Name	Symbol	Atomic Number	Atomic Weight	Foot-notes
Neodymium	Nd	60	144.24*	c
Neon	Ne	10	20.179*	d
Neptunium	Np	93	237.0482	a
Nickel	Ni	28	58.70	
Niobium	Nb	41	92.9064	
Nitrogen	N	7	14.0067	
Nobelium	No	102	(259)	
Osmium	Os	76	190.2	c
Oxygen	O	8	15.9994*	b
Palladium	Pd	46	106.4	c
Phosphorus	P	15	30.97376	
Platinum	Pt	78	195.09*	
Plutonium	Pu	94	(244)	
Polonium	Po	84	(209)	
Potassium	K	19	39.0983*	
Praseodymium	Pr	59	140.9077	
Promethium	Pm	61	(145)	
Protactinium	Pa	91	231.0359	a
Radium	Ra	88	226.0254	a, c
Radon	Rn	86	(222)	
Rhenium	Re	75	186.207	
Rhodium	Rh	45	102.9055	
Rubidium	Rb	37	85.4678*	c
Ruthenium	Ru	44	101.07*	c
Samarium	Sm	62	150.4	c
Scandium	Sc	21	44.9559	
Selenium	Se	34	78.96*	

Name	Symbol	Atomic Number	Atomic Weight	Foot-notes
Silicon	Si	14	28.0855*	
Silver	Ag	47	107.868	c
Sodium	Na	11	22.98977	
Strontium	Sr	38	87.62	c
Sulfur	S	16	32.06	b
Tantalum	Ta	73	180.9479*	
Technetium	Tc	43	(98)	
Tellurium	Te	52	127.60*	c
Terbium	Tb	65	158.9254	
Thallium	Tl	81	204.37*	
Thorium	Th	90	232.0381	a,c
Thulium	Tm	69	168.9342	
Tin	Sn	50	118.69*	
Titanium	Ti	22	47.90*	
Tungsten (Wolfram)	W	74	183.85*	
(Unnilhexium)	(Unh)	106	(263)	
(Unnilpentium)	(Unp)	105	(262)	
(Unnilquadium)	(Unq)	104	(261)	
Uranium	U	92	238.029	c,d
Vanadium	V	23	50.9415	
Xenon	Xe	54	131.30	c,d
Ytterblum	Yb	70	173.04*	
Yttrium	Y	39	88.9059	
Zinc	Zn	30	65.38	
Zirconium	Zr	40	91.22	c

[a] Element for which the value of A_r is that of the radioisotope of longest half-life.

[b] Element for which known variations in isotopic composition in normal terrestrial material prevent a more precise atomic weight being given; $A_r(E)$ values should be applicable to any "normal" material.

[c] Element for which geological specimens are known in which the element has an anomalous isotopic composition, such that the difference in atomic weight of the element in such specimens from that given in the table may exceed considerably the implied uncertainty.

[d] Element for which substantial variations in A_r from the value given can occur in commercially available material because of inadvertent or undisclosed change of isotopic composition.

QUANTITATIVE ANALYTICAL CHEMISTRY

SECOND EDITION

H.A. Flaschka
Georgia Institute of Technology

A.J. Barnard, Jr.
J.T. Baker Chemical Company

P.E. Sturrock
Georgia Institute of Technology

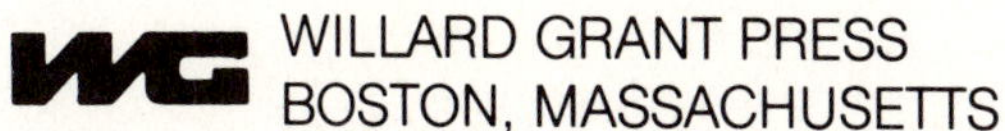

WILLARD GRANT PRESS
BOSTON, MASSACHUSETTS

Statler Office Building, 20 Providence Street
Boston, Massachusetts 02116

Willard Grant Press is a division of Wadsworth, Inc.

Library of Congress Cataloging in Publication Data

Flaschka, Hermenegild Arved
Quantitative analytical chemistry.

Original ed. published in 2 vols. This ed. is a revision of vol. 1, which was published under title: Introduction to principles.
Includes index.
1. Chemistry, Analytic—Quantitative. I. Bernard, Alfred James, joint author. II. Sturrock, P. E., joint author. III. Title.
QD101.F56 1980 545 80-18555
ISBN 0-87150-707-2

Text design by David Chelton, in collaboration with the WGP production staff. Production coordination by Susan London. Copyediting by Janet Wright. Composed in Press Roman by Jay's Publishers Service, Inc. Art drawn by Scientific Illustrators. Cover design by Miriam Recio. Cover printed by John P. Pow Co. Text printed and bound by Maple-Vail Manufacturing Group.

Preface

The central purposes of this introductory textbook and its companion* are to give students a basic understanding of the theory and principles of quantitative analysis, a major area of analytical chemistry, and to provide an introduction to practice. To accomplish these purposes, we believe a textbook should: (1) Introduce quantitative chemistry largely in the framework of inorganic analysis. (2) Provide an understanding of basic analytical methods, operations, and some determinations. (3) Acquaint students with the spectrum of present-day compositional analysis, from gravimetric and titrimetric methods to electrical and optical ones, as well as chromatography. (4) Present this material in separate, but correlated works: one directed to principles and the other to practice.

We believe these aims can be achieved even in the short course required of non-chemistry majors.

A knowledge of concepts and descriptive material covered in a general chemistry course is the only background expected. We have restated some concepts; others we have considered in detail since students may have been exposed to them from a different viewpoint or to an insufficient depth.

We have organized this textbook primarily for courses in which many of the students are enrolled in majors other than chemistry, such as bioscience or engineering. However, sufficient material to allow its use for a more intensive course has been included. This book should also be useful in refresher courses and for in-house training programs in industry and government.

Most of the distinctive features of the earlier edition have been retained. The presentation has again been divided into many chapters and sections rather than a few lengthy ones. We feel this arrangement gives teachers more flexibility in selecting topics and in electing an order of presentation that best suits their course. Some chapters are brief and simple and could be assigned for self study.

We have always felt that questions and problems are important and have included them at the end of each chapter. Students should attempt to answer the questions even if not requested to do so. We have used examples from metallurgical, clinical, environmental, and pharmaceutical analysis as the basis of numerical

---

**Quantitative Analytical Chemistry: An Introduction to Practice*, H.A. Flaschka, A.J. Barnard, P.E. Sturrock, 2nd ed., Willard Grant Press, Boston, 1980.

problems. These problems and questions provide information about important applications and aid student appreciation of the vitality of the principles. For some problems the answers are given; for others, answers are provided in an Instructor's Manual.

We have introduced the principles for most of the chemical methods used in present-day compositional analysis. Potentiometry with ion-selective electrodes has been considered in detail, because of its increasing use in field and process analysis. Both ultraviolet and infrared spectrophotometry have been described, in part because non-chemistry majors may not learn of these techniques in other courses. The treatment of fluorimetry has been broadened because of its importance in inorganic and organic trace analysis and in bioresearch. Atomic absorption spectrometry has come of age, and we have provided a rather detailed exposition. Chromatography is now so important that it must be included in even a short course. The extent, however, to which this subject can be presented remains a problem. Consequently, the treatment has been divided into three chapters with many sections, allowing selection of topics.

We have largely omitted from the text details of the design and operation of instruments. This approach is appropriate for lectures on principles. Instrument operation and design can be described in conjunction with laboratory work and be related to the available models.

The treatment of experimental data, although a topic for practice, is introduced by some teachers during lectures on principles. We have, therefore, included in Chapter 2 the text on the treatment of data and errors that appears in Chapter 7 of the companion work.

Laboratory instruction and experiments have been placed in a separate volume. Students often have only a few hours of lecture before starting laboratory work. Also, special instruments and assemblies may be available in limited number, requiring their use by some students before lecture discussion. In either of these cases, mechanical, "cookbook" performance by students can result. To avoid this, we have prefaced each experiment or group of experiments in the companion work with a summary of their underlying principles. In this way students need not search other sources in order to appreciate an immediate task. We have added more experiments for instrumental methods to include electrometric titrations, polarography, photometry, photometric titrations, fluorimetry, flame photometry, atomic absorption spectrometry and gas chromatography.

We thank many colleagues for their helpful criticisms of the earlier edition and those who commented on the manuscript for the present one. We also thank teachers and reviewers who have communicated their teaching experiences and opinions on how the course should be taught. We welcome comments from teachers as to success or difficulty encountered with our text and will appreciate having errors brought to our attention.

H.A. Flaschka

A.J. Barnard

P.E. Sturrock

Contents

1 Introduction

Analytical chemistry, broadly conceived, underlies and contributes to almost all branches of chemistry as an experimental science. You have already received brief instruction in qualitative analysis. In that field of chemistry, the central question is, *What is present in the sample*? In quantitative analysis, the question is, *How much is present*? Quantitative analysis uses many of the same reactions and phenomena as qualitative analysis. However, quantification has more stringent requirements and demands closer controls.

Consider a reaction that produces a precipitate. When such a reaction is used qualitatively as an identification test, it may be desirable, but not necessary, that the precipitation be complete and the precipitate be pure. In contrast, if the reaction is used for quantitative purposes and the amount of the washed and ignited precipitate is to be related to the amount of one of the components of the precipitate, the precipitation of that component must be essentially complete and the precipitate be freed of all impurities that cannot be removed by either simple washing or another, subsequent treatment—for example, ignition.

The student of qualitative analysis is introduced to only a limited number of the reactions and phenomena that are valuable in analytical chemistry. Quantitative analysis involves many additional reactions and phenomena. Consequently, the student will be learning more about qualitative analysis and also chemistry generally. Although most of the material in this textbook is directed toward inorganic substances and inorganic analysis, the principles introduced also apply to organic analysis.

Quantitative analytical chemistry can, for the purpose of instruction, be separated into two broad areas: (1) the principles and the possibilities they offer, and (2) their application to practice. This volume is directed to the first area. Practice is treated only so far as it shows the vitality of the principles and assures understanding of the various equipment, reagents, instruments, and procedures used.

Many textbooks begin with detailed definition and classification of the subject to be treated. This approach is usually unrewarding to a student encountering the subject for the first time. Such considerations are postponed in this textbook to the terminal pages, Chapter 52, where they can be studied either after some experience has been gained or at the end of the course, with further appreciation and in perspective. Sections devoted to basic chemical and physical concepts, such as chemical equilibria and electrochemical and optical principles, are dispersed throughout the development of the subject.

The student must appreciate the distinction between a determination and an analysis. A determination establishes the amount or content of a *single* component in a sample. An analysis establishes the amounts or contents of a few or all of the components in a material. In an introduction to quantitative analysis, simple determinations must be considered predominantly. Analyses require, beyond a mastery of determinations, additional knowledge, special experience, and detailed attention to the overall composition of the material. The resolution of complex samples is beyond the scope of an introductory course.

2

Experimental Data and Error in Analysis

2.1 *Concept of Significant Figures*

In quantitative analysis, as in most fields of science and technology, numerical values result from measurements. A measurement is restricted in its reliability, that is, in its accuracy and precision (see Section 2.6). When a value is reported that has been based on measurement, the reliability should be reflected in the number of figures retained. A digit that denotes an amount for the place in which it appears and is assigned some reliability is known as a significant figure.* A zero, depending on its position, can be either a significant figure or a mere indicator of the location of the decimal point.

The rule that may be adopted for reporting values based on measurement is to retain only one uncertain figure. In other words, the value should be presented so that the figure before the last one is certain and the last figure is uncertain.† This rule can be appreciated from an example. If a certain amount of a substance is weighed on a pan balance that is known to yield results no more reliable than a few hundredths of a gram, a result might be presented as 5.34 g. To report, say, 5.342 g after an attempted but useless interpolation would not be proper, since the second decimal place is already uncertain. Indeed, the result of an additional weighing might be 5.35 g or 5.33 g. If the same amount of substance is weighed on an analytical balance with a reproducibility of 0.0003 g, the result might properly be written, for example, 5.3434 g. Reporting superfluous figures is not only

*One of the proposals associated with the International System of Units (SI) is the use of *significant digit* for what is here termed *significant figure*. The later term has been retained in this work because of its longtime use and acceptance in diverse fields.

†For intermediate results in various mathematical operations, it is often necessary to retain two or more uncertain figures. This is also the case for the statistical study of measured values (see Section 2.17).

wasted effort but improper presentation, since an incorrect impression is given of the reliability of the value.

In the evaluation of the number of significant figures, care must be taken with terminal zeros. Assume that an object is found to weigh 3.2 g on a crude balance, and that this value is written with the proper number of significant figures. We may want to express this value in milligrams. The expression 3200 mg, although it is numerically correct, is ambiguous, because the two zeros, which locate the decimal point, may or may not be taken subsequently as significant figures. This difficulty is circumvented if the value is written as 3.2×10^3 mg. In contrast, the expression 0.0032 kg, giving the result in kilograms, is unambiguous, because the two zeros to the right of the decimal point only fix its location. Of course, the result can also be represented satisfactorily as 3.2×10^{-3} kg.

If the reliability of a result is known, an even more meaningful way of presenting it can be adopted. Assume that a sample was weighed on a balance that is known to permit no better operation than to ±0.03 g. Then the result might be reported as, for example, 5.34 ± 0.03 g. Here the term ±0.03 g represents the *absolute* uncertainty of the reported result. Another way of expressing the reliability is by the *relative* uncertainty, which is the fraction obtained by dividing the absolute uncertainty by the value of the result. In the present case, the relative uncertainty is 0.03/5.34 = $^3/_{534}$, or about $^1/_{180}$.

The following examples will illustrate the points made so far. Below each of the values is the number of significant figures and the approximate relative uncertainty; the absolute uncertainty is assumed to be ±2 in the last significant figure.

2.34	0.00234	0.0002340	2.34×10^4	23,400	23,400.0
3	*3*	*4*	*3*	*uncertain*	*6*
$^1/_{120}$	$^1/_{120}$	$^1/_{1200}$	$^1/_{120}$		$^1/_{120,000}$

It is sometimes expedient to indicate uncertain figures by special typography, usually italic or subscript digits, for example, 3.5*2* or 3.5_2. These practices are used especially when a 5 would otherwise be rounded.

2.2 Rules for Rounding

When a measured or calculated value is to be adjusted to retain only the proper number of significant figures, superfluous figures must be rejected. The following rules for rounding are recommended.*

1. When the figure next beyond the last place to be retained is *less than 5,* leave

*These are the practices recommended by the American Standards Association and the American Society for Testing and Materials; they are paraphrased from the publications of the latter society.

unchanged the figure in the last place retained. (For example, 452.23 would be rounded to 452.2, and 8.03 to 8.0.)

2. When the figure next beyond the last place to be retained is *greater than 5,* increase by 1 the figure in the last place retained. (For example, 23.67 would be rounded to 23.7, and 0.0699 to 0.070.)

3. When the figure next beyond the last place to be retained is *5,* there are two possible courses. If there are no figures beyond this 5 or only zeros, (1) increase by 1 the figure in the last place retained if it is odd, (2) leave the figure unchanged if it is even. (For example, round 235.5 to 236, 0.6445 to 0.644, and 0.605 to 0.60, and in rounding to the nearest hundred, 2250.0 to 2.2×10^2 and 2350.0 to 2.4×10^3.)

   If there are other figures besides zeros beyond this 5, increase by 1 the figure in the last place retained. (For example, in rounding to the nearest hundred, round 2250.4 to 2.3×10^3 and 2354.0 to 2.4×10^3.)

4. The rounded value should be reached in a single step by direct rounding and not in two or more steps of successive roundings. (Thus, 89,490 rounded to the nearest thousand is at once 89,000 and is best written 8.9×10^4; it would be improper to round first to the nearest hundred, 89,500, and then to the nearest thousand, getting 90,000, that is, 9.0×10^4.)

2.3 Significant Figures in Arithmetic Operations

The rules for rounding can be applied readily to the proper expression of a single measured value. The situation, however, is more complicated when the final result follows from calculations involving several values, each of which may vary in its own reliability. The maxim applies that "no chain is stronger than its weakest link." For significant figures, this means that no calculated value can be more reliable than its least reliable component. Evaluation of the uncertainty of a computed value will depend on the mathematical operations involved.

Addition and Subtraction. In a result from addition or subtraction, or both, the *absolute* uncertainty of the least reliable component determines the reliability of the result. Consequently, only as many places should be retained to the right of the decimal point as there are in the component with the least number of decimal places.

Example 2-1 A formula weight is calculated from the values for the atomic weights recorded in the atomic weight table (see Table L in the Appendix). Thus for lithium fluoride, LiF,

$$
\begin{array}{rl}
\text{Li:} & 6.941 \\
\text{F:} & 18.998403 \\
\hline
\text{LiF:} & 25.939403 = \text{rounded } 25.939
\end{array}
$$

Since the atomic weight of lithium is known reliably to only three decimal places, it is pointless to carry any figures beyond that place, and the formula weight of the salt is properly recorded as 25.939.

Multiplication and Division. The following simple "fewest significant figures rule" can be adopted: A product or a quotient should not contain more significant figures than are contained in the number with the fewest significant figures used in the multiplication or division. The rule holds also for operations involving both multiplication and division.

Example 2-2 Consider the multiplication 34.2051×3.22. The result is 110.140422. The factor with the fewest, namely three, significant figures is 3.22. Consequently, the result is rounded to three significant figures: 110. However, this form of expression is ambiguous; it is not certain whether the zero is significant or only fixes the location of the decimal point. The result is best given as 1.10×10^2.

It can be seen that superfluous digits have been carried through the calculation; it is common sense to round the first factor from the beginning, causing it to have same number of significant figures that the second factor contains. Then the multiplication simplifies to 34.2×3.22.

Example 2-3 Calculate the result for the following expression:

$$\frac{4.6672 \times 12.4}{53.267 \times 0.13862}$$

The factor with the fewest significant figures has three such figures (12.4). Consequently, all the other factors are rounded to three significant figures. Then we get

$$\frac{4.67 \times 12.4}{53.3 \times 0.139}$$

The multiplications in the numerator and the denominator are next performed, and the intermediate results are also written with only three significant figures:

$$\frac{57.9}{7.41}$$

The final result is 7.81.

Logarithms. For converting a number to its logarithm, the following rule applies. Retain as many places in the mantissa of the logarithm (that is, to the right of the decimal point in the logarithm) as there are significant figures in the number itself. For example, $\log 24.5 = 1.389$, and $\log 0.34 = 0.53 - 1$.

When logarithms are used to obtain the results of multiplication, division, calculation of powers, extraction of roots, or a combination of these operations, the logarithms may be noted as given in a table of logarithms and the final result adjusted to the proper number of significant figures according to the rules that have been given. Alternatively, each logarithm in the operations may be written with its proper number of decimals in the mantissa, and the calculation performed with the adjusted logarithms.

2.4 *Significant Figures in Mixed Calculations*

Mixed arithmetic calculations are performed stepwise according to the rules of arithmetic. In addition and subtraction, the absolute uncertainties are considered; and for multiplication and division, the rule of "fewest significant figures" is applied. One has to read twice. The first time "doing likewise" is expected to refer to rounding to three figures.

Example 2–4 Consider the mixed arithmetic calculation

$$\frac{2.357}{0.26} + \frac{1.265}{4.12}$$

For the first term, the numerator is rounded to two significant figures, and the result of the division becomes 2.4/0.26 = 9.2. Doing likewise with the second term yields 1.26/4.12 = 0.306. For the addition according to the rules established, only one decimal place is significant. Consequently, 9.2 + 0.3 = 9.5.

Example 2–5 Consider the mixed arithmetic operations

$$\frac{(248.31 - 248.1) \times 4.012}{(10.01 + 25.12) \times 0.9804}$$

First the subtraction in the numerator is performed. The difference is 0.21; written with the appropriate number of significant figures, it becomes 0.2. This number is the one with the fewest significant figures. Consequently, all other terms, including the result of 35.13 of the addition term in the denominator, are rounded to one significant figure:

$$\frac{0.2 \times 4}{40 \times 1}$$

The result is 0.02, which also is recorded with one significant figure. In such mixed operations, significant figures are lost. From terms with four and five significant figures, a result may have only one significant figure. Such losses always occur when numbers of like size appear in a subtraction term, as in this example.

2.5 Remarks

The preceding treatment is simplified but will suffice the student's need. For multiplication and division, the "fewest significant figures" rule is a pronounced simplification. By this rule, a decision is based solely on the number of significant figures and does not take into account that two numbers with the same number of significant figures may differ in reliability. The number 99, for example, is considered to have two significant figures, but it is so close to 100 that it may well be considered to have a "hidden" third significant figure. In a more rigorous treatment, it is not the number of significant figures but the relative uncertainties that should be taken into account. The decision how many significant figures are to be properly retained in the value resulting from multiplication or division, or both, should be based on the following consideration: The *relative* uncertainty of the result and of the least certain component should be of the same order. The rule can be adopted that the relative uncertainty of the result should be between twice and two-tenths (that is, one-fifth) of the relative uncertainty of the least certain component.

Example 2-6 In the following multiplication, each factor is assumed to be uncertain to the extent of ±1 in the last figure. (The approximate relative uncertainties are expressed as italic fractions.)

$$\underset{\mathit{1/40}}{4.3} \times \underset{\mathit{1/7000}}{6.893} \times \underset{\mathit{1/5000}}{0.5372} =$$

So that superfluous numbers will not be carried, the numbers are rounded according to the simple rule, but one more figure is retained than in the number with the fewest figures. Thus

$$4.3 \times 6.89 \times 0.537 =$$

Multiplying the first two factors and again retaining one additional figure yields

$$29.6 \times 0.537 = 15.90$$

The result should have no better certainty than $\frac{1}{40}$, which is the certainty of the least certain factor. Then $\frac{1}{40} \times 15.90 = 0.4$. Consequently, only one decimal place is to be retained, and the result is properly written 15.9. The uncertainty of the result is $\frac{1}{160}$, which is within twice and two-tenths of the least certain factor: $2 \times \frac{1}{40} = \frac{1}{20}$, and $0.2 \times \frac{1}{40} = \frac{1}{200}$. Note that the simplified approach would lead to expressing the result as 16, which has a relative uncertainty of $\frac{1}{16}$, or approximately $\frac{1}{20}$, which falls outside or just at one limit of the uncertainty range established by the rigorous treatment.

This rigorous treatment is used in all calculations throughout this textbook, and it is assumed for simplicity that the uncertainty of all given values is ±1 in the last figure. This practice has no bearing on the mode of calculation adopted.

Implicit in this discussion of significant figures has been the assumption that knowledge is at hand about how many figures are uncertain in a value obtained by measurement. We shall subsequently consider how to obtain this knowledge.

2.6 *Accuracy and Precision**

When the measurement of a single quantity is repeated, even with the same measuring device and by the same operator, small differences in the values will occur. With improvement of the device and the technique, such differences may decrease but will never vanish completely. The smaller these differences, the more precise the measurements are said to be however, a highly precise measurement is not necessarily an accurate one. A tape measure that has been stretched through long use yields results of a certain precision, but each measurement departs from the "true" value, that is, it is low in accuracy. It can therefore be appreciated that at least two parameters are needed to describe the quality of measurements—accuracy *and* precision.

Accuracy is related directly to how well a measurement agrees with the "true" value; the closer the agreement with the true value, the higher the accuracy. Precision is related to the difference between the values obtained when the same quantity is measured repeatedly. If such differences are small, the precision is high. The term *true value,* used in describing the concept of accuracy, must be treated as an ideal. In most real situations, the true value is either unknown or not fully attainable. In practice it is often necessary to substitute the "most probable" value, or in other terms, the value known to the least uncertainty or having the greatest number of significant figures. In student experiments, the true value often reduces to the value obtained by the instructor (or a competent analyst), who has used perfected techniques and methods; that is, it reduces to an "assumed," "expected," or "accepted" value.

2.7 *Errors*

Errors are conveniently grouped into two classes, systematic errors and random errors. The extent of a systematic error affords a measure of accuracy and may be taken as the difference between the "true" value and either an in-

*Sections 2.6 through 2.21 are almost identical with Chapter 7 of the companion work, *The Practice of Quantitative Analytical Chemistry* (Willard Grant Press, Boston, 1980). The material is included for the convenience of courses that are not using that volume or in which the treatment of experimental data is considered in lectures rather than in the introduction to practice. A few examples and problems assume an elementary knowledge of such terms as *titration volume, pH,* and *solubility product.* For students unfamiliar with such terms, the relevant examples and problems can be deferred to a later point in the study.

dividual measurement or an average of a series of measurements of the same quantity. A systematic error, which is also called a determinate error, is unidirectional; that is, it is either positive or negative and is traceable, at least theoretically, to a definite cause.

Random errors are related to the precision, that is, to the repeatability or reproducibility of the measurement of a quantity.* Random errors are present, regardless of the presence or absence of systematic error, and their consideration does not require knowledge of the true value. For evaluating random errors, the laws of probability are evoked; that is, such errors are treated statistically (see below).

The absolute error ϵ of a single measurement is the difference between the measured value X_i and the true value X_T. Dividing this error by the true value yields the relative error ϵ_{rel}:

$$\epsilon = X_i - X_T \qquad \epsilon_{rel} = \frac{X_i - X_T}{X_T} \tag{2-1}$$

Multiplying the relative error by 100 yields the percentage relative error, and multiplying by 1000, the parts per thousand relative error.

2.8 Normal Distribution Curve

If a measurement is repeated many times, the values can be grouped into small intervals, and the number of values falling into each group (that is, the frequency) plotted versus the group value; a so-called histogram is thus obtained. Such a graph in idealized form is shown as Fig. 2-1(a). If the number of measurements is made infinite and the size of the intervals made increasingly smaller, a curve such as that shown as Fig. 2-1(b) is obtained. This is known variously as the normal distribution curve, Gauss curve, or probability curve. It can be seen (1) that the curve is symmetrical around a central value, the average ($\bar{X}$), (2) that a positive deviation from this central value is as probable as a negative deviation, (3) that small deviations are more frequent than large ones, and (4) that the curve has two points of inflection. The distance from either of these points to the average is known as the "true" standard deviation, designated by the Greek letter sigma, σ. For a normal distribution (infinite number of observations), 68.3% of the measurements will fall within the interval $\bar{X} \pm \sigma$, 95.4% within $\bar{X} \pm 2\sigma$, and 99.7% within $\bar{X} \pm 3\sigma$. These intervals are shown on the curve in Fig. 2-1(b). The smaller the value of the standard deviation, the greater the precision. How the average is determined and the standard deviation estimated is considered below.

---

*The term *repeatability* is best reserved for the precision attained by a single, experienced analyst and the term *reproducibility* for the aggregate precision secured by different laboratories and analysts in cooperative studies. When reproducibility is far poorer than repeatability, a "standard" procedure is subject to review.

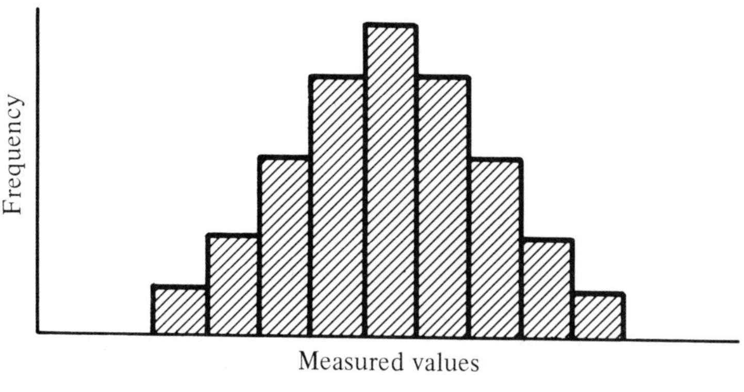

(a)

$\bar{X}$
Frequency
$\bar{X} - \sigma$ 68.3% $\bar{X} + \sigma$
$\bar{X} - 2\sigma$ 95.4% $\bar{X} + 2\sigma$
$\bar{X} - 3\sigma$ 99.7% $\bar{X} + 3\sigma$

(b)

Fig. 2-1. (a) Histogram; (b) normal distribution (Gauss) curve (see the text for explanation).

2.9 *Average*

If there is no valid basis for assuming that one observed value of a series is more reliable than another, that is, if all the values are of equal statistical weight, it can be appreciated without mathematical derivation that the average (also called the simple average, arithmetic mean, or mean) is a more reliable value than any single value. The average is the numerical value resulting from dividing the sum of the measured values of a quantity by the number of measurements. If the N individual values are represented by $X_1, X_2, \ldots, X_{N-1}, X_N$, the average $\bar{X}$ is given by

$$\bar{X} = \frac{X_1 + X_2 + \cdots + X_{N-1} + X_N}{N} = \frac{1}{N} \sum_{i=1}^{N} X_i \qquad (2\text{-}2)$$

The last expression is read "the sum of values of X_i from $i = 1$ through $i = N$."

It can be shown that the average of the values of N equally reliable measurements is $\sqrt{N}$ times as reliable as any single measurement. Hence, the average of four values tends to be twice as reliable as a single value, and the average of nine values, three times as reliable. Thus, the advantage of duplicate or triplicate observations is seen, but the gain in reliability secured by obtaining additional values may not always justify the added effort.

2.10 *Variance and Standard Deviation*

The variance V_x of a series of values X_i is obtained by applying the following formula:

$$V_x = \frac{\sum_{i=1}^{N} (X_i - \bar{X})^2}{N - 1} = \frac{\sum_{i=1}^{N} d_i^2}{N - 1} \qquad (2\text{-}3)$$

where $\bar{X}$ is the average and d_i the difference between a value and the average; that is, $d_i = X_i - \bar{X}$. The variance, in words, is the sum of the squares of the deviations from the average divided by one less than the number of measurements. Variance as a measure of dispersion has the advantage of being additive; that is, the total variance for a multistep process is the sum of the individual variances for the steps.

For the usual purposes of analytical chemistry, the square root of the variance, which is known as the standard deviation s, is of greater interest, because it has the same units as the data from which it is derived. The standard deviation is given by*

$$s = \sqrt{V_x} = \sqrt{\frac{\sum_{i=1}^{N} (X_i - \bar{X})^2}{N - 1}} = \sqrt{\frac{\sum_{i=1}^{N} d_i^2}{N - 1}} \qquad (2\text{-}4)$$

---

*If the denominator of the formula for the standard deviation were N, this measure would be the root mean square of the deviations from the average. The use of $N - 1$ rather than N in this formula (and the formula for the

What is called the standard deviation, and denoted by s, should more precisely be termed the estimate of the standard deviation; the reason is that the "true" standard deviation, designated by the Greek letter sigma, σ, in the consideration of the normal distribution curve is calculable only from an infinite number of values. In addition, by the term *standard deviation* is actually meant the "standard deviation of a single measurement"; this expression is warranted whenever confusion is possible with "standard deviation of the average."

2.11 Standard Deviation of the Average

Occasionally, and especially in the intercomparison of two or more sets of measurements, it is desirable to express (the estimate of) the standard deviation of the average, s_m. This is related to the standard deviation of a single measurement s by

$$s_m = \frac{s}{\sqrt{N}} = \sqrt{\frac{\sum_{i=1}^{N} d_i^2}{N(N-1)}} \tag{2-5}$$

Note that this measure of dispersion demands the words *of the average* in its designation so that it can be distinguished from the standard deviation (of a single measurement).

2.12 Range

When fewer than ten measurements are secured, the range w is often used as a measure of dispersion. This is simply the difference between the largest and the smallest value in the set:

$$w = X_{\text{largest}} - X_{\text{smallest}} \tag{2-6}$$

2.13 Average Deviation

If the arithmetic average of the differences between the values and the average is calculated without regard to the algebraic sign of the differences, the

---

variance) is associated with the loss of one of the N degrees of statistical freedom in calculating the average. As the number of values increases, the distinction between $N - 1$ and N becomes less significant.

result is known as the average deviation (or mean deviation) α. The average deviation can therefore be expressed

$$\alpha = \frac{|d_1| + |d_2| + \cdots + |d_{N-1}| + |d_N|}{N} = \frac{1}{N} \sum_{i=1}^{N} |d_i| \tag{2-7}$$

The vertical lines indicate that the enclosed entity is to be taken without regard to algebraic sign; that is, only the absolute value is to be considered.

The average deviation is a less efficient estimate of dispersion than the other measures mentioned. As the number of measurements increases, the relation between the standard deviation and the average deviation approaches $s = \frac{5}{4}\alpha$.

2.14 Confidence Limits

When we considered the normal distribution curve (Section 2.8), we said that 68.3% of the values fall within the interval $\bar{X} - \sigma$ and $\bar{X} + \sigma$, where σ is the "true" standard deviation. Such an interval is known as a confidence interval and its limits as confidence limits.

When only a limited number of measurements of a quantity are available, the magnitude of the estimate of the standard deviation s depends on the number of measurements, and consequently the calculated confidence interval also depends on that number. For N measurements, the confidence interval for a single measurement can be expressed as $\bar{X} \pm ts$. The values of the parameter t, which depends on N, can be calculated from considerations beyond the scope of the present treatment. The values for percentage probabilities 95% and 99% are given in Table 2-1.

Table 2-1. Values of t for Probability Levels of 95% and 99%

	Number of values N											
	2	3	4	5	6	7	8	9	10	11	15	∞
$t_{95\%}$	12.71	4.30	3.18	2.78	2.57	2.45	2.36	2.31	2.26	2.23	2.14	1.96
$t_{99\%}$	63.66	9.92	5.84	4.60	4.03	3.71	3.50	3.36	3.25	3.17	2.98	2.58

Assume that a single operator titrated five identical samples. The average volume of titrant needed was 15.32 mL and the standard deviation was calculated to be ±0.06 mL. Since $N = 5$, the 95% confidence limits for a single measurement are given by $15.32 \pm 2.78 \times 0.06 = 15.32 \pm 0.17$ mL. This means that if the determination were repeated 100 times, 95 results might be expected to fall within 15.32 + 0.17 and 15.32 - 0.17 mL.

More interesting is the confidence interval of the average. Extending the above reasoning, we can clearly see that this interval is given by $\bar{X} \pm ts_m$ or by its equivalent, $\bar{X} \pm ts/\sqrt{N}$. The confidence limits of the average define an interval about the experimental average within which the "true" *average* value would lie with the stated probability. The true average, the value of which is unknown, is the average that would be obtained from a large (theoretically infinite) number of measurements.

For the example given above, the 95% confidence limits of the average are given by $15.32 \pm 2.78 \times 0.06/\sqrt{5} = 15.32 \pm 0.07$ mL. This means simply that the true average would be expected to lie with 95% probability within 15.32 + 0.07 and 15.32 - 0.07 mL. A more meaningful interpretation is that if 100 sets of five titrations were performed, 95 of the sets would be expected to have an average that would fall within 15.32 + 0.07 and 15.32 - 0.07 mL.

2.15 *Relative Measures of Dispersion*

It is also possible to place the various measures of deviation on a relative basis by dividing by the value of the average. For example,

$$\text{Relative standard deviation} = \frac{s}{\bar{X}} \tag{2-8}$$

$$\text{Relative average deviation} = \frac{\alpha}{\bar{X}} \tag{2-9}$$

$$\text{Relative range} = \frac{w}{\bar{X}} \tag{2-10}$$

By multiplying by 100 or 1000, these relative measures can be placed on a percentage or a parts per thousand basis. The percentage relative standard deviation is also termed the coefficient of variation.

2.16 *Number of Significant Figures in Reporting Experimental Data*

In many analytical applications, one can readily determine how many significant figures are appropriate in presenting a value. In weighing a substance, for example, the reproducibility of the balance is the primary factor, and it can be assessed from the range of successive weighings of a single object. If an analytical balance has a reproducibility of ±0.5 mg, recording a weighing to 0.01 mg is useless. Similarly, the reading of an ordinary 50-mL buret can hardly be better than ±0.02 mL; consequently, recording three decimal places for a volume delivered is unwarranted.

The situation is more complicated when a value is a composite of more than one type of measurement (for example, a sample weight and a volume delivered by a buret). Statistics provides formulas that allow evaluating "error propagation" (their consideration is deferred to Section 2.21). Although the uncertainty of one measurement may be so great that it can be taken as the dominant consideration in selecting the appropriate number of significant figures for the result to be reported, this is not necessarily so.

There is another problem in selecting a proper number of significant figures for the presentation of the average and the individual values of a series of measurements. Calculating the standard deviation allows this situation to be resolved. For many purposes, an average can be reported so that its last significant figure corresponds to the first place of significance in the deviation measure adopted. Thus, if an unrounded average were 11.09 and the calculated standard deviation ±0.11, the average could be reported as 11.1.

When it is important that rounding errors will not be unduly large compared with random errors, a more exacting procedure has merit:

1. Calculate the standard deviation.
2. Divide the value of the standard deviation by 4.
3. Round the resulting number to the nearest multiple or submultiple of 10.
4. Report the relevant result to this place.

When an average is to be rounded, the standard deviation of the average is used with this procedure.

Example 2-7 Assume that the calculated standard deviations of three series, each containing four results, have the following values: (i) 0.36, (ii) 1.6, and (iii) 2.4. (a) In each series, to what number of decimal places should a single result be reported? (b) If an individual result is 19.11, how should it be reported for each of the three series? (c) If the average is 19.223, how should it be reported for each of the three series?

(a) (i) 0.36/4 = 0.09 ~ 0.1. Express a single result to one decimal place.
(ii) 1.6/4 = 0.4 ~ 0.1. Express a single result to one decimal place.
(iii) 2.4/4 = 0.6 ~ 1. Express a single result to the units place only.

(b) Report the single measurement as (i) 19.1, (ii) 19.1, (iii) 19.

(c) (i) $\frac{0.36}{4\sqrt{4}} = 0.045 \sim 0.01$. Report the average as 19.23.
(ii) $\frac{1.6}{4\sqrt{4}} = 0.2 \sim 0.1$. Report the average as 19.2.
(iii) $\frac{2.4}{4\sqrt{4}} = 0.3 \sim 0.1$. Report the average as 19.2.

2.17 Calculating Measures of Dispersion

The mechanics of calculating the various measures of dispersion (that is, of precision) can best be demonstrated by examining fully an example involving the milliliters of titrant required in titrating *identical* amounts of a sample.

Two or even more figures than what seem to be significant from merely inspecting the experimental data are carried through the calculations. This is mandatory because the statistical calculations treat the data as a set of numbers. The purpose of the treatment is to evaluate the extent of random errors. The proper rounding of the final results must be withheld until the evaluation is complete.

Example 2-8

i	X_i(mL)	$d_i = X_i - \bar{X}$(mL)	d_i^2(mL^2)
1	24.32	−0.022	0.000484
2	24.38	+0.038	0.001444
3	24.45	+0.108	0.011664
4	24.29	−0.052	0.002704
5	24.30	−0.042	0.001764
6	24.31	−0.032	0.001024
$N = 6$	$\Sigma X_i = 146.05$ mL	$\Sigma\lvert d_i\rvert = 0.294$ mL	$\Sigma d_i^2 = 0.019084$ mL^2

$$\bar{X} = \frac{\Sigma X_i}{N} = \frac{146.05}{6} = 24.341666\ldots = 24.342 \text{ mL}$$

$$\sigma = \frac{\Sigma|d_i|}{N} = \frac{0.294}{6} = 0.0490 \text{ mL}$$

$$V_x = \frac{\Sigma d_i^2}{N-1} = \frac{0.019084}{6-1} = 0.003817 \text{ mL}^2$$

$$s = \sqrt{V_x} = \sqrt{0.003817} = \pm 0.06178 \text{ mL}$$

$$s_m = \frac{s}{\sqrt{N}} = \frac{0.06178}{\sqrt{6}} = \pm 0.02522 \text{ mL}$$

$$95\% \text{ Confidence limits for average} = \pm ts_m = \pm 2.57 \times 0.02522 = \pm 0.0648 \text{ mL}$$

$$99\% \text{ Confidence limits for average} = \pm ts_m = \pm 4.03 \times 0.02522 = \pm 0.1016 \text{ mL}$$

$$w = 24.45 - 24.29 = 0.16 \text{ mL}$$

How many significant figures should be given in reporting the average?

$$\frac{s_m}{4} = \frac{0.02522}{4} = 0.006305 \sim 0.01$$

Report the average to two decimal places.

$$\text{Relative standard deviation} = \frac{0.06178}{24.342} = \pm 0.00254$$

$$\text{Relative average deviation} = \frac{0.0490}{24.342} = 0.00201$$

$$\text{Relative range} = \frac{0.16}{24.342} = 0.00657$$

Final presentation (rounded):

$\bar{X} = 24.34$ mL
$V_x = 0.0038$ mL^2
$s = 0.06$ mL
$s_m = 0.025$ mL
$\alpha = 0.05$ mL
$w = 0.16$ mL
95% Confidence limits of average = ±0.06 mL
99% Confidence limits of average = ±0.10 mL
Relative standard deviation = $\pm 0.002_5$ or $\pm 0.2_5\%$
Relative average deviation = 0.002 or 0.2%
Relative range = 0.007 or 0.7%

In reporting values with their precision, the particular measure of dispersion should be made clear. It is undesirable, for example, to report an average and its standard deviation simply as 24.34 ± 0.06 mL without further explanation. Knowing the total number of values on which an average is based often helps in interpreting it. Thus a more extended form of reporting is desirable, such as 24.34 mL ($s = \pm 0.06$ mL; $N = 6$); or if the convention is understood and agreed on, 24.34 ±0.06 mL (s; 6).

Note that only the precision is evaluated by applying measures of dispersion, not the accuracy. In Example 2-8, the true value may be close to 24.34 mL or far removed, depending on how large the systematic errors are.

2.18 *Rejecting a Discordant Value*

In a series of repeated measurements, a value may be obtained that differs markedly from the other values. If it is clear that a gross departure from the prescribed procedure or a personal blunder (say loss of material or misreading an instrument) is involved, a value should be rejected whether or not it agrees with the other values. When no cause can be ascertained for a large deviation from other values, however, the question arises whether the outlying value should be rejected or whether it should be retained, thereby markedly affecting the value of the average. In general, a cautious attitude should be adopted. When possible, additional measurements should be made as an aid in resolving the problem or at least in reducing the influence of the single "doubtful" value if it is retained.

In practical analysis, usually only four to eight values at most are available. In such cases, two different approaches may help in deciding whether an outlying value may be rejected. One approach is based on calculating the average and the average deviation with the doubtful value omitted; the other involves the so-called Q test.

For the first approach, with the *single* suspect value omitted, calculate the average and the *average* deviation. If the suspect value differs from this average by more than four times that average deviation, discard the value.

For the Q test, arrange the values in decreasing or increasing order of magnitude. Then calculate the value of Q, which is the absolute value of the ratio of the difference d between the suspect value and its nearest neighboring value to the range w:

$$Q = \frac{|(\text{suspect value} - \text{nearest value})|}{|(\text{largest value} - \text{smallest value})|} = \frac{|d|}{|w|} \tag{2-11}$$

If the calculated value for Q exceeds the tabulated value for the *total* number of values N and the selected confidence level (usually 90% or 95%), reject the suspect outlying value (see Table 2-2).

Table 2-2. Critical Values of Q for Estimating Gross Errors

	Number of values N				
	3	*4*	*5*	*6*	*7*
$Q_{80\%}$	0.781	0.560	0.451	0.386	0.344
$Q_{90\%}$	0.886	0.679	0.557	0.482	0.434
$Q_{95\%}$	0.941	0.765	0.642	0.560	0.507
$Q_{99\%}$	0.988	0.889	0.780	0.698	0.637

Example 2-9 For titrating four identical samples, the following volumes of titrant, (twice the measure given), were needed: 11.12, 10.90, 11.02, and 11.40 mL. Before the average volume is calculated, it has to be decided whether an outlying value should be rejected as discordant. To help you decide, apply the rule based on average deviation and also the Q test at the 90% confidence level.

Ranking the four values yields 10.90, 11.02, 11.12, and 11.40 mL. By inspection, the suspect value is 11.40 mL.

You will find that with that doubtful value omitted, the average is 11.01_3 mL and the average deviation 0.076 mL. The deviation of the doubtful value from the average is 0.39 mL, which is greater than 4 × 0.076 = 0.30 mL. On this basis, the value would be rejected.

The Q value is

$$Q = \frac{(11.40 - 11.12)}{(11.40 - 10.90)} = \frac{0.28}{0.50} = 0.56$$

The calculated Q value, 0.56, does not exceed the $Q_{90\%}$ value for $N = 4$ given in Table 2-2, 0.679. On this basis, the doubtful value would be retained.

This example delineates the difficulty in rejecting an outlying value when only a few experimental values have been secured. The average and the standard deviation of the average calculated with and without the rejection are noteworthy: 11.01 ± 0.05 mL (s_m;3) and 11.11 ± 0.11 mL (s_m;4).

In ending the discussion of the evaluation of dispersion, it is appropriate to note that statistical methods are not a remedy for "sick" experimental data; rather, such methods are a way of assessing the "state of health."

2.19 Systematic Errors in Chemical Analysis

In the following discussion, it is assumed that obvious "blunders" are absent—such as a leaking buret, faulty instruments, loss of sample through spattering, or mathematical mistakes. Errors can be classified in various ways depending on the reason for the classification. Possibly the broadest classification is systematic errors and random errors. This distinction has been drawn in Section 2.7, and the statistical evaluation of random errors has been considered (Section 2.17). Even this classification assumes idealized conditions, since each systematic error will actually be accompanied by a random error. In practice it is usually difficult to separate the two components, especially if the systematic error is relatively small.

Some prominent systematic errors can be differentiated on the basis of their origin: personal errors, instrumental errors, and methodic errors.

Personal (or Operative) Errors. An operator may have a personal idiosyncrasy or prejudice that leads to an incorrect judgment and thereby introducing an error. Thus, an operator may always adjust the liquid level in a pipet either higher or lower than what has been assumed in its calibration, and therefore cause either too large or too small a volume to be regularly delivered. Or an operator may have an abnormal color response (for example, partial color blindness) and place the color change of an indicator in a visual titration at an improper color shade. Hence, in one sense, personal errors relate to the operator as an observer; indeed they are sometimes termed observational errors. But the errors referred to in our study as personal are unidirectional and hence systematic errors; the unbiased operator will sometimes make a reading slightly high and sometimes slightly low, causing fluctuations that are random errors.

When the operator, functioning as such, for example, ignites a precipitate within an improper temperature range, this is not considered a personal error. If the proper ignition temperature has been given in the procedure being followed, the operator has blundered and a "blunder" is said to have occurred, rather than a personal error. On the other hand, if the temperature is incorrectly specified in the procedure, the error is considered methodic (see below).

Personal errors can sometimes be eliminated by carefully checking a single operator's performance and by comparing results obtained by different operators. An operator who becomes aware of personal prejudice may be able to eliminate it.

Instrumental Errors. Many devices and instruments used in analytical procedures require calibration, and any fault in this operation may introduce a systematic error. This holds for the calibration of burets, pipets, and volumetric flasks, which may actually deliver or contain a volume different from that certified. A weight may be incorrectly calibrated or may have changed from its original correct weight due to corrosion, deposition of dust, and so on. Such errors can be eliminated or at least reduced by ensuring that the device is correctly calibrated or applying a correction when such an error is recognized.

Methodic (or Procedural) Errors. Analytical procedures have shortcomings that lead to systematic errors known as methodic, or procedural, errors. Some examples will show the diverse and potentially serious nature of such errors. In a titration, a certain volume of titrant may be required for a reaction with the indicator; when this volume is not taken into account, the result may be incorrect. An indicator selected for an acid-base titration may change color at a pH value far removed from the equivalence point. The presence of carbonate in a solution of a strong base may introduce a significant methodic error. In gravimetric analysis, the finite solubility of the precipitate on extensive washing will cause a negative error; on the other hand, insufficient washing may not remove all contaminants from the precipitate, and a positive error results. Further, the weighing form may not be fully stoichiometric. Of grave consequences are impurities in reagents or interfering substances in the sample.

Systematic errors may be constant or proportionate, or vary in a complex manner. This allows a further differentiation of systematic errors.

Constant Errors. Constant errors are also known as additive errors; in their absolute value they are independent of the amount of the substance to be determined. For example, the indicator in a titration may require 0.15 mL of titrant to bring about the color change (methodic error), or the operator may through prejudice constantly read the volume delivered from the buret 0.15 mL high (personal error). If all other conditions are equal and only the quantity of substance titrated is altered, such an error will be constant in its absolute value (except for the unavoidable random fluctuations superimposed).

A constant error can often be detected by a trend in the results when different amounts of the sample are taken. Assume that a 10.00-mL volume of a sample solution requires a 10.00-mL volume of the titrant solution. It would be expected that 20.00-mL and 30.00-mL volumes of the sample solution would need 20.00 mL and 30.00 mL of the titrant solution. However, if the same amount of indicator is added in all titrations and a volume of 0.15 mL of the titrant is required to change the color of the indicator, only 9.85 mL of the titrant is actually consumed by the 10.00-mL sample volume. Hence, for 20.00-mL and 30.00-mL volumes of the sample solution, the volumes of titrant delivered to the end point would be $2 \times 9.85 + 0.15 = 19.85$ mL and $3 \times 9.85 + 0.15 = 29.70$ mL respectively.

Proportionate Errors. Proportionate errors depend in their absolute value on the amount of the substance to be determined; they decrease and increase proportionately with that amount. Assume that a substance A is precipitated in the presence of a substance B and that the latter is always coprecipitated to the extent of 0.5% of the weight of A. Then 0.5 mg of B is coprecipitated with 0.1 g of A, and 2.5 mg of B with 0.5 g of A. Note that although the absolute error has increased fivefold, the relative error is unchanged; hence, no trend is established. Finding a proportionate error is therefore difficult. In practice, proportionate errors are not always strictly proportional to the amount of the substance present, but show at least some progressive increase.

2.20 Detecting Systematic Errors; The Blank

Detecting a systematic error is not so simple as the above brief considerations might imply, since more than one such error may be present, and also, random errors are superimposed. Personal errors and instrumental errors can sometimes be discovered by checking operator performance and by carefully calibrating all instruments and devices used. Constant errors can often be discovered from a trend in the results as the amount of the sample is varied.

When the true value is known, there is no difficulty in detecting errors except when several errors compensate for each other. Hence, frequently a procedure is checked with the use of samples of exactly known content. The U.S. National Bureau of Standards provides a wide variety of certified samples for this purpose. If using such a sample establishes the presence of a systematic error, attempts can be made to modify the procedure to obviate the error; less satisfactorily, an appropriate correction can be applied to results obtained by the procedure. These approaches are not always possible, and a standard sample may be used in a different way. Both the unknown sample and a standard closely similar in composition and content of the substance to be determined are subjected to the same procedure. Then by simple proportion, the result for the unknown sample is calculated from the known content of the standard. This method is extensively used, for example, in photometric determinations, flame photometry, and emission spectroscopy.

Systematic errors can often be eliminated, or at least reduced, by use of a blank. A blank determination consists in performing all steps of the procedure without introducing the sample, or in the case of a sample solution, by adding an equal volume of the solvent. Thus in a titration, the volume of titrant required by the indicator can be established by titrating the specified amount of the indicator dissolved in either pure water or a solution of inert salts. "Blank" solutions are frequently used in photometric determinations to set the instrument to 100% transmittance, compensating for any impurities in both the reagents and the solvent.

A frequent gross misconception is that a blank is a sure remedy for methodic errors. The following example will show how false this view is. Aluminum is often determined gravimetrically via precipitation of its hydrous oxide by the addition of ammonia, filtration, washing, and ignition to the anhydrous oxide. Ammonia solutions, unless freshly prepared will contain varying amounts of silica dissolved from the glass bottle. This silica is coprecipitated with the aluminum, causing high results. A blank determination effected using water in place of the aluminum salt solution will show a blank value of zero; the reason is that the dissolved silica passes through the filter, since there is no aluminum oxide precipitate to act as carrier. In such a situation, the only remedy is to precipitate an exactly known amount of aluminum, thus establishing the amount of silica present in a certain volume of ammonia, and to apply a correction. Of course, a contaminated reagent is best avoided.

2.21 *Error Propagation*

In considering systematic (determinate) and random (indeterminate) errors, we have emphasized a series of "identical" measurements. The computation, however, of the final result of an analysis almost always involves more than one type of experimental quantity; at issue are weight, volume, electrical or optical response, and the like. Each of the experimental quantities is affected to a different degree by both systematic and random errors. It is often interesting to know how these errors *propagate* in the computation and how the final result is affected. If only a single quantity involves an exceptionally large error, it may be the only one requiring consideration; further mathematical treatment is then a sterile exercise. If, however, the various errors are of the same order of magnitude, such treatment can be revealing.

A general equation for the propagation of errors can be applied.* However, simple formulas can be derived from it for the common mathematical operations of analytical chemistry. These formulas are given in Table 2-3. You see that a

---

*If you are familiar with calculus, you will appreciate the general formula for the propagation of error for a result R, that is a function of independent variables $A, B, C, \ldots$ is given by

$$R = f(A, B, C, \ldots)$$

The general formula for propagating systematic (determinate) errors takes the form

$$r = \frac{\delta R}{\delta A}a + \frac{\delta R}{\delta B}b + \frac{\delta R}{\delta C}c + \cdots$$

where the symbols have the meaning assigned in Table 2-3 and $\delta R/\delta A$, for example, is the partial derivative of R with respect to A.

Table 2-3. Formulas for Evaluating Error Propagation

	Class of Error			
	Systematic (Determinate)		Random (Indeterminate)	
Operation: Addition and subtraction ($R = A + B - C$)	$r = a + b - c$	(2-12)	$\sigma_R = \sqrt{\sigma_A^2 + \sigma_B^2 + \sigma_C^2}$	(2-16)
Multiplication and division ($R = AB/C$)	$\frac{r}{R} = \frac{a}{A} + \frac{b}{B} - \frac{c}{C}$	(2-13)	$\frac{\sigma_R}{R} = \sqrt{\frac{\sigma_A^2}{A^2} + \frac{\sigma_B^2}{B^2} + \frac{\sigma_C^2}{C^2}}$	(2-17)
Exponentials ($R = A^n$, where n is constant)	$\frac{r}{R} = n\frac{a}{A}$	(2-14)	$\sigma_R = nA^{n-1}\sigma_A$	(2-18)
Logarithms ($R = \log A$)	$r = 0.434\frac{a}{A}$	(2-15)	$\sigma_R = \frac{0.434\sigma_A}{A}$	(2-19)

NOTE:
R = calculated result
A, B, C = quantities from which R is calculated
r, a, b, c = absolute signed error of R, A, B, C
$\sigma_R, \sigma_A, \sigma_B, \sigma_C$ = standard deviation of R, A, B, C

distinction has been drawn between systematic and random errors. In brief, for systematic errors, the absolute signed values are propagated for addition and subtraction and the relative ones for multiplication and division. For random errors, the variances (that is, the square of the standard deviations) are added to yield the total variance for addition and subtraction and the relative variances are added for multiplication or division.

When a result is arrived at by addition or subtraction, or both, or by multiplication or division, or both, the equations in the table are immediately applicable. When more complex mixed calculations are involved, addition or subtraction, or both, is considered first, then multiplication or division, or both. Some simple examples show how these error propagation equations can be applied.

The corresponding general formula for random (indeterminate) errors is

$$\sigma_R = \sqrt{\left(\frac{\delta R}{\delta A}\right)^2 \sigma_A^2 + \left(\frac{\delta R}{\delta B}\right)^2 \sigma_B^2 + \left(\frac{\delta R}{\delta C}\right)^2 \sigma_C^2 + \cdots}$$

If you are familiar with calculus, you should derive the equations of Table 2-3 from these general formulas as an exercise.

Example 2-10 Two objects were each weighed the same number of times, and the mean weights and standard deviations were calculated as 5.240 ± 0.003 and 7.608 ± 0.004 g. What is the standard deviation of the sum (or the difference) of the weights? How could the sum and the difference of the weights be reported?

For a simplified calculation, the standard deviations can be expressed in milligrams. Applying Equation (2-16) from Table 2-3 yields

$$\sigma_R = \sqrt{3^2 + 4^2} = \pm 5 \text{ mg}$$

$$R_{\text{sum}} = 5.240 + 7.608 = 12.848 \text{ g}$$

$$R_{\text{diff}} = 7.608 - 5.240 = 2.368 \text{ g}$$

The sum of the weights can be reported as 12.848 ± 0.005 g (s) and the difference as 2.368 ± 0.005 g (s).

Example 2-11 What is the systematic error in the percentage of a component in a mixture when there are relative weighing errors of +0.1% for the component and -0.3% for the total mixture?

$$\% \text{ component} = \frac{\text{wt of component}}{\text{wt of mixture}} \times 100$$

Equation (2-13) of Table 2-3 applies:

$$\frac{r}{R} = \frac{a}{A} + \frac{c}{C} = +0.1 - (-0.3) = +0.4\%$$

Example 2-12 A result is to be calculated, as indicated, taking into account the absolute systematic errors, as given within the parentheses:

$$R = \frac{5.24(-0.03) \times 0.145(+0.002)}{2.16(-0.04)} = 0.351759$$

Calculate the absolute error in the result.
Equation (2-13) of Table 2-3 applies:

$$\frac{r}{R} = \frac{r}{0.351759} = \frac{-0.03}{5.24} + \frac{+0.002}{0.145} - \frac{-0.04}{2.16}$$

$$= -0.00572 + 0.01379 + 0.0185 = 0.0380$$

The absolute error of the result is

$$r = R \times 0.0380 = 0.351759 \times 0.0380 = 0.01338$$

The results can therefore be expressed as 0.35 or 0.35_2.

Example 2–13 Assume the same mixed calculation as in Example 2–12 but consider the errors within the parentheses to be standard deviations (using the signs ±). Calculate the standard deviation of the result.

Equation (2–17) of Table 2–3 applies:

$$\frac{\sigma_R}{R} = \sqrt{\left(\frac{0.03}{5.24}\right)^2 + \left(\frac{0.002}{0.145}\right)^2 + \left(\frac{0.04}{2.16}\right)^2}$$

$$= \sqrt{3.278 \times 10^{-5} + 1.902 \times 10^{-4} + 3.429 \times 10^{-4}}$$

$$= 2.379 \times 10^{-2}$$

and $$\sigma_R = 0.351759 \times 2.379 \times 10^{-2} = \pm 0.00837$$

From the standpoint of random error, the result can be expressed 0.352 ± 0.008 (s).

Example 2–14 A result R is calculated from the relation $R = (A - B)K/W$. (a) Derive an equation for the relative random error in R in relation to the random error for the four parameters. (b) For A, B, K, and W that have the values and standard deviations 50.00 ± 0.02, 37.00 ± 0.02, 1.000 ± 0.002, and 0.1335 ± 0.0002, calculate the value and the standard deviation of R.

Subtraction must be considered first. A symbol D for the term $A - B$ can be introduced:

$$D = A - B$$

Applying Equation (2–16) yields

$$\sigma_D = \sqrt{\sigma_A^2 + \sigma_B^2}$$

Then Equation (2–17) takes the form

$$\frac{\sigma_R}{R} = \sqrt{\frac{\sigma_D^2}{D^2} + \frac{\sigma_K^2}{K^2} + \frac{\sigma_W^2}{W^2}}$$

Substitution of the data and solution is left to the student.

$$\frac{\sigma_R}{R} = 3.3133 \times 10^{-3}$$

$$R = 97.378 \quad \text{and} \quad \sigma_R = \pm 0.3226$$

The result can therefore be expressed as 97.4 ± 0.3 (s).

For logarithmic functions, the propagation of error is also important in considering, for example, concentrations (pH, and the like) and calculations involving equilibrium constants (pK, log K). It is noteworthy that according to Equation (2-19), the *absolute* random error for a logarithm is determined by the *relative* random error of the corresponding number, and vice versa.

Example 2-15 The formal concentration and standard deviation of a hydrochloric acid solution, HCl, is $(2.20 \pm 0.03) \times 10^{-3}F$. For this solution calculate the pH and its standard deviation.

The relation between pH and hydrogen ion concentration, as given in Section 8.4, is pH = $-\log [H^+]$. Substituting in Equation (2–19) yields

$$\sigma_R = \frac{0.434 \times 0.03}{2.20} = \pm 5.92 \times 10^{-3}$$

$$\text{pH} = -\log 2.2 \times 10^{-7} = 2.6576$$

The pH can therefore be reported as 2.678 ± 0.006 (s).

Example 2-16 The solubility product K_{sp} of silver bromide, AgBr, has the value 5.2×10^{-13}. Assume a standard deviation of $\pm 0.3 \times 10^{-13}$ for this value and calculate the formal solubility S^f for this salt (see Section 5.4) and its standard deviation.

$$S^f = \sqrt{K_{sp}} = \sqrt{5.2 \times 10^{-13}} = 7.2 \times 10^{-7}F$$

The function obviously has the form

$$R = A^{1/2}$$

From Equation (2–18)

$$\sigma_R = nA^{n-1}\sigma_A = \tfrac{1}{2}A^{-1/2}\sigma_A = \frac{\sigma_A}{2\sqrt{A}}$$

Substituting the value assumed for σ_A gives

$$\sigma_R = \frac{0.3 \times 10^{-13}}{2\sqrt{5.2 \times 10^{-13}}} = \pm 2.08 \times 10^{-8}F$$

The formal solubility can therefore be presented as $(7.2 \pm 0.2) \times 10^{-7}F$ (s).

2.22 *Questions*

2-1 Which of the following statements are true? Which false? Which need elaboration? Explain.

(a) The average of some experimental values is highly accurate when the values show high precision.

(b) High precision always affords high accuracy.

(c) Low accuracy always excludes high precision.

(d) Low precision is inconsistent with high accuracy.

(e) Low precision always affords low accuracy.

2-2 Indicate what is ambiguous about the following statement: The result to be reported for a titration is 6.03 ± 0.03 mL.

2-3 Comment on the validity of the following statements and qualify them or elaborate.

(a) The mean of five experimental values is more precise than the mean of three such values.

(b) The error in an experiment is either random or systematic.

(c) High precision excludes the presence of a systematic error.

(d) A systematic error can be detected and evaluated solely by a statistical treatment.

(e) A systematic error is always positive.

(f) A random error always involves a negative deviation from the "true" value.

(g) The "true" value must be known to determine the standard deviation.

2-4 In the following circumstances, classify the error and indicate in which direction it will affect the result.

(a) Calcium is coprecipitated in the gravimetric determination of iron(III) based on precipitation of the hydrous oxide and ignition to iron(III) oxide.

(b) The "25.00-mL" pipet actually delivers 25.30 mL.

(c) The color change of an indicator in a visual acid-base titration occurs at the proper pH but is very indistinct; hence it is difficult to locate the end point.

(d) The color change of an indicator is sharp but occurs before the equivalence point is reached.

(e) The color change of an indicator is indistinct and occurs after the equivalence point is reached.

(f) A precipitation reaction does not go to completion, but proceeds reproducibly and exactly to the extent of 96%.

(g) During a titration a few drops of titrant were spilled.

(h) Water containing calcium was used in the solution of the samples to be analyzed for their calcium content.

2-5 Examine the following statement critically: The average weight was found to be 16.342 g.

2-6 Discuss the following statement: The method shows poor precision as a gravimetric determination because the precipitate is relatively soluble.

2-7 Explain the error in the following student statement: I realize that the concentration of my titrant has not been determined *accurately,* but I can perform the titration with this solution many times, average the results, and in that way still obtain a satisfactory answer.

2-8 Discuss the statement: A small value for the range expresses high precision.

2-9 The calculated result of an analysis often is reduced to the fraction (or percentage) R of the determined substance related to some measured value M for that substance and the amount W of the sample by $R = KM/W$, where K is an empirical factor. Write expressions for the systematic and random errors in R due to errors in K, M, and W. Frequently, the errors in M and W are insignificant; then to what do the two expressions reduce? Elaborate on what the various expressions imply for practical analysis.

2-10 In relation to assessing error propagation, elaborate on the advantages and disadvantages of variance instead of standard deviation as a measure of random error.

2-11 The values for S and T are calculated from $S = K^{1/2}$ and $T = K^{1/3}$. Decide whether the absolute and relative systematic and random errors will be greater or smaller for S than for T.

2-12 In assessing propagated errors, it is often permissible to neglect an *absolute* determinate error or a standard deviation that is smaller than other ones. Is the same simplification allowed when *relative* determinate errors and standard deviations are involved? Elaborate.

2-13 Refer to Problem 2-17 and elaborate on why systematic errors may cancel even if they are large and standard deviations never do.

2.23 *Problems*

2-1 The velocity of light in a vacuum is 29,979,300,000 cm/s. How should this value be written to express unambiguously that the measured value is known to seven significant figures? *Ans.:* 2.997930×10^{10} cm/s

2-2 Perform the indicated arithmetic operations and express the result to the proper number of significant figures, assuming that each value is uncertain to the extent of ±1 in the last figure.

(a) 3.67 + 34.236 + 863.6
(b) 521.5 + 3.77 − 8.09
(c) 3.6 × 48.1 × 0.216
(d) (9.8 × 6.44)/1.21

Ans.: (a) 901.5; (b) 517.2; (c) 37; (d) 52

2-3 What is faulty with the presentation of a value as 646.52 ± 2?

2-4 Write the logarithms of the following data and retain the correct number of significant figures: (1) 24.6; (b) 2.6×10^{-3}; (c) 0.0643.
Ans.: (a) 1.391; (b) 0.42 − 3 or −2.58; (c) 0.808 − 2 or −1.192

2-5 Give the correct number of significant figures for the following numbers.

(a) 231.4
(b) 0.10004
(c) 0.002300
(d) 620.04
(e) 3.23×10^{-3}
(f) 0.044×10^4
(g) 202.0
(h) 0.02020
(i) 0.00202
(j) 202

2-6 Give the logarithms of the following values with the correct number of significant figures.

(a) 21.4
(b) 22.301
(c) 1.00×10^4
(d) 1×10^{-3}
(e) 0.023

2-7 Give the numbers, with the correct number of significant figures, corresponding to the following logarithms.

(a) 0.30
(b) 1.301
(c) 2.3010
(d) −4.66
(e) 8.4563

2-8 Round each of the following values to three significant figures.

(a) 88.55
(b) 8.345
(c) 3.2547×10^{-6}
(d) 213.4
(e) 699.9

2-9 Round each of the following values to four significant figures.

(a) 33.352
(b) 32.356
(c) 32.3549
(d) 0.49996
(e) 5,632,445.4

2-10 Perform the indicated addition, giving the correct number of significant figures in the results.

(a) $301.0 + 36.54 + 423 =$
(b) $2.56 + 0.732 + 1.6 =$
(c) $4.367 + 0.486 + 16.3 =$
(d) $0.001 + 0.16 + 1.009 =$

2-11 Perform the indicated arithmetic operations and express the results with the correct number of significant figures, assuming that each value has an uncertainty of ±1 in the last place.

(a) $88.3 + 4.22 - 1.457 =$
(b) $(3.21 \times 10^{-4}) \times 21.3 =$
(c) $4.22 \times 9.22 =$
(d) $7.89 + 3.456 - 6.775 =$
(e) $(2.335 + 7.892) \times 3.678 =$
(f) $\dfrac{6.63 \times 2.98}{4.6997} =$
(g) $\dfrac{88.14 + 24.1}{17.99 - 34.64} =$

2-12 (a) For the given set of values, determine the average, average deviation, range, and a standard deviation for a single measurement and the standard deviation of the average: 52.33, 52.28, 52.35, and 52.30 mL.

Ans.: 52.32, 0.02_5, 0.07, ±0.03, and ±0.02 mL

(b) Decide whether an additional value 52.46 mL could be rejected on the basis of the rule of "four times" average deviation and the $Q_{90\%}$ test.

Ans.: Yes; yes

(c) How should the result be reported in terms of the average and of the standard deviation of a single measurement?

Ans.: 52.32 ± 0.03 mL (*s*; 4)

(d) Express the relative error of the average of the set in parts per thousand when the true value is 52.36 mL. *Ans.:* 0.8 ppt

(e) Calculate the relative range. *Ans.:* 0.0013

(f) Calculate the percentage standard deviation of a single measurement.

Ans.: ±0.06%

2-13 From the given set of results—15.30, 15.45, 15.40, 15.45, and 15.50—calculate (a) the average, and the standard deviation; and (b) the standard deviation of the average. Decide to how many decimal places (c) individual results and (d) the average are properly reported.

Ans.: (a) 15.42 ± 0.08 (*s*; 5); (b) ±0.04; (c) two; (d) two

2-14 The solubility product K_{sp} of a salt of the type AB_2 is related to its formal solubility S^f by $S^f = \sqrt[3]{K_{sp}/4}$ (see Section 5.4). For a given salt, the K_{sp} value of 3.0×10^{-14} is secured, with a systematic error of -0.3×10^{-14}. Calculate the relative and the absolute systematic errors for the value of S^f. The standard deviation of the K_{sp} value is $\pm 0.3 \times 10^{-14}$; calculate the corresponding relative error and absolute random error for S^f.

2-15 A solution of a strong acid was found to have a pH of 3.85, with a standard deviation of ±0.05. What is the standard deviation of the hydrogen ion molar concentration calculated from the pH value?

2-16 Assume the situation of Problem 2-15 with the exception that a determinate error of −0.05 pH unit is involved. Calculate the relative and absolute determinate error in the calculated hydrogen ion concentration.

2-17 When a sample was weighed, the container alone weighed 9.3356, and the container with the sample, 12.3452 g. (a) It is known that one of the weights involved in both operations was in error by −0.5 mg; calculate the systematic error in the sample weight. (b) Assume that the standard deviation in each of the weighings is ±0.5 mg; calculate the standard deviation of the sample weight and also the relative standard deviation in percentage.

Ans.: (a) zero; (b) ±0.7 mg and ±0.023%

2-18 For the following expressions, calculate R and its absolute and relative determinate errors. The absolute systematic error for each value is given within parentheses.

(a) $R = 15.0(-0.2) + 12.0(+0.3)$
(b) $R = 15.0(-0.2) - 12.0(+0.3)$
(c) $R = 23.25(+0.06) \times 0.0143(0.0006)$

(d) $R = \dfrac{23.25(+0.06)}{0.0143(-0.0006)}$

(e) $R = [16.2(-0.5) + 31.8(+0.4)] \times [0.66(+0.05)]$
(f) $R = [16.2(-0.5) - 31.8(+0.4)] \times [0.66(+0.05)]$

(g) $R = \dfrac{16.2(-0.5) + 31.8(+0.4)}{0.66(+0.05)}$

(h) $R = \dfrac{16.2(-0.5) - 31.8(+0.4)}{0.66(+0.05)}$

2-19 Use the expressions of Problem 2-18, taking the errors within the parentheses to be the standard deviations (with the signs ±). Calculate the absolute and the relative standard deviation for each R value.

3

Expression of Concentration and Content

3.1 *Introduction*

Quantitative analysis by its nature involves measurement and thus units. The units encompassed by the International System (SI) are summarized in Table K of the Appendix; added information on SI electrical units is given in Section 22.1.

The analytical chemist is confronted by mixtures of two or more components, and especially by homogeneous mixtures, that is, solutions. Solutions may be liquid, solid, or gaseous; however, in analytical chemistry a solution is assumed to be *liquid* unless otherwise qualified. In considering a solution of a solid (or a gas) in a liquid, the former is called the solute and the latter the solvent. It is necessary to express unambiguously the composition of a mixture and especially the content of the solute, which for a solution is termed *concentration.* Concentration and content are expressed in various ways, depending on the data available and the required purposes. The diverse modes of expression must be mastered to appreciate the calculations and methods of quantitative analysis. These expressions can be grouped into two categories: those involving physical units only and those involving chemical units.

3.2 *Dimensions and Units*

A word on the concepts *units* and *dimensions* is appropriate. Physical terms can be expressed in basic dimensions such as length, time, and mass. Speed, for example, has the dimensions of length per time, that is, $l \cdot t^{-1}$. For practical use, units must be attached to a term. Speed, for example, can be given in centimeters per second, kilometers per hour, and so on.

Sometimes a term is dimensionless and yet has units attached. An example is given in Section 3.4 in the consideration of formula weight. The term *formula weight* is dimensionless, yet has the attached units of grams per mole. There is no contradiction since *grams* and *mole* have the same dimension, namely mass.

A mathematical equality must have identical dimensions on each side of the equals sign. Achieving this may call for introducing a *proportionality constant,* which will have the dimensions required by the equality and the units that follow from whatever units are assigned to the variables. An equality between mass (weight) W and volume V, for example, is secured by introducing density d, so that $W = dV$. The density has the dimensions $m \cdot l^3$; the units will depend on whatever units are assigned to mass and volume.

A *conversion factor* is used to convert from one unit or group of units to another. A conversion factor is a dimensionless number. For converting kilograms to grams, for example, the value is multiplied by 10^{-3} $g \cdot kg^{-1}$. A conversion can be more involved, but the principle of operation is the same. For example, °C = (°F - 32)5/9. Degrees Fahrenheit and degrees Celsius have the same dimension (temperature). The fraction 5/9 in the equality is dimensionless and has the units of °C/°F.

The student can be confused when a logarithmic term is involved, such as log C, where C is a concentration having the dimensions $m \cdot l^{-3}$. The student often, correctly, recalls that the logarithm can be found for a "pure" number but not for a dimensioned one. The trick is to subtract zero. Recall that the common logarithm of zero is 1 and that log a - log b = log (a/b). For a concentration of, say, $10^{-2} m \cdot l^3$,

$$\log C = \log C - 0$$
$$= \log\left(\frac{10^{-2} m \cdot l^3}{1 m \cdot l^3}\right)$$
$$= \log 10^{-2}$$
$$= 2$$

You are well advised to keep dimensional and unit analysis in mind when considering the examples and problems throughout this textbook.

3.3 Concentration and Content in Physical Units

Percentage Weight by Weight. Percentage weight by weight, expressed as % w/w, and often termed *weight percent,* refers to the parts by weight of a component per hundred parts by weight of the mixture. Thus, a binary magnesium-aluminum alloy stated to be 12% w/w in aluminum contains 12 parts by weight (grams, pounds, and so on) of aluminum in 100 parts by weight (grams, pounds, and so on) of alloy. Since it is a binary alloy, it is by difference 88% w/w in magnesium. It is noteworthy that percentage weight by weight is independent of temperature expansion or contraction, in contrast to percentage weight by volume. Percentage weight by weight is common for expressing the composition of alloys, the purity (assay) of chemical substances, and the strength of concentrated acids and bases. When the density of a solution is known, that is, the weight per unit volume, a percentage-weight-by-weight value can be converted to percentage weight by volume.

Percentage Volume by Volume. Percentage volume by volume, expressed as % v/v, and often termed *volume percent,* refers to the parts by volume of a component per hundred parts by volume of the mixture. This form of expression is largely limited to mixtures of gases or of liquids. Thus, a 55% v/v mixture of methanol with water contains 55 volumes (milliliters, pints, and the like) of methanol in 100 volumes (milliliters, pints, and the like) of the methanol-water mixture. Since the volume of a liquid is temperature-dependent, it is usually necessary to specify a temperature when reporting a percentage-volume-by-volume value. It should be emphasized that the mixing of different liquids is often attended by a marked expansion or contraction in volume. A 55% v/v mixture of methanol and water is *not* obtained by mixing 55 mL of methanol with 45 mL of water, but rather by adding water to 55 mL of methanol with mixing, to obtain a total solution volume of 100 mL. If 45 mL of water and 55 mL of ethanol were mixed, the volume of the mixture would be slightly less than 100 mL as a result of contraction caused by intermolecular action of the two components.

Percentage Weight by Volume. Percentage weight by volume, expressed as % w/v, and often termed *weight by volume percent,* refers to the parts by weight of a component in 100 parts by volume of the mixture. The weight and volume units should be compatible. Thus, a 10% w/v sodium chloride solution contains 10 g of sodium chloride in 100 mL of solution (not in 100 mL of solvent!). Since the volume of a solution is temperature-dependent, the temperature must usually be specified in reporting a percentage-weight-by-volume value. If the density of a solution is known for the specified temperature, a percentage-weight-by-volume value can be converted to percent weight by weight, and vice versa. For example, the density of a 10.0% w/w aqueous sodium chloride solution at 20.0°C is 1.07 g/mL; hence, the % w/v = $10.0 \times 1.07 = 10.7$% w/v. For very dilute aqueous solutions, w/w and w/v percentages do not differ much in value, since the density is close to 1 g/mL; the percentages are for practical purposes often taken as identical.

Parts per Thousand and per Million. All the percentage expressions of concentration have analogous parts per thousand (ppt) and parts per million (ppm) expressions. However, a weight-by-weight basis is usually used and is assumed unless it is otherwise stated. With very dilute aqueous solutions, the distinction between w/w and w/v is negligible. An alloy 1 ppt in silver contains 1 part by weight of silver in 1000 parts by weight of the alloy. A water sample 2 ppt in sulfate contains, for example, 2 g of sulfate per 1000 g of sample solution, or since a dilute aqueous solution is involved, per 1000 mL (that is, per liter) of sample solution. A value in parts per million is equivalent to milligrams per kilogram or micrograms per gram, and in the case of dilute aqueous solutions, to milligrams per liter or micrograms per milliliter.

Miscellaneous Expressions. In laboratory practice, the concentration of aqueous solutions of acids and ammonia is frequently expressed, for example, as 1:8 HNO_3 or 1 + 8 HNO_3; either expression implies that one volume of commercial concentrated nitric acid is mixed with eight volumes of water. This special notation has the advantage that no calculation is necessary and only simple measurement of volumes is needed when preparing such a dilute solution. This mode of expression might seem ambiguous at first glance; however, most concentrated acids and ammonia are marketed in the same concentration by all reagent suppliers. Thus, reagent-grade nitric acid contains nominally 70% w/w HNO_3. For a concentrated acid commonly available in two different concentrations, this form of expression is best avoided; an example is perchloric acid, which is available in the reagent grade as 70–72% w/w and 60–62% w/w products. The concentration of solvent mixtures can be expressed similarly; thus, 1:1 methanol-water implies that one volume of methanol is mixed with one volume of water.

Density and Specific Gravity. The density of a substance or of a mixture is the weight per unit volume and is usually expressed in grams per milliliter. The density is temperature-dependent, since the volume varies with the temperature. The specific gravity of a substance or of a mixture is the ratio of its density at a specified temperature to the density of a reference substance, commonly water, at a specified temperature. The two specified temperatures need not be identical. Specific gravity values can be converted to density values if the density (absolute) of the reference substance is known for the specified temperatures. Textbooks on physics and handbooks presenting physical data can supply further details.

For *dilute* aqueous solutions, since the density of water remains close to 1.0 g/mL over a wide range of temperature, specific gravities and densities can be used interchangeably, often with an error less than 1 part in 5000. For simple solutions (that is, one solute and one solvent), the concentration of the solute at a given temperature can be related empirically to the density of the solution. This serves for the assay of some simple solutions, including commercial mineral acids. Common examples are estimating the sulfuric acid content of the electrolyte in lead storage batteries and estimating the ethylene glycol or methanol content of automotive radiator antifreeze solutions using specially calibrated hydrometers.

3.4 Concentration and Content in Chemical Units

The expressions for concentration can be viewed as involving physical units. For the purposes of analytical chemistry, expression in terms of chemical units is often advantageous.

Atomic Weight. In the SI concepts, the atomic weight, also known as the relative atomic mass A_r, is defined as the ratio of the average mass per atom of the element A with its natural isotopic composition to one-twelfth of the mass of an atom of carbon 12. In other words, the atomic weight is the mass ratio of one average atom of the element A to one atom of the carbon 12 isotope, to which a value of exactly 12 is assigned by the international atomic weight scale (see Table L in the Appendix). The atomic weight is consequently a dimensionless number.

Formula Weight. The formula weight of a substance, also known as the relative molecular mass, is the ratio of the average mass "per formula" of the substance with the constituent atoms in their natural isotopic composition to one-twelfth of the mass of an atom of carbon 12. In other words, the formula weight of a substance is the sum of the atomic weights of each element taken as many times as indicated by the formula. The formula weight is consequently also a dimensionless number. (Unless specifically stated otherwise, the formula weight and molecular weight are taken as identical.)

For purposes of description, a species can be stated to correspond to a stated number of "atomic mass units," abbreviated "amu." This assignment of a unit is arbitrary; the formula weight and the atomic weight are still dimensionless. The statements "Chlorine gas, Cl_2, has a formula weight of 70.906" and "Chlorine gas, Cl_2, corresponds to 70.906 amu" are equivalent.

Mole. The mole, abbreviated "mol," is the SI base unit for amount of substance and is defined as the amount of substance of a system that contains as many elementary entities as there are atoms in exactly 0.012 kg of carbon 12. When the mole is used, the elementary entities must be specified, and they may be atoms, molecules, ions, electrons, other particles, or specified groups of such particles.

If you have been exposed to other courses and textbooks not yet conforming to SI practices, you will note that this definition of the mole is broader than older ones. Terms such as *gram atom* and *gram ion* are no longer needed because the species must be declared. Another change, which is of greater consequence, is that the term *amount* in the SI definition of the mole has no stated units and can be interpreted in various ways. For our analytical purposes, the term *amount* can usually be treated traditionally, namely as the formula weight in grams.* The mole is frequently too large a unit, and submultiples find use in analytical chemistry (1 mmol = 1×10^{-3} mol; 1 μmol = 1×10^{-6} mol). For example, the atomic weight of calcium is 40.08 and one mole corresponds to 40.08 g (1 mmol = 40.08 mg and 1 μmol = 40.08 μg). The compound sodium chloride, NaCl, has the formula weight 22.98977 + 35.453 = 58.443. One mole of sodium chloride corresponds to 58.443 g (1 mmol = 58.443 mg and 1 μmol = 58.443 μg).

---

*With the restriction of *amount* to "mass" in the definition of the mole, the SI concept of *molar mass* need not be considered. The molar mass is the mass of a substance divided by the amount of the substance. When molar mass is expressed in the practical units of $g \cdot mol^{-1}$, the "pure" number is identical with the value for the atomic or the formula weight.

The number of moles of a substance corresponding to a certain weight of it in grams is given by the relation

$$\text{Number of moles} = \frac{\text{weight in grams}}{\text{formula weight}}$$

Rearranging this expression yields

$$\text{Formula weight} = \frac{\text{weight in grams}}{\text{number of moles}}$$

The formula weight consequently has the *units* grams per mole. The formula weight by definition, however, is a dimensionless number. Any contradiction is only apparent. *Gram* and *mole* (as used in this textbook) have the same dimension, namely mass. The formula weight, consequently, even with the units "grams per mole" attached, is still dimensionless.

It is important that you fully appreciate these definitions and the distinctions drawn, since you may have been exposed to different ones that are equally valid and may possess advantage for some area of science.

Mole Fraction and Mole Percent. If in a given amount of a binary mixture of two substances A and B, a moles of A are present and b moles of B, then the mole fraction of substance A present is given by $a/(a + b)$ and the mole fraction of B by $b/(a + b)$. The fractions are commonly expressed decimally. The sum of the mole fractions of A and of B is unity. That the concept can be extended to mixtures of three or more components is obvious. If a mole fraction is multiplied by 100, a mole percent is obtained. Consider a mole of sodium chloride; it contains one mole of sodium ion and one mole of chloride ion. The mole fraction and mole percent of sodium present are 0.5 and 50%.

Formality. Formality as an expression of chemical concentration is defined as the moles of a substance per liter of solution (not per liter of solvent). Since one mole corresponds to the amount of substance in grams numerically equal to the formula weight, the formula of the substance must either be given or understood when a formality is stated. With the concentration of a solution expressed in formality, it is possible to relate chemical units to volumes, which are more easily measured than weights. However, since the volume of a solution is temperature-dependent, temperature must be specified when the formality of a solution is stated. If no temperature is given, room temperature is assumed. Formality is denoted by a capital F (italic in printing, and best underlined in handwritten text and problems). Thus a $1.30F$ aqueous sodium chloride solution, which may be further abbreviated as $1.30F$ NaCl (water is assumed as the solvent unless otherwise specified), contains $1.30 \times 58.44 = 76.0$ g of sodium chloride dissolved in water and diluted with water to a total volume of exactly 1 L. (Note that the solution is not obtained by dissolving 76.0 g of sodium chloride *in* 1 L of water.)

In mathematical expressions in this textbook, formal concentration is denoted by the capital letter *C*, with subscripts where necessary to differentiate various species. Thus the expressions C_A and C_{HCl} denote the formal concentration of substance A and of hydrochloric acid.

Molarity and Its Relation to Formality. A solution of a desired formality can be prepared by dissolving the necessary number of moles of the substance and diluting to 1 L. The statement of the formality in no way implies that this substance is present as such, that is, as added. It may dissociate, associate, form complexes with other substances added or present, or undergo some reaction. However, all these changes do not alter the fact that the solution is of the stated formality.

It is often desirable to have a special label for the concentration of any species that forms as a result of these changes. The concentration of a species *actually* present is conveniently expressed in molarity, that is, as the moles of that species per liter of solution. In mathematical expressions, molarity is denoted by placing the designation for the species within brackets. Thus the expressions [A] and $[Cl^-]$ denote the molar concentrations of substance A and of chloride ion.

To bring the distinction between formality and molarity into sharp relief, a simple example will be helpful. Consider that 0.100*F* solutions of hydrochloric acid and acetic acid are prepared. These solutions might be prepared by dissolving in water 1/10 mole of each of hydrogen chloride and acetic acid and diluting each solution to 1 L with water. From elementary chemistry, you will appreciate that the hydrogen chloride completely dissociates in the aqueous solution into hydrogen ion and chloride ion and that no undissociated hydrogen chloride is present. You also recall that the acetic acid solution contains only small amounts of hydrogen ion and acetate ion and that most of the acetic acid is present in its undissociated form. Regardless of these chemical facts, each solution is 0.100*F* in its acid. However, the molarity of undissociated hydrogen chloride is *zero;* in contrast, the molarity of the undissociated acetic acid has a finite value. Because of the complete dissociation of hydrogen chloride, the following relations hold for 0.100*F* HCl:

$$\underset{0.100F}{C_{HCl}} = \underset{0.100M}{[H^+]} = \underset{0.100M}{[Cl^-]}$$

Because of the incomplete dissociation of acetic acid, HOAc, the following relations hold for 0.100*F* acetic acid:

$$\underset{0.100F}{C_{HOAc}} = \underset{98.66 \times 10^{-3}M}{[HOAc]} + \underset{1.34 \times 10^{-3}M}{[OAc^-]}$$

Since one mole of hydrogen ion is formed for each mole of acetate ion, the molarity of hydrogen ion is also $1.34 \times 10^{-3}M$. (The molarities given for the acetic acid solution are calculated by means to be described in a later section.)

The distinction between total analytical concentration and actual concentration expressed in moles per liter, termed, respectively, formality and molarity, is extremely helpful for the unambiguous consideration of chemical equilibria. This distinction is not always made, however, what is here denoted as formality is sometimes also denoted as molarity. (For example, the International System of Units (SI) does not use formality.) When molarity only is used, the intended sense must be recognized from the context on the basis of the relevant chemical facts and in mathematical relations by the frequent use of C with a subscript, versus brackets around a formula or symbol.

Equivalents and Normality. When two substances interreact, they need not do so in equal *molar* amounts. One mole of one substance may be "chemically equivalent" to two or more moles of the other substance (for example, one mole of sulfuric acid requires two moles of sodium hydroxide for complete neutralization). For many purposes it is desirable to have a unit that implicitly incorporates the chemical relation existing in the reaction. This unit is known as the chemical equivalent, or more briefly, the equivalent. The number of equivalents of substance is obtained by multiplying the number of moles by the equivalence number, which is derived from the molar combination ratio, or reaction ratio, of the two substances. The equivalent weight of a substance is obtained by dividing the formula weight by the equivalence number. Since the latter is a dimensionless number, the equivalent and the equivalent weight have the same dimensions and units as the mole and the formula weight, respectively. The normality of a solution is defined as the equivalents of a substance in one liter of solution. Normality is denoted by a capital N (in italic in printing and best underlined in handwritten text and problems).

The terms *equivalent, equivalent weight,* and *normality* must be related to the reactions to which they apply; otherwise, ambiguity often results. For a solution used as an active participant in various reactions involving different numbers of reactive species per molecule, it is best to label concentration in formality. (Further discussion of these various terms is deferred to the consideration of titrimetry, Chapters 15, 19, and 26.)

Titer. Another way of expressing chemical concentration is by titer. A titer expresses the relation of the solute present in a unit volume of its solution to the amount of a stated substance for a particular reaction of the solute with that substance. Thus, the "strength" of a hydrochloric acid might be stated as 1.00 mL = 4.08 mg of Na_2CO_3. The value 4.08 mg is the sodium carbonate titer of the particular hydrochloric acid under the actual reaction conditions. The use of titers gained early recognition in titrimetry, since no knowledge of the chemical reaction (or reactions) involved is needed. The titer method simplifies calculations. It is especially advantageous in the routine analysis of many similar samples; and indeed when "standard" samples are used to establish the titer under the actual conditions of the determination, some errors may be compensated. When the reaction is known to proceed quantitatively, a titer can be calculated from the stoichiometry of the reaction. The titer concept will be more fully understood when we examine the theory and practice of titrimetry (see Chapters 19 and 26).

3.5 *Illustrative Examples: Concentration and Content*

In dilution problems, unless it has been otherwise stated, assume that volumes are additive, that is, that contraction or expansion in volume on mixing either does not exist or is neglected.

Example 3–1 What weight of NaCl is present in 1.50 kg of a NaCl–KCl mixture 16.2% w/w in NaCl?

Since exactly 100 kg of the mixture contains 16.2 kg of NaCl, then 1.50 kg contains x kg; hence by proportion,

$$\frac{100}{16.2} = \frac{1.50}{x}$$

and $$x = \frac{16.2 \times 1.50}{100} = 0.243 \text{ kg} = 243 \text{ g of NaCl}$$

Example 3–2 What weight of the mixture of Example 3–1 must be taken to provide exactly 400 g of KCl?

Since exactly 100 g of the mixture contains 100 – 16.2 = 83.8 g of KCl, then x g contains 400 g; hence,

$$\frac{100}{83.8} = \frac{x}{400}$$

and $$x = \frac{100 \times 400}{83.8} = 477 \text{ g of the mixture}$$

Example 3–3 What weight of water must be used to dissolve 25.0 g NaCl to obtain an 8.00% w/w solution?

Since exactly 100 g of the solution contains 8.00 g of NaCl, then x g contains 25.0 g; hence,

$$\frac{100}{8.00} = \frac{x}{25.0}$$

and $$x = \frac{100 \times 25.0}{8.00} = 312 \text{ g of solution}$$

Therefore, weight of water needed = 312 – 25 = 287 g

Example 3–4 A 45.0% w/v aqueous solution of substance A has a density of 1.62 g/mL. Calculate the % w/w of this solution.

The weight of 100 mL of the solution is 100 × 1.62 = 162 g; this volume contains 45.0 g of substance A. Hence,

$$\% \text{ w/w of substance A} = \frac{45.0 \times 100}{162} = 27.8\%$$

Example 3-5 Express the composition in mole percent of a mixture of NaCl (*58.4*) and K_2SO_4 (*174.3*) that is 65.0% w/w in NaCl.*

Exactly 100 g of the mixture contains 65.0 g of NaCl or 65.0/58.4 = 1.113 mol of NaCl.

Exactly 100 g of the mixture contains 100 − 65.0 g = 35.0 g of K_2SO_4 or 35.0/174.3 = 0.2008 mol of K_2SO_4.

Hence, the total moles in exactly 100 g of the mixture = 1.113 + 0.201 = 1.314 mol and the mole percent of NaCl = 1.113 × 100/1.314 = 84.7 mol % of NaCl.

By difference, the mole percent of K_2SO_4 = 100.0 − 84.7 = 15.3 mol %.

Example 3-6 What weight of a 14.0% w/w aqueous solution of substance B must be added to 35.0 g of a 6.0% w/w aqueous solution of B to obtain a 7.0% w/w solution?

Initially the amount of B present = 35.0 × 0.060 = 2.10 g of B. After the addition, the weight of solution will be 35.0 + x, where x is the number of grams of the 14.0% solution to be added.

The total amount of B in the final solution is 2.10 + x × 0.14 g. Hence the percentage of B in the final solution is given by

$$\frac{(2.10 + x \times 0.14) \times 100}{35.0 + x} = 7.0\%$$

Solving for x yields x = 5.0 g of the 14.0% w/w solution.

Example 3-7 What volume of 0.12*F* H_2SO_4 must be added to exactly 500 mL of 0.09*F* H_2SO_4 to obtain 0.10*F* H_2SO_4?

By reasoning like that used in solving Example 3-6, we arrive at the following equation, where x stands for milliliters of 0.12*F* H_2SO_4 to be added:

$$\frac{500 \times 0.09 + x \times 0.12}{500 + x} = 0.10$$

Solving for x yields $x = 2.5 \times 10^2$ mL of 0.12*F* H_2SO_4.

---

*Here, as often throughout this textbook, formula weights (sometimes rounded) are given parenthetically in italic numerals; these values are to be used in the calculations of the *particular* problem so that the arithmetic effort is reduced.

Example 3-8 What volume of water must be added to exactly 300 mL of $0.250F$ HCl to secure $0.200F$ HCl?

Reasoning like that used in solving Example 3-6 yields

$$\frac{300 \times 0.250 + x \times 0}{300 + x} = 0.200$$

Solving for x yields $x = 75.0$ mL of water to be added.

Note that the term $x \times 0$, which is included to make the reasoning fully analogous to that in Examples 3-6 and 3-7, corresponds to the fact that the concentration of hydrogen chloride in the water is zero.

Example 3-9 What is the formality of a sulfuric acid solution containing 62.0% w/w of H_2SO_4 (*98.0*) and having a density of 1.520 g/mL?

One liter (= 1000 mL) of the acid weighs $1000 \times 1.520 = 1520$ g and contains $1520 \times 0.620 = 942$ g of H_2SO_4; this amount of H_2SO_4 represents $942/98.0 = 9.61$ mol. Since the calculation is based on a volume of 1 L, the result is $9.61F$ H_2SO_4.

In consolidated form and with the units shown, the solution can be given as

$$\frac{10^3 \text{ mL/L} \times 1.520 \text{ g/mL} \times 62.0 \text{ g/100 g}}{98.0 \text{ g/mol}} = 9.61 \frac{\text{mol}}{\text{L}} = 9.61F$$

Example 3-10 What will be the formality of an ammonia solution secured by diluting 20.0 mL of concentrated aqueous ammonia (density, 0.90 g/mL; 28% w/w NH_3; mol wt NH_3, 17.0) to a volume of 50.0 mL with water? (Note that this dilution represents the addition of 30.0 mL of water.)

By reasoning like that in Example 3-6, the formality of the concentrated aqueous ammonia is calculated:

$$\frac{1000 \times 0.90 \times 0.28}{17.0} = 14.8F$$

Then reasoning like that in Example 3-8 yields the formality of the final solution:

$$\frac{20.0 \times 14.8 + 30.0 \times 0}{50.0} = 5.9_2F \text{ in } NH_3$$

Example 3-11 What volume of 40% w/w HCl (*36.5;* density, 1.198 g/mL) must be diluted with water to obtain exactly 250 mL of a $1.0F$ solution?

The formality of the initial acid is

$$\frac{1000 \times 1.198 \times 0.40}{36.5} = 13.1F$$

Hence, if x denotes the milliliters of 40% w/w HCl to be taken,

$$\frac{x \times 13.1}{250} = 1.0$$

and x = 19 mL of the 40% w/w HCl to be taken.

Example 3-12 What is the formality of H_2O in pure water? Water has the rounded formula weight of 18.01, and 1 L of water at 25°C weighs 997.04 g. Hence the formality of H_2O in pure water is

$$\frac{997.04}{18.01} = 55.36F$$

3.6 *Questions*

3-1 Define the terms *mixture, solution, solute,* and *solvent.*

3-2 Assume that a 0.100*F* aqueous solution has been prepared at 20.0°C. Will the formality be greater or smaller at 30.0°C? Explain your answer.

3-3 Discuss the advantages and disadvantages of concentration expressions based on weight by weight compared with expressions based on weight by volume.

3-4 Define the terms *mole, formula weight, formality,* and *molarity.*

3.7 *Problems*

3-1 An alloy is made by melting a mixture of 36.0 g of aluminum, 14.0 g of magnesium, and 2.00 g of manganese. Calculate the percentage composition of the alloy. *Ans.:* 69.2% Al, 26.9% Mg, and 3.8_5% Mn

3-2 Calculate the composition in mole percent of the alloy of Problem 3-1, when the atomic weights of Al, Mg, and Mn are taken as 27.0, 24.3, and 54.9. *Ans.:* 68.5, 29.6, and 1.9 mol%

3-3 At a certain temperature, NaCl shows a solubility of 35.0 g in 100 mL of water. Calculate the % w/w of this saturated solution. *Ans.:* 25.9% w/w

3-4 Solution A is 25% w/w in NaCl and solution B 35% w/w in NaCl. Calculate the percentage of NaCl in the solution obtained by mixing 50 g of solution A and 43 g of solution B. *Ans.:* $29._6$% w/w

3-5 A solution of a certain substance is prepared by mixing 15 g of a 6.3% w/w solution and 23 g of a 7.2% w/w solution. How much pure water should be added to the mixture to obtain a 5.0% w/w solution? *Ans.:* 14 g (= 14 mL) of water

3-6 Calculate the formality of a 26.8% w/w KOH (*56.1*) solution that has a density of 1.255 g/mL. *Ans.:* 6.00*F*

3-7 What volume of concentrated HNO_3 (density, 1.513 g/mL; 100% w/w; mol wt HNO_3, 63.0) is needed to prepare 400 mL of 1.50*F* HNO_3?

Ans.: 25.0 mL

3-8 Express the concentration in parts per million of chloride (*35.5*) of a $1.6 \times 10^{-5} F$ HCl solution.

Ans.: 0.57 ppm of chloride

3-9 Calculate the formality of each of the solutions formed when solution A is mixed with solution B.

	Solution A		Solution B	
	Volume, mL	Formality	Volume, mL	Formality
(a)	300.0	0.120	200.0	0.150
(b)	50.0	0.1015	100.0	0.1250
(c)	80.0	3.2	50.0	2.5
(d)	500.0	1.0	300.0	2.0

3-10 Calculate the formality of each of the following solutions.

(a) 5.00 g of H_2SO_4 (*98.08*) in 400.0 mL of solution
(b) 4.00 g of H_3PO_4 (*98.00*) in 100.0 mL of solution
(c) 2.00 g of HCl (*36.46*) in 50.00 mL of solution
(d) 3.00 g of HNO_3 (*63.01*) in 200.0 mL of solution
(e) 33.5 g of H_2SO_4 (*98.08*) in 1.50 L of solution

3-11 What percentage (w/w) of metal is present in each of the following compounds?

	Compound	Metal
(a)	NaCl (*58.44*)	Na (*22.99*)
(b)	$AgNO_3$ (*169.88*)	Ag (*107.87*)
(c)	Na_2SO_4 (*142.04*)	Na (*22.99*)
(d)	K_2CrO_4 (*194.20*)	K (*39.10*)
(e)	K_2CrO_4 (*194.20*)	Cr (*52.00*)
(f)	K_2CrO_4 (*194.20*)	K (*39.10*) and Cr (*52.00*)

3-12 Calculate the percentage (w/w) of H_2O (*18.02*) in each of the following substances.

(a) $BaCl_2 \cdot 2H_2O$ (*244.28*)
(b) $Na_2CO_3 \cdot 10H_2O$ (*286.14*)
(c) $K_2SO_4 \cdot Al_2(SO_4)_3 \cdot 24H_2O$ (*948.8*)
(d) $FeCl_3 \cdot 6H_2O$ (*270.3*)

3-13 Calculate how many milliliters of pure water must be added to each of the specified solutions to obtain solutions of the desired formalities. Assume that the volumes are additive.

	Original formality	Volume, mL	Desired formality
(a)	0.1012	250.0	0.1000
(b)	0.250	80.0	0.1000
(c)	1.00	125.0	0.250
(d)	0.200	75.0	0.150
(e)	0.200	20.0	0.080

3-14 Calculate the grams of solution II that must be added to the indicated amount of solution I to obtain solution III.

	Grams	Solution I % w/w	Solution II % w/w	Solution III % w/w
(a)	500.0	12.0	20.0	16.0
(b)	32.0	30.0	9.5	16.9
(c)	20.0	24.0	18.4	20.0
(d)	120.0	18.0	39.0	25.0
(e)	200.0	25.0	16.0	19.0
(f)	160.0	38.0	29.0	31.0

3-15 You have 500.0 mL of a 30.0% v/v solution of glycerol in water. How many milliliters of that solution must you remove and replace with an 80.0% v/v solution to obtain 500.0 mL of a 40.0% v/v solution?

3-16 Express the composition of each of the binary alloys in percentage weight by weight.

(a) 350 g of Pb, 450.0 g of Cu
(b) 36 g of Zn, 88 g of Cd
(c) 620 g of Na, 44 g of K
(d) 235 g of Hg, 300.0 g of Ag

3-17 Calculate the number of moles of each of the compounds present in the specified volumes of its solution.

	Compound		Solution density, g/mL	% w/w	Volume, mL
(a)	KOH	*(56.11)*	1.43	43.0	250.0
(b)	NaOH	*(40.00)*	1.28	25.1	450.0
(c)	HCl	*(36.46)*	1.19	37.2	100.0
(d)	HNO_3	*(63.01)*	1.41	70.1	400.0
(e)	H_2SO_4	*(98.08)*	1.83	95.5	150.0
(f)	H_3PO_4	*(98.00)*	1.69	85.0	300.0
(g)	$HClO_4$	*(100.46)*	1.66	69.8	600.0

3-18 Calculate the formality of each of the following solutions of acids and bases.

	Solute		% w/w	Density, g/mL
(a)	HNO_3	*(63.01)*	43.7	1.27
(b)	NaOH	*(40.00)*	25.1	1.275
(c)	HCl	*(36.46)*	10.5	1.05
(d)	NH_3	*(17.03)*	27.3	0.900
(e)	H_3PO_4	*(98.00)*	85.5	1.70
(f)	$HClO_4$	*(100.46)*	70.5	1.67
(g)	H_2SO_4	*(98.08)*	30.2	1.22
(h)	KOH	*(56.11)*	21.4	1.20
(i)	H_3PO_4	*(98.00)*	10.0	1.05

3-19 Calculate the milliliters of each of the concentrated solutions that must be mixed with water to prepare a dilute solution of the specified volume and formality.

Solute	Concentrated solution		Dilute solution	
	Density, g/mL	% w/w	Volume	Formality
(a) $HClO_4$ *(100.46)*	1.66	70.0	500.0 mL	0.500
(b) HNO_3 *(63.01)*	1.42	70.0	1.00 L	1.00
(c) HNO_3 *(63.01)*	1.27	43.7	2.00 L	0.200
(d) NaOH *(40.00)*	1.43	40.0	500.0 mL	0.286
(e) NH_3 *(17.03)*	0.900	28.0	2.00 L	0.100

4

Chemical Equilibrium

4.1 The Law of Mass Action

Many chemical reactions can be made to proceed in either direction by a suitable choice of conditions; that is, they are reversible. The reversibility can be expressed in a chemical equation by a double arrow:

$$aA + bB + cC + \cdots \rightleftharpoons rR + sS + tT + \cdots \tag{4-1}$$

When a reversible reaction has reached the stage at which there is no further change in the relative amounts of the species involved, an equilibrium exists. For such a reaction, it can be derived that the concentrations of the constituents in the equilibrium mixture are governed by the relation

$$K = \frac{[R]^r[S]^s[T]^t \cdots}{[A]^a[B]^b[C]^c \cdots} \tag{4-2}$$

where the constant K is the equilibrium constant. This formulation of a chemical equilibrium is termed the law of mass action, or sometimes the law of Guldberg and Waage.

In writing the above expressions, certain conventions are observed. The reaction equation is written with the actual or assumed reactants on the left side and the products on the right. The expression for the equilibrium constant is written with the concentrations of the products of the reaction as factors in the numerator and the concentrations of the reactants in the denominator; in addition, the coefficient of each participant in the balanced chemical equation becomes the exponent of the corresponding concentration.

It is important to realize that the equilibrium is dynamic. Although on a gross scale there may be no observable change in the relative amounts of the substances involved, the forward and reverse reactions continue on a molecular and an ionic scale at equal rates. The dynamic situation can be readily demonstrated by labeling one of the reactants with a radioactive tracer, as in the following example. Solid silver chloride, in which some of the atoms of silver are radioactive, is introduced into a saturated aqueous solution of inactive silver chloride. After some time, the solid silver chloride is removed by filtration and the filtrate placed in a radiation counter. The previously inactive solution is found to be radioactive. The reaction equilibrium can be represented by the equation

$$\underline{AgCl} \rightleftharpoons Ag^+ + Cl^- \tag{4-3}$$

where the underlining of a formula indicates that the substance, here silver chloride, is present as a solid phase.

Example 4-1 A chemical system is known to react according to the balanced reaction $2A + B \rightleftharpoons 2C$. When an equilibrium reaction mixture was analyzed, the following concentrations were found: $[A] = 1.0 \times 10^{-2} M$, $[B] = 4.0 \times 10^{-9} M$, and $[C] = 2.0 \times 10^{-5} M$. What is the value of the equilibrium constant?

Applying the expression for the law of mass action, (4–2), yields

$$K = \frac{[C]^2}{[A]^2[B]} = \frac{(2.0 \times 10^{-5}\ \text{mol/L})^2}{(1.0 \times 10^{-2}\ \text{mol/L})^2 \times (4.0 \times 10^{-9}\ \text{mol/L})}$$

$$= 1.0 \times 10^3\ \text{L/mol}$$

4.2 *Interpreting the Law of Mass Action*

Some important aspects and consequences of the law of mass action that are often confusing deserve brief mention.

The equilibrium constant describes the *position* of the equilibrium and says nothing about the *rate* at which the equilibrium is attained. Thus, the position of the equilibrium between elemental hydrogen and oxygen, and water, $2H_2 + O_2 \rightleftharpoons 2H_2O$, lies extremely far to the right side, that is, toward the position of water. But it is possible to store a mixture of hydrogen and oxygen gases for months or longer with no significant conversion to water. If, however, a catalyst (say finely divided platinum) is brought into contact with the hydrogen-oxygen mixture, the reaction speeds up tremendously and may even become explosive. Equilibrium is then attained in a fraction of a second.

The expression for the equilibrium constant is valid only for a dilute system and for a *reversible reaction under equilibrium conditions.* Reversibility is a topic suitable for more advanced courses. Here it suffices to state that a reaction is considered reversible if under the condition of interest it can proceed readily in either direction and if equilibrium is rapidly reestablished after a small (infinitesimally small) change in the conditions is imposed. Fortunately, most reactions encountered in inorganic quantitative analysis are reversible and go rapidly to equilibrium; thus, the law of mass action can be properly applied to their study.

Although it is common to write an equilibrium constant without explicit statement of its units, usually the quantity is not dimensionless. Thus, for the reaction $2A + B \rightleftharpoons R + W$, the equilibrium constant has the dimensions of (mass/

volume)$^{-1}$. By the general form of the law of mass action, (4–2), the equilibrium constant has the dimensions of (mass/volume) raised to the power $(r + s + t + \cdots) - (a + b + c + \cdots)$. It follows that it is impossible to compare two equilibria on the basis of the values of their respective constants unless the constants have identical dimensions (and units).

The *numerical* value of the equilibrium constant will change with the units used for expressing concentration. In analytical chemistry, molar concentrations are common. For reactions involving gases, partial pressures, which are proportional to concentrations, are convenient.

4.3 *Effect of Conditions on Chemical Equilibrium*

Some of the factors that influence either the position of a chemical equilibrium or the value of the equilibrium constant, or both, are the nature of the reactive system, concentration, activity, temperature, and catalysis.

Nature of the Reaction System. Obviously each reversible reaction involves a particular set of substances and is unique; this is reflected in the numerical value of its equilibrium constant.

Concentration. If the concentration of any component of an equilibrium mixture is changed, the position of the equilibrium will be shifted. However, the numerical value of the equilibrium constant, to a first approximation, will remain unchanged; rather, the concentrations of the reactants and products will be so altered that the quotient of the products of their concentrations, with the relevant exponents considered, is unchanged.

Activity. Implicit in the consideration of the law of mass action is the assumption that each species involved is not affected by the other substances present. This condition is met only in extremely dilute systems. When an equilibrium constant, as we have defined it, is determined closely, slightly different values are obtained at different concentration levels, even though temperature and pressure are kept constant. For electrolytes in aqueous solution, the attraction between oppositely charged ions becomes more pronounced as the concentration is increased and the effective concentration differs from the actual concentration. If the effective concentration, called activity, is substituted for the actual concentration, the so-called activity constant, or the thermodynamic constant, is obtained. The value of this constant is not affected by the concentration level and varies only with temperature. It should be appreciated that the numerical value of the concentration-based equilibrium constant can change with the concentration of species that do not even participate in the equilibrium.

The activity a of a species A can be related to the molar concentration of that species:

$$a_{\mathrm{A}} = \gamma[\mathrm{A}] \tag{4-4}$$

The activity coefficient is designated by the lowercase Greek letter gamma, γ. The values of such coefficients can be determined experimentally or estimated from theoretical considerations (Debye-Hückel equation). The second approach yields satisfactory results only for solutions of low ionic concentration (say, below about 0.5*M*). For such solutions, the activity coefficients usually have numerical values less than unity and approach unity at infinite dilution.

The value of the activity coefficient of an ion depends on the size and charge of that ion, and most important, on the concentration and charges of other ions in the solution. The following data will give a rough idea of the size of some activity coefficients. In an aqueous solution 0.1*F* in potassium chloride, the activity coefficients of the ions H^+, Ca^{2+}, Al^{3+}, and Th^{4+} are approximately 0.9, 0.7, 0.4, and 0.3. These values are valid for 25°C and if the concentration of each ion named is negligibly small compared with the 0.1 formality of the potassium chloride. The data show that neglecting activity coefficients can lead to sizable errors.

Where equilibrium constants are involved, a practicable way exists to ease the problem. Activity coefficients do not change considerably within the ionic-strength region 0.01 to 1. Therefore, the constants may be determined at an ionic strength within this region, for example, at 0.1. If these "concentration constants" are used in connection with concentrations (rather than activities), the errors will be greatly diminished. Ionic strength in this context is a function of the concentrations and charges of *all* ions present in the solution. All these concepts are properly considered in the physical chemistry realm, and our brief examination here merely points out the importance of activity. Throughout this textbook, concentrations will be used and you should know that the results obtained are often only approximate.

Temperature. For a given chemical reaction, temperature is the only factor that can essentially influence the value of the thermodynamic equilibrium constant. The value may be either increased or decreased by a change in temperature (see below).

Catalysis. The equilibrium constant does not reveal anything about the rate at which the equilibrium is attained. A catalyst affects only the rate of a reaction. In a reversible reaction, the rate of the reaction in either direction is affected similarly; therefore the position of an equilibrium is unaffected by the presence of a catalyst.

4.4 *The Principle of Le Châtelier*

The so-called principle, or theorem, of Le Châtelier permits a qualitative decision about the direction of the shift in the equilibrium when a change in conditions is imposed. This principle may be expressed in the following form: If a stress (by a change in temperature, concentration, or pressure, for example) is applied to a system at equilibrium, the equilibrium will be displaced in the direction that tends to diminish the effect of the stress.

Consider the equilibrium A + B ⇌ S + T. If more S is added, the equilibrium position will shift to the left. This can be reasoned from the law of mass action. The same result would be predicted from the principle of Le Châtelier—that the system shifts to the left to diminish the effect of the stress imposed by adding more S.

This principle can also indicate the influence of a temperature change if the reaction equation is written to include the heat that is involved as either a reactant or product. (Whether the reaction is exothermic or endothermic, that is, whether heat is evolved or consumed, must be established experimentally.) Consider the reversible, *exothermic* reaction of A and B to yield R. This may be written A + B ⇌ R + n calories. An increase in temperature may be considered to add the reagent "calories" to the right side. By the Le Châtelier principle, it would be surmised that with an increase in temperature, the equilibrium would shift to the left, thus diminishing the applied stress. Since calories do not occur in the expression for the equilibrium constant, this shift causes that constant to have a smaller numerical value at higher temperature. The use of calories as a reactant produces only a first approximation, because "caloric" energy may be only a portion of the total energy involved in the reaction. However, this approximation is good enough for many purposes. For a more rigorous treatment of chemical equilibrium and its thermodynamic basis, textbooks on physical chemistry should be consulted.

4.5 *Questions*

4-1 What is the meaning of the statement: A chemical equilibrium is dynamic?

4-2 In what units can you express the rate of a chemical reaction?

4-3 How does temperature affect the rate at which an equilibrium is attained?

4-4 What effect does temperature have on the value of an equilibrium constant?

4-5 What would be the dimensions of a rate constant for a reaction? Give an example.

4-6 Are quantitative predictions possible from the principle Le Châtelier?

4-7 How will an increase in temperature shift the equilibrium of an endothermic reaction?

4-8 What is the thermodynamic equilibrium constant?

4-9 Will the value of the equilibrium constant of an uncatalyzed reaction change when a catalyst is added?

4-10 How can you prove experimentally that a chemical equilibrium is dynamic?

4-11 When silver chloride, AgCl, is prepared by mixing solutions of its ions, heat is evolved. Would you predict silver chloride to be more soluble in cold water or hot water? Explain your conclusion.

4-12 Write the expression for the equilibrium constant of the reaction 3A + B ⇌ 2U + V.

4-13 What is a concentration constant? When is it used? What are its advantages?

4-14 The reaction A + B $\rightleftharpoons$ C is reversible and endothermic, and at 20°C has the equilibrium constant $K = 4.0 \times 10^3$. At 46.5°C, would the numerical value of the constant be the same, larger, or smaller? Explain.

5

Solubility and Solubility Product

5.1 *Solubility and Its Expression*

All solutions involving solids dissolved in liquids can be thought of as saturated, unsaturated, or supersaturated. The solid is called the solute and the liquid the solvent. A saturated solution is one containing the maximum amount of solute that can be dissolved in the solvent under equilibrium conditions.

For most solutes, solubility increases with temperature, although no direct proportionality exists. For example, 100 g of water dissolves 13 g of potassium nitrate at 0°C and 246 g at 100°C; in contrast, the same quantity of water dissolves 35.7 g of sodium chloride at 0°C and 39.8 g at 100°C. Substances that decrease in solubility with an increase in temperature are less common; well-known examples include aqueous solutions of calcium acetate, calcium hydroxide, and sodium sulfate. The temperature must therefore be specified when a solubility value is given for a particular solute-solvent system. For a dissolved gas, pressure must also be specified.

It is important to appreciate that solubility refers to equilibrium conditions. A solution saturated at a high temperature may be cooled, and although the solute may be far less soluble at a lower temperature, precipitation may not occur even over an extended period. Such a system is not in equilibrium but is, it is said, supersaturated; that is, it contains a greater amount of solute than what is allowed by the solubility equilibrium. In most cases, if a supersaturated solution is agitated by either shaking or stirring, some solute crystallizes promptly; adding solute crystals also helps in initiating crystallization.

If enough solute is placed in contact with a solvent, eventually a saturated solution will be attained. A dynamic equilibrium is thereby established, and the amount of solute dissolving in a given interval of time equals the amount leaving the solution. As stressed in Section 4.1, in the general consideration of chemical equilibria, the rate of attainment of an equilibrium in no way reflects the position of the equilibrium. Because of the inapt use of such expressions as "readily soluble," the beginner often fails to appreciate the difference between the solubility of a substance and its rate of dissolution. A substance may dissolve slowly or rapidly, depending on its state of subdivision (for example, large crystals or a fine powder) and on the rate of agitation of the solvent; but this has no relation to the solubility of the substance. Substances that are of quite low solubility are often called insoluble. More precise terms are *essentially, sparingly,* or *practically insoluble.*

Solubility can be expressed in many different ways. Tables of solubility data most often use the grams of solute per 100 g of *solvent* at a definite temperature, and the amount usually appears as a superscript number. The statement that the water solubility of sodium chloride is 35.7^{0} and 39.1^{100} is understood to mean 35.7 g of sodium chloride per 100 g of water at 0°C and 39.1 g of sodium chloride per 100 g of water at 100°C.

Solubility may be related to the solution, rather than the solvent, by expressing the concentration of a saturated solution in percentage (%w/w, %w/v, %v/v), or for substances of low solubility, as parts per thousand or parts per million. (These modes of expression were discussed in Section 3.2.) The solubility may also be expressed as the formality of the saturated solution, that is, moles of solute present in 1 L of the saturated solution. This is referred to as the formal solubility, S^f, and is an especially convenient term for the theoretical treatment of some problems. (The formal solubility is also often termed the *molar solubility.*)

5.2 Solubility Product

If the solute is a strong electrolyte, that is, a substance that in aqueous solution is completely (or almost completely) dissociated into ions, an equilibrium exists that involves only the ions in the saturated solution and the undissolved solute in contact with the solution. For a saturated aqueous solution of silver chloride, for example, the equilibrium may be written as follows (the underrule denotes the solid state):

$$\underline{AgCl} \rightleftharpoons Ag^+ + Cl^- \tag{5-1}$$

The amount of undissolved solute in contact with the saturated solution has no effect on the concentration of that solution. If the solid silver chloride were separated by filtration, the filtrate (at the same temperature) would still be a saturated solution; and any more solid silver chloride added would remain undissolved in dynamic equilibrium with the dissolved salt. Consequently, giving the concentration of the ions of the solute is enough to describe the equilibrium. The equilibrium constant in this case is the product of the molar concentrations of the silver ion and chloride ions and is known as the solubility-product constant, or simply the solubility product K_{sp} of silver chloride in water:

$$K_{sp} = [Ag^+][Cl^-] \tag{5-2}$$

The concept can be generalized with the restriction that it applies to only sparingly soluble, strong electrolytes. Thus, the solubility equilibrium of the strong electrolyte M_mA_a in water is represented by the following (where charges on the ions are omitted to simplify the notation):

$$\underline{M_mA_a} \rightleftharpoons mM + aA \tag{5-3}$$

The solubility-product expression then takes the form*

$$K_{sp} = [M]^m [A]^a \tag{5-4}$$

The solubility-product principle can, of course, be applied to electrolytes that yield more than two ionic species in water solution. For example, the expression for the solubility product of magnesium ammonium phosphate, $Mg(NH_4)PO_4$, is $K_{sp} = [Mg^{2+}] [NH_4^+] [PO_4^{3-}]$.†

5.3 *Ion Product*

Assume that an aqueous solution of silver chloride is unsaturated. The product of the molar concentrations of the silver and chloride ions in the solution, which is known as the ion product, IP, will have a value less than that of the solubility product. If more silver chloride is dissolved, the value of the ion product is increased. Eventually, when a saturated solution is reached, the value of the ion product will equal the value of the solubility product. In a supersaturated solution, the ion product will exceed the solubility product in value. From the values of the ion product and the solubility product, it can be judged whether a solution is saturated or not, and how far the solution is removed from that equilibrium condition.

5.4 *Formal Solubility and the Solubility Product*

It is possible to calculate the formal solubility S^f from the solubility product, and the reverse.

Example 5-1 The solubility product of silver chloride in water is about 1×10^{-10}. Calculate the formal solubility of this salt. Since the solution contains one

*Values for the solubility product of some common electrolytes at 25 °C are given in Table C in the Appendix.

†The solubility of an electrolyte and hence the value of the solubility product may change if the solid phase undergoes modification; this modification could be the formation of a different crystalline form, a change in the hydration, or where very small particles are involved, a change in the particle size.

silver ion for every chloride ion, the molar concentrations of these two ions are identical and equal the formal solubility S^f of the salt:

$$[Ag^+] = [Cl^-] = S^f$$

$$1 \times 10^{-10} = [Ag^+][Cl^-] = S^f \times S^f = (S^f)^2$$

and

$$S^f = \sqrt{1 \times 10^{-10}} = 1 \times 10^{-5} F$$

Example 5-2 What is the formal solubility of silver chromate, Ag_2CrO_4, when its solubility product has the value 2.0×10^{-12}? The solubility-product expression is

$$K_{sp} = 2.0 \times 10^{-12} = [Ag^+]^2[CrO_4^{2-}]$$

Since on dissolution of the salt two silver ions are formed for each chromate ion, and since the formal solubility of the salt S^f equals the molar concentration of the chromate ion, it follows that

$$[CrO_4^{2-}] = S^f$$

$$[Ag^+] = 2[CrO_4^{2-}] = 2S^f$$

$$2.0 \times 10^{-12} = (2S^f)^2 \times S^f$$

$$2.0 \times 10^{-12} = 4(S^f)^3$$

$$S^f = \sqrt[3]{\frac{2.0 \times 10^{-12}}{4}} = \frac{10^{-4}}{\sqrt[3]{2.0}} = 8 \times 10^{-5} F$$

The general relation between the formal solubility S^f and the solubility product K_{sp} of a strong electrolyte M_mA_n in water is given by

$$S^f = \sqrt[m+n]{\frac{K_{sp}}{m^m n^n}} \qquad (5\text{-}5)$$

You should derive this formula by generalizing the reasoning of Example 5-2.

Superficially comparing the values of the solubility products of silver chloride ($K_{sp} = 1 \times 10^{-10}$) and of silver chromate ($K_{sp} = 2.0 \times 10^{-12}$) might lead you to think that silver chloride is more soluble than silver chromate. However, comparing the corresponding formal solubilities ($1 \times 10^{-5} F$ and $7.9 \times 10^{-5} F$) reveals that the opposite is true. The explanation is simply that solubility products of electrolytes of different types cannot be compared. Values of equilibrium constants are generally recorded without their units explicitly stated (Section 4.1). The solubility products of silver chloride and silver chromate have different units—(moles/liter)^2 and (moles/liter)^3—hence, direct comparison is impossible.

For both silver chloride and barium sulfate, the solubility product has a value of about 1×10^{-10}. Since the two substances are of the same type, MA, their formal solubility is identical, even though one involves the singly charged ions Ag^+ and Cl^- and the other the doubly charged ions Ba^{2+} and SO_4^{2-}.

5.5 Common-Ion Effect

Adding a participant to a chemical system at equilibrium shifts the position of the equilibrium (see Section 4.3). Consequently, adding a further amount of an ion involved in the solubility equilibrium should lead to further precipitation of the solute from the saturated solution. This phenomenon, known as the common-ion effect, finds important application either for reducing the concentration of an ionic species or for maintaining its concentration at a low level. The effect of adding a common ion can be evaluated from the expression for the solubility product.

Example 5-3 In 1 L of a saturated water solution of AgCl ($K_{sp} = 1 \times 10^{-10}$), KCl is dissolved to the extent of 10 mmol. What is the effect on the silver ion concentration?

The initial silver ion concentration is obtained from the expression for the solubility product:

$$K_{sp} = 1 \times 10^{-10} = [Ag^+][Cl^-]$$

The reasoning and calculation related to the solution with the KCl added takes the following form. The potassium ion is not a common ion and hence has no effect on the equilibrium. Chloride ion now stems from two sources, the AgCl and the KCl. Thus, we can write for the total chloride concentration,

$$[Cl^-] = [Cl^-]_{AgCl} + [Cl^-]_{KCl}$$

The concentration of chloride from the KCl equals 10 mmol/L, or $1.0 \times 10^{-2} M$. The concentration of the chloride from the AgCl equals the concentration of the silver ion, since one silver ion is produced from each chloride ion. Hence,

$$[Cl^-] = [Ag^+] + 1.0 \times 10^{-2}$$

Substituting this in the solubility-constant expression yields

$$1 \times 10^{-10} = [Ag^+]([Ag^+] + 1.0 \times 10^{-2})$$

This is a quadratic equation, and its solution yields an exact value. However, it is possible to simplify matters. It is known that adding KCl causes a repression of the silver concentration. Since this concentration without repression is $10^{-5} M$, it is permissible to neglect the term $[Ag^+]$ in the parentheses because 10^{-5} is small compared with 10^{-2}.* Then

$$1 \times 10^{-10} = [Ag^+] \times 1 \times 10^{-2}$$

---

*As a rule, in equilibrium calculations a term can be neglected if it is 5% or less of the numerical value of the term to which it is to be added or from which it is to be subtracted. If the value is unknown or cannot be estimated, the simplified calculation should be made. If examination of the result thus obtained indicates that the simplification is invalid, the more exact expression must then be solved.

and $$[Ag^+] = 1 \times 10^{-8} M$$

By increasing the chloride ion concentration by a factor of 10^3, then, the silver ion concentration is decreased one thousand times.

Example 5-4 To 1 L of a saturated solution of AgCl, let KCl be added until its concentration is $1 \times 10^{-5} F$. What will the silver ion concentration be when equilibrium is reestablished?

Proceeding as in Example 5-3, we obtain the following equation:

$$1 \times 10^{-10} = [Ag^+]([Ag^+] + 1 \times 10^{-5})$$

Neglecting the term $[Ag^+]$ in the parentheses and solving for $[Ag^+]$ as in the preceding example yields

$$[Ag^+] = \frac{1 \times 10^{-10}}{1 \times 10^{-5}} = 1 \times 10^{-5} M$$

On checking the result, it becomes obvious that the neglected term is not much smaller than 10^{-5} and that an approximation analogous to that in the preceding example is not allowed here. Consequently, the full equation must be solved. The quadratic equation, as obtained on rearrangement, can be solved by using the quadratic formula:

$$[Ag^+]^2 + 1 \times 10^{-5} [Ag^+] - 1 \times 10^{-10} = 0$$

$$[Ag^+] = \tfrac{1}{2}(-1 \times 10^{-5} \pm \sqrt{1 \times 10^{-10} + 4 \times 10^{-10}})$$

$$= 6 \times 10^{-6} M$$

Only the positive root is taken, because a negative concentration has no physical meaning.

5.6 *Diverse-Ion Effect*

The solubility product, as considered above, is a "concentration" constant and its numerical value will change, even at constant temperature, when further electrolytes are dissolved that have no ion in common with the insoluble substance. If activities (Section 4.3) were used in place of molar concentrations, the solubility-product expression would yield a true constant, influenced only by temperature. The solubility of a sparingly soluble, strong electrolyte in water generally will increase with increasing concentrations of other electrolytes. This phenomenon is variously known as the inert-electrolyte effect, diverse-ion effect, or salt effect. For example, barium sulfate is increasingly soluble at increasing concentrations of such salts as KNO_3 and $Mg(NO_3)_2$. This enhanced solubility depends not only on the concentration of the diverse ions but also on their charges. Thus magnesium chloride, $MgCl_2$, will increase the solubility of barium sulfate in water more than an equimolar amount of potassium chloride.

5.7 *Complications with Solubility-Product Principle*

The solubility-product expression assumes that the ions of the dissolved strong electrolyte undergo no other secondary reactions than the formation of hydrated species, that is, aquo complexes. However, the anion or cation or both may undergo hydrolysis. An example of this is a solution of metal sulfide, MS, for which $K_{sp} = [M^{2+}][S^{2-}]$. The sulfide ion, S^{2-}, undergoes hydrolysis to the hydrogen sulfide ion, HS^-, according to

$$S^{2-} + H_2O \rightleftharpoons HS^- + OH^- \qquad (5\text{-}6)$$

The solubility-product expression involves the ion S^{2-}, but much of the ionic sulfur present in the solution is in the form of the ion HS^-. If this hydrolysis is neglected, the value for the formal solubility calculated from the solubility product and the value established experimentally do not agree. (The influence of hydrolysis on solubility is deferred until Section 8.13.)

Complex formation is a further type of secondary reaction that may affect a solubility equilibrium. Silver ion, for example, forms strong, soluble complexes with ammonia, and hence silver chloride is soluble in ammoniacal solution. (You are familiar with this reaction, which is used in qualitative analysis.)

A complex may even form with an ion directly involved in the solution equilibrium. Thus, silver ion forms the soluble chloro complex, $AgCl_2^-$, and hence silver chloride is soluble in solutions containing a high concentration of chloride ion. If chloride is added, in the form of hydrochloric acid, for example, to a saturated solution of silver chloride in water, initially silver chloride precipitates, from the common-ion effect. Then, as more hydrochloric acid is added, the effect of the formation of soluble complexes predominates; the precipitate redissolves progressively, to ultimate completion. For the present purpose, it suffices to mention that hydrolysis and complex formation may occur, to a degree that they cannot be neglected in calculations based on the solubility-product principle.

5.8 *Calculations Using the Solubility-Product Principle*

Some additional examples will help you to become familiar with calculations based on the solubility-product principle.

Example 5–5 A saturated aqueous solution of AgI($K_{sp} = 1 \times 10^{-16}$) is given. (a) What is the concentration of silver ion in this solution? (b) What concentration of iodide ion must be established (say by adding KI) to reduce the silver ion concentration to $1 \times 10^{-12}\,M$?

(a) $$1 \times 10^{-16} = K_{sp} = [Ag^+][I^-]$$

and since $[Ag^+] = [I^-]$,

$$[Ag^+] = \sqrt{1 \times 10^{-16}} = 1 \times 10^{-8}\,M$$

(b) Because the concentration of iodide ion furnished by the AgI can be safely neglected, it follows that

$$1 \times 10^{-16} = 1 \times 10^{-12}[I^-]$$

and

$$[I^-] = \frac{1 \times 10^{-16}}{1 \times 10^{-12}} = 1 \times 10^{-4}M$$

Example 5-6 Solid Ag_2CrO_4($K_{sp} = 2 \times 10^{-12}$) is equilibrated with $1 \times 10^{-3}F$ $AgNO_3$. What is the equilibrium concentration of chromate ion in the solution?

$$2 \times 10^{-12} = K_{sp} = [Ag^+]^2[CrO_4^{2-}]$$

The concentration of silver is $1 \times 10^{-3}M$ (from the $AgNO_3$) plus twice the chromate concentration (since two silver ions are produced for each chromate ion):

$$[Ag^+] = 1 \times 10^{-3} + 2[CrO_4^{2-}]$$

Substitution in the solubility-product expression yields

$$2 \times 10^{-12} = (1 \times 10^{-3} + 2[CrO_4^{2-}])^2[CrO_4^{2-}]$$

Neglecting $2[CrO_4^{2-}]$ within the parentheses gives the approximation

$$2 \times 10^{-12} \cong (1 \times 10^{-3})^2[CrO_4^{2-}]$$

and

$$[CrO_4^{2-}] = \frac{2 \times 10^{-12}}{1 \times 10^{-6}} = 2 \times 10^{-6}M$$

Finally, the approximation must be checked. The value 1×10^{-3} is indeed much larger than $2 \times 2 \times 10^{-6}$, so neglecting the parenthetical term is permissible.

Example 5-7 The value of the ion product of an unsaturated solution of $BaSO_4$ ($K_{sp} = 1 \times 10^{-10}$) is 1×10^{-12}. (a) What is the concentration of barium ion and sulfate ion in this solution? (b) What weight (in milligrams) of 100% w/w of H_2SO_4(*98*) can be added to 2 L of this solution before precipitation of sulfate commences?

(a)

$$\text{IP} = 1 \times 10^{-12} = [Ba^{2+}][SO_4^{2-}]$$

$$[Ba^{2+}] = [SO_4^{2-}] = \sqrt{1 \times 10^{-12}} = 1 \times 10^{-6}M$$

(b) The concentration of barium remains $1 \times 10^{-6}M$ up to the point of saturation. Therefore,

$$1 \times 10^{-10} = K_{sp} = [Ba^{2+}][SO_4^{2-}] = 1 \times 10^{-6} \times [SO_4^{2-}]$$

and

$$[SO_4^{2-}] = \frac{1 \times 10^{-10}}{1 \times 10^{-6}} = 1 \times 10^{-4}M$$

The difference in $[SO_4^{2-}]$ between the saturated and unsaturated solutions corresponds to

$$1 \times 10^{-4} - 1 \times 10^{-6} = 99 \times 10^{-6} \cong 1 \times 10^{-4} M$$

Hence for 2 L, 2×10^{-4} mol, or 0.2 mmol, of sulfate can be added before precipitation commences. This amount is furnished by $0.2 \times 98 = 20$ mg of 100% w/w H_2SO_4.

The next example demonstrates one possible way in which a formal solubility and the corresponding solubility product can be determined.

Example 5-8 Solid $SrSO_4$(*184*) was added to about 2 L of distilled water. The mixture was shaken for several hours at constant temperature and then filtered. Exactly 1.5 L of the clear filtrate was evaporated to dryness in a platinum dish. The residue of $SrSO_4$, after ignition and cooling, was found to weigh 0.174 g. Calculate the solubility product of $SrSO_4$.

The saturated solution contains 0.174 g of $SrSO_4$ in 1.50 L. The formal solubility of this salt is therefore

$$S^f = \frac{0.174}{184} \times \frac{1}{1.50} = 6.30 \times 10^{-4} F$$

Hence

$$K_{sp} = [Sr^{2+}][SO_4^{2-}] = (6.30 \times 10^{-4})^2 = 39.7 \times 10^{-8}$$
$$= 3.97 \times 10^{-7}$$

5.9 *Questions*

5-1 Distinguish between the formal solubility and the solubility product of a strong electrolyte and explain their relation.

5-2 Explain why the solubility-product concept can be applied only to strong electrolytes of low solubility.

5-3 Why is it unnecessary to include the concentration of the undissolved solute in the solubility-product expression?

5-4 Explain what is meant by the statement that a solubility equilibrium is dynamic?

5-5 In equilibrium calculations a concentration term is frequently neglected in additions or subtractions. What general rule is adopted for deciding whether the resulting approximation is valid?

5-6 What is the common-ion effect? Explain it in terms of the solubility equilibrium.

5-7 How is the solubility equilibrium of an electrolyte affected by adding a salt that does not contain a common ion?

5-8 Describe one way in which the value of a solubility product can be determined.

5-9 Elaborate on how the solubility of an electrolyte can be influenced by the presence of other substances.

5-10 What is the inert-electrolyte effect?

5.10 Problems

5-1 Calculate the formal solubilities of the salts MA ($K_{sp} = 7.0 \times 10^{-10}$) and M_3A_2 ($K_{sp} = 1.0 \times 10^{-31}$) and state which is more soluble.
Ans.: $2.6 \times 10^{-5} F$ and $2.5 \times 10^{-7} F$; MA is more soluble

5-2 When solid AgBr ($K_{sp} = 5.0 \times 10^{-13}$) is equilibrated with $0.020F$ KBr, what will be the silver ion concentration in the solution?
Ans.: $2.5 \times 10^{-11} M$

5-3 Calculate the formal solubility of the salt M_2A_3 for which K_{sp} has the value 5.0×10^{-96}.
Ans.: $3.4 \times 10^{-20} F$

5-4 How many milligrams of barium (*137*) will be present in 500 mL of a solution obtained by equilibrating solid $BaSO_4$ ($K_{sp} = 1 \times 10^{-10}$) with $0.1F$ H_2SO_4?
Ans.: 7×10^{-5} mg

5-5 Calculate the value of the solubility product of the salt from its given formal solubility. Neglect hydrolysis.

Salt	Formal solubility
(a) AgCl	1.3×10^{-5}
(b) AgBr	7.2×10^{-7}
(c) $BaSO_4$	1.1×10^{-5}
(d) $Ca(C_2O_4)$	4.8×10^{-5}
(e) Ag_2SO_4	$1.7_5 \times 10^{-2}$
(f) Ag_3PO_4	1.6×10^{-5}

5-6 Calculate the formal solubility of the salt from the given value of its solubility product. Neglect hydrolysis.

Salt	Solubility product
(a) AgI	8.3×10^{-17}
(b) $SrSO_4$	3.2×10^{-7}
(c) $PbCl_2$	1.6×10^{-5}
(d) AgSCN	1×10^{-12}

5-7 From the given value for the solubility product, calculate the amount, in milligrams, of the salt present in 50.0 mL of its saturated solution.

Salt		Solubility product
(a) $BaSO_4$	(*233.4*)	1.3×10^{-10}
(b) $Ba(IO_3)_2$	(*487.2*)	2×10^{-9}
(c) AgI	(*234.8*)	3×10^{-7}
(d) SrF_2	(*125.6*)	1.2×10^{-5}

5-8 The solubility of a salt AB_2 (FW = 100.0) is 0.60 g per 100 mL of water. A volume of 30.0 mL of a 0.300*M* solution of A^{2+} is mixed with 30.0 mL of a 0.200*M* solution of B^-. Show by calculation whether or not a precipitate of AB_2 will form.

5-9 What concentration of KCl (*74.6*) in milligrams per liter must be established in a saturated solution of AgCl ($K_{sp} = 1.8 \times 10^{-10}$) to adjust the silver ion concentration to $5.0 \times 10^{-7} M$?

5-10 A volume of 16.0 mL of 0.040*F* $AgNO_3$ is mixed with 30.0 mL of 0.030*F* NaCl. Calculate the amount, in milligrams, of Ag^+ (*107.9*) remaining in solution. The K_{sp} for AgCl is 1.8×10^{-10}.

5-11 To 50.0 mL of a saturated solution of AgCl ($K_{sp} = 1.8 \times 10^{-10}$) is added 50.0 mL of a solution containing 1.70 g of $AgNO_3$ (*169.9*). Calculate the chloride ion molarity in the resulting solution.

5-12 To 10.0 mL of 0.010*F* $AgNO_3$ is added 10.0 mL of 0.020*F* Na_2SO_4. Show by calculation whether or not a precipitate of Ag_2SO_4 ($K_{sp} = 2.1 \times 10^{-5}$) will form.

5-13 What molar concentration of chloride ion is needed for starting the precipitation of the metal chloride of the given solubility product if the solution contains 10 mg of the metal per milliliter?

Metal chloride	Solubility product
(a) AgCl	1.8×10^{-10}
(b) $PbCl_2$	1.6×10^{-5}
(c) Hg_2Cl_2	1.3×10^{-18}

5-14 A saturated solution of AgBr is shaken with an excess of AgCl until equilibrium is established. Calculate the molar concentration of bromide ion in the solution.

5-15 To a saturated solution of AgI is added an excess of solid AgCl. The system is shaken until equilibrium is attained. What is the ratio of the molar concentration of iodide ion in the resulting solution to that originally present?

5-16 Two compounds, AB and M_2N, have solubility products of 4.0×10^{-10} and 4.0×10^{-12}, respectively. Calculate the ratio of their formal solubilities.

Ans.: 0.2

6

Principles of Gravimetric Analysis

6.1 General Considerations

Gravimetric analysis depends on measuring weights or weight changes and using these data as the basis for calculating the amount of a sought-for substance. Although in analytical chemistry the expression *weight* is used, it is actually *mass* that is determined. The usage stems from the fact that a balance is employed, which establishes the mass of an object by comparing the weight of the object with the known weight of a reference mass. It is convenient to distinguish three different types of gravimetric determinations: precipitation methods, electrogravimetric methods, and evolution methods.

Precipitation Methods. Precipitation methods typically involve the following steps:

1. Weighing of a certain amount of material to be analyzed.
2. Dissolving the weighted sample.
3. Adding an appropriate reagent to form a sparingly soluble compound with the sought-for substance.
4. Isolating the precipitate formed.
5. Purifying the precipitate.
6. Weighing either the precipitate after drying or a compound formed from it by an appropriate conversion.

From the final weight, the amount of sought-for substance can be calculated, or in conjunction with the sample weight, the content of sought-for substance in the sample material can be established.

For example, if sodium chloride is to be determined by a gravimetric method in a mixture with sodium carbonate, a weighed amount of the mixture is dissolved in dilute nitric acid and a solution of silver nitrate is added. The silver chloride precipitated is separated by filtration; it is then washed, dried, and finally weighed. From the weight of the silver chloride, the corresponding weight of sodium chloride can be calculated and consequently the percentage of this substance in the original mixture.

Electrogravimetric Methods. Electrogravimetric methods are based on the deposition of a substance on an electrode. The weight of the deposit is calculated as the difference in the weights of the electrode with and without the dried deposit. (Electrogravimetric methods are considered in Chapter 30.)

Evolution Methods. Evolution methods are based on the volatilization of a component of the sample, usually by heating. The amount of substance volatilized is the difference in the sample weights before and after the volatilization step. This simple type of a gravimetric method is frequently used for determining water by heating the sample for some time at 105 to 110°C and for carbon dioxide by ignition at a higher temperature. It is also possible to absorb the volatilized compound (water or carbon dioxide in the examples above) in an appropriate medium, which is weighed before and after the absorption. The amount of the volatilized compound is obtained from the weight increase. (Various evolution methods are considered in Chapter 48.)

6.2 *Computations in Gravimetric Analysis*

You are already familiar with "chemical arithmetic" from an introductory course in chemistry; consequently, the calculations in gravimetric analysis require only brief treatment.

The balanced reaction equation

$$Ag^+ + Cl^- \rightarrow \underline{AgCl} \tag{6-1}$$

expresses not only qualitatively that silver and chloride ions react to form silver chloride, but also quantitatively that one silver ion reacts with one chloride ion to yield one molecule of silver chloride. Since one mole of silver ion contains the same number of particles as one mole of chloride ion or one mole of silver chloride, the equation also implies that one mole of silver ion reacts with one mole of chloride to yield one mole of silver chloride. Since one mole of a substance is its formula weight expressed in grams, the reaction equation also implies that 107.868 g of silver ion reacts with 35.453 g of chloride ion to yield 107.868 + 35.453 = 143.321 g of silver chloride, provided, of course, that the substances are pure and the reaction goes to completion.

This quantitative relation permits the calculation of the amount of silver or chloride, in a given amount of silver chloride, that is obtainable from a given amount of silver or chloride ion on complete reaction with the ion of the other species.

Example 6–1 What weight of Ag (*107.87*) is present in 100.0 g of AgCl (*143.32*)?

One mole of AgCl contains one mole of Ag; consequently 143.32 g of AgCl contains 107.87 g of Ag, and 100.0 g of AgCl contains x g of Ag. By proportion:

$$\frac{143.32}{107.87} = \frac{100.0}{x}$$

and

$$x = \frac{107.87}{143.32} \times 100.0 = 75.27 \text{ g of Ag}$$

The student oriented to solving problems by "ratio" rather than by "proportion" will achieve the result directly by writing the last line of the above solution of the problem (and similarly in the following illustrative problems).

Example 6–2 What amount of Na (*22.99*) is present in 50.00 g of Na_2SO_4 (*142.04*)?

Since one mole of Na_2SO_4 contains two moles of Na, 142.04 g of Na_2SO_4 contains 2 × 22.99 g of Na, and 50.00 g of Na_2SO_4 contains x g of Na. By proportion,

$$\frac{142.04}{2 \times 22.99} = \frac{50.00}{x}$$

and

$$x = \frac{2 \times 22.99}{142.04} \times 50.00 = 16.19 \text{ g of Na}$$

Example 6–3 What amount of NaCl (*58.44*) is present in a sample that yields, on precipitation of the chloride with $AgNO_3$, 0.8342 g of AgCl (*143.32*)?

Since one mole of NaCl yields one mole of AgCl, 58.44 g of NaCl corresponds to 143.32 g of AgCl, and x g of NaCl corresponds to 0.8342 g of AgCl:

$$\frac{58.44}{143.32} = \frac{x}{0.8342}$$

and

$$x = \frac{58.44}{143.32} \times 0.8342 = 0.3402 \text{ g of NaCl}$$

Example 6–4 How much $BaCl_2$ (*208.24*) is present in a solution if on addition of a sufficient amount of $AgNO_3$, the weight of the isolated AgCl (*143.32*) is 1.3456 g?

One mole of $BaCl_2$ contains two moles of Cl and therefore yields two moles of AgCl; consequently 208.24 g of $BaCl_2$ corresponds to 2 × 143.32 g of AgCl, and x g of $BaCl_2$ to 1.3456 g of AgCl:

$$\frac{208.24}{2 \times 143.32} = \frac{x}{1.3456}$$

and $$x = \frac{208.24}{2 \times 143.32} \times 1.3456 = 0.9776 \text{ g of } BaCl_2$$

In Examples 6-3 and 6-4 it is not necessary to calculate the amount of chloride ion and then to convert this result into sodium chloride and barium chloride, respectively. The calculating is done directly through the weight relations.

6.3 *Gravimetric Factors*

In each of the preceding examples, the left side of the proportion, or the ratio, immediately following the equals sign in the last line of the calculation scheme involves only formula weights and is independent of the actual amounts of the substances involved in the particular example. These terms are known as chemical factors, and in reference to gravimetric analysis, as gravimetric factors. Such a factor can be represented by simply writing the relevant chemical formulas (commonly in italic in printed works), which are then understood to imply the formula weights.

For determining sodium chloride by the precipitation of silver chloride (Example 6-3), the following general proportion can be written:

$$\frac{NaCl}{AgCl} = \frac{x}{w} \tag{6-2}$$

where w is the weight of the precipitate and x is the corresponding weight of sodium chloride. By rearrangement,

$$x = \frac{NaCl}{AgCl} \times w \tag{6-3}$$

Here the quotient $NaCl/AgCl$ is a gravimetric factor and needs to be calculated only once. This factor actually expresses the amount of sodium chloride corresponding to a unit amount of silver chloride and is a dimensionless number; w can therefore be expressed in any desired unit of weight (grams, milligrams, pounds, etc.) and x will and must (!) have the same unit.

The gravimetric factor for sulfate determined by precipitating and weighing barium sulfate corresponds to $SO_4^{2-}/BaSO_4$. For certain applied purposes, the sulfate content is expressed as sulfur trioxide; the factor then is $SO_3/BaSO_4$.

Phosphate can be determined by precipitating magnesium ammonium phosphate hexahydrate, which can be dried and weighed. The gravimetric factor for the orthophosphate ion will be $PO_4^{3-}/MgNH_4PO_4 \cdot 6H_2O$. If the phosphate content is to be given in terms of phosphorus pentoxide (as is common in the analysis of ores, fertilizers, and so on), the factor is

$$P_2O_5/2MgNH_4PO_4 \cdot 6H_2O$$

Observe that the coefficient 2 in the denominator compensates for the fact that two phosphorus atoms appear in the formula of the oxide but only one in the phosphate salt formula.

By ignition, magnesium ammonium phosphate can be readily converted to magnesium pyrophosphate, which can then be weighed. In this case, the factors for the orthophosphate ion and for the pentoxide become

$$2PO_4^{2-}/Mg_2P_2O_7 \quad \text{and} \quad P_2O_5/Mg_2P_2O_7$$

In the first of these factors, the coefficient 2 is required in the numerator to compensate for the fact that only one phosphorus atom appears in the formula of the phosphate ion but two in that of the pyrophosphate salt. In the second factor, two phosphorus atoms appear in both formulas; consequently, both coefficients are unity and need not be written.

The coefficients are governed solely by the relation between the "linking" species (the phosphorus atom in the above examples) in the sought-for substance and the weighed substance. Mistakes of the novice are often associated with failure to recognize the linking atom.

The result of a determination is frequently expressed as the percentage of a constituent in the sample. The calculations then involved are best explained by examples.

Example 6–5 A mixture of NaCl (*58.44*) and Na_2SO_4 is to be analyzed for its NaCl content. A sample weighing 0.9532 g is dissolved in water and the chloride precipitated by adding an excess of $AgNO_3$. The dried precipitate of AgCl (*143.32*) weighs 0.7033 g. What is the percentage of NaCl in the sample?

The calculation is evident from the following (compare Example 6–3):

$$x = \frac{NaCl}{AgCl} \times w$$

x ↑	$\frac{NaCl}{AgCl}$ ↑	w ↑
g of NaCl present in sample	gravimetric factor	g of AgCl found in precipitate

$$= \frac{58.44}{143.32} \times 0.7033$$

The percentage of NaCl in the mixture is given by the following, where W_s is the sample weight in grams.

$$\% = \frac{x \times 100}{W_s} = \frac{58.44 \times 0.7033 \times 100}{143.32 \times 0.9532} = 30.09\% \text{ NaCl}$$

Example 6–6 When a 1.3260-g sample of a substance was heated at 110°C, water was volatilized; after the product was cooled under anhydrous conditions, the weight was 1.0112 g. What is the percentage of volatizable water in the substance?

$$\%H_2O \text{ lost at } 110^\circ C = \frac{\overbrace{(1.3260 - 1.0112)}^{\text{loss in weight = g of } H_2O} \times 100}{\underbrace{1.3260}_{\text{sample weight in g}}} = 23.74\%\ H_2O$$

A common error is using different units in expressing the weight of the final substance and the weight of the sample. If the sample weight is expressed in grams or milligrams, then the weight of the final substance must also be expressed in grams or milligrams.

6.4 *Calculating Results on a "Dry Basis"*

Samples of a material often vary considerably in their moisture content, and these differences cause changes in the percentage composition relative to other constituents. Sometimes analyses are performed to compare various samples, to check a material to learn whether it complies with certain specifications, or for similar purposes. It then becomes necessary to refer to the composition of the material on a "dry basis." Although detailed consideration of moisture determination is deferred, the calculations can be appreciated from examples.

Example 6–7 A sample of an iron ore is found to have a moisture content of 1.56% and an "as is" iron content of 26.24%. What is its iron content on a "dry basis"?

A 100.00-g sample of ore contains 1.56 g of H_2O and 26.24 g of Fe. When the material is dried, the weight remaining is 100.00 – 1.56 = 98.44 g, of which 26.24 g is still Fe. Hence,

$$\frac{26.24 \times 100}{98.44} = 26.66\% \text{ Fe (dry basis)}$$

Example 6–8 One portion of a sample of a copper ore weighing 1.5653 g is dried to a constant weight of 1.4920 g. A second portion weighing 1.0075 g is analyzed for copper, and 320.3 mg of copper is found. Calculate the moisture content of the ore and the copper content on both the "as is" basis and the dry basis.

The moisture content is

$$\frac{(1.5653 - 1.4920) \times 100}{1.5653} = 4.68\%\ H_2O$$

The copper content is

$$\frac{0.3203 \times 100}{1.0075} = 31.79\% \text{ Cu (``as is'' basis)}$$

and

$$\frac{31.79 \times 100}{100.00 - 4.68} = 33.35\% \text{ Cu (dry basis)}$$

6.5 Precipitation

Precipitation is a technique used in quantitative analysis for various reasons. It allows separating one substance from other substances, as the basis for a precipitation titration (Chapter 18), and is one of the initial operations in a gravimetric determination. For the last purpose, it is unnecessary for the precipitate to be of the same composition as the compound finally weighed. Hence in gravimetric analysis, it is convenient to differentiate between the precipitation form and the weighing form. The requirements to be fulfilled by the two forms are different.

Precipitation Form. The precipitation form should have a low solubility in the medium used; it should be pure, or at least contain only impurities that can be readily removed before the weighing step; preferably, it should be coarsely crystalline. The compound precipitated should also be such that it can be readily dried, ignited, or otherwise converted to an appropriate weighing form.

Weighing Form. It is highly desirable that the compound representing the weighing form be strictly stoichiometric, because the gravimetric factor can then be derived readily from stoichiometric considerations. If the composition is not stoichiometric, an empirical factor can be applied, but this practice is not favored; moreover, it requires that the composition be the same in every determination. An additional practical requirement is that the weighing form not be subject to ready attack by moisture or by carbon dioxide or oxygen of the air. These attacks can be prevented by weighing in a closed container, but such a procedure is inconvenient. It is desirable that the compound used as a weighing form afford a gravimetric factor small in its numerical value; that is, the amount of weighing form corresponding to a unit weight of the sought-for substance should be large. In such a case, for a given amount of the sought-for substance, the influence of weighing errors, impurities, uptake of small amounts of moisture or carbon dioxide, and the like, is greatly decreased.

6.6 Mechanism of Precipitation

Sooner or later a precipitate must form when a solution contains more of a particular solute than what is allowed by the solubility under the prevailing conditions. Consequently, the first step in the precipitation is establishing supersaturation. Then, as the second step, crystal nuclei make their appearance; that is, nucleation takes place. Crystal nuclei are exceedingly minute and may even consist of only a few unit cells of the crystal lattice. The rate at which such nuclei form gets larger with the degree of supersaturation existing.

In the third, and final, step, these nuclei are centers for the deposition of more material, and larger crystalline aggregates are formed. Besides forming individual aggregates, two or more established crystals may cluster to form a larger particle. These processes act to achieve what is called crystal growth.

More nuclei may form while crystal growth proceeds. Consequently, the rate of nucleation must be kept low and crystal growth encouraged to get a small number of large crystals rather than many small crystals. Towards this goal, the supersaturation must be kept low, because as it becomes greater, the rate of nucleation is increased much more than the rate of crystal growth. Supersaturation can be kept low by adding the precipitant slowly, under vigorous stirring, and as a dilute solution. But the degree of dilution allowed for the precipitant solution is limited. The more dilute this solution, the more of it that must be used to bring the necessary amount of precipitant into the sample solution. Thus the latter solution will be diluted, possibly to such a degree that precipitation becomes incomplete. Even when a dilute precipitant solution is used, a region of local supersaturation develops at the site where a drop enters the sample solution. While local supersaturation can be greatly reduced by adequate stirring, it is never avoided completely.

Even when the precipitation has been effected according to these findings, the improvement in the character of the precipitate for some systems is inadequate. Such unfavorable cases include various metal sulfides and the hydrous oxides of aluminum, iron(III), and certain other polyvalent metal ions. In these cases, the rate of nucleation or other unfavorable factors predominate and crystal growth is inadequate. As a result, a gelatinous network of barely crystalline material forms. In some of these stubborn cases and in others, crystallinity can be improved, or at least the particle size increased, by aging the precipitate (see below).

6.7 Contamination of Precipitates

A precipitate, as it forms, is rarely pure. For gravimetric purposes, it should be possible to remove the impurities readily or at least to reduce them to such an extent that the accuracy of the determination is not affected. This removal should take place when the precipitate is washed or during its subsequent conversion to the weighing form.

A precipitate can become contaminated through various processes; examining the important ones will throw some light on how to conduct a precipitation for gravimetric purposes.

Coprecipitation. In the broadest sense, coprecipitation is the concurrent precipitation of two or more substances. Thus, for example, adding silver nitrate to a solution containing both sodium chloride and sodium bromide leads to the precipitation of both silver chloride and silver bromide. In analytical chemistry and especially in studying the contamination of a precipitate, the term *coprecipitation* is conveniently used in a restricted sense. Here it refers to the carrying down of one or more substances that if present alone would not be precipitated by the reagent used. For example, calcium is partially coprecipitated when iron(III) is precipitated as its hydrous oxide by neutralizing an acidic solution to pH 4 to 5. Under the same conditions without the iron, calcium remains in solution.

Coprecipitation can involve several mechanisms, including solid solution formation, surface adsorption, and occlusion. These mechanisms are considered separately below, but what we stress here is that contamination often results from the concurrent action of more than one of these processes.

Solid Solutions. Just as two liquids may be miscible to form a liquid solution, two solids may be sufficiently soluble in each other to form a solid solution, which should be differentiated from a mixture obtained by merely grinding the substances together. Two substances may form "mixed crystals", wherein one substance is incorporated into the crystal lattice of the other. Such mixed crystals are frequently, but not exclusively, formed when the two substances are isomorphous, that is, when their crystals are of the same crystallographic system. Such incorporation occurs even if the contaminating compound is soluble in the liquid solution under the conditions of the precipitation. For example, owing to the close similarity of structure, size, charge, and electronic configuration of chromate ion and sulfate ion, barium chromate can be incorporated into the precipitate of barium sulfate. This coprecipitation occurs even when barium sulfate is precipitated from a strongly acidic solution in which the barium chromate is soluble. The precipitate is then more or less yellow, depending on the chromate content. Permanganate ion can be incorporated somewhat into barium sulfate, although barium permanganate is quite soluble and the permanganate ion has a charge differing from that of the sulfate ion. The barium sulfate that precipitates is pink. The minor constituent of a solid solution is scattered throughout the crystal and firmly held there by the lattice forces; washing the precipitate is therefore ineffective in removing the contaminant.

Adsorption. On the surface of the particles of a precipitate, active sites are present, capable of attracting and holding species that by themselves are not precipitated. Obviously, contamination by this mechanism increases with the surface area of the precipitate. Consequently, in a gravimetric determination, a coarse, crystalline precipitate with its smaller relative surface area is preferred over a finely divided precipitate. Adequate washing of a precipitate contaminated through adsorption is often a satisfactory remedy. However, such contamination occurs frequently with gelatinous precipitates; here washing fails because the wash liquid does not penetrate the precipitate fully within a reasonable time, and filtration proceeds very slowly.

Occlusion. The carrying down of impurities in the interior of primary particles is known as occlusion. This process would include in a broad sense the formation of solid solutions but can be used in the more restricted sense of mechanical occlusion, including the inclusion of mother liquor and ion entrapment—the growth of a precipitate around an adsorbed ion. Retention of mother liquor is especially pronounced with gelatinous precipitates. The degree to which occlusion takes place depends somewhat on the rate of the precipitation. Washing of the precipitate will never adequately remove occluded impurities.

Postprecipitation. In postprecipitation, an initially pure precipitate is contaminated by the subsequent deposition of a second solid phase formed by another substance that is only slightly soluble in the solution. Frequently this deposition proceeds from a supersaturated solution of the contaminant. Consider the precipitation of calcium as the oxalate from a magnesium-containing solution. If the calcium oxalate precipitate is not separated promptly by filtration, some magnesium oxalate is adsorbed on the surface and induces the precipitation of more magnesium oxalate, which once deposited does not redissolve. However, in the absence of calcium oxalate a supersaturated solution of magnesium oxalate can be kept for a considerable period without precipitation.

Various metal sulfides offer interesting examples of postprecipitation. When copper(II) is precipitated from a zinc-containing solution by saturation of an acidic solution with hydrogen sulfide, zinc-free copper(II) sulfide is formed initially. When the copper(II) sulfide is left in contact with the mother liquor, it increases progressively in its zinc content. It may be reasoned that sulfide ion is adsorbed on the surface of the precipitate, so that the sulfide ion concentration is raised locally to an extent that zinc sulfide deposits. Postprecipitation can usually be avoided, or at least reduced, by prompt separation of the precipitate from the mother liquor.

6.8 Purification of Contaminated Precipitates

Every precipitate is washed before being converted to the weighing form. The wash liquid should at least remove the mother liquor adhering to the precipitate. Since washing does not always result in a complete removal of impurities, additional steps may be needed for adequate purity.

Reprecipitation (Double Precipitation). Reprecipitation is often a satisfactory, but tedious, method to ensure a pure precipitate. The extent of the contamination of a precipitate depends largely on the concentration of the contaminating material in the solution. Hence, if an impure precipitate is redissolved and the precipitation repeated, the resulting precipitate is often far less contaminated. When hydrous iron(III) oxide is precipitated from a calcium-containing solution, some calcium is coprecipitated. The contaminated hydrous iron(III) oxide can be separated by filtration, washed and dissolved in dilute hydrochloric acid, and precipitated again by adding ammonia. Since the concentration of calcium is far less in the second solution than in the first, the calcium content of the second precipitate is usually negligible. Reprecipitation is impossible, or at least extremely difficult and tedious, when the precipitate cannot be redissolved readily. Barium sulfate represents a common example of such a case. In addition, reprecipitation may be of little value if the contamination is caused by an impurity of low solubility or by one that is isomorphous, or if the contamination is produced by the precipitant itself.

From the facts and phenomena considered above, it can be appreciated that it is not a simple task to obtain a precipitate of a purity suitable for a gravimetric determination. Since removing a contaminant, once introduced, is especially difficult, it is desirable to perform the precipitation in such a way that a sufficiently pure product is obtained initially.

6.9 *Aging Precipitates (Digestion)*

When a precipitate is allowed to stand in contact with its mother liquor, changes often take place that lead to a product of larger particle size and thereby improved filterability and possibly less contamination. This transformation is called aging and may be the result of many processes. Small crystals have a slightly greater solubility than large ones; consequently they dissolve and their material is then deposited on the larger crystals. Small clusters of crystals may be joined, and imperfections in crystals may be "smoothed." In addition, colloids undergo coagulation; that is, small colloidal particles join and the larger particles that are formed precipitate.

In practice, aging is often effected at an elevated temperature, for example, on a steam bath, so that the processes involved are accelerated. The operation is called aging by digestion, or simply digestion of the precipitate.

Aging, or digestion, may also reduce the contamination of a precipitate, especially that due to adsorption of substances, since the large particles formed have a relatively small surface area. The precipitation of barium sulfate is an outstanding example of how a steam bath digestion greatly improves the filterability of the precipitate and significantly reduces some of the contamination. Aging, of course, fails and is inappropriate when the contamination is due to postprecipitation.

6.10 *Precipitation from Homogeneous Solution*

If the precipitant solution is added to the sample solution, undesirably high supersaturation results, at least locally, at the site at which the drops of the precipitant solution enter. Here extensive nucleation may take place even with slow addition and vigorous stirring. This difficulty can be avoided if the precipitant is not added from without but forms within the sample solution itself. This elegant technique is appropriately termed *precipitation from homogeneous solution;* its principles are best understood by looking at some practical examples. In the precipitation of hydrous iron(III) oxide, ammonia is often used as the precipitating reagent. More realistically, the precipitant is the hydroxide ion, which is formed according to

$$NH_3 + H_2O \rightleftharpoons NH_4^+ + OH^- \tag{6-4}$$

Even if a dilute aqueous solution of ammonia is added slowly and with vigorous stirring, a local excess of hydroxide is unavoidable and a slimy precipitate results. If the hydroxide ion is generated in the solution itself at a low rate, however, the supersaturation is controlled, any local overconcentration is avoided, and a coarse, crystalline precipitate results. Urea is one substance that allows this favorable sequence of events. When urea is heated in slightly acidic aqueous solution, it reacts slowly with water, according to the reaction

$$CO(NH_2)_2 + 2H_2O \rightarrow (NH_4)_2CO_3 \rightarrow CO_2\uparrow + 2NH_3 + H_2O \tag{6-5}$$

The carbon dioxide is evolved from the boiling solution and the ammonia causes formation of hydroxide ion, as indicated. Consequently, throughout the solution, hydroxide ion is generated evenly and slowly, and there is no unduly large local supersaturation. By this method a very coarse product is obtained in the formation of the hydrous oxides of iron(III), aluminum, and certain other metals.

Urea can also be used for precipitating calcium oxalate in a coarse form. Calcium oxalate is insoluble in neutral or alkaline aqueous solution but soluble in acidic solution. Thus, if an aqueous solution of a calcium salt is made acidic and oxalate ion is added, there is no precipitation. If urea is then added and the solution boiled, hydroxide ion is generated, and calcium oxalate is precipitated progressively as the solution is neutralized. Crystals longer than 0.5 mm can be readily obtained.

The sulfides of nickel and cobalt are "problem" precipitates in the sulfide separation of cations when gaseous hydrogen sulfide is used. Colloidal suspensions are formed and they do not coagulate and are not easily separated by filtration. Thioacetamide can serve for the homogeneous precipitation of the sulfides of these and other metals. In alkaline solution, the formation of sulfide ion from thioacetamide can be represented by

$$CH_3CSNH_2 + 2OH^- \rightarrow CH_3COO^- + NH_3 + HS^- \tag{6-6}$$

$$HS^- + OH^- \rightleftharpoons S^{2-} + H_2O \tag{6-7}$$

The decomposition of thioacetamide proceeds slowly even in boiling solution; consequently, the generation of sulfide ion is gradual and leads to the homogeneous precipitation of the metal sulfide. The products of this procedure are coarsely crystalline and are readily separable by filtration.

Precipitation from homogeneous solution is a tedious process, because boiling for 30 min to several hours may be needed. The time spent, however, is often compensated by the ready filtration of the product and its lower degree of contamination that make reprecipitation unnecessary. Contamination is reduced because the large particles have a relatively small surface area, and therefore adsorption is decreased. In addition, contamination caused by all types of occlusion and entrapment is greatly reduced, since the low rate of crystal formation allows time for any foreign material to be displaced from the crystal lattice.

6.11 Separating and Washing Precipitates

The precipitate, no matter whether it is separated from the mother liquor by filtration, centrifugation, or even decantation, must be properly washed, and there are certain criteria for selecting a wash liquid. The final wash liquid should leave in the precipitate no foreign substance that cannot be removed in the subsequent treatment (usually drying or ignition), before the weighing.

Unless the precipitate has an extremely low solubility, it is important to keep the amount of wash liquid as small as possible. The washing will be more efficient if each small portion is completely removed from the precipitate before the next portion is added. If a significantly soluble precipitate is involved, its solubility in the wash liquid can often be reduced by resorting to the common-ion effect (Section 5.5), that is, by having present in the wash liquid an ion common with the precipitate (not, of course, the one to be determined). Another possibility is to use at first a wash liquid saturated with the compound precipitated and in a final washing step to remove the saturated liquid with a minimum amount of water. Organic solvents such as alcohol or acetone, in which most ionic compounds are less soluble, can be used to reduce losses during the washing procedure.

If the precipitate is a coagulated colloid, say hydrous tin(IV) oxide, using pure water as the wash liquid may lead to losses—the precipitate may revert to a colloidal solution that passes through the filter. This reversion, known as peptization, is avoided if the water contains a small amount of electrolyte. To avoid permanent contamination of the precipitate, this electrolyte must be one that can be volatilized in the subsequent ignition (for example, ammonium salts).

6.12 Converting the Precipitation Form to a Weighing Form

Where the precipitation form and the weighing form are essentially identical, the precipitated and washed compound may be merely dried and weighed. Simple drying needs only a brief remark. In some cases the temperature must be maintained within a specified range to ensure that a compound of definite composition forms without any partial decomposition or volatilization. More involved operations are often necessary to ensure an appropriate and suitable weighing form; such operations range from ignition to treatment with one or more chemicals. The object of ignition is to obtain a stable substance of definite composition. The best conditions will depend on the particular determination. Any filter paper used in filtering the precipitate must be burned, and often most carefully, to avoid any reduction of the filtered substance. For example, the carbon formed in the initial, incomplete combustion of the paper may reduce barium sulfate to barium sulfide according to the reaction

$$BaSO_4 + 4C \rightarrow BaS + 4CO\uparrow \tag{6-8}$$

Sometimes for various reasons the ignition must be at a much higher temperature than what is needed to attain a desired stoichiometric composition. For example, hydrous aluminum oxide when heated to 200 to 300°C loses its water completely; however, the aluminum oxide thus produced is extremely hygroscopic at room temperature. The aluminum oxide, on being heated for 10 min at 1100°C or higher, is converted to a nonhygroscopic form that can be weighed in an open crucible after cooling.

The determination of calcium involving calcium oxalate as the precipitation form offers instructive examples of weighing forms. The precipitate can be dried at 100°C to the monohydrate, but extreme care is required in the drying to attain a product of the stoichiometric composition of the monohydrate. Calcium oxalate can be ignited to calcium carbonate, which is a suitable weighing form and is common as such. Controlled ignition (475 to 525°C) is mandatory since at too high a temperature, the calcium carbonate decomposes to calcium oxide. Conversion to this oxide, which is also a possible weighing form, occurs if the temperature is high enough (above 880°C). However, the oxide is extremely hygroscopic and absorbs carbon dioxide; consequently it is not a recommended weighing form.

6.13 *General Use of Precipitation*

Although the considerations above have been directed mainly to precipitation as a step for a gravimetric determination, this method of separation has many other applications in both qualitative and quantitative analysis. The phenomena and facts already considered are relevant to this broader use of precipitation.

Coprecipitation can be used to advantage to achieve the concentration and recovery of trace amounts of a substance. Such a substance may be deliberately coprecipitated with another compound, which is then known as a trace carrier, collector, or scavenger. As you might suppose, gelatinous (colloidal) precipitates are especially suitable for this purpose.

6.14 *Questions*

6–1 Define *weight* and *mass* and elaborate on their relation, emphasizing their common usage in gravimetric analysis.

6–2 In your own words define *gravimetric analysis.* Outline the general principles on which the technique is based; give diverse classifications of gravimetric analysis as they are possible from various points of view.

6–3 What is *coprecipitation* in the broad and restricted senses of the term?

6-4 List the main steps usually involved in a gravimetric determination.

6-5 Differentiate between *precipitation* and *weighing* forms and name the requirements for the two forms that are similar, different, mandatory, and desirable.

6-6 What three steps lead to the formation of a precipitate?

6-7 What is "digestion" of a precipitate and what is its purpose?

6-8 By what mechanisms can a precipitate be contaminated? How can such contaminations be avoided or at least reduced?

6-9 What is a gravimetric factor? Should its numerical value be high or low? Explain.

6-10 How can crystal growth be favored over nucleation?

6-11 Compare the usual precipitation technique with precipitation from homogeneous solution. Discuss requirements in time and apparatus; also, purity, composition, and appearance of the precipitates.

6-12 For what purpose besides gravimetry might precipitation be performed in analytical laboratories?

6-13 What is coagulation?

6-14 What is a colloid?

6-15 Discuss advantages and disadvantages of separation techniques such as filtration, decantation, and centrifugation.

6-16 What means can you name that can improve the situation in a gravimetric analysis when a precipitate is involved that is not essentially insoluble?

Note: Problems related to gravimetry appear in Section 7.5.

7 Applied Gravimetric Analysis

7.1 *Selectivity and Sensitivity*

Almost all of the common elements can be accurately determined by one or more gravimetric methods. Although the methods differ both in their chemistry and complexity, it is generally not difficult to determine one element when it is present alone. In practical analysis, however, the sought-for-substance is usually accompanied by other substances. Hence, getting a sufficiently sensitive determination is often overshadowed by the greater problem of securing adequate selectivity. In other words, the precipitant and the procedure selected must not only guarantee a complete precipitation but also a "clean" separation from other constituents of the solution. A beginner, not recognizing this difficulty, frequently thinks that gravimetric analysis is an easy task, requiring for each element or ion nothing more than just applying one method. In analytical practice, a large variety of methods must be at hand and an appropriate selection must be made according to the amount and nature of the substances to be determined and also the amount and nature of accompanying materials.

For example, the gravimetric determination of barium via the precipitation of barium sulfate is satisfactory when barium is present in a solution as the only cation. In contrast, the method is useless in the presence of strontium or calcium, or both, because these other alkaline earths also form sparingly soluble sulfates. In such a case barium is precipitated under carefully adjusted conditions as barium chromate. The precipitation of iron(III) as the hydrous oxide by the addition of ammonia followed by ignition and weighing as iron(III) oxide, Fe_2O_3, furnishes an additional example of the problem of analytical selectivity. When iron is present as the only cation, the method is satisfactory. In the presence of aluminum, chromium(III), and titanium(IV), the method fails because these metals (and some others) are also precipitated when ammonia is added. It is difficult to achieve a sufficiently "clean" separation of iron from these metals that a gravimetric determination is feasible. Even the presence of calcium (Section 6.7) may complicate the determination of iron because of coprecipitation. The difficulties increase when a gravimetric determination is contemplated for a substance that is present either in small or trace amounts. In a complex mixture, separations then become even much more involved. The use of organic precipitants in many cases greatly improves the situation. In contrast to inorganic precipitants, the organic precipitants represent a greater variety of compounds, and many of the reagents themselves show a high degree of selectivity. Also, the properties of organic compounds can be changed within certain limits by modifying their structure. Often, favorable gravimetric factors are obtained with organic reagents of large formula weights. You may have used dimethylglyoxime in the qualitative test for nickel. This oxime is also useful in quantitative analysis and under appropriate conditions is a highly selective precipitant for nickel. (See Section 7.2).

An introductory course cannot cover the many possibilities and practical considerations for gravimetric determinations. But you will do well to remember the limitations that have been described when you are studying the following important and instructive examples of gravimetric determinations for inorganic substances.

7.2 *Selected Examples, Gravimetric Determinations*

Many trivalent and tetravalent cations can be readily precipitated as hydrous oxides by ammonia; this affords a possibility for their separation from monovalent and some divalent ions, but reprecipitation is often necessary because of coprecipitation. The conversion to the anhydrous oxide as a weighing form is achieved simply by ignition. The sum of the sesquioxides of iron and aluminum (and chromium, if present) is often designated R_2O_3 (that is, "residual" oxides) when determined in this way and is often reported as such in the analysis of rocks, minerals, glass, and the like, especially where further differentiation is not required. The precipitate will also contain any titanium present as TiO_2, and also manganese as MnO_2 if an oxidant such as hydrogen peroxide is present during the ammonia precipitation.

The precipitation of hydrous oxides with ammonia will be familiar from the qualitative separation of inorganic cations for their detection. Many other precipitation reactions of the qualitative scheme are applied for quantitative purposes. Some cations can be precipitated and weighed as sulfides, including cations of lead, mercury(II), cadmium, zinc, arsenic(V), antimony(V), bismuth, cobalt(II), nickel, and manganese(II). Since an excess of sulfur is present in certain of the precipitates, however, special measures are needed to obtain a suitable weighing form. Because sulfide precipitations are unselective, they are seldom used as the basis of a gravimetric determination, but hydrogen sulfide still is an important reagent for "group" separations.

Chloride ion is a familiar precipitant for silver ion and is favored for the gravimetric determination of that metal. This precipitation reaction can also be applied in the determination of chloride, using silver nitrate as the precipitant. Such dual use of a given precipitation reaction is a common practice. The precipitation of barium sulfate serves for the separation and gravimetric determination of both barium and sulfate, and the precipitation of barium chromate for both barium and chromate.

Dimethylglyoxime in slightly alkaline solution is a highly selective precipitant for nickel(II). The precipitate can be dried to the stoichiometric composition $Ni(C_8H_{14}N_4O_4)_2$, which affords a very favorable gravimetric factor (0.2032). The gravimetric determination of even small amounts of nickel is therefore possible. The reaction of dimethylgyloxime with nickel involves the accompanying structural arrangement. Note that a single nickel ion combines with two molecules of the

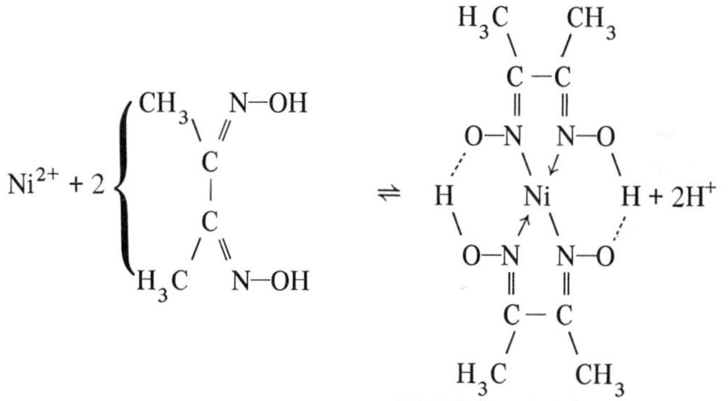

```
                                        H3C     CH3
                                           \   /
                                           C — C
               CH3     N—OH                ‖   ‖
                  \   //                 O—N   N—O
                    C                   .·   \ ↙   \
Ni2+ + 2 {          |          ⇌      H      Ni      H + 2H+
                    C                  \    ↗ \    .·
                  /   \\                O—N   N—O
               H3C     N—OH                ‖   ‖
                                           C — C
                                          /     \
                                       H3C       CH3
```

Nickel dimethylglyoximate

reagent, forming a neutral compound in which the nickel participates in stable chelate rings (see Chapter 20 for chelation).

The gravimetric determination of calcium based on the precipitation of calcium oxalate has been described in Section 6.12, including some of the possible weighing forms. Magnesium can be precipitated from ammoniacal solution as magnesium ammonium phosphate hexahydrate, $MgNH_4PO_4 \cdot 6H_2O$, and can be weighed as such after drying. Solutions of this salt show a pronounced tendency for supersaturation. The vessel walls can be scratched to initiate the precipitation, and prolonged standing is necessary to assure its completeness, especially when only a small amount of magnesium is present. Well-crystallized precipitates are obtained by adding orthophosphate to the acidic sample solution and then adding a large excess of ammonia to the boiling solution slowly, with stirring. The precipitate can be ignited at a high temperature, whereby according to the reaction

$$2MgNH_4PO_4 \cdot 6H_2O \rightarrow Mg_2P_2O_7 + 2NH_3\uparrow + 7H_2O\uparrow \qquad (7\text{-}1)$$

magnesium pyrophosphate is formed. This compound is preferred as the weighing form since its stoichiometry is more readily assured than the stoichiometry of the hexahydrate salt.

The same precipitation reaction can be used for the gravimetric determination of orthophosphate; magnesium ion is the precipitant. An ammonium salt (chloride or nitrate) must be added to prevent such a high alkalinity that magnesium hydroxide will be precipitated.

The important alkali metals, sodium and potassium, form only a few compounds that are sparingly soluble in water. Their gravimetric determination is therefore often accomplished after all other cations are completely removed by precipitation with reagents like ammonia, sulfide ion, and oxalate ion, which are readily volatilized or destroyed. The final solution, after addition of sulfuric acid, is evaporated to dryness and the residue ignited. The alkali metal sulfates remaining are weighed. Frequently, as in the analysis of rocks and minerals, only the sum of the alkali metals is thus established.

Potassium can be determined in the sulfate mixture by dissolving the mixture and precipitating potassium as the sparingly soluble perchlorate, $KClO_4$. Ethanol is added to decrease the solubility sufficiently and is also used in washing the precipitate. Sodium is then found by difference. Sodium tetraphenylborate, $NaB(C_6H_5)_4$ is an excellent precipitant for potassium. Potassium tetraphenylborate, $KB(C_6H_5)_4$, of all known potassium compounds, has the lowest solubility in water. It is stable in air, can be dried to the stoichiometric composition, and has a very favorable gravimetric factor.

7.3 Indirect Gravimetric Analysis

When both components of a binary mixture are to be determined, it is preferable to determine each of them individually, performing a separation if necessary. This is called direct analysis. Another approach involves determining one of the components and establishing the other by difference. For example, the percentage of aluminum in a binary aluminum-magnesium alloy can be determined and the magnesium content found by subtraction from 100. A third approach is known as indirect analysis. Two measurements, of which one is usually the weight of the sample, are secured and used to set up a pair of simultaneous equations. Their algebraic solution establishes the amount of each component.

One type of indirect analysis involves a mixture in which the components have an element or a radical in common. From the determination of this element or radical and the weight of the sample, the amounts of the individual components can be established.

Example 7-1 A sample, known to contain only AgCl (*143.32*) and AgI (*234.77*), weighs 1.5000 g. It is reduced quantitatively to metallic silver (*107.87*), which amounts to 0.8500 g. What is the weight of AgCl and AgI in the sample?

Let

$$x = \text{grams of AgCl in the sample}$$

$$y = \text{grams of AgI in the sample}$$

Then by the description of the sample

$$x + y = 1.5000$$

Now

$$\text{grams of Ag from the AgCl} + \text{grams of Ag from the AgI} = 0.8500$$

and by the application of the chemical factors

$$\text{grams of Ag from the AgCl} = x \times \frac{Ag}{AgCl}$$

$$\text{grams of Ag from the AgI} = y \times \frac{Ag}{AgI}$$

Hence,

$$x \times \frac{Ag}{AgCl} + y \times \frac{Ag}{AgI} = 0.8500$$

Substituting the values for the chemical factors gives

$$0.7527x + 0.4595y = 0.8500$$

This constitutes the second equation of the required pair. Multiplying the first equation of the pair by 0.7527 yields

$$0.7527x + 0.7527y = 1.1290$$

Subtracting the second equation of the pair from this equation gives

$$0.2932y = 0.2790$$

and $$y = 0.952 \text{ g of AgI in sample}$$

From the first equation of the pair

$$x = 1.5000 - y = 0.548 \text{ g of AgCl in sample}$$

A second type of indirect analysis involves converting a mixture in which the components have an element or a radical in common to a second mixture in which the components have a different element or radical in common.

Example 7–2 The mineral dolomite is a mixture of $CaCO_3$ (*100.07*) and $MgCO_3$ (*84.32*), with small amounts of impurities neglected for the present purposes. A 1.000-g sample of a dolomite is ignited, volatilizing CO_2 and leaving a residue of CaO (*56.08*) and MgO (*40.31*); the residue is found to weigh 0.5200 g. What is the approximate percentage composition of the dolomite?

Let

$$x = \text{grams of } CaCO_3 \text{ in the 1-g sample}$$

$$y = \text{grams of } MgCO_3 \text{ in the 1-g sample}$$

Then $x + y = 1.000$ as the first equation of the pair. Applying chemical factors yields

$$x \times \frac{CaO}{CaCO_3} + y \times \frac{MgO}{MgCO_3} = 0.5200$$

and substituting the values for the factors gives

$$0.5604x + 0.4781y = 0.5200$$

as the second equation of the pair. Multiplying the first equation by 0.5604 gives

$$0.5604x + 0.5604y = 0.5604$$

and on solving,

$$x = 0.509 \text{ g} \qquad \text{and} \qquad y = 0.491 \text{ g}$$

Hence the approximate composition is 51% $CaCO_3$ and 49% $MgCO_3$.

Other types of indirect analysis might be illustrated. It will suffice, however, to state that if *n* constituents are to be determined, *n* independent measurements are needed, and *n* simultaneous equations must be somehow formulated and solved. When more than two constituents are to be determined, indirect analysis is seldom practical, since the accuracy and precision are poor. Indeed, even when the necessary mathematical treatment is applied to only two components, it leads to the loss of significant figures and hence a lowering of the precision.

Best results come from an indirect analysis if all the substances involved are present in reasonable amounts. If only a small amount of one constituent is present, then even a slight absolute deviation in the measurement will lead to a large relative deviation in the result for that constituent. In Example 7-2, in the indirect analysis of dolomite, the conditions are favorable since the components of the mixtures are present in similar amounts. From looking at the examples, it is evident that the tendency to lose significant figures is less if there is a greater difference in the numbers on the right side of the simultaneous equations and a greater difference in the conversion factors on the left side of the equations.

7.4 Questions

7-1 What two forms of the separated product can be differentiated in gravimetric analysis?

7-2 In your own words, summarize the mechanisms and phenomena that lead to coprecipitation.

7-3 In gravimetric analysis, why are ammonium salts present in many wash liquids?

7-4 Elaborate on the terms *solid solution, occlusion, peptization, postprecipitation,* and *precipitation from homogeneous solution.*

7-5 What is the significance of a gravimetric factor? Of an empirical gravimetric factor?

7-6 How would you weigh a hygroscopic substance?

7-7 How can the solubility of a precipitate be reduced?

7-8 Is a large or a small value for a gravimetric factor preferred? Explain your answer.

7-9 For a sought-for substance A, there are two possible precipitants, B and C. The formula weights of these precipitants are 100 and 300. The precipitates obtained are AB and A_2C. If all other factors are assumed equal (solubility, ease of handling, and so on), which precipitant would you select?

7-10 In your own words, state some of the considerations involved in selecting a wash liquid for a precipitate.

7-11 What is indirect analysis? What are its advantages and disadvantages?

7-12 Elaborate on the merits of dimethylglyoxime as a precipitant for the gravimetric determination of nickel, and also of sodium tetraphenylborate as a precipitant in the determination of potassium.

7.5 Problems

7-1 A volume of 50.00 mL of a solution containing 0.460 mmol of Na_2SO_4 (*142.04*) per mL is evaporated to dryness. What is the weight of the residue in grams? *Ans.:* 3.27 g

7-2 How many grams of AgCl (*143.32*) will be formed when the chloride present in a solution containing 0.2435 g of NaCl (*58.44*) is completely precipitated? *Ans.:* 0.5972 g

7-3 A sample of KI (*166.01*) weighing 0.2463 g is treated with H_2SO_4, the excess of acid evaporated, and the residue ignited. What is the weight in milligrams of the residue of K_2SO_4 (*174.27*)? *Ans.:* 129.3 mg

7-4 A sample of Ag_2CrO_4 (*331.73*) weighing 0.1000 g is so treated with an excess of $BaCl_2$ (*157.34*) that the precipitation reaction

$$Ag_2CrO_4 + BaCl_2 \rightarrow \underline{2AgCl} + \underline{BaCrO_4}$$

goes to completion. The precipitate containing AgCl (*143.32*) and $BaCrO_4$ (*253.33*) is filtered, washed, dried, and weighed. What will it weigh in grams? *Ans.:* 0.1628 g

7-5 Calculate the gravimetric factors for

Substance sought		Substance weighed		Answer
SO_3	(*80.06*)	$BaSO_4$	(*233.40*)	0.34302
As_2O_3	(*197.84*)	Ag_3AsO_4	(*462.53*)	0.21387
$HClO_4$	(*100.459*)	AgCl	(*143.323*)	0.70093
$FeSO_4$	(*152.908*)	Fe_2O_3	(*159.692*)	1.91504

7-6 In a mixture, $CaCO_3$ (*100.09*) is the only component that loses weight on ignition, by volatilization of CO_2 (*44.01*). When a sample of this mixture weighing 0.4532 g was ignited, the weight of the residue was 428.9 mg.

(a) What was the percentage of loss in weight on ignition? (b) What was the percentage of $CaCO_3$ in the mixture? *Ans.:* (a) 5.36%; (b) 12.19% $CaCO_3$

7-7 Analyzing a mixture has established that it contains 32.55% Fe_2O_3 and that the percentage of loss in weight on drying is 1.25%. Calculate the percentage of Fe_2O_3 on a "dry" basis. *Ans.:* 32.96% Fe_2O_3 (dry basis)

7-8 The formal solubility of AgCl in water is $1.0 \times 10^{-5} F$. A precipitate of AgCl (*143*) is washed with 50 mL of water. If we assume that the wash liquid becomes 80% saturated in AgCl, how many milligrams of the precipitate will be "lost" by this washing? *Ans.:* 0.057 mg

7-9 A mixture containing only NaCl (*58.44*) and KCl (*74.56*) and weighing 1.2328 g, was treated with H_2SO_4 and then heated to dryness. The residue of Na_2SO_4 (*142.04*) and K_2SO_4 (*174.27*) was found to weigh 1.4631 g. What is the percentage composition of the original mixture? (*Caution!* Note that only one mole of an alkali metal sulfate is obtained from two moles of the corresponding chloride.) *Ans.:* 39% NaCl, 61% KCl

7-10 With the data of Example 7-1, consider that all possible pairings of 0.5-mg errors are made in the two weighings, that is, +0.5, +0.5; +0.5, −0.5; −0.5, +0.5; −0.5, −0.5 mg. Observe the effect of these errors on the calculated results. Do likewise with the data of Problem 7-9.

7-11 Exactly 1 g of an alloy gave a weight of 0.24 g of Al_2O_3 (*102*) after proper treatment. Calculate the % w/w of Al (*27*) in the alloy.

7-12 A sample of 1.116 g of an alloy gave a weight of 1.280 g of Fe_2O_3 (*159.7*) after proper treatment. Calculate the % w/w of iron in the alloy.

7-13 What amount in grams of a 1.00% w/w solution of $BaCl_2$ (*208.25*) is required to react quantitatively with 0.100 g of Na_2SO_4 (*142.04*)?

7-14 The sulfur in 0.2280 g of thiourea, $(NH_2)_2CS$ (*76.13*), is converted to sulfate and precipitated as barium sulfate (*233.4*). How many grams of $BaSO_4$ are obtained?

7-15 From qualitative analysis a coin was known to contain only silver (*107.87*) and copper (*63.5*). The coin was dissolved in nitric acid and the silver precipitated as AgCl (*143.32*), of which an amount of 0.2868 g was obtained. The sample taken weighed 0.3083 g. Calculate the % w/w of silver and copper in the coin.

7-16 Sodium tetraphenylborate, $NaB(C_6H_5)_4$, is used as a precipitant for potassium. What is the % w/w of K_2O (*94.20*) in a fertilizer when a 0.4320-g sample gives 0.2138 g of $KB(C_6H_5)_4$ (*358.2*)?

7-17 The arsenate present in a solution is precipitated as Ag_3AsO_4 by adding 40.00 mL of 0.100*F* $AgNO_3$. The precipitate is separated by filtration, and the silver in the filtrate is precipitated as AgCl, yielding 0.1434 g. Calculate the amount of arsenate in the original solution, expressing it as milligrams of As_2O_3.

7-18 Calculate the percentage loss on ignition for the following substances from which CO_2 (*44.01*) is volatilized:

(a) $CaCO_3$ (*100.09*)
(b) $MgCO_3$ (*84.32*)
(c) $BaCO_3$ (*197.35*)
(d) $MnCO_3$ (*114.95*)
(e) $SrCO_3$ (*147.63*)

7-19 A pure dolomite [only $CaCO_3$ (*100.1*) and $MgCO_3$ (*84.32*) present] shows a 47.29% w/w loss on ignition. Calculate the % w/w of $CaCO_3$ in the dolomite.

7-20 A mineral as received contains 10.12% w/w of lithium (*6.94*) and has a moisture content of 4.5% w/w. What is the lithium content on a "dry basis"?
Ans.: 10.60% w/w

7-21 For surface waters, *suspended solids* and *dissolved solids* are defined as the milligrams of solids that, respectively, are retained by and pass thorough a stated glass-fiber disk per liter of sample. For these determinations, a filter is placed in a holder, washed with distilled water, and dried. A 100.0-mL portion of the sample is filtered under suction. The filter and holder are dried to constant weight at 105°C. The filtrate is transferred to a dish and evaporated to constant weight at 180°C. For a river-water sample, the filter and holder weighed originally 27.2442 g and after use 27.4323 g. The dish weighed originally 64.3261 g and with the residue 64.8457 g. Calculate the suspended and dissolved solids for the sample.

7-22. For surface waters, the determination of *total solids* involves evaporating a measured volume of the sample in a dish to constant weight at 103–105°C. By heating the dried total solids to 550°C in a muffle furnace and reweighing the dish, a value for *volatile solids* is then secured. Calculate the milligrams per liter of total and volatile solids from the following data: original weight of dish, 59.6920 g; weight of dish and the residue after the evaporation of 100.0 mL of lake water and drying at 105°C, 61.2149; weight of dish and residue after heating to 550°C, 60.3442 g.

8

Acid-Base Equilibria

8.1 *Introduction*

According to the classical concepts, developed by Arrhenius, acids are substances that yield in aqueous solution hydrogen ion, H^+, and bases are substances that yield hydroxide ion, OH^-. An acid-base reaction is therefore the combination of hydrogen ion with hydroxide ion to yield water. These classical concepts are useful, chiefly when applied to acid-base reactions in aqueous solution. Although they allow a mathematical treatment of the equilibria involved, they fail, however, to explain adequately many of the phenomena encountered, especially in non-aqueous solutions.

In 1923, Brønsted introduced a more general acid-base theory. According to his concepts, an acid is a substance, electrically neutral or ionic, that can give up a proton, H^+. A base is a substance, electrically neutral or ionic, that can take up a proton. In other words, acids are defined as proton donors and bases as proton acceptors. Lowry independently recognized the ideas underlying Brønsted's definitions; consequently, the phrase "Brønsted-Lowry acid-base concepts" is often used.

Consider hydrogen chloride, for which the following ionization equilibrium can be written:

$$HCl \rightleftharpoons H^+ + Cl^- \tag{8-1}$$

In the forward reaction, hydrogen chloride gives up a proton; consequently it is an acid. In the reverse reaction, chloride ion accepts a proton; consequently it is a base. Hydrogen chloride and chloride ion are said to form a conjugate acid-base pair; HCl is the conjugate acid of Cl^-, and Cl^- is the conjugate base of HCl.

Such an equilibrium can be written in general terms:

$$\text{acid} \rightleftharpoons \text{proton} + \text{base} \tag{8-2}$$

Acids need not be uncharged molecules (molecular or electrically neutral acid), such as hydrogen chloride; they can equally well be a cation (cationic acid) or an anion (anionic acid). Identical considerations apply to bases. The following selected examples illustrate the situation:

$$HBr \text{ (molecular acid)} \rightleftharpoons \text{proton} + Br^- \text{ (anionic base)} \quad (8\text{-}3a)$$

$$NH_4^+ \text{ (cationic acid)} \rightleftharpoons \text{proton} + NH_3 \text{ (molecular base)} \quad (8\text{-}3b)$$

$$H_2SO_4 \text{ (molecular acid)} \rightleftharpoons \text{proton} + HSO_4^- \text{ (anionic base)} \quad (8\text{-}3c)$$

$$HSO_4^- \text{ (anionic acid)} \rightleftharpoons \text{proton} + SO_4^{2-} \text{ (anionic base)} \quad (8\text{-}3d)$$

$$Al(H_2O)_6^{3+} \text{ (cationic acid)} \rightleftharpoons \text{proton} + Al(H_2O)_5(OH)^{2+} \text{ (cationic base)} \quad (8\text{-}3e)$$

In this list, the hydrogen sulfate ion, HSO_4^-, appears both as an anionic acid and an anionic base; species like these, which can both donate and accept a proton, are called ampholytes and are said to be amphiprotic or amphoteric.

Acids and bases can also be classified according to the number of protons they can donate or accept per molecule. If an acid can donate only one proton, it is variously known as monoprotic, monofunctional, monobasic, or monoequivalent. The prefixes *di-*, *tri-*, *tetra-*, and so on are substituted for *mono-* if two, three, four, and so on protons can be donated or accepted. For an acid (or base) capable of donating (or accepting) more than one proton but with no particular number of protons specified, the prefix *poly-* is used.

The electrical conductance of benzene is virtually zero and remains so when hydrogen chloride is dissolved in this solvent. This finding indicates that neither the solvent nor the solution contains any appreciable concentration of ions. In contrast, the electrical conductance of pure water is extremely low but increases tremendously if hydrogen chloride is dissolved in it. Consequently, it must be concluded that the aqueous solution, that is, hydrochloric acid, contains a large concentration of ions; in other words, the originally molecular hydrogen chloride has ionized.

The extreme difference in the behavior of hydrogen chloride in the two solvents indicates that whether or not ionization takes place does not depend on the acid alone; the solvent plays an active role.

According to Brønsted's concepts, a proton is given up by an acid only if a base is present that can accept the proton. Since benzene does not possess basic properties, no proton transfer takes place and consequently no ions are formed. Water, however, can act as a base and accept protons. In water, therefore, ionization of the acid occurs. Consequently, the classical "dissociation" of an acid in terms of Brønsted's concepts is actually an acid-base reaction between the solute and the solvent.

The present example can be formulated as

$$HCl + H_2O \rightleftharpoons Cl^- + H_3O^+ \quad (8\text{-}4)$$

The dissociation of hydrogen chloride in water is a protolytic reaction involving two conjugate acid-base pairs, HCl–Cl^- and H_3O^+–H_2O. The monohydrated proton, H_3O^+, is known as the oxonium ion; however, where an indefinite degree of hydration is actually intended, the term *hydronium ion* can be used. The dissociation of an acid may be formulated as a proton-transfer reaction of the following general type:

$$\text{acid}_1 + \text{base}_2 \rightleftharpoons \text{base}_1 + \text{acid}_2 \qquad (8\text{–}5)$$

where commonly base_2 is the solvent.

8.2 Acid-Base Strength

The strength of an acid or a base can be described in terms of the intrinsic strength of the substance, that is, its tendency to donate or accept a proton. This tendency can be measured only on a relative basis, however, because the acid donates a proton only if a base is present to accept it and the base can accept a proton only if an acid is present that donates it. Consequently, the strength of an acid (or a base) can be evaluated only in relation to a base (or an acid), commonly the solvent itself. In aqueous solution, hydrochloric, nitric, and perchloric acids, for example, all behave as strong acids of essentially equal strength. All three acids donate their protons virtually completely to the base water to form the hydronium ion, which is the strongest acid that can exist in water. In water-free acetic acid, however, perchloric acid acts as a far stronger acid than nitric or hydrochloric acids. The greater intrinsic strength of perchloric acid (that is, its greater tendency to donate protons) comes into fuller play in the acetic acid medium because this solvent is less basic than water, that is, it accepts a proton less readily.

It can be concluded that the position of the equilibrium in reaction (8–5) will be shifted more to the right when acid_1 and base_2 are stronger and base_1 and acid_2 are weaker. In other words, if an acid-base reaction takes place, the weaker acid and the weaker base of the two conjugate acid-base pairs will always form.

If an acid is strong, that is, if it has a great tendency to donate a proton, then obviously its conjugate base has a small tendency to accept a proton. Since hydrochloric acid (in water) is a strong acid, chloride ion is an extremely weak base. This conclusion can be presented as a general statement: The stronger an acid (or base), the weaker its conjugate base (or acid). The strengths of an acid and its conjugate base can be related mathematically (see Section 8.5).

8.3 Autoprotolysis of Water

The "dissociation" of water, after Brønsted, is actually an acid-base reaction, involving two conjugate acid-base pairs that, in this special case, have one component in common. The process can be written

$$\underset{\text{acid}_1}{H_2O} + \underset{\text{base}_2}{H_2O} \rightleftharpoons \underset{\text{base}_1}{OH^-} + \underset{\text{acid}_2}{H_3O^+} \tag{8-6}$$

One molecule of water acts as the proton donor (acid) and the other as the proton acceptor (base). The hydronium ion is the strongest acid that can exist in aqueous solution and the hydroxide ion is the strongest base. Since an acid-base reaction always goes in the direction of forming the weaker acid and base, it can be surmised that the position of the equilibrium is such as to favor water. Indeed, the experimental evidence shows that the equilibrium position is extremely far to the left. In pure water, the concentrations of hydronium and hydroxide ions are each only 10^{-7} mol/L.

As stated earlier, writing in H_3O^+ is not to imply that this species is actually present as such. Solvation definitely takes place and species such as $H(H_2O)_2^+$ or $H(H_2O)_3^+$ may persist. The "hydrogen ion" molar concentration always equals the total molar concentration of solvated protons regardless of the species considered present. Consequently, it is possible, whenever explicit recognition of the solvation of protons is not required, to simplify the notation by writing H^+. This common practice is followed in this textbook.

If the law of mass action is applied to the autoprotolysis reaction of water, reaction (8-6), the expression for the equilibrium constant follows:

$$K_{eq} = \frac{[H_3O^+][OH^-]}{[H_2O]^2} = \frac{[H^+][OH^-]}{[H_2O]^2} \tag{8-7}$$

Since the extent of ionization is exceedingly small, the concentration of water can be considered to remain essentially unaltered and can be incorporated into the equilibrium constant. Then

$$K_w = [H_3O^+][OH^-] = [H^+][OH^-] \tag{8-8}$$

This new constant, K_w, is known as the ion product of water. Its numerical value at 25°C is 1.00×10^{-14}. Since the autoprotolysis of water is strongly endothermic, the value of the constant increases greatly with temperature, reaching, at 60°C, ten times the value for 25°C.

The autoprotolysis reaction yields equimolar concentrations of hydrogen and hydroxide ions. Consequently, for pure water

$$[H^+] = [OH^-] \tag{8-9}$$

Hence

$$[H^+]^2 = [OH^-]^2 = K_w = 1.00 \times 10^{-14} \tag{8-10}$$

$$[H^+] = [OH^-] = \sqrt{K_w} = 1.00 \times 10^{-7} M \text{ (at 25°C)} \tag{8-11}$$

Pure water or any aqueous solution that affords equimolar concentrations of hydrogen and hydroxide ions is said to be neutral. Pure water in this context means chemically pure water, in contrast to distilled or deionized water in equilibrium with the air. The latter, so-called equilibrium water, contains a small amount of carbon dioxide, which forms carbonic acid. Consequently, such water will be slightly acidic; that is, the hydrogen ion concentration will be greater than $1.00 \times 10^{-7} M$. Depending on the carbon dioxide content of the air, this concentration ranges from 1×10^{-6} to $3 \times 10^{-6} M$.

8.4 The Concept of pH

It is sometimes inconvenient to present small molar concentrations in an exponential manner. Søfrensen therefore proposed that such concentrations and the small values of equilibrium constants should be expressed by the negative of the logarithm of the numerical value. This mode of expression is indicated by a prefixed letter "p."

Accordingly, for the molar concentrations of hydrogen and hydroxide ions, the expressions pH and pOH are used; they are defined by

$$pH = -\log [H^+] \quad \text{and} \quad [H^+] = 10^{-pH} \tag{8-12}$$

$$pOH = -\log [OH^-] \quad \text{and} \quad [OH^-] = 10^{-pOH} \tag{8-13}$$

The present-day definition of pH, however, involves the hydrogen ion activity rather than concentration. But as in previous cases, for dilute solutions, substituting concentration for activity is made as an approximation. (For the establishment of a practical, that is, an operationally defined pH scale, see Chapter 23.)

The expression for the ion product of water, using Søfrensen's convention, can be written

$$pK_w = pH + pOH = 14.00 \tag{8-14}$$

Consequently, for pure water or a neutral solution, from equation (8-11), pH = pOH = 7.00.

Unless it is stated otherwise, the temperature in connection with pH values is assumed to be 25°C. In the numerical expression of either pH or pOH, usually only two decimal figures are given, because in the common experimental determination of the hydrogen or hydroxide ion concentration it is difficult to attain a precision greater than that corresponding to two significant figures (see Section 2.3 for significant figures in logarithmic expressions).

The relation of the hydrogen and hydroxide ion concentrations to the acidity or alkalinity of an aqueous solution at 25°C can be summarized:

$$\text{Alkaline solution:} \quad pH > 7.00 > pOH \tag{8-15a}$$

$$\text{Neutral solution:} \quad pH = 7.00 = pOH \tag{8-15b}$$

$$\text{Acidic solution:} \quad pH < 7.00 < pOH \tag{8-15c}$$

Note that the numerical value of the ion product is not changed when an acid or base is added. Only the hydrogen and hydroxide ion molar concentrations are altered and in such a manner that their product still equals K_w.

A few numerical examples show practical aspects of the concepts developed.

Example 8–1 What is the pH of a solution 0.030*M* in hydrogen ion?

$$[H^+] = 0.030 = 3.0 \times 10^{-2} M$$

$$pH = -\log [H^+] = -\log (3.0 \times 10^{-2}) = -\log 10^{-2} - \log 3.0$$

$$= 2 - 0.48 = 1.52$$

Example 8–2 Calculate the pH and pOH of a solution 0.0020*M* in hydroxide ion.

$$[OH^-] = 0.0020 = 2.0 \times 10^{-3} M$$

$$pOH = -\log [OH^-] = -\log (2.0 \times 10^{-3}) = -\log 10^{-3} - \log 2.0$$

$$= 3 - 0.30 = 2.70$$

$$pH = 14.00 - pOH = 14.00 - 2.70 = 11.30$$

This last calculation can be checked by applying equation (8–8).

$$[H^+] = \frac{1.00 \times 10^{-14}}{[OH^-]} = \frac{1.00 \times 10^{-14}}{2.0 \times 10^{-3}} = 0.50 \times 10^{-11}$$

$$= 5.0 \times 10^{-12} M$$

$$pH = -\log (5.0 \times 10^{-12}) = 12 - \log 5.0$$

$$= 12 - 0.70 = 11.30$$

Example 8–3 What is the molar concentration of hydrogen ion in a solution having a pH value of 4.30?

$$[H^+] = 10^{-pH} = 10^{-4.30} = 10^{-5.00+0.70} = 10^{-5} \times 10^{0.70}$$

$$= 10^{-5} \times \text{antilog } 0.70 = 5.0 \times 10^{-5} M$$

Expressing the negative exponent as the algebraic sum of a positive and a negative sum is necessary, since an antilogarithm must be a positive decimal number to be found in a table of logarithms.

Example 8–4 What is the hydroxide ion concentration in a solution of pH 10.15?

$$pOH = 14.00 - pH = 14.00 - 10.15 = 3.85$$

$$[OH^-] = 10^{-pOH} = 10^{-3.85} = 10^{-4.00+0.15} = 10^{-4} \times 10^{0.15}$$

$$= 10^{-4} \times \text{antilog } 0.15 = 1.4 \times 10^{-4} M$$

8.5 Acidity and Basicity Constants

Equation (8-5) can be written in a brief form,

$$a_1 + b_2 \rightleftharpoons b_1 + a_2 \tag{8-16}$$

where *a* and *b* designate an acid and a base. If the molar concentrations of the various species are denoted by brackets, the equilibrium-constant expression for reaction (8-16) can be written

$$K_{prot} = \frac{[b_1][a_2]}{[a_1][b_2]} \tag{8-17}$$

The equilibrium constant K_{prot} is called the protolysis constant.

If the discussion is restricted to the solution of an acid in water, b_2 and a_2 become H_2O and H_3O^+, respectively. The reaction taking place when an acid *a* is dissolved in water can be written

$$a + H_2O \rightleftharpoons b + H_3O^+ \tag{8-18}$$

The equilibrium constant for this reaction is given by

$$K_{a,prot} = \frac{[b][H_3O^+]}{[a][H_2O]} \tag{8-19}$$

Since the concentration of water remains essentially unchanged in a dilute solution and since $[H^+]$ can be substituted for $[H_3O^+]$, relation (8-19) reduces to

$$K_a = [H_2O]\, K_{a,prot} = \frac{[H^+][b]}{[a]} \tag{8-20}$$

The constant K_a is variously known as the dissociation, ionization, or acidity constant of the acid *a*. The larger the numerical value of this constant, the stronger the acid.

Equation (8-18) seems to be unbalanced in its charge. As it has been pointed out, the essential criterion for an acid is its ability to donate protons, regardless of whether the acid is electrically neutral, anionic, or cationic. Consequently, in the general formulation, charges are omitted on *a* and *b*. With actual species written, either *a* or *b* or both can carry charges and in such a way that the charges on the left side of the equation balance the charges on the right.

If a base *b* is dissolved in water, the protolysis reaction taking place is

$$b + H_2O \rightleftharpoons a + OH^- \tag{8-21}$$

The protolysis constant for this equilibrium is

$$K_{b,prot} = \frac{[a][OH^-]}{[b][H_2O]} \tag{8-22}$$

Again the concentration of water remains essentially unchanged and the value can be incorporated into a new constant, which is variously known as the hydrolysis, dissociation, ionization, or basicity constant K_b of base b.

$$K_b = [H_2O]\,K_{b,\text{prot}} = \frac{[a]\,[OH^-]}{[b]} \tag{8-23}$$

Multiplying equation (8-23) by (8-20) yields

$$K_aK_b = \frac{[H^+]\,[b]}{[a]} \times \frac{[a]\,[OH^-]}{[b]} = [H^+]\,[OH^-] = K_w \tag{8-24}$$

The statement made previously that the stronger an acid the weaker its conjugate base can now be made quantitative. The acidity constant of an acid is inversely proportional to the basicity constant of its conjugate base, and vice versa. The proportionality constant is the ion product of the solvent. Since hydrochloric acid is a strong acid with a large acidity constant, its conjugate, chloride ion, is an extremely weak base. In fact, chloride is so weak a base that it does not influence the acidity of a solution (for example, a sodium chloride solution is neutral because neither the chloride ion nor the sodium ion affects the pH). The base character of the conjugate of a strong acid does not generally need to be considered in calculations related to acid-base equilibria.

8.6 *Monofunctional Strong Acids and Bases*

An acid or a base is defined as strong if its protolytic reaction with the solvent (water in the present context) goes essentially to completion, even in moderately concentrated solutions. Calculating the pH value of the solution of a strong monofunctional acid (or base) is simple. The molar concentration of hydrogen ion (or hydroxide ion) equals the formality of the acid (or base) as long as the concentration of the solute is large enough to make the contribution of hydrogen (or hydroxide) ion from the autoprotolysis of water insignificant.

Example 8-5 What is the pH of a 0.0040F solution of the strong acid HCl?

$$C_{HCl} = [H^+] = 0.0040 = 4.0 \times 10^{-3}$$

$$pH = 3 - \log 4.0 = 3.00 - 0.60 = 2.40$$

Example 8-6 What is the pH of a 0.020F solution of the strong base NaOH?

$$C_{NaOH} = [OH^-] = 2.0 \times 10^{-2}$$

$$pOH = 2 - \log 2.0 = 1.70$$

$$pH = 14.00 - 1.70 = 12.30$$

Sodium hydroxide, NaOH, according to the Brønsted concepts, is strictly not a base but a salt. This compound has an ionic lattice and on dissolution in water simply dissociates to yield sodium ion, Na^+, and hydroxide ion, OH^-. The sodium ion undergoes solvation, but this is not a protolytic reaction and is of no further consequence. The hydroxide ion is actually the base. According to older concepts, sodium hydroxide was described as a base; and even when the Brønsted concepts are applied, this compound is still loosely termed a base.

The simple approach used in Examples 8-5 and 8-6 fails if the concentration of either of the ions produced in the autoprotolysis of water cannot be neglected. For example, if the simple approach were applied to $1.0 \times 10^{-7} F$ hydrochloric acid, the calculated result would be pH = 7.00. This value would imply that adding a strong acid to pure water yields a neutral solution. This is nonsense. Example 8-7 shows how to proceed in this case.

Example 8-7 What is the pH of a solution $1.0 \times 10^{-7} F$ in HCl? Hydrogen ion is produced by two reactions, the autoprotolysis of water and the protolysis of the hydrogen chloride. Consequently,

$$[H^+] = [H^+]_{water} + [H^+]_{HCl}$$

From the HCl, one hydrogen ion is obtained for every chloride ion formed. Since the protolysis of the HCl goes to completion, the molar concentration of Cl^- equals the formality of the acid, and this in turn equals the molar concentration of the hydrogen ion produced from the HCl. For the autoprotolysis of water, it is known that for each hydrogen ion one hydroxide ion is formed. Consequently, the molar concentration of hydrogen ion must equal the sum of the molar concentrations of the hydroxide ion and chloride ion:

$$[H^+] = [OH^-] + [Cl^-]$$

The chloride concentration is known to equal $1.0 \times 10^{-7} M$. The hydroxide ion concentration can be expressed in terms of $[H^+]$ from the ion product of water; this leads to

$$[H^+] = \frac{1.0 \times 10^{-14}}{[H^+]} + 1.0 \times 10^{-7}$$

Rearrangement gives the following quadratic equation:

$$[H^+]^2 - 1.0 \times 10^{-7} [H^+] - 1.0 \times 10^{-14} = 0$$

Application of the quadratic formula yields

$$[H^+] = \tfrac{1}{2} [1.0 \times 10^{-7} \pm \sqrt{(1.0 \times 10^{-7})^2 + 4.0 \times 10^{-14}}]$$

$$= 1.6 \times 10^{-7} M$$

Only the positive root has physical meaning. The pH is then

$$pH = 7 - \log 1.6 = 7 - 0.20 = 6.80$$

Example 8-7 serves only to demonstrate that under certain conditions the autoprotolysis of water cannot be neglected and to indicate how to proceed in such a case. The numerical value obtained has little practical significance, because such a dilute acid solution is extremely susceptible to small amounts of impurities (for example, carbon dioxide from the air).

This refinement in the calculation has to be made only if very small concentrations of an acid or a base are involved; that is, the pH resulting is close to that of the neutral point. When the pH is slightly removed from that point, the differences between the concentrations of hydrogen and hydroxide ions become so great that the concentrations of one of these species can be neglected. For example, if the acidity of $1.00 \times 10^{-6} F$ HCl is calculated according to the refined approach, the result is $[H^+] = 1.01 \times 10^{-6} M$ (the actual calculation is left to you), whereas the approximate approach yields $1.00 \times 10^{-6} M$. Consequently, the simple calculation is quite satisfactory for concentrations of monofunctional acids (and bases) greater than $10^{-6} M$.

8.7 *Charge Balance*

The equation $[H^+] = [OH^-] + [Cl^-]$ of Example 8-7 was derived by appropriate reasoning; however, this equation can be obtained more directly.

A solution must be electrically neutral; that is, the number of positive charges must equal the number of negative charges. Consequently, for any solution, the sum of the molar concentrations of all positive charge must equal the sum of the molar concentrations of all negative charge. The mathematical formulation of this fact is the "charge-balance equation," or simply the "charge balance." Thus the charge balance for Example 8-7 can be immediately written as given above.

The novice frequently fails to differentiate between the concentration of *charge* and the concentration of *species.* For example, the molar concentration of barium ion in a solution is written $[Ba^{2+}]$; the molar concentration of the positive charge for this ion requires the term $2[Ba^{2+}]$, since each ion of barium carries two positive charges and consequently one mole of barium ion corresponds to two moles of positive charge. Similarly, the orthophosphate ion in a charge balance requires the term $3[PO_4^{3-}]$, since the ion carries three negative charges. Another common mistake is not realizing that the concentration term for a species occurs only once, even when the species is produced by different solutes. For example, the charge-balance equation for a solution obtained by dissolving sodium sulfate and sodium chloride in water is

$$[H^+] + [Na^+] = [OH^-] + [Cl^-] + [HSO_4^-] + 2[SO_4^{2-}]$$

The coefficient of the sodium term is unity, although sodium ion stems from two sources, one of which even provides two sodium ions per molecule. Also note that the concentration term for the hydrogen sulfate ion occurs but not the term for the electrically neutral sulfuric acid. Of course, every charge balance for an aqueous solution contains the autoprotolysis products of water. The mistakes mentioned happen partly from the failure of the beginner to differentiate clearly between a charge-balance equation and the equation describing a dissociation equilibrium.

The charge balance is not restricted to problems like those considered here but is of general value. It often provides the additional equation required to make the number of equations equal the number of unknowns.

To reduce the mathematical effort in solving problems, it is always good practice to determine whether any terms in the charge-balance equation are significantly smaller than others and can therefore be neglected. For example, in Example 8-5, the charge balance could be taken as $[H^+] = [OH^-] + [Cl^-]$. It could then be reasoned that in a strongly acidic solution, the hydroxide concentration is much smaller than the hydrogen ion concentration. Thereby the equality $[H^+] = [Cl^-] = C_{HCl}$ could be obtained directly.

One additional example of pH calculations should clarify the point.

Example 8-8 What is the pH of a solution $2.0 \times 10^{-7}F$ in the strong diacidic base $Ba(OH)_2$? The charge balance is

$$[H^+] + 2[Ba^{2+}] = [OH^-]$$

Appropriate substitutions yield

$$[H^+] + 2 \times 2.0 \times 10^{-7} = \frac{1.0 \times 10^{-14}}{[H^+]}$$

On rearrangement

$$[H^+]^2 + 4.0 \times 10^{-7}[H^+] - 1.0 \times 10^{-14} = 0$$

Solving the quadratic equation yields

$$[H^+] = \tfrac{1}{2}(-4.0 \times 10^{-7} \pm \sqrt{16 \times 10^{-14} + 4 \times 10^{-14}})$$
$$= 2.5 \times 10^{-8}M$$

and $$\text{pH} = 8 - \log 2.5 = 7.60$$

The remark made after Example 8-7 also applies here.

8.8 *General Treatment, Monofunctional Strong Acids and Bases*

It is simple to derive a general formula that allows the acidity (alkalinity) of a solution of a strong acid (base) to be calculated from its concentration. The reasoning involved is merely a generalization of that used in Example 8-7. The hydrogen ions stem from two sources, the ionization of the acid and the autoprotolysis of water. Since the acid is strong and thus ionizes completely and since the acid is monofunctional, the hydrogen ion concentration from this source equals the formal concentration of the acid C_a. The hydrogen ion concentration from the autoprotolysis of the water equals the hydroxide concentrations. Consequently,

$$[H^+] = C_a + [OH^-] \tag{8-25}$$

From the expression of K_w, $[OH^-]$ is given as

$$[OH^-] = \frac{K_w}{[H^+]} \tag{8-26}$$

Substitution of this yields

$$[H^+] = C_a + \frac{K_w}{[H^+]} \tag{8-27}$$

Since the value of K_w is known, this equation suffices to calculate the acidity. The quadratic equation obtained on rearranging equation (8-27) is

$$[H^+]^2 - [H^+]C_a - K_w = 0 \tag{8-28}$$

Solution yields

$$[H^+] = \tfrac{1}{2}(C_a + \sqrt{C_a^2 + 4K_w}) \tag{8-29}$$

In almost all practical cases, the solution of a strong acid has an acidity such that $[H^+]$ is much larger than $[OH^-]$. The $[OH^-]$ term in equation (8-25) can then be neglected, and the simple relation obtained is

$$[H^+] \cong C_a \tag{8-30}$$

and

$$pH = -\log C_a \tag{8-31}$$

For a strong monofunctional base at formal concentration C_b, the reasoning is analogous and can be summarized as

$$[OH^-] = C_b + [H^+] \tag{8-32}$$

Proceeding as before, we arrive at the following equation for $[OH^-]$:

$$[OH^-] = \tfrac{1}{2}(C_b + \sqrt{C_b^2 + 4K_w}) \tag{8-32}$$

For a strong base, $[OH^-]$ in almost all practical cases is much larger than $[H^+]$, and the $[H^+]$ term can be neglected in equation (8-32):

$$[OH^-] \cong C_b \tag{8-34}$$

Substituting $K_w/[H^+]$ for $[OH^-]$ and rearranging yields

$$[H^+] = \frac{K_w}{C_b} \tag{8-35}$$

and

$$pH = pK_w - \log C_b = 14 - \log C_b \tag{8-36}$$

It is of interest to investigate the situation in which a strong monofunctional acid and a strong monofunctional base are dissolved simultaneously. At equilibrium, the sum of the concentrations of *all* bases must equal the sum of the concentrations of *all* acids. Consequently, with the equalities already previously stated,

$$C_a + [OH^-] = C_b + [H^+] \tag{8-37}$$

Substituting $K_w/[H^+]$ for $[OH^-]$ and solving the quadratic equation obtained gives

$$[H^+] = \tfrac{1}{2}\left[(C_a - C_b) + \sqrt{(C_a - C_b)^2 + 4K_w}\right] \tag{8-38}$$

If $C_a = C_b$ the equation yields $[H^+] = \sqrt{K_w} = 10^{-7}$; that is, as expected, the solution is neutral. If either the acid or the base is in slight excess, the pH is far removed from that of the neutral point and then either $[OH^+]$ or $[H^+]$ can be neglected in equation (8-37). For an acidic solution (acid in excess, $C_a > C_b$),

$$[H^+] = C_a - C_b \tag{8-39}$$

For an alkaline solution (base in excess, $C_b > C_a$),

$$[OH^-] = C_b - C_a \tag{8-40}$$

or

$$[H^+] = \frac{K_w}{C_b - C_a} \tag{8-41}$$

The formulas for a strong acid, a strong base, or both are also valid when more than one acid or base is present if C_a and C_b are replaced by the sum of the concentrations, that is, ΣC_a and ΣC_b.

8.9 Monofunctional Weak Acids and Bases

An acid or base is defined as weak if the protolysis reaction with the solvent, here water, does not go essentially to completion, unless the solution is extremely dilute. The calculation of the acidity and alkalinity, that is, of the pH, of solutions containing weak protolytes can best be explained by examples. To simplify the notation, we can denote the weak acid as HA. For example, for acetic acid, CH_3 COOH, the symbol A^- is the anion CH_3COO^-. The protolysis reaction can now be written

$$HA + H_2O \rightleftharpoons A^- + H_3O^+ \tag{8-42}$$

The acidity constant of the acid according to equation (8–20) is given as

$$K_a = \frac{[A^-][H^+]}{[HA]} \tag{8–43}$$

Since the ionization does not go to completion, the acid, when equilibrium is established, will be present partly as non-ionized acid and partly as the anion, that is, as the conjugate base. The formal concentration of the acid C_a will equal the sum of the molar concentrations of the non-ionized and ionized portions,

$$C_a = [HA] + [A^-] \tag{8–44}$$

This equation is called a material-balance equation, or simply a material balance. The charge balance provides one more equation,

$$[H^+] = [A^-] + [OH^-] \tag{8–45}$$

Rearrangement yields

$$[A^-] = [H^+] - [OH^-] \tag{8–46}$$

Combining this equation with equation (8–44) yields an expression for [HA]:

$$[HA] = C_a - ([H^+] - [OH^-]) \tag{8–47}$$

Substituting the relevant terms in the expression of the constant (8–43) gives

$$K_a = \frac{[H^+]([H^+] - [OH^-])}{C_a - ([H^+] - [OH^-])} \tag{8–48}$$

When $[OH^-]$ is replaced by $K_w/[H^+]$, an expression is obtained that can be solved for $[H^+]$; with K_a, K_w, and C_a known, rigorous calculation of the acidity of the solution is possible. Unfortunately, the equation is third-order in $[H^+]$ and thus not easily solved. However, for practical cases simplification is almost always possible. Unless the acid is extremely weak and the solution very dilute, the acidity will always be high enough to permit neglecting $[OH^-]$ with respect to $[H^+]$; this amounts to neglecting the contribution due to the autoprotolysis of water. Then the expression simplifies to

$$K_a \cong \frac{[H^+]^2}{C_a - [H^+]} \tag{8-49}$$

For many practical cases only a moderate acidity results, and then the term $[H^+]$ in the denominator can be dropped, since it is much smaller than C_a. Then equation (8–49) simplifies to

$$K_a \cong \frac{[H^+]^2}{C_a} \tag{8-50}$$

The hydrogen ion concentration can therefore be obtained from the simple equation

$$[H^+] \cong \sqrt{K_a C_A} \tag{8-51}$$

The critical question arises: When is it allowable to neglect a term? The values of the acidity constants are commonly reliable only to a small percentage. Thus the rule of thumb can be established that a term can be dropped if its value is 5% or less of the value of the term with which it is compared. Hence the first approximation, that is, neglecting $[OH^-]$ in equation (8–48), can be made if $[OH^-]$ is 5% of $[H^+]$ or less. The second approximation, that is, neglecting $[H^+]$ in the denominator of equation (8–49), is permissible if $[H^+]$ is 5% of C_a or less. Approximations like this are possible only with sums and differences and *never* with products or quotients; beginners often overlook this restriction.

One cannot decide offhand whether an approximation is allowable or not. For example, the decision whether $[H^+]$ can be dropped in the denominator of equation (8–49) calls for knowing the value of $[H^+]$, and it is this value that must be calculated. This difficulty is resolved in the following way. The approximation is made and the approximate value thus obtained is used for a check. If the 5% requirement (or whatever other limit is considered appropriate) is not fulfilled, the calculation must be repeated, using the more involved formula without the approximation that has proved unallowable. Here are some examples.

Example 8–9 What is the pH of a 0.10*F* solution of acetic acid when $K_a = 1.8 \times 10^{-3}$?

Applying the approximate formula, (8–51), yields

$$[H^+] \cong \sqrt{1.8 \times 10^{-3} \times 1.0 \times 10^{-1}} = \sqrt{1.8 \times 10^{-6}} = 1.3 \times 10^{-3} M$$

This value is now used to test the approximation. The hydrogen ion concentration is definitely large enough to drop $[OH^-]$, that is, to neglect the contribution of H^+ by the autoprotolysis of water. Thus the first approximation is valid. The value of $[H^+]$ obtained amounts to $(1.3 \times 10^{-3} \times 100)/10^{-1}$ = 1.3% the value of C_a. Consequently, the second approximation is also permitted. It is now safe to go ahead and calculate the pH of the solution:

$$pH = -\log (1.3 \times 10^{-3}) = 3 - \log 1.3 = 3 - 0.11 = 2.89$$

Example 8–10 What is the pH of a 0.0050*F* solution of formic acid with $K_a = 1.8 \times 10^{-4}$?

Applying the approximate formula, (8–51), yields

$$[H^+] \cong \sqrt{1.8 \times 10^{-4} \times 5 \times 10^{-3}} = \sqrt{9 \times 10^{-7}}$$

$$= \sqrt{90 \times 10^{-8}} = 9.5 \times 10^{-4} M$$

and $$pH = 3.02$$

Testing the approximations reveals that the ionization of water does not contribute significantly to the acidity. However, comparing $[H^+]$ with C_a shows that $[H^+]$ is

$$\frac{9.5 \times 10^{-4} \times 100}{5 \times 10^{-3}} \cong 20\%$$

of C_a. Consequently, the second approximation is not permitted according to the 5% requirement, and the calculation must involve the quadratic equation, (8–49).

Substituting the numerical values yields

$$1.8 \times 10^{-4} = \frac{[H^+]^2}{5 \times 10^{-3} - [H^+]}$$

Rearranging gives

$$[H^+]^2 + 1.8 \times 10^{-4}\,[H^+] - 9 \times 10^{-7} = 0$$

and on solving,

$$[H^+] = \tfrac{1}{2}(-1.8 \times 10^{-4} + \sqrt{3.2 \times 10^{-8} + 36 \times 10^{-7}})$$
$$= 8.6 \times 10^{-4} M$$

Finally,

$$\text{pH} = 3.07$$

Comparing this result with the pH value obtained by the approximate formula shows the difference to be only 0.05 pH unit; we see, then, that the 5% requirement is fairly stringent.

Example 8-11 What is the pH of a $5 \times 10^{-6} F$ solution of hydrocyanic acid with $K_a = 5 \times 10^{-10}$?

Application of the approximate formula, (8–51), yields

$$[H^+] \cong \sqrt{5 \times 10^{-6} \times 5 \times 10^{-10}} = \sqrt{25 \times 10^{-16}}$$
$$= 5 \times 10^{-8} M$$

The result is obviously wrong. This hydrogen ion concentration corresponds to an alkaline solution. Weak as an acid may be, its solution can never be alkaline. Obviously, with so weak an acid and at so small a concentration, the contribution of ions by the water cannot be neglected, and even the first approximation is not permitted. It might seem necessary to apply the third-order equation, (8–48). In fact, however, the calculation is simple. The pH is obviously close to 7. Consequently, $[H^+]$ and $[OH^-]$ are therefore very small compared with C_a and can be neglected in the denominator of (8–48). Then

$$[H^+] \cong \sqrt{K_a C_a + K_w} = \sqrt{25 \times 10^{-16} + 10^{-14}}$$
$$= \sqrt{1.25 \times 10^{-14}} = 1.12 \times 10^{-7} M$$

and $$\text{pH} = 6.95$$

However the calculation is practically meaningless, because such a solution would be so susceptible to influences of trace impurities (like carbon dioxide from the air) that the pH actually measured would probably differ significantly from the pH calculated. The example has the sole purpose of showing the importance of testing approximations and how to proceed with evaluating possibilities of dropping terms. Later, we shall have examples that have practical significance.

Deriving a formula for calculating the alkalinity of a solution that contains a weak monofunctional base, B, is fully analogous. The base protolyzes according to

$$B + H_2O \rightleftharpoons BH^+ + OH^- \tag{8-52}$$

The basicity constant is given as

$$K_b = \frac{[OH^-]\,[BH^+]}{[B]} \tag{8-53}$$

Material balance yields

$$C_b = [B] + [BH^+] \tag{8-54}$$

Charge balance yields

$$[H^+] + [BH^+] = [OH^-] \tag{8-55}$$

By rearrangement and substituting in the expression for the basicity constant, we get

$$K_b = \frac{[OH^-]([OH^-] - [H^+])}{C_b - ([OH^-] - [H^+])} \tag{8-56}$$

This formula could have been written without derivation by simply replacing in the formula for a weak acid, (8-48), $[H^+]$ by $[OH^-]$ and vice versa, and substituting C_b and K_b for C_a and K_a. The expression obtained by neglecting the contributions of ions by water (here equivalent to dropping the terms $[H^+]$) is

$$K_b \cong \frac{[OH^-]^2}{C_b - [OH^-]} \tag{8-57}$$

If the base is not too strong and the solution not too dilute, $[OH^-]$ is small enough to be neglected in comparison with C_b, and the further approximate formula,

$$[OH^-] \cong \sqrt{K_b C_b} \tag{8-58}$$

is obtained. Applying these equations and the reasoning, and testing the approximations are analogous to these procedures for a weak acid. An example will delineate the point.

Example 8-12 What is the pH of a $0.020F$ solution of ammonia (protolysis corresponding to $NH_3 + H_2O \rightleftharpoons NH_4 + OH^-$) if $K_b = 1.8 \times 10^{-5}$?

Applying the approximate formula, (8-58), yields

$$[OH^-] = \sqrt{1.8 \times 10^{-5} \times 2 \times 10^{-2}} = \sqrt{3.6 \times 10^{-7}}$$
$$= \sqrt{36 \times 10^{-8}} = 6.0 \times 10^{-4}M$$

Testing the approximations shows that neglecting the protolysis of water is permitted. Since $[OH^-]$ is only $(6 \times 10^{-4} \times 100)/(2 \times 10^{-2}) = 3\%$ of C_b, the second approximation is also permitted. Going on to the calculation of the pH, we get

$$pOH = 4 - \log 6.0 = 3.22$$

and

$$pH = 14 - 3.22 = 10.78$$

It can be surmised that for acid-base equilibria, the region about pH 7 is the critical one for either of the terms $[H^+]$ and $[OH^-]$ to be neglected in the relevant equations. This region can be defined by the 5% requirement introduced earlier in this section. If $[OH^-]$ is 5% of $[H^+]$, then by the expression for K_w, (8-24), $K_w = [H^+] \times 0.05\ [H^+]$, and therefore $[H^+] = \sqrt{10^{-14}/0.05}$ and the corresponding pH is 6.35. Because of the symmetry of the equation for K_w, the other limit of the region is given by $pK_w - pH = 14 - 6.35 = 7.65$. Within the pH range 6.35 to 7.65, consequently, use approximations with caution.

8.10 General Treatment, Monofunctional Weak Acids and Bases

The derivations given in Section 8.9 involve molecular acids and bases only and may have recalled to you facts from previous courses. Analogous derivations are possible for ionic acids and bases. It is more convenient, however, to have general formulas available that apply to all types of monofunctional acids and bases. Regardless of whether an acid is molecular or ionic, the protolysis equilibrium in water yielding the conjugate base proceeds according to

$$a + H_2O \rightleftharpoons b + H_3O^+ \tag{8-59}$$

The formal concentration of the acid C_a is, according to material balance,

$$C_a = [a] + [b] \tag{8-60}$$

Next we write a scheme that shows the reactants and products of all protolytic reactions:

Initial protolytes	Bases formed	Acids formed
a of formal concentration C_a	b of concentration $[b]$	
H_2O	OH^- of concentration $[OH^-]$	H_3O^+ of concentration $[H^+]$

The sum of the concentrations of all acids formed must equal the sum of the concentration of all bases formed. (This condition is analogous to the charge balance used in the previous derivation.) Consequently,

$$[H^+] = [OH^-] + [b] \tag{8-61}$$

and on rearrangement,

$$[b] = [H^+] - [OH^-] \tag{8-62}$$

Combining (8-60) and (8-62) yields

$$[a] = C_a - [b] = C_a - ([H^+] - [OH^-]) \tag{8-63}$$

Substitution in the expression for the acidity constant, (8-20), yields the desired formula:

$$K_a = \frac{[H^+]([H^+] - [OH^-])}{C_a - ([H^+] - [OH^-])} \tag{8-64}$$

This expression is identical with the one derived for a molecular acid, (8-48). The arguments leading to approximations apply identically here, and the simplified formulas are

$$K_a \cong \frac{[H^+]^2}{C_a - [H^+]} \tag{8-65}$$

and

$$[H^+] \cong \sqrt{K_a C_a} \tag{8-66}$$

The derivation for weak bases is analogous and is left as an exercise for you. The formulas obtained are

$$K_b = \frac{[OH^-]([OH^-] - [H^+])}{C_b - ([OH^-] - [H^+])} \tag{8-67}$$

$$K_b \cong \frac{[OH^-]^2}{C_b - [OH^-]} \tag{8-68}$$

and

$$[OH^-] \cong \sqrt{K_b C_b} \tag{8-69}$$

Example 8-13 What is the pH of a $0.015F$ solution of ammonium chloride if K_b for ammonia is 1.8×10^{-5}?

Ammonium chloride is a strong electrolyte and on dissolution in water completely dissociates, according to

$$NH_4Cl \rightarrow NH_4^+ + Cl^-$$

The chloride ion is the conjugate base of the strong acid HCl and is such a weak base that its basicity is of no consequence. The ammonium ion is a cationic acid. Thus the problem reduces to finding the pH of a solution containing the weak acid NH_4^+ in a concentration of 0.015*F*. Recall that NH_4^+ is the conjugate acid of the base ammonia. By the relation (8–24) ($K_a K_b = K_w$), the acidity constant of NH_4^+ is found to be

$$K_a = \frac{1.0 \times 10^{-14}}{1.8 \times 10^{-5}} = 5.6 \times 10^{-10}$$

Now the formula for a weak acid, (8–66), is applied:

$$[H^+] = \sqrt{5.6 \times 10^{-10} \times 1.5 \times 10^{-2}} = \sqrt{8.4 \times 10^{-12}}$$

$$= 2.9 \times 10^{-6} M$$

and $\text{pH} = 5.54$

The tests for the validity of the approximations are left to you.

Example 8–14 What is the pH of a 0.080*F* solution of sodium benzoate if K_a for benzoic acid is 6.3×10^{-5}?

If the benzoate ion is denoted by R^-, the dissociation of the strong electrolyte sodium benzoate on dissolving in water can be described as NaR $\rightarrow Na^+ + R^-$. The sodium ion solvates but does not show any acid-base properties; it is therefore of no further interest. The benzoate ion is the conjugate base of benzoic acid and thus the problem reduces to finding the pH of a solution containing this weak base in a concentration of 0.080*F*. The basicity constant of the benzoate ion is readily found to be

$$K_b = \frac{1.0 \times 10^{-14}}{6.3 \times 10^{-5}} = 1.6 \times 10^{-10}$$

Applying the approximate formula, (8–69) yields

$$[OH^-] = \sqrt{1.6 \times 10^{-10} \times 8 \times 10^{-2}} = \sqrt{12._8 \times 10^{-12}}$$

$$= 3.6 \times 10^{-6} M$$

$$\text{pOH} = 5.44$$

and $\text{pH} = 8.56$

You should test the validity of the approximations.

Example 8–15 Calculate the pH of a solution obtained by dissolving 4.92 g of sodium acetate (*82.0*) in water to a volume of 400 mL. The value of K_a for acetic acid is 1.8×10^{-5}.

When sodium acetate dissolves in water, it fully dissociates into sodium ion (which is of no further consequence) and acetate ion. The acetate ion is the conjugate base of acetic acid and therefore has a basicity constant of

$$K_b = \frac{1.0 \times 10^{-14}}{1.8 \times 10^{-5}} = 5.6 \times 10^{-10}$$

Applying the approximate formula, (8–69), requires knowing the formal concentration of the base; in this case, this value equals the concentration of the salt. This concentration is given by

$$C_b = C_a = \frac{4.92}{82.0} \times \frac{1000}{400} = 0.150F$$

By substitution,

$$[OH^-] = \sqrt{5.6 \times 10^{-10} \times 1.5 \times 10^{-1}} = \sqrt{8.4 \times 10^{-11}}$$

$$= 9.2 \times 10^{-6}M$$

$$pOH = 5.04$$

and $$pH = 8.96$$

Example 8–16 The solution of the sodium salt of a weak monofunctional acid has a pH of 9.00. What is the pH if the solution is diluted with pure water to twice its volume?

The anion of the weak acid is a weak base with K_b and at concentration C_b. Applying the approximate formula to the original solution yields

$$[OH^-]_{orig} = \sqrt{K_b C_b}$$

On dilution to twice the volume, the relation holds:

$$[OH^-]_{dil} = \sqrt{\frac{K_b C_b}{2}}$$

Dividing one equation by the other gives

$$\frac{[OH^-]_{orig}}{[OH^-]_{dil}} = \sqrt{2}$$

Substituting $K_w/[H^+]$ for $[OH^-]$ yields

$$\frac{K_w/[H^+]_{orig}}{K_w/[H^+]_{dil}} = \sqrt{2}$$

and finally

$$[H^+]_{dil} = [H^+]_{orig}\sqrt{2}$$

Substituting the numerical value gives

$$[H^+]_{dil} = 1.0 \times 10^{-9} \times \sqrt{2} = 1.4 \times 10^{-9}M$$

and $$pH = 8.85$$

8.11 Solution of a Monofunctional Weak Acid (or Base) and Its Salt

The situation becomes more involved if a weak acid (or base) is present in a solution together with its salt, that is, its conjugate. While it is always possible to establish enough independent relations to obtain a number of equations equal to the number of unknowns, the final general expression for $[H^+]$ (or $[OH^-]$) is always of third order or higher. Fortunately, in most of the cases that are of practical importance, reasonable approximations are allowed, and if applied lead to expressions that can be handled more readily. At least an approximate value can always be reached that permits the rough evaluation of the situation at hand.

No full derivations will be given in this section. Rather, the approximations will be made a priori and only approximate expressions will be derived. You should, however, take a cautious look at the numerical results whenever such expressions are used—especially in cases in which the acid or base is either extremely weak or relatively strong, either a very dilute or a relatively concentrated solution is involved, or the pH value is very high or very low.

First the situation will be examined for a weak acid present in a solution with its sodium salt. The acid undergoes protolysis according to

$$a + H_2O \rightleftharpoons b + H_3O^+ \tag{8-70}$$

The acidity constant is

$$K_a = \frac{[H^+]\,[b]}{[a]} \tag{8-71}$$

The salt dissociates and the sodium ion is of no further consequence. The conjugate base, however, undergoes hydrolysis according to

$$b + H_2O \rightleftharpoons a + OH^- \tag{8-72}$$

Neither the protolytic reaction of the acid nor the protolytic reaction of the conjugate base is assumed to proceed to any great extent. Protolysis of the acid is restricted because its conjugate base b is present in significant concentration from the addition of the salt. Analogously, the hydrolysis of the base is restricted because its conjugate acid a is present in significant amounts. Consequently, as an approximation, the molar concentration of the acid $[a]$ can be taken as equal to the concentration of the acid added [acid]. Analogously, the molar concentration of the conjugate base $[b]$ can be set equal to the concentration of the salt [salt]. Substituting in the expression for the acidity constant and rearranging yields the approximate formula for calculating the acidity of the solution secured:

$$[H^+] \cong K_a \frac{[\text{acid}]}{[\text{salt}]} \tag{8-73}$$

Taking negative logarithms of this formula and substituting pH and pK_a yield

$$\mathrm{pH} = \mathrm{p}K_a + \log \frac{[\mathrm{salt}]}{[\mathrm{acid}]} \tag{8-74}$$

This logarithmic form of Equation (8-73) is often found in the biosciences; it is known as the Henderson-Hasselbach equation. Note that the argument of the logarithm's term, [salt]/[acid], is the reciprocal of the ratio in Equation (8-73); the logarithmic term can then carry a positive sign.

You can see from Equation (8-73) that for a given system, that is, with K_a given, the acidity is solely a function of the ratio of the formal concentrations of acid and salt. This simple approach holds only for acids with K_a in the approximate range 10^{-3} to 10^{-10} and at reasonable concentrations, say not below 10^{-3} to $10^{-4}F$. The approach cannot be applied to very highly concentrated solutions; here the calculations are invalid anyway because of the failure to consider activities. Also, the simple approach does not apply if the salt involved introduces any other species, besides the conjugate base, that shows significant acid-base properties.

These restrictions hold analogously for solutions containing a weak base and its conjugate acid. From the analogy between acids and bases already stressed several times, the formula for calculating the hydroxide in concentration of a solution containing a weak base and its salt can be written immediately as

$$[\mathrm{OH^-}] \cong K_b \frac{[\mathrm{base}]}{[\mathrm{salt}]} \tag{8-75}$$

The derivation is recommended to you as an exercise.

Example 8-17 What is the pH of a solution 0.050F in sodium acetate and 0.025F in acetic acid ($K_a = 1.8 \times 10^{-5}$)?

Substituting in (8-73) yields

$$[\mathrm{H^+}] = \frac{1.8 \times 10^{-5} \times 2.5 \times 10^{-2}}{5.0 \times 10^{-2}} = 9.0 \times 10^{-6}F$$

and

$$\mathrm{pH} = 6.00 - \log 9.0 = 6.00 - 0.95 = 5.05$$

Example 8-18 Calculate the pH of a solution prepared by dissolving 3.0 g of acetic acid (*60.0*; $K_a = 1.8 \times 10^{-5}$) and 4.9 g of sodium acetate (*82.0*) in water and diluting with water to a volume of exactly 200 mL.

The moles of acid and salt present are

$$\text{Moles of acetic acid} = \frac{3.0}{60.0} = 0.050$$

$$\text{Moles of sodium acetate} = \frac{4.9}{82} = 0.060$$

The corresponding concentrations are

$$[\text{acid}] = 0.050 \times \frac{1000}{200}$$

$$[\text{salt}] = 0.060 \times \frac{1000}{200}$$

Substituting in (8–73) yields

$$[H^+] = 1.8 \times 10^{-5} \times \frac{0.050 \times 1000/200}{0.060 \times 1000/200} = 1.5 \times 10^{-5}$$

and $$pH = 4.82$$

You can see in Example 8–18 that the volume of the solution cancels. In many applications of formulas (8–73) and (8–75) this is the case, and it is unnecessary to carry the volume through the calculations; that is, the millimoles of the species can be used instead of their concentrations.

8.12 *Polyfunctional Acids, Bases, and Their Salts*

Polyfunctional acids (bases) are capable of donating (accepting) more than one proton per molecule. Here we consider only diprotic acids. Diacidic bases can be treated analogously, replacing $[H^+]$ by $[OH^-]$ and the acidity constant by the basicity constant.

A molecular diprotic acid, H_2A, in water undergoes protolysis in steps:

$$H_2A + H_2O \rightleftharpoons HA^- + H_3O^+ \tag{8-76}$$

$$HA^- + H_2O \rightleftharpoons A^{2-} + H_3O^+ \tag{8-77}$$

The corresponding expressions for the two acidity constants are

$$K_{a,1} = \frac{[H^+]\,[HA^-]}{[H_2A]} \tag{8-78}$$

and

$$K_{a,2} = \frac{[H^+]\,[A^{2-}]}{[HA^-]} \tag{8-79}$$

Generally, the value of the acidity constant for each successive protolytic reaction is smaller than the preceding value; that is, $K_{a,1} > K_{a,2}$. This is largely explained by the following. In the first step, the acid loses a proton and the conjugate base has one less positive charge. This base is the acid in the second step and with one less negative charge, it holds remaining protons more strongly; in other words, it has a smaller tendency to donate a proton.

For sulfuric acid, the first step corresponds to the protolysis of a strong acid and goes essentially to completion. The acidity constant for the second step is fairly large (1.1×10^{-2}). Thus, frequently the approximation is made that ionization is also complete here. Hence, the hydrogen ion molar concentration is taken as twice the formality of the acid. However, it is interesting to test this approximation by calculating the actual acidity of a sulfuric acid solution.

Example 8-9 (a) Derive a formula for the calculating of the acidity of a sulfuric acid solution, assuming that the first protolytic step goes to completion and that the second step is for an acid having $K_a = 1.1 \times 10^{-2}$. (b) Use the formula to calculate the hydrogen ion concentration of a 0.010F sulfuric acid.

To derive the requested formula, we must state a number of equations that together equal the number of unknowns. These equations will then be combined according to the rules of algebra, so that eventually $[H^+]$ is obtained as a sole function of C_a and K_a.

The first ionization step, as that of a strong acid, goes to completion. For the second step, the acidity constant expression can be written

$$K_a = \frac{[H^+][SO_4^{2-}]}{[HSO_4^-]}$$

The charge balance is

$$[H^+] = 2[SO_4^{2-}] + [HSO_4^-] + [OH^-]$$

which can definitely be simplified by neglecting $[OH^-]$ since the solution is strongly acidic:

$$[H^+] = 2[SO_4^{2-}] + [HSO_4^-]$$

If the formal concentration of the sulfuric acid is denoted by C_a, the material balance yields

$$C_a = [SO_4^{2-}] + [HSO_4^-]$$

By rearrangement,

$$[HSO_4^-] = C_a - [SO_4^{2-}]$$

Combining this equation with the simplified charge balance yields

$$[H^+] = 2[SO_4^{2-}] + C_a - [SO_4^{2-}] = [SO_4^{2-}] + C_a$$

or after rearrangement,

$$[SO_4^{2-}] = [H^+] - C_a$$

Combining the two preceding equations gives

$$[HSO_4^-] = C_a - [H^+] + C_a = 2C_a - [H^+]$$

Substituting in the acidity constant expression yields

$$K_a = \frac{[H^+]([H^+] - C_a)}{2C_a - [H^+]}$$

Rearrangement gives

$$[H^+]^2 - [H^+](C_a - K_a) - 2C_aK_a = 0$$

and solution secures the requested formula,

$$[H^+] = \tfrac{1}{2}[(C_a - K_a) + \sqrt{(C_a - K_a)^2 + 8C_aK_a}\ ]$$

(b) Substituting the numerical value gives

$$[H^+] = \tfrac{1}{2}(1.0 \times 10^{-2} - 1.1 \times 10^{-2} + \sqrt{1.0 \times 10^{-6} + 8 \times 1.1 \times 10^{-4}}\ ) = 1.4 \times 10^{-2} M$$

Thus the hydrogen ion concentration is $1.4 \times 10^{-2} M$, which is 30% less than the value of $2.0 \times 10^{-2} M$ obtained under the assumption that both steps go to completion. You should evaluate results for more concentrated and more dilute solutions, say $0.10F$ and $0.0010F$, and observe the differences.

If both protolytic steps correspond to weak acids, the following reasoning is applied in evaluating the acidity of a solution containing the acid. Since the second acidity constant is usually much smaller than the first, ionization according to the second step can be assumed to be completely repressed by the relatively large hydrogen ion concentration produced in the first step. In other words, all the hydrogen ion is assumed to stem from the first step, and calculating the acidity of the solution reduces to the case of a weak monoprotic acid having an acidity constant equal to $K_{a,1}$.

The calculation of the pH of solutions containing salts of difunctional acid and bases is deferred until Section 11.9.

8.13 *Mixed Equilibria: Acid-Base and Salt Solubility*

When we talked about dissolving a sparingly soluble salt (Chapter 5), it was in terms of the solubility equilibrium only. This approach is inadequate or at most is only a first approximation if either the cation or the anion undergoes protolytic reaction, or both do. If protolysis is extensive, and is neglected in the calculation, the value of the formal solubility thus computed can be grossly in error (low). How to reach a correct result by taking protolysis into account will be explained for a few simple cases in examples.

Example 8–20 Calculate the formal solubility in pure water of the sparingly soluble salt MA (K_{sp} = 2.5 × 10^{-27}), composed of the metal M^+ and the anion A^- that is the conjugate base of weak acid HA (K_a = 1.2 × 10^{-14}). The metal ion is assumed not to undergo any protolytic reaction.

The competitive solubility and protolysis equilibria can be displayed as

$$\underline{MA} \rightleftharpoons M^+ + A^-$$

$$+$$

$$H_2O \rightleftharpoons HA + OH^-$$

From this reaction scheme, you see that the protolysis of A^- shifts the solubility equilibrium to the right, that is, increases the solubility of the salt.

It is evident that the following material balance holds:

$$[M^+] = [A^-] + [HA]$$

Rearranging the expression for the acidity constant yields

$$[HA] = \frac{[A^-][H^+]}{K_a}$$

By substitution,

$$[M^+] = [A^-] + \frac{[A^-][H^+]}{K_a} = [A^-]\left(1 + \frac{[H^+]}{K_a}\right)$$

Regardless of protolysis, for a saturated solution the solubility-product expression for the salt (5–2) still holds. This expression, however, involves only the concentrations of M^+ and A^- and does not consider the portion of the anion that is transferred by hydrolysis to HA. Hence

$$[A^-] = \frac{K_{sp}}{[M^+]}$$

Combining the last two equations yields

$$[M^+] = \frac{K_{sp}}{[M^+]}\left(1 + \frac{[H^+]}{K_a}\right)$$

And the formal solubility, since it here equals the molar concentration of M^+, becomes

$$S^f = [M^+] = \sqrt{K_{sp}\left(1 + \frac{[H^+]}{K_a}\right)}$$

For calculation of the formal solubility, the pH of the solution must be known. Since the salt is of very low solubility, however, the approximation can be introduced that the hydroxide ion concentration due to the hydrolysis of the anion is negligibly small compared with that concentration due to the autoprotolysis of water. Then the hydrogen ion concentration is $10^{-7}M$. Substituting this and the other numerical values yields

$$S^f = \sqrt{2.5 \times 10^{-27}\left(1 + \frac{1.0 \times 10^{-7}}{1.2 \times 10^{-14}}\right)} = 1.4 \times 10^{-10}F$$

Testing the approximation involves the following reasoning. Since hydroxide ion is produced by the dissolution and protolysis of molecular MA, the hydroxide ion concentration stemming from this process can at most equal the value of the formal solubility and is thus negligible compared with the value 10^{-7} from the autoprotolysis of water. It is interesting to contrast the values of the solubility obtained with and without considering hydrolysis. The value without hydrolysis is

$$S^f = \sqrt{2.5 \times 10^{-27}} = 5.0 \times 10^{-14}F$$

Due to the protolysis of the anion, the formal solubility of the salt is increased by a factor of about 10,000.

In the equation for the solubility of a sparingly soluble salt of type $M^+ A^-$ developed in Example 8-20,

$$S^f = [M^+] = \sqrt{K_{sp}\left(1 + \frac{[H^+]}{K_a}\right)} \tag{8-80}$$

the term $1 + ([H^+]/K_a)$ can be denoted α_H. For a particular acid, the value of this term is a function only of the acidity of the solution. The product $K_{sp} \times \alpha_H$ can be denoted K'_{sp} and be termed the *effective* or *conditional* solubility product. Either term is appropriate since this constant is the one *effective* under the particular acidity *condition* prevaling. The values of α_H or of log α_H can be tabulated or plotted as a function of pH. Then for any sparingly soluble salt of the particular acid, the effective solubility product can be calculated readily and used in any other kind of calculations by using the effective constant exactly in the same way as K_{sp} itself.

By reasoning as in Example 8-20, you can derive the general formula for the formal solubility of a salt $M^{(2+)} A^{(2-)}$, where the anion A^{2-} of a weak diprotic acid H_2A undergoes hydrolysis. The derivation starts with the equality

$$[M^{2+}] = [A^{2-}] + [HA^-] + [H_2A]$$

The result is

$$S^f = \sqrt{K_{sp}\left(1 + \frac{[H^+]}{K_{a,2}} + \frac{[H^+]^2}{K_{a,1}K_{a,2}}\right)} = \sqrt{K_{sp}\alpha_H} \tag{8-81}$$

To gain deeper insight, you are strongly advised to calculate α_H in equation (8-81) for the weak diprotic acid hydrogen, sulfide ($K_{a,1} = 1.0 \times 10^{-7}, K_{a,2} = 1.2 \times 10^{-14}$) as a function of pH for pH values of values of 0, 1, 2, 3, and so on up to pH 10, and to plot log α_H versus pH. You should observe when the various terms in the expression for α_H contribute significantly and when they can be neglected. It is also recommended that you apply the values obtained to calculate the solubilities of the sulfides of lead, cadmium, and zinc in acidic solutions ranging from pH 0 to pH 6. The results should be related to the qualitative sulfide separation scheme for cations.

A further example will demonstrate additional reasoning involved in this kind of problem.

Example 8-21 Calculate the formal solubility of calcium oxalate (a) in pure water and (b) in an aqueous solution buffered to pH 2.00. The following data are given: $K_{sp} = [Ca^{2+}][C_2O_4^{2-}] = 2.3 \times 10^{-9}; K_{a,1} = 5.4 \times 10^{-2}; K_{a,2} = 5.1 \times 10^{-5}$.

(a) Since both dissociation constants of oxalic acid are relatively large, hydrolysis of oxalate ion in pure water can be neglected and the formal solubility of the salt calculated from the simple expression for the solubility product,

$$S^f = [Ca^{2+}] = \sqrt{K_{sp}} = \sqrt{2.3 \times 10^{-9}} = 4.8 \times 10^{-5} F$$

By substituting the value $[H^+] = 1.0 \times 10^{-7} M$ in the full expression, (8-81), you can confirm that protolysis can indeed be neglected in pure water.

(b) For pH = 2.00, the hydrogen ion concentration $[H^+] = 1.0 \times 10^{-2} M$. Substituting this and the other relevant numerical values in equation (8-81) yields

$$S^f = \sqrt{2.3 \times 10^{-9}\left(1 + \frac{1.0 \times 10^{-2}}{5.1 \times 10^{-5}} + \frac{(1.0 \times 10^{-2})^2}{5.4 \times 10^{-2} \times 5.1 \times 10^{-5}}\right)}$$

$$= 7.3 \times 10^{-4} F$$

Calcium oxalate is therefore about 15 times more soluble at pH 2 than at pH 7.

8.14 Graphical Treatment of Acid-Base Equilibria

The algebraic treatment of acid-base equilibria presented in the preceding sections can be applied to any situation, however complex, if the necessary data are available. For the general treatment of n acid-base equilibria, the equation in $[H^+]$ will be of order $n + 1$. By simplification and neglecting minor contributions (5% requirement), approximate solutions for particular problems can be achieved (cf. Section 8.8). But even then, the calculations provide the concentrations of the species for a single set of equilibrium conditions—with no comprehensive view of the acid-base system. In contrast, a graphical approach can ensure such a view and implicitly make use of the allowed simplifications. The basis of such a graphical treatment of acid-base equilibria (and of ionic equilibria generally) can be appreciated by constructing two simple graphs step by step and interpreting them. The completed graphs are presented as Figs. 8-1 and 8-2. You will, however, gain a better understanding by preparing your own graphs as you follow the development of the topic. For this purpose, draw on ordinary graph paper a square for which 14 spaces on a

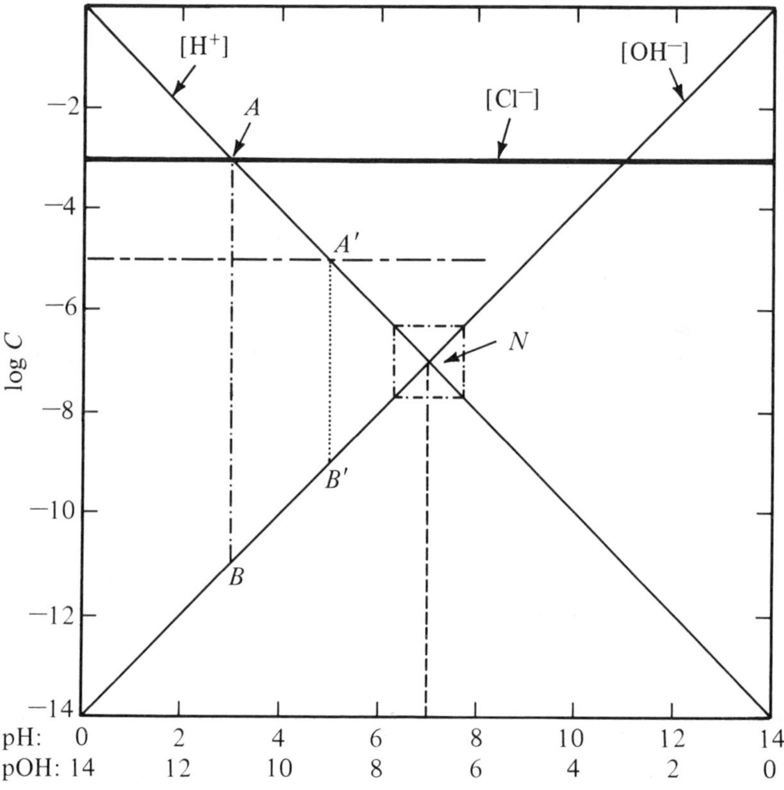

Fig. 8-1. Logarithmic graph for $10^{-3}F$ hydrochloric acid.

side are designated. The abscissa is marked to represent pH (from left to right 0 to 14) and also pOH (from left to right over the same space 14 to 0; see Fig. 8-1). The ordinate is labeled "log C" (from 0 to -14 downwards). Here C is the *molar* concentration of any species involved in the equilibria and should be designated by its symbol in brackets adjacent to its curve.

For aqueous solutions at 25°C, the value $K_w = 1.00 \times 10^{-14}$. The algebraic relation between hydrogen ion concentration and pH ($\log [H^+] = -pH$) corresponds to a diagonal line between the corners having the coordinates log $C = 0$, pH = 0 and log $C = -14$, pH = 14. The relation between the hydroxide ion concentration and the pH is obviously a diagonal line from log $C = 0$, pH = 14 to log $C = -14$, pH = 0. This line corresponds to the equation $\log [OH^-] = -pOH = pH - K_w$. The two lines should be labeled "$[H^+]$" and "$[OH^-]$." The two lines intersect at pH = $\frac{1}{2} pK_w$ = 7.00, and the intersection should be labeled N (for neutrality).

Consider now a solution of a monoprotic strong acid, for example, $10^{-3}F$ hydrochloric acid. The line for $[Cl^-]$ is slope zero, that is, a horizontal line, and for this example passes along the ordinate log $C = -3$. This line should be labeled "$[Cl^-]$." This line implies that the chloride molarity is constant (the acid is completely dissociated) and independent of pH.

Since a strong acid is involved, the intersection of the $[Cl^-]$ and $[H^+]$ lines at point *A* defines the pH of the solution, namely, pH 3. The corresponding hydroxide ion concentration is determined from the $[OH^-]$ line at point *B* as $10^{-11}M$.

Consider next $10^{-5}F$ hydrochloric acid. The $[Cl^-]$ line moves downward to an ordinate value of -5; in Fig. 8-1, it is drawn as a dot-dash line. Points *A* and *B* become points *A′* and *B′*.

Now consider $10^{-7}F$ hydrochloric acid. The $[Cl^-]$ line passes through point *N*, and points *A* and *B* coincide with that point. Obviously something is incorrect. The "acidic" solution cannot be neutral. To this point the 5% rule has been applied tacitly (see Section 8.9). In other words, so long as the contribution of a species from one source, here water, was less than 0.05 times the contribution from the other source (here hydrochloric acid), the smaller contribution was neglected. For $10^{-7}F$ hydrochloric acid, the rule no longer applies. (This situation has been considered algebraically in Section 8.6.) An analogous situation exists for the hydroxide ion.

Consequently, a "warning" area may be defined within which the 5% requirement cannot be applied and for which a refined treatment is demanded. This area corresponds to a square with sides of 1.3 units (since log 0.05 = -1.3). In other words, a distance of 1.3/2 = 0.65 unit is measured to the left, right, up, and down from point *N*, and the square is drawn in dotted lines.

Finally, consider $10^{-9}F$ hydrochloric acid. The $[Cl^-]$ line now intersects the $[H^+]$ line at a point corresponding to pH 9. This is obviously nonsense—the "acidic" solution cannot be alkaline. The acid does provide a hydrogen ion concentration of $10^{-9}M$. The water, however, contributes about 100 times as great a hydrogen ion concentration, that is, $10^{-7}M$. The total contribution of hydrogen ion from the two sources is roughly $1.01 \times 10^{-7}M$, which differs only minutely from $1.00 \times 10^{-7}M$, the value for pure water. In other words, the pH of the solution is below 7, but only so slightly that the departure is negligible. The last two examples show the danger of the graphical treatment. In using such graphs, exercise common sense and accept answers only when the implicit simplifications are valid.

The complete graph appears as Fig. 8-1. You will profit by exploring the graphical treatment for $10^{-3}F$, $10^{-5}F$, and $10^{-7}F$ sodium hydroxide, which is analogous to the treatment explored for hydrochloric acid.

The graphical treatment of the acid-base equilibrium for a monofunctional strong acid or base is simple, and the results and predictions are relatively obvious if the algebraic treatment has been mastered. The merit of the graphical approach can be better appreciated by considering a slightly more advanced situation—the solution of a monoprotic weak acid.

Consider a weak monoprotic acid HA with $K_a = 10^{-5}$ at a concentration of $10^{-2}F$. The finished graph for this example is shown as Fig. 8-2. Again you will benefit by preparing your own graph step by step as the construction is described. As a help, the lines and certain points in the figure are marked by circled numbers that correspond to the numbered steps below. The basic graph of Fig. 8-1 should first be drawn, that is, a square 14 units to a side with the diagonal lines drawn and marked "$[H^+]$" and "$[OH^-]$" and the "warning" square drawn about the intersection of these lines.

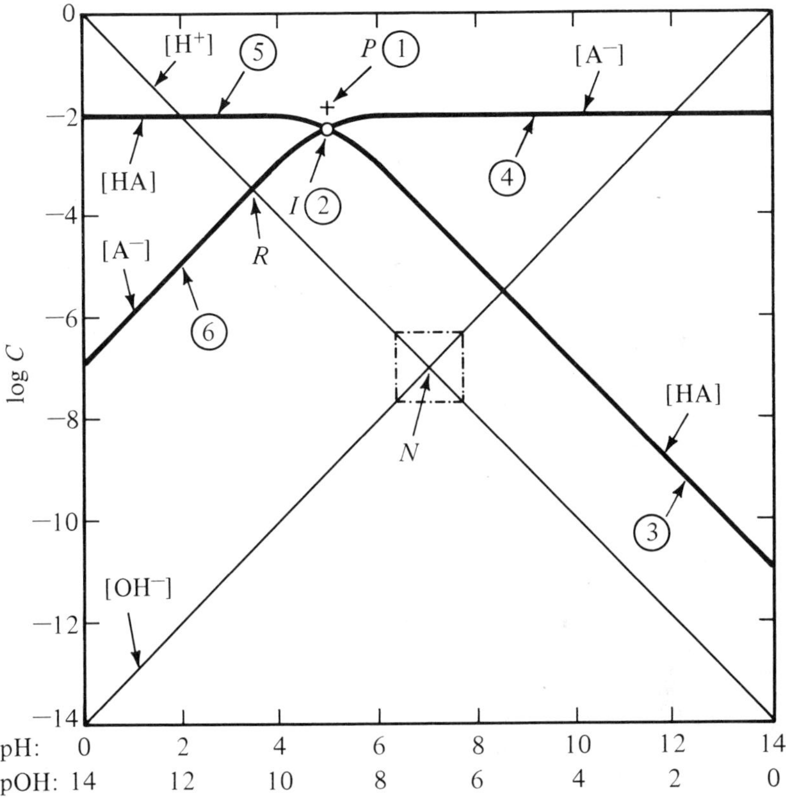

Fig. 8-2. Logarithmic graph for weak monoprotic acid HA with $K_a = 10^{-5}$ at a concentration of $10^{-2}F$.

Step 1: Mark a point P having the coordinates pH = pK_a = 5 and log C = log C_a = -2. The significance of this point will become clear from the considerations to follow.

The following equations are at hand:

$$\textit{Acidity constant:}\quad K_a = \frac{[H^+][A^-]}{[HA]} \qquad (8\text{-}82)$$

$$\textit{Material balance:}\quad C_a = [HA] + [A^-] \qquad (8\text{-}83)$$

Solving equation (8-83) for $[A^-]$ and substituting for $[A^-]$ in equation (8-82) yields

$$K_a = \frac{[H^+](C_a - [HA])}{[HA]} \qquad (8\text{-}84)$$

Solving this equation for [HA] provides

$$[\mathrm{HA}] = \frac{[\mathrm{H}^+]C_a}{K_a + [\mathrm{H}^+]} \tag{8-85}$$

Likewise, solving equation (8-83) for [HA], substituting for [HA] in equation (8-82), and solving for $[\mathrm{A}^-]$ yields

$$[\mathrm{A}^-] = \frac{K_a C_a}{K_a + [\mathrm{H}^+]} \tag{8-86}$$

To establish the [HA] and $[\mathrm{A}^-]$ curves, we must examine three regions, which are defined by the following relations: (a) $[\mathrm{H}^+] = K_a$; (b) $[\mathrm{H}^+] \ll K_a$; and (c) $[\mathrm{H}^+] \gg K_a$.

For case (a), equations (8-85) and (8-86) reduce to

$$[\mathrm{HA}^+] = \frac{C_a}{2} \tag{8-87}$$

$$[\mathrm{A}^-] = \frac{C_a}{2} \tag{8-88}$$

or in logarithmic form

$$\begin{aligned} \log\,[\mathrm{HA}] &= \log\,[\mathrm{A}^-] \\ &= \log \frac{C_a}{2} \\ &= \log C_a - \log 2 = \log C_a - 0.3 \end{aligned} \tag{8-89}$$

Step 2: From Equation (8-87), the [HA] and $[\mathrm{A}^-]$ curves intersect at a point I having an ordinate 0.3 unit below point P and at the same abscissa value (by assumption $[\mathrm{H}^+] = K_a$ and pH = pK_a). Mark this point.

For case *b*, equations (8-85) and (8-86) reduce to

$$[\mathrm{HA}] = \frac{C_a\,[\mathrm{H}^+]}{K_a} \tag{8-89}$$

$$[\mathrm{A}^-] = C_a \tag{8-90}$$

or in logarithmic form

$$\log\,[\mathrm{HA}] = -\mathrm{pH} + \mathrm{p}K_a + \log C_a \tag{8-91}$$

$$\log\,[\mathrm{A}^-] = \log C_a \tag{8-92}$$

Step 3: From equation (8-91), note that the [HA] line in this region has slope -1. For the condition pH = pK_a (abscissa of point *P*), this equation reduces to log [HA] = log C_a (ordinate of point *P*). Consequently, the [HA] line passes through that point. Draw this line parallel to the $[H^+]$ line and in the direction of point *P*, ending about 1 pH unit before it.

The line drawn in Step 3 indicates that as the pH increases, the concentration of the undissociated acid decreases; that is, more of the acid is present as the conjugate base.

Step 4: According to equation (8-92), in this region the $[A^-]$ line is independent of pH and corresponds to a horizontal line with a constant ordinate value of -2. Draw such a line from the right of the graph toward point *P*, terminating about 1 pH unit from it.

The line drawn in Step 4 implies that the acid is almost all converted to the conjugate base at high pH values.

For case *c*, equations (8-85) and (8-86) reduce to

$$[HA] = C_a \tag{8-93}$$

$$[A^-] = \frac{K_a C_a}{[H^+]} \tag{8-94}$$

or in logarithmic form,

$$\log\,[HA] = \log C_a \tag{8-95}$$

$$\log\,[A^-] = pH - pK_a + \log C_a \tag{8-96}$$

Step 5: Equation (8-95) implies that in a reasonably acidic solution, independent of pH, the acid is almost all in its undissociated form. Consequently the [HA] line in this region is horizontal, starting at log *C* = -2 and going toward point *P*, ending about 1 pH unit before it. Draw this line.

Step 6: From equation (8-96), the relation between $[A^-]$ and pH in this region is represented by a line of slope +1; and because pH = pK_a is the ordinate of point *P*, the line passes through that point. Draw such a line, parallel to the $[OH^-]$ line toward point *P*, and ending about 1 pH unit from it.

Step 7: Draw the remaining curved portions of the [HA] and $[A^-]$ curves freehand, dotted, to intersect at point *I*.

The complete graph shown in Fig. 8-2 provides a comprehensive view of the species involved in the equilibrium for a weak acid and the conjugate base. Furthermore, the concentrations of the relevant species can be derived readily for any pH value. At pH 4.50, for example, the graph indicates

$$\begin{aligned} \text{pH} &= 4.50, \text{ therefore } [\text{H}^+] = 3.2 \times 10^{-5} M \\ \text{pOH} &= 9.50, \text{ therefore } [\text{OH}^-] = 3.2 \times 10^{-10} M \\ \log [\text{HA}] &= -2.10, \text{ therefore } [\text{HA}] = 7.9 \times 10^{-3} M \\ \log [\text{A}^-] &= -2.50, \text{ therefore } [\text{A}^-] = 3.2 \times 10^{-3} M \end{aligned}$$

As a check, the material balance of equation (8-83) can be applied:

$$C_a = [\text{HA}] + [\text{A}^-] = (7.9 + 3.2) \times 10^{-3} = 1.1 \times 10^{-2} F.$$

Considering the precision of reading the graph, we find that the value is sufficiently close to the correct value of $1.0 \times 10^{-2} F$.

The construction is the same for either a molecular acid or an ionic acid (cf. Section 8.10).

In Fig. 8-2 it is instructive to consider point *R*, the intersection of the $[\text{H}^+]$ and $[\text{A}^-]$ lines. This point moves toward the neutrality point *N* as the concentration of the acid becomes smaller or the value of the acidity constant becomes smaller, or both. When point *R* reaches the "warning" square, the approximations inherent in the treatment no longer apply and a more refined treatment becomes necessary (see Example 8-11).

8.15 *Questions*

8-1 What is an acid? a strong acid? a weak acid? a salt?

8-2 What fundamental relation exists between the hydrogen ion concentration and the formality of a monoprotic strong acid?

8-3 What are the approximate formulas describing the acidity of a weak acid solution and the basicity of a weak base solution?

8-4 Elaborate on the restriction existing for the application of the approximate formulas mentioned in question 8-3.

8-5 What is stated by the electroneutrality rule?

8-6 Discuss the advantages of the pH concept.

8-7 How does the value of the ion product of water change with temperature?

8-8 Elaborate on the meaning of a negative pH value.

8-9 What does "protolysis of an ion" mean?

8–10 What is an amphiprotic substance? Give examples.

8–11 Explain why SO_4^{2-} is a weak base in aqueous medium.

8–12 State whether the solution of the following substances in pure water will be acidic, neutral, or alkaline; explain your answers: (a) NaCl; (b) NH_4Cl; (c) ammonium acetate ($K_a = K_b$); (d) potassium acetate.

8–13 Is the statement "HCl is a strong acid" complete? If not, what might be added to make it complete?

8–14 Elaborate on the fact that according to the strict Brønsted concept KOH is a salt. Why may it loosely be called a base?

8–15 What relation exists between the acidity constant of a weak acid and the basicity constant of its conjugate base?

8–16 Why is the second dissociation constant of a diacidic weak base always smaller than the first dissociation constant?

8–17 What is an effective solubility-product constant?

8–18 Why does a solution of KCN show an alkalinity close to that of NaOH?

8–19 Elaborate on the conditions under which the autoprotolysis of water can be neglected in dealing with acid-base problems.

8–20 Two acids are of equal strength in aqueous solution. Are these two acids necessarily of equal strength in another solvent? Elaborate.

8–21 In the *strict* sense of the Brønsted concepts, which of the following substances would be considered acids, bases, ampholytes, and salts in aqueous medium?

HBO_2	KCN	CN^-	HSO_4^-	$Ba(OH)_2$	CH_3COONa
1	*2*	*3*	*4*	*5*	*6*
$Fe(H_2O)_6^{3-}$	KCl	HCN	NH_3	NH_4^+	
7	*8*	*9*	*10*	*11*	

Ans.: acids: *1, 4, 7, 9,* and *11;* bases: *3, 4,* and *10;* ampholyte: *4;* salts: *2, 5, 6,* and *8*

8–22 Write for each of the substances in Question 8–21 the protolysis reaction occurring on their contact with water and discuss the type of species formed according to the Brønsted concepts.

8–23 Draw a graph like Fig. 8–2 for a weak monoprotic acid of $pK_a = 10^{-4}$ at a formality of 10^{-2} and derive the concentration of the relevant species at pH 3, 5, and 8. For each pH value, check the material balance for the calculated concentrations.

8–24 Draw a graph for a base of $pK_b = 10^{-4}$ at a formality of 10^{-3}. For pH 4, 10, and 12, secure the concentrations of the relevant species and make a material balance check.

8.16 Problems

8-1 Calculate the pH corresponding to the following conditions in an aqueous solution at room temperature:

(a) $[H^+] = 1.0 \times 10^{-9}M$
(b) $[H^+] = 2.35 \times 10^{-2}M$
(c) $[H^+] = 0.00321M$
(d) $[OH^-] = 1.2 \times 10^{-3}M$
(e) $[OH^-] = 0.012M$
(f) $[OH^-] = 1 \times 10^{-5}M$

Ans.: (a) 9.0; (b) 1.63; (c) 2.49; (d) 11.08; (e) 12.08; (f) 9.0

8-2 Convert the following pH values to the corresponding values of $[H^+]$ and $[OH^-]$: (a) 7.63; (b) 1.48.

Ans.: (a) $[H^+] = 2.3 \times 10^{-8}M$ and $[OH^-] = 4.3 \times 10^{-7}M$; (b) $[H^+] = 3.3 \times 10^{-2}M$ and $[OH^-] = 3.0 \times 10^{-13}M$

8-3 Calculate the pH of $2.0 \times 10^{-7}F$ HCl. *Ans.:* pH 6.62

8-4 Calculate the pH of $1.5 \times 10^{-7}F$ NaOH. *Ans.:* pH 7.30

8-5 Calculate the pH of a 0.050F solution of a weak monoprotic acid for $K_a = 1.0 \times 10^{-7}$. *Ans.:* pH 4.15

8-6 What is the pH of a 0.020F solution of a weak monoprotic base when $K_b = 5 \times 10^{-6}$? *Ans.:* pH 10.5

8-7 What is the pH of a 0.040F solution of a weak monoprotic acid when $K_a = 1.0 \times 10^{-3}$? *Ans.:* pH 2.23

8-8 What is the pH of a 0.10F solution of a weak monoprotic base when $K_b = 5.0 \times 10^{-3}$? *Ans.:* pH 12.30

8-9 What is the pH of an equimolar mixture of a weak base, B^+, of $K_b = 1.8 \times 10^{-5}$ and its nitrate salt, BNO_3? *Ans.:* pH 9.26

8-10 Calculate the pH of a solution obtained by dissolving 50 mL of concentrated aqueous ammonia (*17.0*) (density 0.90 g/mL; 28% w/w NH_3) and 2.0 g of NH_4Cl (*53.5*) in water and diluting to exactly 250 mL ($K_b = 1.8 \times 10^{-5}$). *Ans.:* pH 10.55

8-11 What is the pH of the solution obtained by mixing 50 mL of a 0.15F solution of the sodium salt of a weak monoprotic acid ($K_a = 4.0 \times 10^{-6}$) with 40 mL of 0.10F HCl? *Ans.:* pH 5.34

8-12 The pH of a 0.050F solution of sodium acetate was found to be 8.70. What is the value of the dissociation constant of acetic acid based on this experiment? *Ans.:* $K_a = 2.0 \times 10^{-5}$

8-13 A volume of 50 mL of 1.120F HCl was diluted with 120 mL of pure water. What is the pH of the resulting solution? *Ans.:* pH 0.48

8-14 Calculate the formal solubility of a salt MR in (a) pure water, (b) a solution buffered to pH 3.00, and (c) 0.100F HCl, when $K_{sp} = 4.0 \times 10^{-8}$ and $K_{a,HR} = 8.0 \times 10^{-4}$.

Ans.: (a) $2.0 \times 10^{-4}F$; (b) $3.0 \times 10^{-4}F$; (c) $2.2 \times 10^{-3}F$

8-15 Write the complete charge-balance equation for the solution obtained by dissolving NaCl, Na_2HPO_4, and $ZnSO_4$, and HCl in water.

Ans.: $[H^+] + [Na^+] + 2[Zn^{2+}] = [OH^-] + [Cl^-] + [HSO_4^-] + 2[SO_4^{2-}] + [H_2PO_4^-] + 2[HPO_4^{2-}] + 3[PO_4^{3-}]$

8-16 Which of the terms in the charge-balance equation of Problem 8-15 can be neglected in good approximation if the concentrations of each of the solutes present is about 0.1*F*? *Ans.:* $[OH^-]$, $3[PO_4^{3-}]$, $2[HPO_4^{2-}]$

8-17 What is the basicity constant in water of the anion of a weak acid HA with $K_a = 1 \times 10^{-11}$? *Ans.:* 1×10^{-3}

8-18 Aniline is an organic base, $C_6H_5NH_2$. It undergoes protolysis in aqueous media to form the anilinium ion, $C_6H_5NH_3^+$. The value K_b for aniline is 4.0×10^{-10}. Calculate the acidity constant of the anilinium ion.

Ans.: 2.5×10^{-5}

8-19 Calculate the pH value of a 0.040*M* solution of anilinium chloride in water. (*Hint:* The anilinium ion is an acid; the chloride ion is such a weak base that its influence can be neglected.) *Ans.:* pH 3.0

8-20 The hydroxopentaaquoaluminate ion, $Al(H_2O)_5(OH)^{2+}$, is a base with a K_b value in water of 3.5×10^{-10} at 15°C. Calculate the acidity constant of the hexaaquoaluminate ion at this temperature, at which pK_w has the value 14.35.

Ans.: 1.3×10^{-5} (Note that this ion has about the same acid strength as acetic acid.)

8-21 Calculate the pH value of the solutions of the following molar hydrogen ion concentrations.

(a) 5.2×10^{-3}
(b) 3.0×10^{-5}
(c) 6.5×10^{-5}
(d) 5.0×10^{-9}
(e) 9.3×10^{-6}
(f) 1.2
(g) 2.9
(h) 4.5×10^{-14}
(i) 1.9×10^{-15}
(j) 4.0×10^{-15}

8-22 Calculate the molar hydrogen ion concentration of the solutions with the following pH values.

(a) 6.53
(b) 9.82
(c) 4.85
(d) −1.22
(e) 7.20
(f) 14.57
(g) −1.52
(h) 14.20
(i) 8.98
(j) 7.91

8-23 Calculate the molar hydroxide ion concentration of the solutions with the following pH values.

(a) 1.51
(b) 13.62
(c) 3.67
(d) 6.29
(e) 2.70
(f) 1.32
(g) 3.89
(h) 6.73

8-24 The following solutions of strong acids are mixed. Calculate the pH of each resulting solution.
(a) 70.0 mL of 0.010*F* HCl and 30.0 mL of 0.015*F* $HClO_4$
(b) 25.0 mL of 0.050*F* HNO_3 and 50.0 mL of 0.025*F* HCl
(c) 25.0 mL of 0.0080*F* $HClO_4$ and 20.0 mL of 0.011*F* HNO_3

8-25 What is the pH of $3.00 \times 10^{-7}F$ HCl?

8-26 What is the pH of $4.00 \times 10^{-7}F$ HCl?

8-27 What is the pH of $1.00 \times 10^{-7}F$ $Ba(OH)_2$?

8-28 What is the pH value of each of the following solutions of a weak acid or base having the indicated K_a or K_b value:

Solution	Dissociation constant
(a) 0.40*F* $HC_2H_3O_2$	1.8×10^{-5}
(b) 0.15*F* HClO	2.8×10^{-8}
(c) 0.01*F* HSCN	1.4×10^{-1}
(d) 0.05*F* HF	6.0×10^{-4}
(e) 0.02*F* weak base	4.2×10^{-3}
(f) 0.18*F* weak base	6.0×10^{-4}

8-29 Calculate the value of the dissociation constant of each of the weak monobasic acids listed below from the pH values of their respective solutions:

Solution	pH
(a) 0.010*F* $HC_2H_3O_2$	3.37
(b) 0.050*F* HCN	5.35
(c) 0.20*F* $HCHO_2$	2.22
(d) 0.40*F* HNO_2	1.85

8-30 Calculate the pH value of each of the following solutions. The solution is
(a) 0.0200*F* in acetic acid ($K_a = 1.8 \times 10^{-5}$) and 0.100*F* in sodium acetate.
(b) 0.0200*F* in ammonia ($K_b = 1.8 \times 10^{-5}$) and 0.600*F* in ammonium chloride.
(c) 0.100*F* in formic acid ($K_a = 1.8 \times 10^{-4}$) and 0.150*F* in sodium formate.
(d) 0.0100*F* in boric acid ($K_a = 5.9 \times 10^{-10}$) and 0.0500*F* in sodium borate.
(e) 0.0500*F* in hydrofluoric acid ($K_a = 6.0 \times 10^{-4}$) and 0.150*F* in sodium fluoride.
(f) 0.300*F* in tartaric acid ($K_{a,1} = 9.2 \times 10^{-4}$) and 0.0200*F* in potassium acid tartrate.

8-31 Calculate the pH value of each of the solutions obtained by mixing
(a) 25.0 mL of 0.100*F* acetic acid ($K_a = 1.8 \times 10^{-5}$) and 20.0 mL of 0.100*F* NaOH
(b) 30.0 mL of 0.100*F* KCN and 25.0 mL of 0.100*F* HCl (for HCN, $K_a = 5.0 \times 10^{-10}$)
(c) 20.0 mL of 0.100*F* formic acid ($K_a = 1.8 \times 10^{-4}$) and 20.0 mL of 0.150*F* sodium formate

(d) 50.0 mL of 0.200*F* acetic acid (K_a = 1.8 × 10^{-5}) and 50.0 mL of 0.100*F* NaOH

(e) 30.0 mL of 0.100*F* ammonia (K_b = 1.8 × 10^{-5}) and 20.0 mL of 0.100*F* HCl

(f) 35.0 mL of 0.300*F* hydrofluoric acid (K_a = 6.0 × 10^{-4}) and 65.0 mL of 0.100*F* sodium fluoride

8-32 What is the formal solubility of LiF (K_{sp} = 4.0 × 10^{-3}) in (a) pure water and (b) 1.0*F* HF (K_a = 6.0 × 10^{-4})?

8-33 At what pH value will the formal solubility of CaF_2 (K_{sp} = 4.0 × 10^{-11}) be 10 times the solubility in pure water? $K_{a,HF}$ = 6.0 × 10^{-4}.

8-34 The dissociation constant of HF has the value 6.0 × 10^{-4}. What will be the solubility, in moles per liter, of CaF_2 (K_{sp} = 4.0 × 10^{-11}) (a) in pure water; and in solutions buffered to (b) pH 3.50, (c) pH 2.50, and (d) pH 1.50?

8-35 What amount, in milligrams, of lead (*207.2*) is present in 500 mL of a saturated solution of PbF_2 (K_{sp} = 3.7 × 10^{-8}) (a) in pure water; in a solution buffered to (b) pH 2.00 and (c) pH 1.00; and (d) in a solution buffered to pH 1.50 and made 0.50*F* in NaF?

9

Review Questions on Chemical Equilibria and Gravimetric Analysis

9–1 Elaborate on the difference between an ion product and a solubility product.

9–2 Elaborate on the factors that govern the rate of precipitation.

9–3 What is peptization? What difficulty may it cause? What can be done to avoid it?

9–4 Elaborate on the possibilities of obtaining a coarsely crystalline precipitate relatively free of impurities.

9–5 What advantages and disadvantages does reprecipitation have?

9–6 When a certain substance is dissolved in water, a temperature decrease is observed. From this fact predict whether its solubility will increase or decrease with increasing temperature.

9–7 Give the units of the solubility products for lead(II) sulfate and lead(II) iodide.

9–8 Strontium oxalate is slightly soluble in water. Name several ions that would influence the solubility of this salt through the common-ion or diverse-ion effects.

9–9 By applying the solubility-product concept, establish whether a medium $0.1F$ in $AgNO_3$ or one $0.1F$ in K_2CrO_4 would be more effective in repressing the solubility of Ag_2CrO_4.

9–10 The following statement is made: The formal solubility of a salt MA is 1×10^{-6}; consequently, the solubility product is 1×10^{-12}. Elaborate on this statement and discuss what qualifications must be made before the statement can be considered correct.

9–11 Is it safe to assume that drying a substance at 110°C would remove all the water? Elaborate.

9–12 In quantitative analysis, samples are usually dried before they are weighed. What is the merit of this procedure?

9–13 How can the common-ion effect be used to ensure essentially complete precipitation?

9–14 In gravimetric analysis, it is necessary to ignite some precipitates rather than to simply dry them. Elaborate on this statement and give examples.

9–15 How does adding a catalyst affect the position of an equilibrium?

9–16 When barium sulfate is precipitated in the presence of sodium ions, some of the sodium is coprecipitated as sodium sulfate. Will this cause a low or a high result in the gravimetric determination of sulfate?

9–17 In the gravimetric determination of lead, the $PbSO_4$ precipitate is first washed with a dilute solution of sulfuric acid instead of water. Explain the merit of this practice.

9–18 Elaborate on the mechanisms by which impurities can be carried down with a precipitate.

9–19 Two substances have identical numerical values for their solubility products. One substance is of the type MA; the other, MA_2. When hydrolysis is negligible, which substance is more soluble?

9–20 Consider the situation of question 9–19 when the substances are of the types M_2A and MA_2.

9–21 How would you treat a sample solution containing NaCl and Na_2S before the gravimetric determination of chloride as AgCl? What would happen if the solution were used without a pretreatment?

9–22 Why is it important in many cases to refer to the "dry basis" in reporting an analytical result?

9–23 During the ignition of hydrous iron(III) oxide to Fe_2O_3, some Fe_3O_4 may form if improper conditions prevail. Will the result of the gravimetric determination of iron then be low or high?

9–24 Elaborate on the requirements for a substance to be used (a) as a precipitation form and (b) as a weighing form.

9–25 Consider the precipitation of the following substances: $BaSO_4$, CaC_2O_4, CaF_2, Ag_2CrO_4, $BaCO_3$, and AgCl. For which, do you assume, is the acidity of the solution during the precipitation an important factor? Explain your answer.

9–26 Explain the experimental finding that sparingly soluble salts of weak acids, but not of strong ones, are commonly dissolved by strong acids.

9–27 What is the meaning assigned the "p" in terms such as p*K*?

9–28 Explain the alkaline reaction of a solution of the sodium salt of a weak acid.

9–29 What is coagulation?

9–30 In a calculation related to the pH of an acid-base equilibrium, approximate formulas were used and the value for the pH was found to be close to 7. What should you check in that case?

9–31 An acid-base equilibrium is to be considered at pH 3. Is it necessary to take the hydroxide ion concentration into account when formulating the charge balance? Explain your answer.

9–32 What is the effect of temperature on the ionization equilibrium of water?

9–33 Distinguish between strong and weak electrolytes.

9–34 Distinguish between monoprotic and polyprotic acids.

9–35 What will be the pH of a solution of a salt formed from a weak acid and a weak base if $K_a = K_b$?

9–36 Compare the pH values of aqueous solutions of hydrochloric acid, sulfuric acid, and acetic acid equal in *formality*. Explain the differences.

9–37 Distinguish between the members of the following pairs: common-ion effect and foreign-ion effect, coprecipitation and postprecipitation; molarity and formality, protolyte and ampholyte, dry basis and "as is" basis, weak base and strong base, and charge balance and material balance.

9–38 Is the pAg value larger or smaller in a saturated silver chloride solution than in a saturated silver bromide solution? Explain.

9–39 A reaction in which heat is produced is said to be exothermic. Heat (H) in such a case can be treated formally as a reactant and the process can be formulated, for example, as

$$A + B \rightleftharpoons C + D + H$$

In such a case, what will be the effect of an increase in temperature on the position of the equilibrium?

9–40 In an endothermic reaction, heat is absorbed. Write an equation like that of question 9–39 and discuss the influence of raising and lowering the temperature on the composition of the equilibrium mixture.

9–41 Besides shifts in the position of an equilibrium, what other effects may be encountered on increasing the temperature at which a reaction proceeds?

9–42 Commonly the ion product of water is taken as exactly 10^{-14} if the situation is close to room temperature. The true values of K_w at the temperatures given within parentheses are 1.29×10^{-15} (0°C), 3.55×10^{-5} (10°C), 8.51×10^{-15} (20°C), 1.86×10^{-14} (30°C), 5.62×10^{-14} (50°C), and 7.41×10^{-13} (100°C). In the light of these data, criticize the definition (which is incorrect) of a neutral solution as a solution having a pH of 7.0. What is a correct definition of a neutral aqueous solution? What is the pH of "pure" water at 100°C? Does the degree of dissociation of water increase or decrease when the temperature is increased?

9–43 In your own words, omitting mathematical details, summarize the steps to be taken in applying the graphical treatment to the acid-base equilibrium for a weak monoprotic acid.

10

Titrimetry: General Considerations

Here we introduce the subject of titrimetry, unify it, and relate it to facts already known from previous studies. The full meaning and nuances of many of the statements will be better appreciated with the more detailed knowledge of titrimetric methods that a thorough study of subsequent material in chapters and laboratory experience will bring. We strongly recommend that you return to this section and relate its content to each type of titration presented.

10.1 *Concepts and Definitions*

Titrimetric analysis, or *titrimetry,* is a term for quantitative methods in which the amount of sought-for substance is calculated from the known concentration and measured amount (usually volume) of reagent solution added to the sample solution until essentially all the sought-for substance has been consumed in a definite reaction with this reagent. The process is known as a titration. The reagent is called the titrant and its solution is known as the titrant solution, or standard solution. The concentration of this solution is either calculated from the known amount of titrant dissolved and diluted to an exact volume, or established by a determination known as a standardization (see Section 10.3). If standardization is used, the solution may be called a standardized solution when this distinction is significant. The species (or solution) titrated is sometimes termed the titrand. However, this term is not encouraged, because the close similarity between *titrant* and *titrand* readily leads to confusion.

The following requirements must be fulfilled for a reaction to be the basis of a titrimetric determination: (1) The reaction must proceed rapidly so that the titration can be performed within a reasonable period; (2) the reaction must be unambiguous, or at least sufficiently "well-behaved," so that a definite equivalence of the reactants exists; and (3) the reaction should go essentially to completion.

There are titrimetric procedures in which the reaction is not fully complete, side reactions occur, or other factors adversely affect the stoichiometry. In such cases, an empirical factor is used in calculating the result. The preferred situation, however, is that the computation be based on the strictly stoichiometric relation between the sought-for substance and the titrant; this requires that the relevant balanced reaction equation be known. The reaction on which the titration is based is the titration reaction; and it can be expressed by a chemical equation, which is termed the titration equation.

The point in the process of a titration at which the titrated species and the titrant are present in exactly equivalent amounts is called the equivalence point. Unless an empirical factor is used in the computation (see page 131), this point is unambigously defined by the relations implicit in the titration equation. However, some substances can be titrated with the same titrant according to different titration equations, and then two or more equivalence points are defined.

It is necessary to establish the point at which sought-for substance and titrant are present in equivalent amounts. This process is called indication. Ideally the indication should locate the equivalence point. In practice, however, it is difficult to accomplish this and the point actually established will be near but not coincident with the equivalence point. This experimental point is termed the end point of the titration. Indication can be achieved visually or by instrumental methods. In some cases, the end point can be calculated from data taken during the titration or be established from a graphical presentation of such data.

Some titrations proceed so that visual indication is afforded by the titration system itself; this is termed self-indication. If a system is not self-indicating, an indicator can be added to the titration medium. Visual indicators, here under consideration, show a change in color related to the change of some property of the titration system.

The titration curve is a curve obtained as a plot on a rectilinear coordinate system. On the abscissa, the volume of the titrant solution is plotted, usually in milliliters. On the ordinate is plotted the numerical value of some parameter that is related to the concentration of one or more species participating in the titration reaction. If the value of this parameter varies during the titration over many orders of magnitude, it is often plotted as a logarithmic function. The titration curve may be calculated from the stoichiometry of the titration equation and is then called the theoretical titration curve. Alternatively it may be obtained from plotting actual experimental data and may then be called the experimental titration curve whenever such a distinction is warranted.

10.2 Classification of Titrations

Titrimetric methods can be classified from different points of view. They may be grouped into four broad categories according to the nature of the titration reaction: acid-base titrations, precipitation titrations, compleximetric titrations, and redox titrations. They may also be classified according to the titrant used, for example, acidimetry, alkalimetry, permanganatometry, argentimetry, and iodimetry. Another approach is to arrange titrations according to the method of end-point detection; thus visual titrations, electrometric titrations, photometric titrations, and the like can be differentiated. An additional criterion is the size of the sample taken, giving macro, semimicro, micro, or submicro titrations. All these classification schemes have merit, and the one selected will depend on the purpose to be served. In this book, the considerations are arranged according to the nature of the titration reaction involved; and visual titrations are dealt with before titrations involving instrumental methods of end-point detection.

One important description of titrations is according to mode of performance. When the standard solution is run directly into the sample solution until the end point is established, the process is called a direct titration. Sometimes a known amount of titrant solution in excess of that required to react completely with the sought-for substance is added to the sample solution, and the excess is titrated with a second titrant, called the back-titrant. The process is called a back-titration.

10.3 Standardization and Primary Standards

It is not always possible to prepare a standard solution so that its concentration is known to the necessary degree of reliability (usually four significant figures) from the amount of titrant dissolved and diluted to definite volume. This happens, for example, where the starting material is of unknown purity, contains variable amounts of water, or presents difficulties in handling (say it absorbs moisture or carbon dioxide from the air). In such a case, a solution of approximately the desired concentration is prepared and the exact concentration is then established by a determination known as a standardization. The reference substance used in the standardization is often known as the primary standard. Such a substance must fulfill certain requirements: (1) It must be of known purity, preferably 100% or very close to it; (2) its reaction with the relevant component of the solution to be standardized must be unequivocal and stoichiometric so that the reaction can be used as the basis of the calculations; (3) it should present no difficulties in handling; and (4) it should have a high equivalent weight, so that the weighing error has a negligible effect. Also, the substance should be readily available.

11

Acid-Base Titration Curves

A titration curve is drawn by plotting the numerical values of some parameter of the titration system that changes as a function of the amount of titrant added versus the volume of titrant solution. For an acid-base titration in an aqueous medium, the hydrogen ion concentration can be this parameter. However, since this concentration changes over many orders of magnitude, it is expedient to use a logarithmic function and to plot pH.

Titration curves help in establishing the feasibility of a titration, in predicting the sharpness of the end point, and in picking a suitable indicator. These curves are often derived from calculations based on the equilibria involved. We shall now examine such calculations and evaluate the curves.

11.1 *Titrating a Strong Acid with a Strong Base*

The calculations necessary to obtain data for constructing the titration curve of a strong acid with a strong base, both monofunctional, are straightforward because the acid, the base, and the salt formed are all completely dissociated. The calculations are best explained by examples.

Assume that a volume of exactly 100 mL of a 0.100*F* solution of a monoprotic strong acid is titrated with a 0.100*F* solution of a strong monoequivalent base (for example, hydrochloric acid titrated with sodium hydroxide).

The pH at the start of the titration is calculated as given in Section 8.4:

$$[H^+] = \text{formality of the acid} = 0.100M$$

and

$$pH = 1.00$$

On adding 5.00 mL of the base, for example, calculation of the pH takes the following form. The amount of acid initially present is 100 × 0.100 = 10.0 mmol.*

*The relation between the amounts of acid and base reacting could be expressed in milliequivalents; since both reactants are monofunctional, however, the number of milliequivalents and millimoles are identical.

The amount of base added is 5.00 × 0.100 = 0.500 mmol; consequently, the amount of acid remaining unreacted is 10.0 - 0.5 = 9.5 mmol. The solution volume is now 100 + 5 = 105 mL. Consequently, the pH of the solution is calculated as follows:

$$[H^+] = \frac{9.5}{105}\,M$$

and

$$pH = \log 105 - \log 9.5 = 2.02 - 0.98 = 1.04$$

Analogous calculations are applicable as further volumes of base are added; the pH values obtained are plotted on the ordinate and the milliliters of base added on the abscissa (see Fig. 11-1).

The mode of calculation can readily be put in general terms by the following formula, which should be understandable without detailed derivation:

$$\frac{\overbrace{\underbrace{mL_a \times F_a}_{\text{initial millimoles acid}} - \underbrace{mL_b \times F_b}_{\text{millimoles base added}}}^{\text{millimoles acid remaining unreacted}}}{\underbrace{mL_a + mL_b}_{\text{total volume}}} = \text{formality of unreacted acid}$$

$$= \text{molarity of hydrogen ion}$$

$$= [H^+] \qquad (11\text{-}1)$$

Here mL_a and mL_b are the volumes and F_a and F_b the formalities of the acid titrated and the base added.

How the formula is applied can be shown in an example in which we calculate the pH after 99.99 mL of base is added.

$$\frac{100.0 \times 0.100 - 99.99 \times 0.100}{100.0 + 99.99} = \frac{10.00 - 9.999}{199.99} \cong \frac{0.001}{200} \cong 5 \times 10^{-6} M$$

and

$$pH \cong 5.3$$

For the equivalence point, equation (11-1) yields $[H^+] = 0$, which is an impossible value. At this point, the pH value is 7.00. The value of zero is obtained because (11-1) is only an approximation. The formula neglects the hydrogen ion resulting from the dissociation of water. However, applying this formula to points removed from the equivalence point is appropriate. At such points, the pH is essentially governed only by the concentration of unreacted acid, and the hydrogen ion concentration is so large that the contribution from the dissociation of water can safely be neglected. Water can at most provide a concentration of $[H^+] = 10^{-7}$; actually it

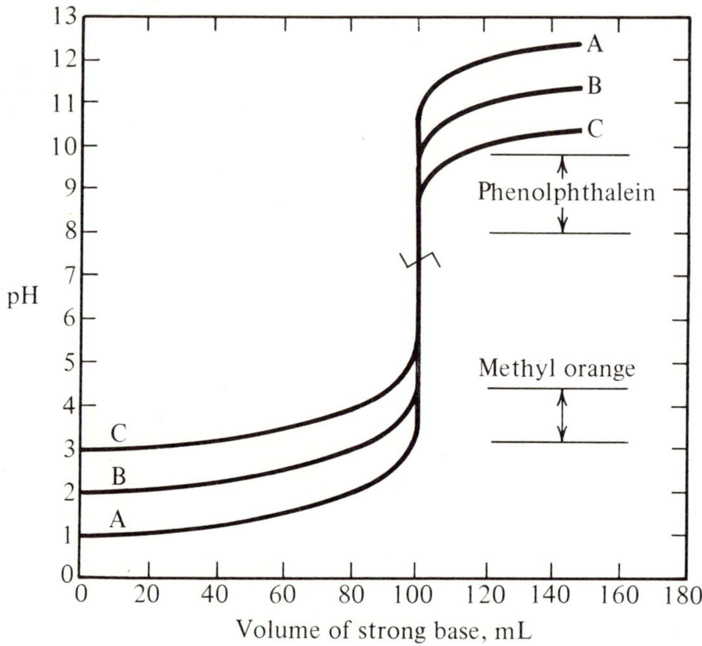

Fig. 11-1. Calculated titration curves for the titration of a monoprotic strong acid with a monoequivalent strong base, with the pH transition intervals of methyl orange and phenolphthalein shown. Curve A: 100 mL of $0.100F$ acid titrated with $0.100F$ base. Curve B: 100 mL of $0.0100F$ acid titrated with $0.0100F$ base. Curve C: 100 mL of $0.00100F$ acid titrated with $0.00100F$ base.

contributes much less because its dissociation is repressed by the presence of free acid. Even after the addition of 99.99 mL of base, the pH value, as calculated above, is still only about 5.3, far removed from pH 7.0 (see Section 8.4).

Points on the titration curve beyond the equivalence point are calculated in the following manner. Suppose that the total volume of base added is 101.00 mL. From this volume exactly 100 mL is consumed for the neutralization of the acid; exactly 1 mL of base remains as unreacted excess. This amount of base in excess corresponds to $1.00 \times 0.100 = 0.100$ mmol, and is present in a volume of $100 + 101 = 201$ mL. Hence,

$$[OH^-] = \frac{0.100}{201} M$$

$$pOH = \log 201 - \log 0.100 = 2.30 + 1.00 = 3.30$$

and

$$pH = 14.00 - pOH = 14.00 - 3.30 = 10.70$$

You should derive a general formula analogous to (11-1) for points beyond the equivalence point. (*Hint:* The signs of the terms in the numerator are reversed, and the result is the formality of the unneutralized base.)

With increasing amounts of base, the curve levels off and approaches pH = 13.00 asymptotically, since with the addition of an infinite volume of the base the solution would assume the pH of the titrant.

Inspection of the complete titration curve in Fig. 11-1, curve A, reveals that adding about 90% of the volume of titrant solution necessary to reach the equivalence points fails to change the pH by as much as adding only a fraction of a drop of base near that point. It is this sharp rise in pH on addition of a very small amount of titrant that makes indication of the end point possible.

Curves B and C given in Fig. 11-1 apply to the analogous titration of 0.0100*F* and 0.00100*F* strong acid with a strong base of corresponding formality. It can be seen that the shape of the curve remains unchanged, but the "break," that is, the nearly vertical rise in the region around the equivalence point, becomes progressively smaller as the concentrations of the acid and the titrant are decreased.

11.2 *Titration of a Strong Base with a Strong Acid*

The titration curve of a monoequivalent base with a strong monoprotic acid is the reverse of the curve just described and is calculated in the same way, except that $[H^+]$ is replaced by $[OH^-]$. It is good practice for you to derive the general formula, perform the calculations, and plot such a curve. It can be readily appreciated that each point on the titration curve for the base can be obtained by subtracting the corresponding pH values of the titration curve for the acid from 14.00. In other words, the titration curve for the base is the mirror image of the corresponding curve for the acid, and is obtained by reflection of the latter curve along the line passing through pH 7.00 and parallel to the abscissa.

11.3 *Titration of a Weak Acid with a Strong Base*

The formulas derived in Sections 8.9, 8.10, and 8.11 for the acid-base equilibria involving a weak acid and its conjugate base, present as its salt, can be applied in calculating the curve for the titration of a weak monoprotic acid with a strong monoequivalent base.

Assume that a volume of exactly 100 mL of a 0.100*F* solution of a weak monoprotic acid, having an acidity constant $K_a = 1.0 \times 10^{-5}$, is titrated with a 0.100*F* solution of a strong monoequivalent base. (The data closely correspond to those in the titration of acetic acid with sodium hydroxide.) The pH at the start of the titration is calculated from equation (8-66) as follows:

$$[H^+] \cong \sqrt{K_a C_a} = \sqrt{1.0 \times 10^{-5} \times 1.00 \times 10^{-1}}$$

$$= \sqrt{1.0 \times 10^{-6}} = 1.0 \times 10^{-3} M$$

and $$pH = 3.00$$

After the addition of 10.00 mL of the strong base, for example, the calculation of the pH is based on equation (8-74) for the hydrogen ion concentration of a solution containing a weak acid and its conjugate base, present as the alkali metal salt. As pointed out, in this approximate formula the concentrations of the acid and its salt appear in a fraction, the solution volume cancels, and the calculation can be based simply on the millimoles of acid and salt present.

Initially the acid present amounted to 100 × 0.100 = 10.0 mmol. The base added amounts to 10.00 × 0.100 = 1.00 mmol and transforms an equivalent amount of acid to the conjugate base, that is, to the salt; consequently, the amount of salt present is 1.00 mmol and the amount of the acid, 10.0 - 1.0 = 9.0 mmol. Hence

$$[\mathrm{H}^+] = K_a \frac{[\text{acid}]}{[\text{salt}]} = 1.0 \times 10^{-5} \frac{9.0}{1.0} = 9.0 \times 10^{-5} M$$

and

$$\mathrm{pH} = 5 - \log 9.0 = 5 - 0.95 = 4.05$$

By this mode of calculation, the pH values corresponding to the addition of 20, 30, and more milliliters of strong base can be calculated and plotted (see curve A, Fig. 11-2).

At the equivalence point, the acid is completely neutralized and the solution is identical to that containing the equivalent amount of conjugate base as the alkali salt in the prevailing volume. The concentration of the base or the salt can be calculated from the original concentration of the acid by recognizing that the volume has changed during the titration. The amount of salt is 100 × 0.100 = 10.0 mmol present in a total of 100 + 100 = 200 mL. Hence C_s at the equivalence point is 10.0/200F. The basicity constant of the conjugate base of the acid is $K_b = K_w/K_a$. Applying equation (8-58) and substituting the numerical values yields

$$[\mathrm{OH}^-] = \sqrt{\frac{1.0 \times 10^{-14} \times 10.0}{1.0 \times 10^{-5} \times 200}} = \sqrt{5.0 \times 10^{-11}}$$

and

$$\mathrm{pH} = 14.00 - \tfrac{1}{2}(11 - \log 5.0) = 8.85$$

Note that the number of millimoles of acid initially present and the number of millimoles of salt present at the equivalence point are identical. But the concentration of the acid at the beginning of the titration and the concentration of the salt at the equivalence point are different, because of the dilution effected during the titration. The novice often fails to appreciate this fact.

Beyond the equivalence point, the strong base is solely responsible for the hydroxide ion concentration and governs the further shaping of the titration curve; consequently, this portion of the curve is identical to the curve derived for the titration of a strong acid with a strong base involving identical volumes and concentrations. The complete titration curve is shown as curve A in Fig. 11-2 and should be compared with the corresponding titration curve for a strong acid (curve A, Fig. 11-1). The curve for the latter remains at a low pH for a large part of the titration,

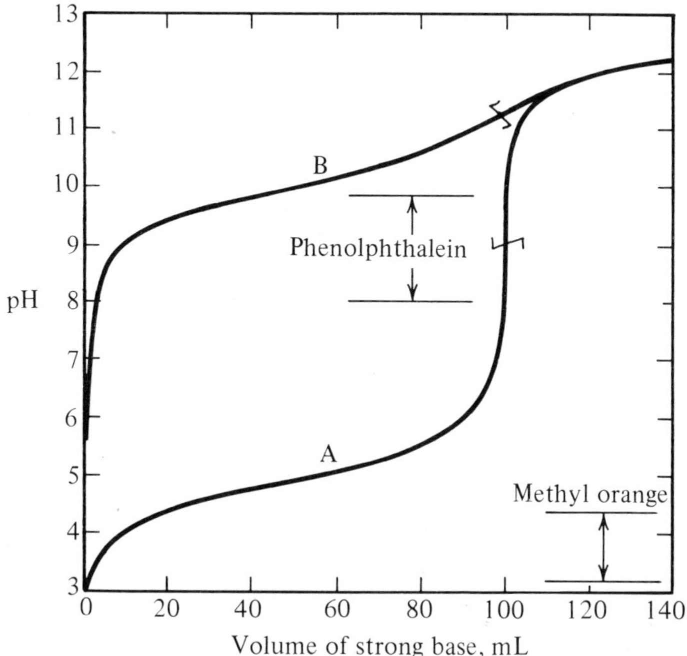

Fig. 11-2. Calculated titration curves for the titration of a monoprotic weak acid with a monoequivalent strong base, with the pH transition intervals of methyl orange and phenolphthalein shown. Curve A: 100 mL of 0.100F acid (K_a = 1.0 × 10^{-5}) titrated with 0.100F base. Curve B: 100 mL of 0.100F acid (K_a = 1.0 × 10^{-10}) titrated with 0.100F base.

then inclines only moderately, exhibits a large break around the equivalence point, and then levels off. In contrast, the curve for the weak acid rises from the start of the titration, then levels off, shows a smaller break around the equivalence point, and then levels off again. The equivalence point for the titration of the weak acid occurs at a pH value greater than 7.

The halfway point in the titration of the weak acid is important—that is, the point at which one-half the volume of base required to reach the equivalence point has been added. At this point (because [acid] = [salt]) the approximation pH ≅ pK_a is fulfilled, and all titration curves for the given weak acid pass essentially through this point, regardless of the initial concentration of the acid. Calculating the pH at the start, the halfway point, and the equivalence point in the titration of a weak acid is often enough to draw a rough approximation of the titration curve and from it to evaluate the feasibility of the titration, the choice of indicator, the sharpness of the end point, and so on (see Section 11.6). Of further interest in the titration curve of a weak acid is the moderately sloped portion around the halfway point. Solutions corresponding to this region resist changes in the pH when base or acid is added. A solution having such behavior is called a buffer (see Chapter 13).

11.4 Titration of a Weak Base with a Strong Acid

The titration curve of a weak monoequivalent base with a strong monoprotic acid can be calculated exactly like the titration curve for the titration of a weak acid; this would be by using the formulas given in Section 8.9 for the acid-base equilibria involving a weak base and its conjugate acid in the form of a salt. The titration curve can also be obtained by a mirror reflection of the curve for the corresponding titration of a weak acid along the line passing through pH 7.00 and parallel to the abscissa, provided that K_b equals K_a and the concentrations involved are the same.

11.5 Feasibility of an Acid-Base Titration

The nature of a titration curve near the equivalence point is important when one is judging the feasibility of an acid-base titration, and is especially helpful in selecting an appropriate acid-base indicator. (Such indicators are considered in Chapter 12.) For the present purpose we need only say that an acid-base indicator changes in color within a definite pH interval characteristic of the particular indicator. In Figs. 11-1, 11-2, and 11-4, the pH transition intervals of two such indicators, methyl orange and phenolphthalein, are designated.

Ideally the titration should be performed to the particular color shown by the indicator at the exact pH of the equivalence point. However, since the human eye is incapable of such subtle color discrimination, the experimental end point usually does not coincide exactly with the equivalence point, even with a satisfactory indicator. With either of the two indicators mentioned, the resulting deviation from the correct result is quite small in the titration of a strong acid (Fig. 11-1), since the titration curve passes nearly perpendicularly through the transition intervals of both indicators. Consequently, either indicator can be applied in titrating a $0.1F$ solution of a monofunctional strong acid with a $0.1F$ solution of a monofunctional strong base, even though the end point signaled is displaced from the pH of the equivalence point.

The situation is different in the titration of a weak acid. From curve A in Fig. 11-2 it can be appreciated that with methyl orange as the indicator, the change in color begins at the start of the titration and is complete when only about one-half of the weak acid has been titrated; for this titration, methyl orange is not suitable as an indicator. In contrast, the steepest part of the titration curve passes through the transition interval of phenolphthalein, and the result is satisfactory when this indicator is used. Your attention is drawn to the importance of making a clear distinction between the equivalence point and the experimentally established end point.

Thus a titration curve permits a judgment about what indicators might be satisfactory and also about whether a sharp end point can be secured. In this context, sharpness denotes the extent of the color change shown when a certain, small amount of titrant is added around the end point. Inspection of the titration curve, in connection with the transition interval and limiting colors of an indicator, also allows an estimate of what color shade should best be taken as the end point.

11.6 Titration of Very Weak Acids and Bases

The method of calculation just outlined can also be applied to approximate the curve for the titration of an extremely weak acid or base that is, one for which K_a or K_b is smaller than about 10^{-9}. Inspection of such a curve reveals that a titration is not feasible; this can be appreciated from an example.

Assume that a volume of exactly 100 mL of a 0.100*F* solution of a weak monoprotic acid having an acidity constant K_a of 1.0×10^{-10} is titrated with a 0.100*F* solution of a strong monofunctional base.

The halfway point of the titration corresponds to pH $\cong$ pK_a $\cong$ 10. The pH value at the equivalence point is found from equation (8-58) with C_s ($=C_b$) being 10.0/200*F* and the basicity constant of the conjugate base given by $K_b = K_w/K_a = 1.0 \times 10^{-14}/1.0 \times 10^{-10} = 1.0 \times 10^{-4}$.

$$[OH^-] \cong \sqrt{1.0 \times 10^{-4} \times \frac{10.0}{200}} = \sqrt{5.0 \times 10^{-6}}$$

and

$$pH = 14.00 - \tfrac{1}{2}(6 - \log 5.0) = 11.35$$

The titration curve must approach pH 13.00 asymptotically, since this is the pH of the 0.100*F* titrant. Hence, between the halfway point of the titration and the ultimate pH, the interval is only 13 - 10 = 3 pH units. Most two-color acid-base indicators have an actual transition interval of about 1.8 pH units (Section 12.1). Even without the calculation of the pH of the equivalence point, you can see that the titration is not feasible. Assume that the midpoint of the transition interval of the indicator approximates the equivalence point; then the color change is very gradual. The basic limit of the color transition is reached only after the titrant is added in an excess of about 50%. Hence, the color change extends over the addition of about 70 mL of titrant. The calculated titration curve, given as curve B in Fig. 11-2, bears out these conclusions.

It should be pointed out that the conjugate base of a very weak acid is a base of considerable basicity.

In the example just given, the basicity constant of the conjugate base is 10^{-4}, which is greater than that of ammonia. Consequently, if the conjugate base is present as a solution, for example, of the alkali metal salt of the acid, such a solution can be titrated with a strong acid. The traditional term for such a titration is a *replacement titration,* because the anion of the very weak acid on addition of the strong acid titrant reacts with hydrogen ion to form the undissociated acid and is "replaced" by the anion of the strong acid.

Analogously, the solution of a salt of a very weak base contains its corresponding conjugate acid, which is fairly strong and can be titrated readily with a strong base.

11.7 Titration of Polyfunctional Acids and Bases

Polyprotic acids furnish more than one hydrogen ion per molecule. It is of interest to investigate the curve for their titration with a monofunctional strong base, such as sodium hydroxide. The treatment will be limited here to a diprotic acid, H_2A; however, extension to acids furnishing more than two hydrogen ions per molecule is readily possible. (Titrations involving triprotic phosphoric acid are considered in Section 14.5.)

The dissociation of a diprotic acid, H_2A, was considered in Section 8.12. The relevant expressions (8-78) and (8-79) are repeated here:

$$H_2A \rightleftharpoons H^+ + HA^-: \quad K_{a,1} = \frac{[H^+][HA^-]}{[H_2A]} \tag{11-2}$$

$$HA^- \rightleftharpoons H^+ + A^{2-}: \quad K_{a,2} = \frac{[H^+][A^{2-}]}{[HA^-]} \tag{11-3}$$

In principle, each of the dissociable hydrogens should be capable of being titrated with a strong base:

$$H_2A + OH^- \rightleftharpoons HA^- + H_2O \tag{11-4}$$

$$H_2A + 2OH^- \rightleftharpoons A^{2-} + 2H_2O \tag{11-5}$$

To obtain two usable equivalence points, that is, two well-defined breaks in the titration curve, the values of $K_{a,1}$ and $K_{a,2}$ must differ sufficiently. If the two constants are too close, the first break is obliterated or poorly defined. Also, since an acid-base indicator has a color transition interval of about 1.8 pH units (Section 12.1), if the two acidity constants are too close in value, the two equivalence points may fall within a single transition interval. Mathematical treatment of the situation reveals that to secure two adequately defined breaks in the titration of 0.1*F* to 0.001*F* solutions, the necessary condition is that $K_{a,1}/K_{a,2} > 10^3$ to 10^4. In addition, the value of the second acidity constant must be large enough to make the titration of the second hydrogen feasible; this limitation is identical to that already considered for a weak monoprotic acid.

Assume that a volume of exactly 100 mL of a 0.100*F* solution of a diprotic acid H_2A, having acidity constants $K_{a,1} = 1.0 \times 10^{-4}$ and $K_{a,2} = 1.0 \times 10^{-8}$, is titrated with a 0.100*F* solution of sodium hydroxide.

At the start, it is "safe" to assume that the second dissociation step will be completely repressed by the high acidity resulting from the first dissociation step. Hence, the expression for a monoprotic acid, (8-51), can be applied:

$$[H^+] \cong \sqrt{K_{a,1}C_a} = \sqrt{1.0 \times 10^{-4} \times 0.100}$$
$$= \sqrt{1.0 \times 10^{-5}}$$

and $\qquad$ $\text{pH} = 2.50$

Even in the further course of the titration of the first hydrogen, the effect of the second dissociation can be neglected up to the vicinity of the first equivalence point. The curve is therefore calculated as described for a monoprotic weak acid. At the halfway point (the first), pH = p$K_{a,1}$ = 4.00.

The situation at the first equivalence point corresponds to that of a solution of the acid salt NaHA. Applying the neutrality rule gives

$$[Na^+] + [H^+] = [OH^-] + [HA^-] + 2[A^{2-}] \tag{11-6a}$$

and expressing the material balance yields

$$C_a = [HA^-] + [A^{2-}] \tag{11-6b}$$

At this equivalence point, by definition, $C_a = [Na^+]$. Substituting this relation and combining equations (11-6a) and (11-6b) yield

$$[OH^-] + [A^{2-}] = [H^+] + [H_2A] \tag{11-6c}$$

For most dibasic acids, the solution at the first equivalence point is still quite acidic and the term $[OH^-]$ can be neglected in equation (11-6c):*

$$[A^{2-}] = [H^+] + [H_2A] \tag{11-7}$$

Replacing terms in this equality by the expressions obtained from equations (11-3) and (11-2) through rearrangement yields

$$\frac{K_{a,2}[HA^-]}{[H^+]} = [H^+] + \frac{[HA^-][H^+]}{K_{a,1}}$$

The result on further rearrangement is

$$[H^+]^2\left(1 + \frac{[HA^-]}{K_{a,1}}\right) = K_{a,2}[HA^-]$$

Solving for $[H^+]$ yields

$$[H^+] = \sqrt{\frac{K_{a,1}K_{a,2}[HA^-]}{K_{a,1} + [HA^-]}} \tag{11-8}$$

---

*A less formal treatment of the situation at the first equivalence point can be achieved. The salt NaHA, as a strong electrolyte, dissociates into Na^+ and HA^-. The anion HA^- can dissociate according to $HA^- \rightleftharpoons H^+ + A^{2-}$. If no further equilibrium were involved, as with the salt of a monoprotic weak acid, the concentration of the anion A^{2-} would equal the hydrogen ion concentration. The anion HA^-, however, can also combine with hydrogen ion and form the undissociated acid: $HA^- + H^+ \rightarrow H_2A$. Consequently, the concentration of the anion A^{2-} equals the sum of the concentrations of the "free" hydrogen ion and the hydrogen ion "hidden" in H_2A. Hence, equation (11-7) follows.

If the amount of H_2A formed is small, and if $K_{a,1}$ is not too great, it is possible as an approximation to take $[HA^-]$ equal to $C_{s,1}$, the formal concentration of the salt NaHA. It then follows that

$$[H^+] \cong \sqrt{\frac{K_{a,1}K_{a,2}C_{s,1}}{K_{a,1}+C_{s,1}}} \qquad (11\text{-}9)$$

If the concentration of the acid being titrated is not extremely low (and consequently $C_{s,1}$ is not too small) and the acid is relatively weak, then $K_{a,1} \ll C_{s,1}$; with this condition, $K_{a,1}$ can be neglected in the denominator, and the factor $C_{s,1}$ cancels, to yield the approximate formula

$$[H^+] \cong \sqrt{K_{a,1}K_{a,2}} \qquad (11\text{-}10)$$

or in logarithmic form,

$$pH_1 \cong \tfrac{1}{2}(pK_{a,1} + pK_{a,2}) \qquad (11\text{-}11)$$

As a first approximation, therefore, the hydrogen ion concentration is independent of the concentration of the salt. It is necessary to check whether the approximations in the above derivation are allowed when the exact pH of a solution of an acid salt is calculated. In any event, the expression permits a rough evaluation of the situation and hence a judgment whether the titration is feasible. In the example, the pH at the first equivalence point is given by

$$pH_1 \cong \tfrac{1}{2}(4.00 + 8.00) = 6.00$$

The calculations for points beyond the first equivalence point are based solely on the second dissociation equilibrium; that is, one proceeds as if a monoprotic acid with $K_{a,2}$ were titrated. Thus at the second halfway point, pH $\cong pK_{a,2} = 8.00$.

The pH of the second equivalence point is calculated as follows. At this point the acid is completely neutralized and present as its conjugate A^{2-} in the form of the sodium salt. Consequently, equation (8-58) can be applied. The concentration of the salt, $C_{s,2}$, required for the calculation is readily obtained. The initial amount of acid was $100 \times 0.100 = 10.0$ mmol. Since all the acid is transferred to the salt, Na_2A, this is also the amount of salt. This amount is present in a volume of 300 mL (100 mL of original acid, 100 mL of base to reach the first equivalence point, and an additional 100 mL to reach the second equivalence point). Consequently, the concentration of the salt is $10.0/300F$. The basicity constant of base A^{2-} is $K_b = K_w/K_{a,2}$. Substituting the numerical values into equation (8-58) yields

$$[OH^-] = \sqrt{\frac{1.0 \times 10^{-14}}{1.0 \times 10^{-8}} \times \frac{10.0}{300}} = \sqrt{\frac{1.0 \times 10^{-7}}{3.0}}$$

and $$pH = 14.00 - \tfrac{1}{2}(7 + \log 3.0) = 10.26$$

Beyond the second equivalence point, the pH is governed solely by the excess of sodium hydroxide. This portion of the curve takes the same shape as in the titration

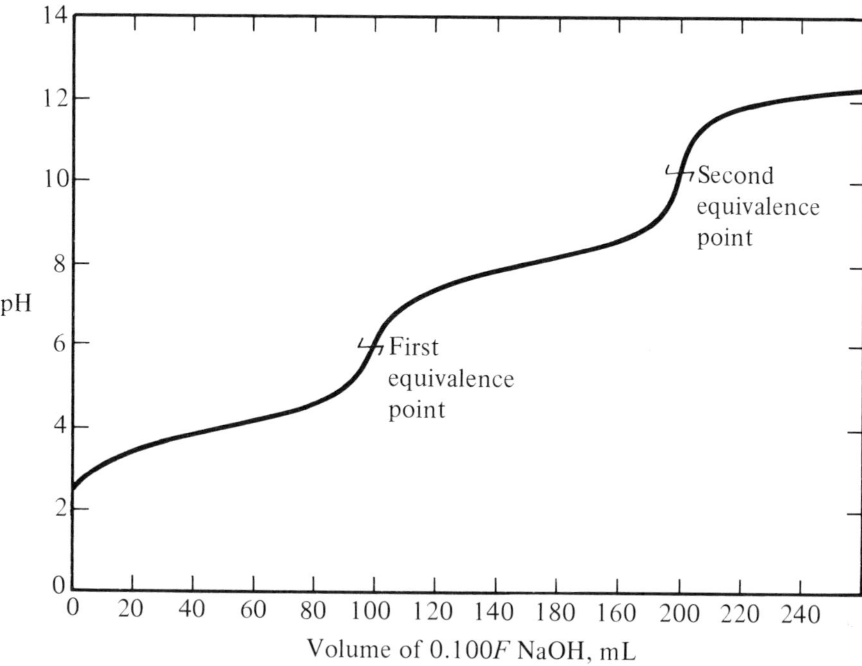

Fig. 11-3. Calculated titration curve for the titration of 100 mL of a 0.100F solution of a diprotic acid ($K_{a,1} = 1.0 \times 10^{-4}$; $K_{a,2} = 1.0 \times 10^{-8}$) with a 0.100$F$ solution of a monoequivalent strong base.

of a strong acid with a strong base. The entire titration curve for the diprotic acid is given in Fig. 11-3.

In deriving equation (11-7), and consequently, equation (11-10), it was assumed that the solution at the first equivalence point is still acidic. When the pH is close to 7, $[H^+] \cong [OH^-]$, equation (11-6b) reduces to $[A^{2-}] = [H_2A]$; and when the expressions for the acidity constants are applied, equation (11-10) directly follows. When the pH is in the alkaline region, $[H^+]$ can be neglected in equation (11-6c) and the $[OH^-]$ term replaced by $K_w/[H^+]$. Substituting the acidity constant expressions for the other terms and solution yields

$$[H^+] = \sqrt{\frac{K_{a,1}(K_{a,2}[HA^-] + K_w)}{[HA^-]}} \tag{11-12}$$

Unless $K_{a,2}$ or the concentration of the acid being titrated is extremely small, or both are, K_w can be neglected, and equation (11-12) reduces to equation (11-10).

The titration of a dibasic acid can be considered a special case of the titration of the mixture of two weak acids of different acidity constants and present at identical initial concentrations. The calculations in the above example reflect this view.

The titration of a polyfunctional base can be treated analogously by substituting OH^- for H^+ and K_b for K_a.

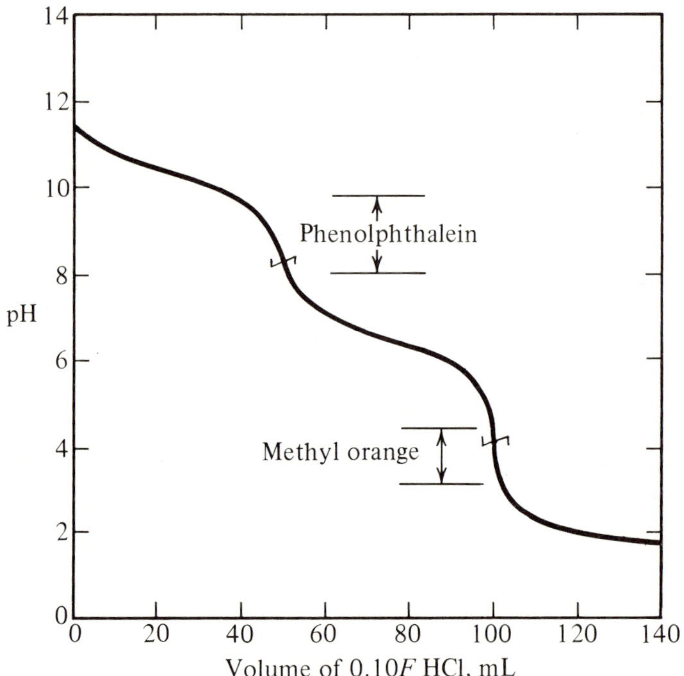

Fig. 11-4. Calculated titration curve for the titration of 100 mL of a 0.050*F* solution of sodium carbonate with 0.10*F* hydrochloric acid, with the pH transition intervals of methyl orange and phenolphthalein shown.

Of special practical interest is the titration of the difunctional base carbonate ion, CO_3^{2-}, present in the solution as an alkali metal or alkaline earth salt. The titration of carbonate with strong acid proceeds in two steps, first to hydrogen carbonate ion, HCO_3^-, and then to carbonic acid:

$$CO_3^{2-} + H^+ \rightarrow HCO_3^-$$

$$HCO_3^- + H^+ \rightarrow H_2CO_3$$

The carbonic acid decomposes according to the equilibrium

$$H_2CO_3 \rightleftharpoons H_2O + CO_2$$

From the acidity constants of carbonic acid, $K_{a,1} = 4.5 \times 10^{-7}$ and $K_{a,2} = 5.6 \times 10^{-11}$, the basicity constants of carbonate ion are readily obtained as $K_{b,1} = 1.8 \times 10^{-4}$ and $K_{b,2} = 2.2 \times 10^{-8}$. Note the reverse order of numbering: $K_{a,1}$ and $K_{a,2}$ correspond to $K_{b,2}$ and $K_{b,1}$.*

*The first acidity constant for carbonic acid and consequently the second basicity constant for carbonate ion are "apparent" constants. It is assumed

The calculated titration curve is shown in Fig. 11-4. The two breaks are clearly defined since $K_{b,1}/K_{b,2} \cong 10^4$. The first equivalence point occurs at pH about 8.3 (the calculation is left to you) and phenolphthalein can be used as the indicator. The second break at pH about 4.0 is not extremely sharp because the second basicity constant is that of a fairly weak base. However, the situation can be greatly improved by heating the solution to volatilize carbon dioxide and thereby shift the position of the second titration reaction to the right. (Further consideration of the carbonate titration is deferred until Section 14.7.)

11.8 Questions

11-1 Discuss the significance of an acid-base titration curve and elaborate on what judgments can be based on this curve.

11-2 Why is it expedient to plot pH rather than $[H^+]$ versus the volume of titrant added?

11-3 What are the significant differences between the titration curves of a strong and a weak acid with a strong base?

11-4 How can the titration curve of an acid be transformed into the curve for the titration of a base under identical conditions? Elaborate on the phrase "identical conditions."

11-5 What special significance has the halfway point when the same amount of a particular weak acid is titrated at different concentrations?

11-6 Why is it impractical or sometimes even impossible to use a weak acid or a weak base as a titrant?

11-7 Explain difficulties met in titrating a very weak acid or base.

11-8 What is a replacement titration?

11-9 Elaborate on the conditions that must be fulfilled to permit the visual titration of a diprotic acid to two equivalence points.

11-10 Elaborate on the possibility of titrating each of two acids in their mixture.

11.9 Problems

11-1 A solution in a volume of 250 mL contains 5.0 mmol of HCl and 2.0 mmol of H_2SO_4. What is the pH of the solution? Assume that both hydrogen ions of sulfuric acid are completely dissociated. *Ans.:* pH 1.44

11-2 Calculate the pH of a solution obtained by mixing 100 mL of 0.10*F* HCl and 150 mL of 0.080*F* NaOH. *Ans.:* pH 11.90

---

that all the carbonic acid is present as such, that is, as undissociated acid. Actually, however, it holds that $C_{H_2CO_3} = [H_2CO_3] + [CO_2]$ according to the above equilibrium.

11–3 Calculate the pH of a solution obtained by mixing 50 mL of 0.00010*F* HCl and 50 mL of 0.10*F* HA ($K_a = 1.0 \times 10^{-4}$). *Ans.:* pH 2.66

11–4 A 25.0-mL portion of a 0.0800*F* solution of a strong monoprotic acid is diluted with water to exactly 100 mL and titrated with 0.100*F* NaOH. Calculate the pH at the starting point and after the addition of 2.0, 5.0, 10.0, 20.0, and 22.0 mL of titrant.

Ans.: pH 1.70, 1.75, 1.84, 2.04, 7.00, 11.21

11–5 A 25.0-mL portion of 0.0800*F* NaOH is diluted to exactly 100 mL with water and titrated with a 0.100*F* solution of a monoprotic strong acid. Calculate the pH at the starting point and after the addition of 2.0, 5.0, 10.0, 20.0, and 22.0 mL of titrant.

11–6 A 25-mL portion of a 0.0800*F* solution of a weak monoprotic acid ($K_a = 5.0 \times 10^{-5}$) is diluted to exactly 100 mL with water and titrated with 0.100*F* NaOH. Calculate the pH for same points as in Problem 11–4.
Ans.: pH 3.00, 3.35, 3.82, 4.30, 8.26, 11.21

11–7 A 25-mL portion of a 0.0800*F* solution of a weak monofunctional base ($K_b = 5.0 \times 10^{-5}$) is diluted to exactly 100 mL with water and titrated with 0.100*F* HCl. Calculate the pH at the same points as in Problem 11–4.
Ans.: pH 11.00, 10.65, 10.18, 9.70, 5.74, 2.79

11–8 A 50.0-mL portion of a 0.1020*F* solution of a diprotic acid ($K_{a,1} = 4.0 \times 10^{-4}$; $K_{a,2} = 6.0 \times 10^{-8}$) is diluted to exactly 150 mL with water and titrated with 0.1200*F* NaOH. Calculate the pH for the starting point, first halfway point, first equivalence point, second halfway point, and the second equivalence point. *Ans.:* pH 2.43, 3.40, 5.31, 7.22, 9.82

11–9 Assume the same numerical data as in Problem 11–8, with the difference that a diprotic base is titrated with a strong acid.
Ans.: pH 11.57, 10.60, 8.69, 6.78, 4.18

11–10 How many milliliters of 0.0500*F* H_2SO_4 must be diluted to exactly 1000 mL with water so that the pH of the resulting solution will equal the pH of a 0.080*F* solution of acetic acid ($K_a = 1.8 \times 10^{-5}$)? *Ans.:* 12 mL

11–11 Calculate what the pH is at the start, halfway to the equivalence point, and at the equivalence point when the solutions of weak acids specified below are titrated with a strong base.
(a) 25.0 mL of 0.100*F* acid ($K_a = 8.0 \times 10^{-6}$) with 0.050*F* NaOH
(b) 50.0 mL of 0.050*F* acid ($K_a = 5.0 \times 10^{-7}$) with 0.100*F* NaOH
(c) 100.0 mL of 0.010*F* acid ($K_a = 5.7 \times 10^{-9}$) with 0.050*F* NaOH
(d) 20.0 mL of 0.030*F* acid ($K_a = 3.6 \times 10^{-7}$) with 0.040*F* NaOH

11–12 A weak acid, HA, has a dissociation constant of 1.0×10^{-6}. A volume of 30.0 mL of 0.100*F* HA is titrated with 0.050*F* NaOH. Calculate what the pH is after each of 0.0, 30.0, 60.0, and 90.0 mL of the NaOH solution has been added.

11–13 Calculate what the pH is at the start, halfway to the equivalence point, and at the equivalence point when solutions of the weak bases as specified below are titrated with a strong acid.

(a) 25.0 mL of 0.150*F* base (K_b = 4.0 × 10^{-6}) with 0.100*F* HCl
(b) 30.0 mL of 0.040*F* base (K_b = 1.0 × 10^{-7}) with 0.040*F* HCl
(c) 100.0 mL of 0.010*F* base (K_b = 5.0 × 10^{-8}) with 0.050*F* HCl
(d) 50.0 mL of 0.050*F* base (K_b = 6.5 × 10^{-6}) with 0.070*F* HCl

11–14 Calculate the approximate pH values for the start, halfway to the first equivalence point, the first equivalence point, midway between the two equivalence points, and the second equivalence point when the following solutions of difunctional weak acids (or bases) are titrated with strong base (or acid).

(a) 10.0 mL of 0.010*F* acid ($K_{a,1}$ = 1.0 × 10^{-6}, $K_{a,2}$ = 1.0 × 10^{-8}) with 0.020*F* NaOH
(b) 25.0 mL of 0.400*F* acid ($K_{a,1}$ = 1.0 × 10^{-4}, $K_{a,2}$ = 1.0 × 10^{-9}) with 0.200*F* NaOH
(c) 25.0 mL of 0.100*F* base ($K_{b,1}$ = 1.8 × 10^{-4}, $K_{b,2}$ = 2.2 × 10^{-8}) with 0.100*F* HCl
(d) 30.0 mL of 0.050*F* base ($K_{b,1}$ = 2.0 × 10^{-10}, $K_{b,2}$ = 1.9 × 10^{-13}) with 0.100*F* HCl

12

Acid-Base Indicators

For visual indication of the end point in a neutralization titration, a so-called acid-base indicator is used. Such an indicator is a weak organic acid (or base) having the special feature that it differs in color from the corresponding conjugate base (or acid). At least one form of the indicator must be colored. When one form is colorless, the substance is described as a one-color indicator; this is in contrast to a two-color indicator. Each indicator changes in color within a definite pH range, called its visual transition interval or its (pH) transition range. Some indicators have more than one such range (see Table F in the Appendix). The location of this range on the pH scale depends on the acidity (or basicity) constant of the indicator. At pH values below and above this range, the indicator is present predominantly in its so-called acid form and base form, respectively. Within the transition range, both forms are present in substantial proportions.

The color change of an acid-base indicator is related to structural changes resulting from either protonation or deprotonation; this can be shown for two important classes. Methyl orange typifies a "two-color" azo indicator. It has a visual transition interval of pH 3.2 to 4.4 (see Section 12.2) and undergoes the structural changes indicated in Equation (12-1). The sulfonate group ($-SO_3^-$) confers water solubility. In acidic media, a proton is accepted by an unshared electron pair of the azo group, $-\ddot{N}=\ddot{N}-$, yielding the red (acid) form, which persists as a resonance (mesomeric) system. Above about pH 5, the indicator is deprotonated and the yellow (base) form results.

Phenolphthalein is a "one-color" indicator and an important member of the phthalein class. (see Section 12.3). As indicated in Equation (12-2), the color change is from a colorless lactone form (acid) present in acidic media, through an intermediate, which is not shown, to a red-violet form (base), which is stabilized by resonance. The relation of structure to color is an important topic for chemistry; such knowledge, however, is unnecessary for understanding the functioning and practical application of acid-base indicators.

The most important requirements for an acid-base indicator can be summarized:

1. At least one form must be intensely colored, so that a clearly visible colora-

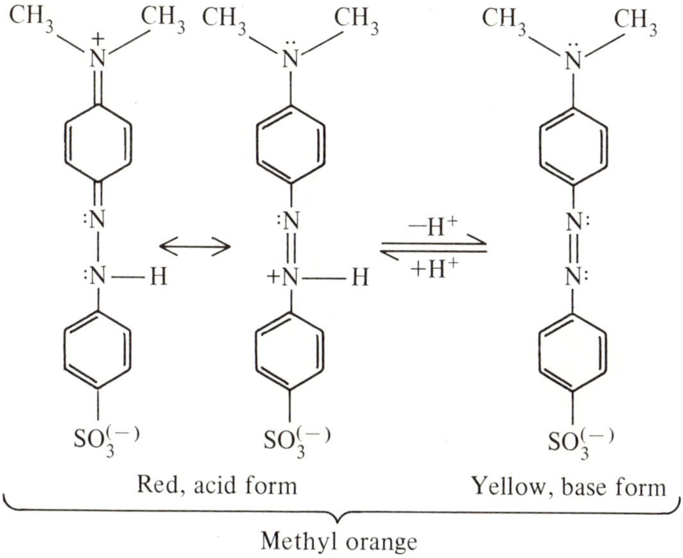
CH3
CH3
CH3
CH3
CH3
CH3
N
:N
:N—H
:N
+N—H
−H+
+H+
N:
N:
SO3(−)
SO3(−)
SO3(−)
Red, acid form
Yellow, base form
Methyl orange

(12-1)

tion is imparted to the solution to be titrated, even at the extremely low indicator concentration usually employed (10^{-4} to 10^{-5} *F*, or lower).

2. The visual transition range should be small, so that an abrupt color change results when a very small amount of titrant is added.
3. For a two-color indicator, the two limiting colors (that is, the colors of the acid and the base forms) should differ markedly in hue so that good color contrast is realized, ideally, the colors should be complementary.

An indicator should be selected so that the pH of the end point signaled is close—ideally, identical—to the pH of the equivalence point. With regard to the titration curve, the transition range should lie on the steepest portion. Since an indicator

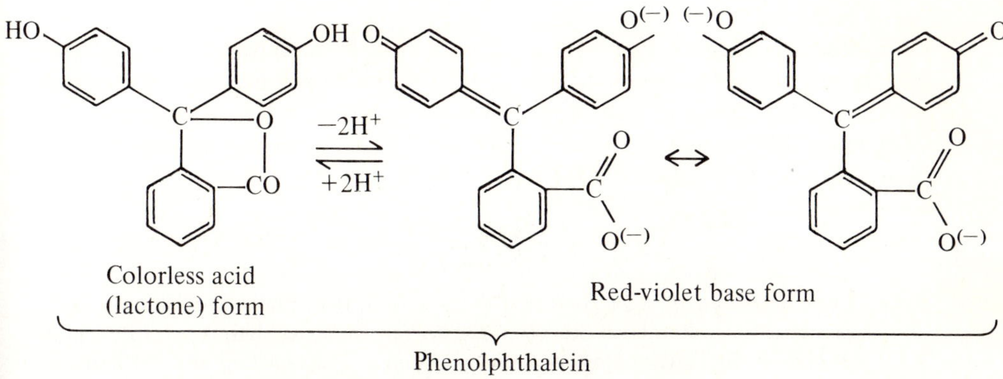
HO
OH
O
C
O
CO
−2H+
+2H+
O(−)
(−)O
O
C
O
C
O(−)
Colorless acid
(lactone) form
Red-violet base form
Phenolphthalein

(12-2)

is an acid or a base, it may require titrant or may neutralize a portion of the substance to be titrated. Consequently, the smallest amount of indicator necessary should be used, and approximately the same indicator concentration should be used in titrating standards and unknowns. Adequate reproduction of the indicator concentration is assured by adding a definite number of drops of a dilute indicator solution to a stated volume of solution to be titrated.

These and several other points will be discussed in detail and become more clear in the following development of the topic (see also Chapter 14).

12.1 Indicator Constant

The ionization reaction of an indicator may be presented as follows:

$$\text{acid form} \rightleftharpoons \text{H}^+ + \text{base form} \tag{12-3}$$

This generalized form has the advantage that it applies to indicator acids and also indicator bases—and regardless of whether they are uncharged, cationic, or anionic. The equilibrium constant, here called the indicator constant, is then given as*

$$K_{In} = \frac{[\text{base form}]\,[\text{H}^+]}{[\text{acid form}]} \tag{12-4}$$

This equation can be rearranged to yield

$$[\text{H}^+] = K_{In}\,\frac{[\text{acid form}]}{[\text{base form}]} \tag{12-5}$$

or in logarithmic form, it becomes

$$\text{pH} = \text{p}K_{In} + \log \frac{[\text{base form}]}{[\text{acid form}]} \tag{12-6}$$

This formula in connection with assumptions about the intensities of indicator colors and the color discrimination of the observer's eye permits a theoretical treatment that should provide insight into how indicators function. Since there are distinctive differences in the theory of two-color and one-color indicators, these substances will be considered separately.

---

*For an indicator acid (like $HIn \rightleftharpoons H^+ + In^-$), the indicator constant equals the acidity constant of the acid K_a. For an indicator base (like $In + H_2O \rightleftharpoons InH^+ + OH^-$), the relation between the indicator constant and the basicity constant is $K_{In} = K_w/K_b$.

12.2 *Two-Color Indicators*

Looking at equation (12-6) shows that with changes in pH, the ratio [base form]/[acid form] changes. In other words, since both forms are colored, a certain hue, due to the mixture of the two colorants, is related to a certain pH value. To assess the situation more closely, we can make the following assumptions, admittedly approximations:

1. To the eye, the color intensity is proportional to the concentration of the colored species, and both forms of the indicator have about the same color intensity.
2. If two colors are mixed, the eye is no longer able to discern a change in hue (that is, color shade) if the concentration ratio of the two colorants exceeds a value of 8:1 to 9:1.

If statement 2 and equation (12-6) are combined, it follows that the eye is able to observe color changes, due to change in the ratio [base form]/[acid form], within a pH range of approximately 1.8 pH units (see Example 12-1). Beyond this visual transition range, no changes in hue can be differentiated, and the eye sees the limiting colors of the indicator. The span of 1.8 pH units can also be obtained from the data in Table F in the Appendix, by averaging the transition intervals of the indicators listed. If [base form] = [acid form], the logarithmic term in equation (12-6) vanishes and pH = pK_{In}. This pH value approximates the midpoint of the transition interval that stretches about 0.9 pH unit above and below the midpoint. If the contrast in the two limiting colors of two two-color indicators is of the same quality, the indicator with the smaller transition range is better, because with a smaller transition range, a larger change in hue is obtained for a certain change in pH.

Some examples will advance your understanding.

Example 12-1 A two-color indicator has pK_{In} = 5.3. (a) Within what approximate pH range will it be possible to discern a color change? Will it be possible to use this indicator to indicate a pH of about (b) 4.5 and (c) 8.5?

(a) As an average the values 8.5:1 and 1:8.5 may be taken for the limiting ratio [base form]/[acid form]. Substituting these values in equation (12-4) yields

$$\text{pH} = 5.3 \pm \log 8.5 = 5.3 \pm 0.93$$

The transition range therefore stretches from pH 4.4 to 6.2.

(b) The required pH of 4.5 is within the transition range. If the color is brought very close to that of the pure acid form, the pH of 4.5 can be well approximated.

(c) The pH of 8.5 cannot be approximated. At any pH above 6.2, only the limiting color of the base form is viewed. Consequently, once this color is reached, no conclusion can be drawn except that a pH higher than about 6.2 prevails.

Example 12-2 At what pH will a two-color indicator with $K_{In} = 1.0 \times 10^{-10}$ be present 60% in the base form?

Sixty per cent base form corresponds to a base-acid form ratio of 60:40. Substitution in equation (12–4) yields

$$\text{pH} = 10.00 + \log {}^{60}\!/_{40} = 10.18$$

It is important to appreciate that for a two-color indicator, although the intensity of the color viewed increases with increasing indicator concentration, it is only a certain hue to which the titration is performed. This hue depends only on the ratio [base form]/[acid form] and is independent of the indicator concentration, at least for indicator concentrations used in actual titrations. Thus the indicator concentration for a two-color indicator is not overly critical. If the concentration is too great, the solution will be so dark that changes in hue can no longer be precisely observed; and if it is too small, there will not be enough color for visual differentiation.

12.3 One-Color Indicators

The considerations of indicator concentration are different for a one-color indicator. The titration with such an indicator is performed to a certain color *intensity* instead of a particular hue. This intensity exists at a certain pH for a given indicator concentration. For an indicator with a colored base form, the same color intensity occurs at a lower pH if the indicator concentration is increased. This is the case with phenolphthalein, for example, the most important one-color indicator. When phenolphthalein is used, an acid is titrated to a certain intensity of pink. In practice, the titration is often stopped when a permanent trace of pink first appears, corresponding to a pH of 8.3 to 8.5. This procedure is permissible in titrating a strong acid with a strong base because this pH range clearly falls on the steep part of the titration curve. In titrating a weak acid, with an equivalent point, say, at pH 9.0 to 9.5, a deeper pink color should be taken (compare Figs. 11–1 and 11–2).

For such a one-color indicator, the transition interval is not a region within which one color changes to another, but a region in which colorless changes to colored with a progressive increase in intensity, until a final color intensity is reached when essentially all the indicator is in the colored form.

Example 12-3 A one-color indicator has $\text{p}K_{In} = 8.0$ and molecular weight 120. It is known that a coloration is just discernible if the indicator is present in its base form at concentration of $2 \times 10^{-5} M$. How many drops of an exactly 0.1% solution of this indicator must be added to the solution to be titrated so that the appearance of the first discernible color indicates a pH of 8.5? Assume the volume at the end point to be 100 mL.

Material balance for the indicator yields

$$C_{In} = [\text{base form}] + [\text{acid form}]$$

Combining this equation with equation (12–4) gives

$$\text{pH} = \text{p}K_{In} + \log \frac{[\text{base form}]}{C_{In} - [\text{base form}]}$$

Substituting the numerical values yields

$$8.5 = 8.0 + \log \frac{2 \times 10^{-5}}{C_{In} - 2 \times 10^{-5}}$$

Solving for C_{In} gives

$$10^{0.5} = 3.2 = \frac{2 \times 10^{-5}}{C_{In} - 2 \times 10^{-5}}$$

and

$$C_{In} = 2.6 \times 10^{-5} F$$

The formality of the 0.1% indicator solution is

$$\frac{0.1}{120} \times \frac{1000}{100} = 8.3 \times 10^{-3} F$$

To establish the above-calculated indicator concentration in 100 mL of the solution to be titrated, the number of milliliters of indicator solution needed is

$$\frac{100 \times 2.6 \times 10^{-5}}{8.3 \times 10^{-3}} = 0.31 \text{ mL}$$

If one drop is assumed to be 0.05 mL, the number of drops required is 0.31/0.05 = 6 drops.

12.4 *Indicator Blank*

Since an acid-base indicator is either a weak acid or a weak base, it consumes either sample or titrant. For titrations involving relatively concentrated sample solutions and titrants, this influence can be neglected, since only a minute amount of the indicator is added. When dilute solutions and small amounts of unknown are involved, an indicator blank should be determined and used to correct the titration result. Assume that an acid is titrated with a base and that an acid-base indicator in its acid form is used. The blank in this case is secured by titrating an amount of the indicator in a volume of water, both substances identical to those used in the principal titration. The small volume of base consumed is then subtracted from the volume required in the principal titration. If in the same titration an indicator in its base form is used, the blank is secured by a similar titration, except that a standard acid solution is used. The volume of acid consumed is then expressed as the equivalent volume of the base used in the principal titration, and this volume is then added numerically to the volume required in that titration.

The value of an indicator constant can be markedly influenced by the presence of salts. Hence, in the determination of an indicator blank, a salt solution of appropriate concentration is preferably substituted for the water. Temperature sometimes markedly influences an indicator constant, especially for a two-color indicator.

For a two-color indicator, an indicator correction can often be avoided by "pretitrating" the indicator. For this purpose, the indicator in the amount to be used in the principal titration is added to a few milliliters of water and by addition of very dilute acid or base is brought to the exact shade of the end-point color. This solution is then added to the solution to be titrated.

12.5 Screened and Mixed Indicators

The purpose of a screened or a mixed indicator is to produce a more pronounced color change at the end point. These kinds of indicators consist of either a mixture of an indicator and an inert dye or a mixture of two indicators. Their functioning will be understood from the following considerations.

Ideally, the two limiting colors of any two-color indicator should be fully complementary, so that a gray develops at some pH within the transition range. Since the indicator concentration is small, the gray content will be so feeble that the solution is virtually colorless. In other words, under ideal conditions, the end point corresponds to the change from one color to its complementary color, passing through colorless. This condition would be realized, for example, if the limiting colors were a red and its exactly complementary green, because red + green yields black, which on dilution is gray. In practice, however, the limiting colors are usually far from complementary, and red and yellow are the most frequent pair (see Table F in the Appendix). If an inert dye, that is, one showing no acid-base properties (at least not in the pH range of interest), is mixed with the indicator, the dye acts as a screen against which or through which the actual color change is seen. Such a mixed indicator is therefore appropriately termed a *screened indicator.* If the inert dye is properly selected for color and mixed in the right proportion with the indicator, the color change is then from one color to its complementary color or close to it. Adding an appropriate blue dye, for example, will screen an end-point color change from red to yellow in the following manner: Red + blue yields violet and yellow + blue yields green. The end-point color is an intermediate gray, since violet + green yields gray. The end point is then more easily detected and more precisely located. For some examples of screened indicators, see Table F in the Appendix.

Screening can also be established by mixing two indicators that undergo color transition in approximately the same pH range. When the two indicators are appropriately chosen for hue and mixed in correct proportions, the solution becomes gray or colorless at the end point, and the color change is sharper than with either indicator alone.

Several indicators selected so that their transition intervals overlap can be mixed, and such a mixture gives continuous color changes as the pH is changed over a wide range. Such mixtures can be used to establish the pH of a solution by comparing the color developed in the test solution with a color chart. Impregnating filter paper with such an indicator mixture produces so-called pH paper.

12.6 *Color-Comparison End Point*

Even when the limiting colors of an indicator afford good color contrast, it is difficult to locate the end point if the slope of the titration curve is small within the transition range of the indicator. In such a case, the color change is so gradual that the change after each addition of a drop of titrant is not pronounced. The operator is limited in ability to remember the color that existed before an increment of titrant solution was added and to distinguish this coior from what was observed after the addition. The operator can improve his color recall as he gains practice with a given end point but only to a limitcd degree. However, the human eye is capable of detecting minute differences in hues or shades when they are presented side by side under identical illumination. This ability is used in performing a titration to a color-comparison end point. This technique can be explained by an example—the titration of orthophosphoric acid, H_3PO_4, with a standard base to the methyl orange end point. Here it is difficult to locate this point exactly and the remedy proceeds as follows. At the equivalence point of this titration, the salt NaH_2PO_4 is present. Therefore, two titration vessels of identical size and shape are selected. In one is placed the sample solution, in the other a solution of sodium dihydrogen phosphate of a concentration and volume close to what will exist at the end point in the titration of the sample. If necessary, a rough preliminary titration is performed of an aliquot of the sample solution, to establish the approximate concentration and volume involved. Then the same number of drops of the indicator solution is added to each vessel and the sample is titrated until the two solutions are identical in color. By this technique, the precision in the location of the end point can be increased by a factor of 10.

12.7 *Questions*

12-1 Define the terms *indicator, acid-base indicator,* and *screened indicator.*

12-2 What substances are used as acid-base indicators?

12-3 What is the significant difference observed in the behavior of a one-color and a two-color acid-base indicator as the indicator concentration is changed?

12-4 Why can only *weak* acids and bases be used as acid-base indicators?

12-5 State two causes for a sluggish color change—one related to the acid-base indicator, the other to the system being titrated.

12-6 Explain why a titration to a color-comparison end point can improve the precision of concentration and volume.

12-7 What is the transition range of a two-color indicator?

12-8 Formerly, pH values were often established colorimetrically. Can you suggest a procedure? Elaborate.

12-9 Elaborate on the possibility of improving the color contrast of indicators by the wearing of colored eyeglasses or the use of colored light to illuminate the titration vessel.

12-10 Explain why any indicator changing its color in the interval pH 4 to 9 gives acceptable results when hydrochloric acid is titrated with sodium hydroxide unless extremely dilute solutions are involved.

12-11 Explain why methyl orange is an unsatisfactory indicator for titrating acetic acid with sodium hydroxide.

12-12 Present in your own words the significant factors involved in choosing an indicator for a particular acid-base titration.

12-13 What significant advantage do you find in the use of the indicator constant rather than the acidity or basicity constant of an acid-base indicator?

12-14 One-color indicators can be screened, too, and the screened preparation behaves like a two-color indicator. If phenolphthalein is to be screened, what color should be shown by the screening dye? What will be the color change by the appropriately screened indicator?

12-15 Some acid-base indicators have two transition ranges (for example, thymol blue; see Table F in the Appendix). Suggest a general ionization scheme and write expressions for the two indicator constants, $K_{In,1}$ and $K_{In,2}$.

12.8 Problems

12-1 An indicator has pK_{In} = 5.3. In a certain solution the indicator is found to be 80% in its acid form. What is the pH of the solution? *Ans.:* pH 4.7

12-2 An indicator with pK_{In} = 9.0 has a color transition from yellow (acid form) to blue (base form). At what pH will the indicator be 75% in the blue form? *Ans.:* pH 9.5

12-3 A volume of 100 mL of a 0.100*F* solution of a triprotic acid is titrated with 0.100*F* NaOH. The acidity constants of this acid have the values 10^{-3}, 10^{-8}, and 10^{-12}. Two-color indicators with the following pK_{In} values are available: *A*, 3.1; *B*, 4.9; *C*, 7.8; *D*, 9.9. Which indicator will be satisfactory for titrating the acid to the (a) first, (b) second, and (c) third equivalence point? State also whether the titration should be performed to the midcolor or to a more alkaline or acid color of the indicator.

Ans.: (a) *B*, quite alkaline color; (b) *D*, close to midcolor; (c) the acidity constant is too small to permit a titration using an acid-base indicator

12-4 A one-color indicator (base form colored) with pK_{In} = 9.30 is used in a titration performed to exactly pH 9.00 and to a certain color intensity. The indicator was present at a concentration of $1.0 \times 10^{-5}F$. What will be the pH of the end point when the same titration is performed to the same color intensity but at an indicator concentration of $1.0 \times 10^{-4}F$? (*Hint:* Apply the equation derived in Example 12-3 of the text to both titrations; combine the two equations through the term [base form] and solve for $[H^+]$.)
Ans.: pH 7.84

12-5 At what pH will a "weak-acid" indicator ($K_{In} = 1.0 \times 10^{-8}$) be 60% dissociated?
Ans.: pH = 8.18

12-6 A "weak-acid" indicator was found to be 60% dissociated at pH 9.20. What will be the percentage of dissociation at pH 9.00?

12-7 From the percentage dissociation and the solution pH, calculate the indicator constant for each of the following two-color acid-base indicators of the "weak-acid" type.

(a) 80% at pH 9.00
(b) 55% at pH 7.90
(c) 60% at pH 5.80
(d) 40% at pH 6.20

12-8 The following two-color indicators (with the pK_{In} values given) are available: *A*, 3.2; *B*, 4.1; *C*, 4.8; *D*, 5.5; *E*, 6.7; *F*, 7.6; *G*, 8.4; *H*, 9.2. Establish which indicator (or more than one) is suitable for use in the titration of each of the following acids or bases. Assume a $1.0 \times 10^{-2}F$ salt concentration at each end point.

	K_1	K_2	K_3
(a) Acid	1.0×10^{-6}	2.0×10^{-8}	–
(b) Acid	6.0×10^{-3}	6.0×10^{-8}	4.4×10^{-13}
(c) Acid	2.0×10^{-3}	–	–
(d) Acid	5.4×10^{-2}	5.1×10^{-5}	–
(e) Base	1.0×10^{-4}	–	–
(f) Base	1.8×10^{-4}	2.2×10^{-8}	–
(g) Base	1.0×10^{-6}	5.0×10^{-8}	–
(h) Base	4.0×10^{-4}	4.0×10^{-12}	–

12-9 The color of a one-color indicator is just perceptible in a $4.0 \times 10^{-5}F$ solution when the indicator is 25% dissociated. If the pK_{In} value is 8.30, at what pH will the color first be visible?

13 Buffers

13.1 Buffering Action and Buffer Capacity

In many chemical operations, it is necessary to establish a certain pH or to maintain the pH at a definite level, or both. A solution that resists changes in pH is called a *buffered solution.* Buffering action in a solution is achieved by adding a reagent or a mixture of reagents; these are called *buffer substances.* They are often added in the form of a solution termed a *buffer solution,* or simply a *buffer.* Two types of buffered solutions can be differentiated by action. Solutions of the first type resist changes in pH on changes in concentration, commonly dilution. Solutions of the second type not only are indifferent to concentration changes, but also resist changes in pH when acid or base is added. The degree of the latter resistance is the *buffer capacity,* which is commonly expressed as the number of moles of strong base that must be added or removed (by addition of acid) from 1 L of solution to change the pH by 1 unit. The buffer capacity depends on the type and concentration of the buffer. The discussion of buffering action and buffer capacity can be based on acid-base titration curves.*

13.2 Strong Acids or Strong Bases as Buffers

Inspection of the titration curve of either a strong acid or a strong base (Fig. 11-1) shows that its initial portion is nearly horizontal. Consequently, it can be expected that a solution of a strong acid or base would not show a marked change in pH when a small amount of a base or acid is added. Since the buffer capacity depends on concentration, a strong acid is an effective buffer only below

*Students familiar with calculus will appreciate the mathematical definition of the buffer capacity, given as $dC_b/d(\text{pH})$, where dC_b is the change in concentration of base. When the buffer composition is described by a point on a titration curve, the buffer capacity is proportional to the reciprocal of the slope of the curve at that point.

about pH 2, and a strong base, above about pH 12. Unless very concentrated buffers are involved, however, dilution affects any strong acid or strong base buffer to the same extent, regardless of the buffer concentration. You can see this in Example 13-1.

Example 13-1 Compare two buffers, one consisting of $0.10F$ HCl and one of $0.0010F$ HCl. (a) Calculate the pH values of the original solutions, (b) calculate the pH increase due to dilution of 100 mL of buffer with 100 mL of water, and (c) calculate the increase in pH when to 100 mL of each buffer is added a volume of 20 mL of $0.0010F$ NaOH.

The calculations involved are straightforward and have been consolidated here.

For the $0.10F$ solution:

(a) $\text{pH} = -\log 1.0 \times 10^{-1} = 1.00$

(b) $[\text{H}^+] = \dfrac{100 \times 1.0 \times 10^{-1}}{200} = 5.0 \times 10^{-2}M$

$\text{pH} = 1.30$

$\Delta\text{pH} = 1.30 - 1.00 = 0.30$ pH unit

For the $0.0010F$ solution:

(a) $\text{pH} = -\log 1.0 \times 10^{-3} = 3.00$

(b) $[\text{H}^+] = \dfrac{100 \times 1.0 \times 10^{-3}}{200} = 5.0 \times 10^{-4}M$

$\text{pH} = 3.30$

$\Delta\text{pH} = 3.30 - 3.00 = 0.30$ pH unit

Note that the dilution affects both buffers to the same degree.

For the $0.10F$ solution:

(c) $[\text{H}^+] = \dfrac{100 \times 1.0 \times 10^{-1} - 20 \times 1.0 \times 10^{-3}}{120}$

$= 8.33 \times 10^{-2}M$

$\text{pH} = 1.08$

$\Delta\text{pH} = 1.08 - 1.00 = 0.08$ pH unit

Note that here the increase in pH is essentially due to the dilution.

For the $0.0010F$ solution:

(c) $[\text{H}^+] = \dfrac{100 \times 1.0 \times 10^{-3} - 20 \times 1.0 \times 10^{-3}}{120}$

$= 6.7 \times 10^{-4}M$

$\text{pH} = 3.17$

$\Delta\text{pH} = 3.17 - 3.00 = 0.17$ pH unit

13.3 *Mixture of Weak Acid and Conjugate as Buffer*

Inspection of the titration curve of a weak monoprotic acid with a strong base (Fig. 11-2) reveals a portion of small slope around the halfway point. For this region, adding small amounts of base or of acid results in only a slight change in pH. The halfway point corresponds to a solution containing equimolar amounts of the weak acid and its conjugate base. The pH of such a solution is obtained by the approximation pH $\cong$ pK_a. For a given total concentration of the acid and its conjugate base, this composition exhibits the greatest buffer capacity.

The buffer system acetic acid–acetate is frequently used in analytical chemistry. The solution is prepared by mixing appropriate amounts of acetic acid and sodium acetate, acetic acid and sodium hydroxide, or sodium acetate and hydrochloric acid. It has a "practical" buffer range about pH 3.7 to 5.7, which corresponds to approximately pK_a ± 1; beyond this region the buffer capacity decreases rapidly, as can be deduced from the shape of the titration curve (Fig. 11–2).

Some calculations involving mixtures of a monoprotic acid and its conjugate base have already been illustrated by Examples 8–10 through 8–13. Several additional examples are worthy of study.

Example 13–2 A buffer solution is prepared by mixing 100 mL of 0.100*F* HA ($K_a = 1.0 \times 10^{-5}$) and 100 mL of 0.100*F* NaA. (a) Calculate the pH of the mixture. (b) Calculate the change in pH when a volume of 20 mL of 0.100*F* HCl is added. (c) Calculate the change in pH when the same volume of 0.100*F* HCl is added to 200 mL of pure water.

(a) The approximate expression for the pH of a mixture of a weak acid and its conjugate base, added as an alkali metal salt of the acid, is given by equation (8–74):

$$\mathrm{pH} = \mathrm{p}K_a + \log \frac{[\mathrm{salt}]}{[\mathrm{acid}]}$$

where [salt] and [acid] are the concentration of the salt (or the conjugate base) and the weak acid. Since here [salt] = [acid],

$$\mathrm{pH} \cong \mathrm{p}K_a = 5.00$$

(b) Adding the strong acid converts an equivalent amount of the conjugate base to the weak acid by the reaction

$$A^- + H^+ \rightarrow HA$$

The amount of strong acid added is 20 × 0.100 = 2.0 mmol. Before this addition, the amounts of weak acid and of its conjugate base present were each 100 × 0.100 = 10.0 mmol. Hence, the amount of that base now present is 10.0 – 2.0 = 8.0 mmol, and of weak acid, 10.0 + 2.0 = 12.0 mmol. As observed in Example 8–11, the value for the ratio [salt]/[acid] can be computed using millimoles instead of concentrations, since the volume cancels:

$$\mathrm{pH} = 5.00 + \log \frac{8.0}{12.0} = 5.00 + \log 8.0 - \log 12.0$$

$$= 5.00 + 0.903 - 1.079 = 4.82$$

Hence the change in pH when the strong acid is added is 5.00 − 4.82 = 0.18 pH unit.

(c) Adding the strong acid to 200 mL of pure water would yield a hydrogen ion concentration of

$$[H^+] = \frac{20 \times 0.100}{200 + 20} = 0.0091M$$

Hence, $$pH = 2.04$$

and the change in pH is 7.00 − 2.04 = 4.96 pH units.

Comparing the results in (b) and (c) delineates the difference in behavior between a well-buffered and an unbuffered system.

Example 13-3 The volumes are the same as in Example 13-2 but the solutions of the weak acid and its alkali metal salt are 0.0500F each. What is the change in the pH on the addition of 20 mL of 0.100F HCl?

The initial pH is again 5.00, since [salt] = [acid]. The calculation of the final pH parallels that of Example 13-2 and is shown in consolidated form.

$$pH = 5.00 + \log \frac{100 \times 0.0500 - 20 \times 0.100}{100 \times 0.0500 + 20 \times 0.100}$$

$$= 5.00 + \log \frac{5.00 - 2.0}{5.00 + 2.0}$$

$$= 5.00 + \log \frac{3.0}{7.0} = 4.63$$

The change in pH is therefore 5.00 − 4.63 = 0.37 unit.

Note that although the buffer of Example 13-3 is one-half the concentration of the buffer of Example 13-2, adding the same amount of strong acid still leads to only a relatively small change in pH (0.37 versus 0.18 pH unit). Compare the results of Examples 3-2 and 3-3 with the results of Example 3-1.

In actual practice, concentrated solutions, containing a weak acid and an alkali metal salt of it in the proper proportion, are commonly held in stock and an appropriate volume of this solution is added to the solution to be buffered.

When the solution to be buffered has a pH far removed from the desired value, it is best to adjust the solution to about that value by adding a strong acid or base, and only then to add the buffer solution. If the buffer were introduced without the previous adjustment, its capacity might be insufficient to secure the desired pH unless a very large amount were added. Too large an addition would be undesirable, or in many instances, intolerable.

13.4 Mixture of Weak Base and Conjugate as Buffer

A mixture of a weak base and its conjugate acid, added as an appropriate salt, exhibits buffer action in the alkaline range. The situation and mathematical treatment resemble those in Section 13.3. The buffer system ammonia-ammonium chloride is frequently used in analytical chemistry and has a "practical" buffer range about pH 8.3 to 10.3.

13.5 Buffers of Extended Range

When polyfunctional weak acids or bases and their conjugates are used or more than one weak acid and base are involved, buffers can be obtained that show high capacity in several pH ranges. In addition, when the dissociation constants are close together, the regions of high buffer capacity overlap and buffers can be obtained that show high capacity over a large pH range. For example, buffers prepared by mixing solutions of sodium hydroxide and citric acid (tribasic) show useful buffer capacity over the interval pH 2 to 7. Phosphate buffers, prepared by mixing aqueous ammonia or an alkali metal hydroxide with phosphoric acid, are widely applied and have a useful range from pH 3.1 to 10.1.

13.6 Acid Salts of Dibasic Acids

The acid salt of the type MHA of a dibasic acid H_2A, where M is an alkali metal, receives special attention for the preparation of a solution of known pH. According to equation (11-11), the pH of a solution of such an acid salt is given as $\text{pH} \cong \frac{1}{2}(\text{p}K_{a,1} + \text{p}K_{a,2})$. The pH remains essentially constant over a relatively wide range of salt concentration, but the buffer capacity is extremely low. Such a solution corresponds to that at the first equivalence point in the titration of a weak dibasic acid (Section 11.9), and at this point the titration curve is far from horizontal. Consequently, adding small amounts of an acid or base will cause a substantial change in pH. Such salt solutions are used as pH standards ("buffers") to standardize pH meters (Section 23.2 and Table E in the Appendix). A $0.050F$ solution of potassium hydrogen phthate provides a pH of 4.01 at 25° C, for example.

13.7 Summary of Buffer Types

Table 13-1 summarizes the various types of buffers and compares their behavior with various changes.

Table 13-1. Buffer Types and Behavior

	Effect of dilution	*Effect of adding small amounts of strong acid or strong base*
Strong acid or strong base	Slight. For solutions up to ~1F, about 1 pH unit for a change in concentration by a factor of 10	Slight, depending on buffer concentration
Weak acid or weak base, and its conjugate	Negligible	Slight, depending on concentration and composition of buffer
Ampholyte	Negligible	Large, regardless of concentration

13.8 *Questions*

13-1 Define *buffer.*

13-2 Elaborate on the correlation of the various types of buffers with the appropriate acid-base titration curves.

13-3 Describe the buffer capacity qualitatively. On what does it depend?

13-4 In what sense can a solution of sodium hydrogen carbonate, $NaHCO_3$, be called a buffer?

13-5 Elaborate on the possibility of preparing buffer solutions by adding hydrochloric acid to a solution of sodium acetate and to a solution of ammonia.

13-6 What acid, base, salt, or mixture of these would you select to get a buffer of pH 9.0? Explain your answer.

13.9 *Problems*

13-1 A buffer solution is prepared by mixing 20.0 mL of 0.10*F* sodium acetate and 8.0 mL of 0.12*F* HCl (K_a for acetic acid, 1.8×10^{-5}). (a) Calculate the pH of this solution. (b) Calculate the change in pH caused by adding 5.0 mL of 0.10*F* HCl. *Ans.:* (a) pH 4.78; (b) 0.47 pH unit

13-2 An NH_3-NH_4Cl buffer is prepared by mixing equal volumes of 0.200*F* NH_3 ($K_b = 1.8 \times 10^{-5}$) and 0.200*F* NH_4Cl. Calculate the pH of this buffer. *Ans.:* pH 9.26

13-3 Calculate the pH of the solution obtained if to 100 mL of buffer described in Problem 13-2 is added 10.0 mL of (a) 0.10*F* NH_3, (b) 0.10*F* NaOH, and (c) 0.10*F* HCl. *Ans.:* (a) pH 9.30; (b) pH 9.34; (c) pH 9.18

13-4 What minimum volume (in millimeters) of a buffer solution that is 0.100*F* in both lactic acid ($K_a = 1.4 \times 10^{-4}$) and sodium lactate must be diluted to exactly 100 mL so that the resulting buffered solution will change its pH by not more than 0.50 pH unit when 10.0 mL of 0.010*F* HCl is added? *Ans.:* 1.9 mL

13-5 How many grams of sodium acetate (*82.0*) must be added to 50.0 mL of 0.080*F* acetic acid ($K_a = 1.8 \times 10^{-5}$) to attain a pH of 4.20? *Ans.:* 0.094 g

13-6 Calculate how many grams of ammonium chloride (*53.49*) must be added to each of the ammonia solutions of specified volume and formality to establish the desired pH.

	Ammonia solution	Desired pH
(a)	100.0 mL of 0.100*F*	9.00
(b)	50.0 mL of 0.120*F*	9.25
(c)	30.0 mL of 0.195*F*	9.05
(d)	25.0 mL of 0.679*F*	10.00
(e)	50.0 mL of 0.500*F*	10.00
(f)	200.0 mL of 0.200*F*	9.75

13-7 The following solutions of weak acids are mixed with solutions of the sodium salt of the same weak acid. Calculate the pH of each mixture.

	Acid solution	K_a	Salt solution
(a)	30.0 mL of 0.020*F*	3.2×10^{-7}	40.0 mL of 0.015*F*
(b)	50.0 mL of 0.015*F*	4.0×10^{-5}	25.0 mL of 0.025*F*
(c)	20.0 mL of 0.100*F*	6.5×10^{-6}	150.0 mL of 0.080*F*
(d)	35.0 mL of 0.040*F*	1.8×10^{-5}	45.0 mL of 0.050*F*

13-8 How many milliliters of the acetic acid solution of indicated formality must be added to each of the NaOH solutions of specified volume and formality to obtain the desired pH?

	NaOH solution	Acetic acid solution	Desired pH
(a)	100.0 mL of 0.100*F*	0.200*F*	4.74
(b)	150.0 mL of 0.215*F*	0.100*F*	5.00
(c)	200.0 mL of 0.075*F*	0.095*F*	4.50
(d)	75.0 mL of 0.095*F*	0.350*F*	4.90

13-9 A buffer is made by adding 25.0 mL of 0.100*F* sodium acetate to 50.0 mL of 0.100*F* acetic acid ($K_a = 1.8 \times 10^{-5}$). What would the pH be if to this buffer were added: (a) 25.0 mL of distilled water? (b) 10.0 mL of 0.100*F* HCl? (c) 10.0 mL of 0.100*F* NaOH?

13–10 Three buffer solutions are prepared from 0.100F ammonia (K_b = 1.8 × 10^{-5}) and 0.100F NH_4Cl solutions.

Buffer	Ammonia solution, mL taken	NH_4Cl solution, mL taken
A	50.0	50.0
B	25.0	75.0
C	75.0	25.0

(a) Calculate the pH of each buffer. (b) Compare the change of pH resulting when 1.0 mL of 1.0F HCl is added to each buffer.

13–11 What must the formal concentrations be of HA (K_a = 1.0 × 10^{-6}) and NaA in a buffer of pH 5.30 so that the pH will be lowered to 5.00 when 10.0 mL of 0.100F HCl is added to 100.0 mL of the buffer solution?

13–12 A buffer solution has been prepared by making the solution 0.100F in acetic acid (K_a = 1.8 × 10^{-5}) and 0.100F in sodium acetate. What volume of this buffer solution must be used so that the addition of 50.0 mL of 0.050F H_2SO_4 will change the pH by exactly 1 unit?

13–13 A buffer solution has been prepared by mixing 25.0 mL of a 0.100F solution of the sodium salt of a weak monoprotic acid (K_a = 1.0 × 10^{-5}) and an equal volume of a solution of that acid, of the same formality. What change in the pH of this buffer will result when 5.0 mL of 0.050F HCl is added?

Ans.: ΔpH = –0.09

14

Applied Acid-Base Titrations

14.1 Standard Acid Solutions

Hydrochloric acid is the acid most frequently used as a titrant in acid-base titrations in aqueous media. Occasionally sulfuric acid and perchloric acid are used. Nitric acid is suitable in some special determinations. Possibilities exist for directly preparing a standard acid solution by weighing a known amount of acid and diluting it to a known volume with water. The use of constant-boiling hydrochloric acid for this purpose is a notable example. The composition of the distillate of the constant-boiling mixture depends only on the atmospheric pressure. Consequently it is possible to distill this acid under exactly known pressure and to read the concentration of the distillate from a table. In practice, however, it is simpler to prepare a solution of approximately the desired strength and to establish its exact concentration by an acid-base titration using a primary standard or by evaluation against a standard solution of a strong base.

Sulfuric acid and perchloric acid are not volatilized when their aqueous solutions, even those of considerable strength, are boiled. Hydrochloric acid and nitric acid are more volatile, and it is important to appreciate this fact when the solution must be boiled during a titration. Note, however, that a $0.1F$ hydrochloric acid solution can be boiled for over 30 min with no detectable loss of hydrogen chloride, provided that the water evaporated is replaced at intervals.

Storing standard acid solutions presents no serious problems. Their strength is maintained in glass vessels that are tightly closed to prevent the evaporation of water or the absorption of basic substances from the laboratory air.

14.2 Primary Standards for Acids

The requirements for primary standards are summarized in Section 10.3. Although many primary standards for acids are known, they often do not function as well as the primary standards for bases. Hence, frequently a solution of a strong base is first standardized, and then the strength of the acid is evaluated against solution of the base.

Sodium Carbonate. Sodium carbonate, Na_2CO_3, is suitable as a primary standard for acids. For accurate work it is necessary to heat the reagent-grade powdered form of the anhydrous compound to about 285 °C for at least 30 min. This treatment removes all water and converts any traces of hydrogen carbonate to carbonate. Since the product is somewhat hygroscopic, it should be weighed promptly after cooling in a desiccator, and its storage for a protracted period is not recommended. A weighed amount of the sodium carbonate is dissolved in water and is directly titrated with the acid to be standardized to the methyl orange end point. (The details of this titration are considered in Section 14.7.) In this titration the equivalent weight of sodium carbonate is one-half its formula weight = ½ × 105.989 = 52.994.

Sodium Tetraborate Decahydrate. Sodium tetraborate decahydrate (borax), $Na_2B_4O_7 \cdot 10H_2O$, is a salt of a very weak acid, and in aqueous solution the tetraborate ion reacts with a strong acid as follows:

$$B_4O_7^{2-} + 2H^+ + 5H_2O \rightarrow 4H_3BO_3 \qquad (14\text{-}1)$$

The equivalent weight of borax is therefore one-half its formula weight = ½ × 381.373 = 190.686. The equivalence point in the standardization of a 0.1*F* solution of strong acid with borax is at about pH 5.1; hence, methyl red is an appropriate indicator. The advantages of borax as a primary standard include its high equivalent weight and its ease of preparation in a highly pure state, simply by recrystallization from water below 55 °C. The product requires special storage to ensure that the exact composition of the decahydrate is maintained; this is best accomplished by storage over deliquescent sodium bromide, which affords the proper humidity.

2-Amino-2-(hydroxymethyl)-1.3-propanediol. The substance 2-amino-2-(hydroxymethyl)-1,3-propanediol, $(HOCH_2)_3CNH_2$, is a weak monoequivalent base having a formula (and equivalent) weight of 121.137. It is often assigned the trivial designation *tris.* The high-purity material available can be dried at 100 to 105 °C. The equivalence point in its titration with a strong acid occurs at about pH 4.7. Methyl red is a suitable indicator for this titration; bromocresol green is even better. For general analytical work, this base is possibly the most suitable one for the standardization of strong acid solutions.

14.3 Standard Base Solutions

Solutions of bases must be prepared, stored, and used in such a way that they are protected from absorption of carbon dioxide from the air, because the presence of carbonate is a frequent source of difficulties in titrations (see Section 14.7). For ordinary work, a small amount of carbonate can be tolerated, but it must be kept as small as possible. Sodium hydroxide is the base most frequently applied as a titrant in acid-base titrations in aqueous media. Brief rinsing of the reagent-grade pellets of this base with water before dissolution in water is one possible approach to removing carbonate. The preferred classical method is to dissolve 50 parts by weight of sodium hydroxide in pellet form in 50 parts of water. In such a solution, sodium carbonate is insoluble and settles out. A portion of the clear supernatant liquid, or of the filtrate obtained by vacuum filtration through a sintered glass filter, is used in the preparation of the dilute solution. This preparation has been simplified by the availability of reagent-grade 50% sodium hydroxide solution low in carbonate furnished in polyethylene bottles; clear portions of this solution can be directly diluted. The water used to prepare the standard solution is either boiled to remove carbon dioxide and then cooled or is taken directly from a mixed bed of ion-exchange resins (Chapter 47). The amount of carbon dioxide in water in equilibrium with the general atmosphere is small ($1.5 \times 10^{-5} F$). Laboratory air, however, may contain substantially more carbon dioxide, and, hence, so may distilled water that has been in contact with such air. Solutions of sodium hydroxide should be stored in containers that are either airtight or fitted with a tube filled with soda-lime or soda-asbestos. Since alkaline solutions attack glass, sodium hydroxide solutions are best stored in polyethylene bottles, although for short periods borosilicate glass bottles can be used.

Sodium hydroxide solutions can be freed of carbonate by being passed through a column of a strongly basic anion-exchange resin in the hydroxide form (Chapter 47).

For very accurate work, barium hydroxide is often recommended, since barium carbonate is insoluble. A barium hydroxide solution of about the desired strength is prepared and allowed to stand in an airtight bottle for several days. The clear supernatant liquid is siphoned off and stored in a polyethylene bottle with precautions to prevent absorption of carbon dioxide.

Potassium hydroxide is sometimes used as a base in acid-base titrimetry. However, potassium carbonate is relatively soluble in concentrated solutions of potassium hydroxide, hence, the classical technique for removing carbonate from sodium hydroxide is not applicable. When an ethanolic solution of a base is needed, potassium hydroxide is often used because it is readily soluble in this organic solvent.

14.4 Primary Standards for Bases

Potassium Hydrogen Phthalate. Potassium hydrogen phthalate is probably the primary standard that is used most often in acid-base titrimetry. This compound is the acid salt of the weak diprotic acid, phthalic acid ($K_{a,2} = 3.1 \times 10^{-6}$), and acts as a weak monoprotic acid. Its equivalent weight in the titration with a strong base equals its formula weight, 204.229. The equivalence point is in the alkaline region, and phenolphthalein is generally used as the indicator. Potassium hydrogen phthalate is stable, may be heated to 130°C without decomposition, and is virtually nonhygroscopic. This salt is commercially available in a primary standard grade.

Sulfamic Acid. Sulfamic acid, NH_2SO_3H, is a strong monoprotic acid of formula weight 97.093; and in its titration with a strong base, any indicator changing in color in the region pH 4 to 9 can be used. The substance is stable in air, but decomposes slowly in aqueous solution, to yield ammonium hydrogen sulfate, NH_4HSO_4. Consequently, the titration is best performed soon after dissolving the weighed amount of sulfamic acid.

14.5 *Determination of Phosphoric Acid*

Orthophosphoric acid, H_3PO_4, is a tribasic acid ($K_{a,1} = 6.0 \times 10^{-3}$, $K_{a,2} = 6.3 \times 10^{-8}$, $K_{a,3} = 4.4 \times 10^{-13}$; $pK_{a,1} = 2.15$, $pK_{a,2} = 7.20$, $pK_{a,3} = 12.36$).

Only the first two of the dissociable hydrogen ions can be titrated directly with a strong base, such as sodium hydroxide, since the third constant is too small. The approximate pH values for the equivalence points are $1/2\ (2.15 + 7.20) = 4.68$ and $1/2\ (7.20 + 12.36) = 9.73$. These pH values correspond to the pH of a solution of monosodium phosphate, NaH_2PO_4, and of disodium phosphate, Na_2HPO_4, respectively.

The titration to the first equivalence point can be performed using methyl orange and titrating almost to its alkaline color. Locating the color is made easier by comparison with a solution containing the indicator and monosodium phosphate in concentrations equal to those in the principal titration (see Section 12.4). Bromocresol green is a better indicator, but again a comparison solution is preferable. The second equivalence point is at a pH value slightly more alkaline than what can readily be indicated by phenolphthalein; hence, thymolphthalein is a better indicator.

Even though a direct, simple titration of the third dissociable hydrogen ion in phosphoric acid is impracticable, phosphoric acid can be titrated as a tribasic acid if the phosphate ion is precipitated under proper conditions. For this purpose, a neutralized solution of calcium chloride is added to the sample solution, and the titration with sodium hydroxide solution is started. As the titration proceeds, the pH increases progressively until a value is reached at which calcium orthophosphate starts to precipitate. Its occurrence, however, is disregarded and the titration completed to the permanent pink color of phenolphthalein. The precipitate has hardly any effect on the recognition of the end-point color but can coprecipitate unreacted material, and thus errors as great as 1 or 2% may be experienced.

14.6 *Boric Acid*

Borate is often separated from interfering substances by distillation of trimethyl borate. $B(OCH_3)_3$, which is formed when the borate-containing sample is boiled with concentrated hydrochloric acid, anhydrous calcium chloride (which binds the water present and formed in the esterification), and methanol, CH_3OH. This borate ester is condensed and collected in a sodium hydroxide solution. Next, this solution is boiled to decompose the ester and to volatilize the methanol. The remaining solution is then neutralized to methyl orange and boiled to remove carbon dioxide. Free boric acid is thus present in the solution, acting as a very weak monoprotic acid ($K_a = 5.9 \times 10^{-10}$) that cannot be titrated.* Boric acid, however, forms complex entities with organic polyhydroxy compounds, such as glycerol or mannitol. These complex compounds are far stronger acids than boric acid itself. With mannitol, the acidity constant of the complex acid formed has a value about 1×10^{-4}, thus the complex acid is stronger than acetic acid and can readily be titrated with a strong base to a sharp end point, using phenolphthalein as indicator. The reaction with mannitol can be represented by the following equation, where mannitol is denoted by *Ma:*

$$HBO_2 + 2Ma \rightarrow H^+ + Ma_2BO_2^- \qquad (14\text{-}2)$$

14.7 Determination of Carbonates

The principles underlying the "replacement" titration of a carbonate salt with a strong acid have been outlined in Section 11.16.

Carbonate can be titrated to hydrogen carbonate ion (pH 8.3). The "break" at this equivalence point is not well defined, however, as you can see from the curve in Fig. 11-4; consequently, when highly reliable results are desired, a comparison solution should be used containing amounts of indicator and sodium hydrogen carbonate like those in the titration solution. This is especially necessary when phenolphthalein is used as the indicator, since its complete decoloration occurs at about pH 8.0, which is beyond the equivalence point and in a region where the titration curve is far from perpendicular.

The hydrogen carbonate can be further titrated to carbonic acid. The titrant is added until the methyl orange indicator color begins to change from yellow to orange. Of course, it is possible to perform the titration of carbonate ion directly to carbonic acid, omitting phenolphthalein and adding methyl orange only. The solution is then boiled to expel carbon dioxide, and after it is cooled, the titration is continued to a red-orange color. Cooling is necessary because the dissociation equilibrium of methyl orange and consequently of the color exhibited at a certain pH strongly depends on the temperature. The small amount of carbon dioxide formed when the last portions of titrant are added does not affect the color change, and a relatively sharp color change occurs.

Alternatively, it is often more convenient to add an excess of acid in known amount, to heat the solution to boiling, then to cool, and to back-titrate with a standard sodium hydroxide solution to the methyl orange end point. Indeed, since all carbonate has been evolved as carbon dioxide and only strong acid remains, it is

*Metaboric acid, HBO_2, has one fewer water molecule than orthoboric acid, H_3BO_3. In aqueous media, the two cannot be distinguished and hence boric acid can be written as either H_3BO_3 or HBO_2.

possible to back-titrate to the phenolphthalein end point. In this case it is unnecessary to wait until the solution has cooled. This back-titration approach is mandatory when a water-insoluble carbonate such as calcium carbonate is to be titrated.

Hydroxide ion can be determined in the presence of carbonate ion if the carbonate ion is first precipitated as barium carbonate. An excess amount of a neutralized solution of barium chloride is added to the hydroxide-carbonate sample; the hydroxide ion is then titrated with hydrochloric acid to the phenolphthalein end point in the presence of the precipitate. (The methyl orange end point cannot be used since the barium carbonate would dissolve in the acidic solution.) This procedure has limitations because the precipitate may carry down some hydroxide ion.

Various titrimetric processes can be combined to effect the resolution of mixtures of carbonate and hydrogen carbonate, or of carbonate and hydroxide. If a solution contains substantial amounts of both carbonate and hydroxide ions, titration with hydrochloric acid first to the phenolphthalein end point neutralizes the hydroxide and converts carbonate to hydrogen carbonate; then, after methyl orange is added, continued titration to a second end point (with the heating step discussed above) converts hydrogen carbonate to carbonic acid. The volume of hydrochloric acid required for the second titration step corresponds to one half of that required to convert carbonate to carbonic acid. Hence, both the hydroxide and the carbonate contents can be calculated. A similar stepwise titration permits the determination of both hydrogen carbonate and carbonate when they are present together. When only a small amount of carbonate is present, the method fails, since then the volume of titrant solution needed to go from the phenolphthalein end point to the methyl orange end point is small and the results are unreliable.

In such a case, a different approach is used. In one aliquot of the sample solution, the total content of base, that is, the sum of hydroxide and carbonate, is determined by titration to the methyl orange end point. In a second aliquot, the carbonate is precipitated by adding barium chloride and the titration performed to the phenolphthalein end point. The difference in consumption of titrant in the two titrations corresponds to the carbonate. This approach also permits determining a small amount of hydrogen carbonate in a carbonate sample. To one aliquot of the sample solution, a known volume of standard sodium hydroxide solution is added, in excess of what is required to convert hydrogen carbonate to carbonate ion. Then barium chloride is added, and the mixture is titrated with hydrochloric acid to the phenolphthalein end point; an identical volume of sodium hydroxide is similarly titrated. The difference in titrant volumes needed in the two titrations corresponds to hydrogen carbonate. In a second aliquot of the sample solution, the total of carbonate and hydrogen carbonate is determined by titration to the methyl orange end point; carbonate is then obtained by difference.

It is convenient now to examine the effect of the presence of carbonate in a base used for the titration of a strong acid. From the above considerations, it is obvious that there will be a difference in the titration results, depending on whether phenolphthalein or methyl orange is used as the indicator. The greater the carbonate content, the greater the difference. When only a small amount of carbonate is present, it is possible to standardize the base using either indicator and to apply the corresponding "phenolphthalein" (P.P.) or "methyl orange" (M.O.) formality as appropriate. If large amounts of carbonate are present, however, this approach fails because further difficulties arise.

When the carbonate-containing base is added to the acidic sample solution, carbon dioxide is formed. As long as this formation is not excessive, the titration to the methyl orange end point will not be affected. If the titration is conducted to the phenolphthalein end point, however, the carbon dioxide is titrated to hydrogen carbonate. But some of it may have already escaped from the solution, and since the amounts lost are uncontrolled, the titrations are not closely reproducible. Furthermore, the reconversion of carbon dioxide to carbonic acid (which is the species that reacts with the base) is not very rapid and leads to a fading end point; that is, the solution becomes pink on addition of one drop of the base, but the color fades within a few seconds. This fading is not to be confused with the slower fading that occurs after possibly 30 s or longer due to the absorption of atmospheric carbon dioxide by the alkaline solution. For the experienced analyst, the rapid fading and sluggish character of an end point in a titration with a strong base to the phenolphthalein end point is a clear signal that an undesirably high amount of carbonate is present in either the standard base or the sample solution or both.

14.8 Determination of Temporary Water Hardness

Natural water may contain significant amounts of the soluble hydrogen carbonate salts of the alkaline earths (principally calcium and magnesium) as well as the sulfates and chlorides of these elements. These are the main constituents contributing to the hardness of water. When hard water is boiled, carbonate precipitates according to the reactions

$$Ca^{2+} + 2HCO_3^- \rightleftharpoons \underline{CaCO_3} + H_2O + CO_2 \qquad (14\text{-}3)$$

$$Mg^{2+} + 2HCO_3^- \rightleftharpoons \underline{MgCO_3} + H_2O + CO_2 \qquad (14\text{-}4)$$

The portion of the hardness corresponding to the hydrogen carbonate salts is termed temporary hardness or carbonate hardness. The remaining hardness is due to the alkaline earth sulfates or chlorides, which do not precipitate on boiling, and is termed the permanent hardness or noncarbonate hardness. The sum of the two types of hardness, expressed in identical units, is known as total hardness. Hardness is usually expressed in milligrams per liter (that is, parts per million) of calcium, calcium oxide, or calcium carbonate, regardless of whether calcium or magnesium is actually present. The determination of water hardness is of practical importance, for example, in boiler-feed water and in water intended for laundry purposes. When hard water is heated in a boiler, calcium and magnesium carbonates deposit in the boiler tubes as so-called boiler scale. Calcium and magnesium ions also form insoluble soaps, and consequently hard water requires the use of much soap before adequate suds can be developed.

Temporary hardness is determined by simply adding methyl orange to a known volume of the water sample and titrating the hydrogen carbonate with hydrochloric acid. Since a low concentration of hydrogen carbonate ion is involved, a blank is run, using an identical volume of carbonate-free distilled water, and the volume of acid required for the blank is subtracted from the volume of acid in the titration of the water sample.

Total hardness was formerly determined by a tedious acid-base titration procedure, but the result can now be attained more rapidly and with higher reliability by a compleximetric titration (Section 20.10).

14.9 Determination of Nitrogen

Where the nitrogen in a substance is present as ammonium ion or can be converted to it, a determination of nitrogen based on an acid-base titration is feasible. The solution of an ammonium salt can be made strongly alkaline and boiled, thereby volatilizing ammonia with water. Ammonia in the distillate is determined by an acid-base titration. Kjeldahl found that the nitrogen in many organic substances can be converted to ammonium ion by heating with concentrated sulfuric acid. After alkalization of the digested mixture, the ammonia can be distilled and determined. Inorganic nitrates, which are not reduced to ammonia by this acid treatment, can be converted to ammonia by an alkaline digestion with a metal or an alloy, such as Devarda's alloy, which has reducing properties under such conditions.

Kjeldahl Digestion. The reactions that take place when an organic substance containing nitrogen is heated with concentrated sulfuric acid are complex and not fully understood. The aggregate process may be described as the oxidation of the substance by sulfuric acid, which is reduced to sulfuric dioxide, and the reduction of nitrogen by the sulfur dioxide to ammonium ion. (With organic matter that has oxidized nitrogen, such as nitro groups, a preliminary reduction is often appropriate.) To hasten the digestion, potassium sulfate is added to elevate the boiling point of the digestion mixture. Various substances can be added to catalyze the reactions occurring during the digestion, including salts and oxides of copper, mercury, or selenium. The digestion is usually carried out in special long-necked flasks, known as Kjeldahl flasks, in a hood or a collecting manifold, since fumes of sulfuric acid and sulfur dioxide are evolved. The digestion is continued until the solution is colorless, indicating the complete destruction of organic matter. The final decolorization may be hastened by adding a few drops of hydrogen peroxide.

Distillation after Kjeldahl Digestion. When the Kjeldahl digestion is complete, the solution is allowed to cool. A distillation assembly may be attached to the flask, or the solution may be transferred to a separate distillation apparatus. An excess of sodium hydroxide is added and the solution is brought to a boil. Ammonia and water vapor are driven off and condensed, and are caught in a flask containing a measured amount of standard acid. After all the ammonia has been distilled, the excess of the acid is back-titrated with standard base to the methyl red end point (since ammonia is a weak base, the equivalence point is in the acidic region).

Treatment after Devarda. The reduction of inorganic nitrate to ammonia can be effected in strongly alkaline solution by Devarda's alloy (45% aluminum, 5% zinc, and 50% copper):

$$3NO_3^- + 8Al + 5OH^- \rightarrow 8AlO_2^- + 3NH_3 \qquad (14\text{-}5)$$

The solution is heated; the ammonia distills with water and is determined as described above.

14.10 Formol Method for Ammonia

Ammonia in ammonium salts can be determined by a more rapid method, which involves the reaction of formaldehyde, CH_2O, with ammonia to form the extremely weak base hexamethylenetetramine, $(CH_2)_6N_4$, and hydrogen ion, which can be titrated with a strong base. This reaction is represented by

$$4NH_4^+ + 6CH_2O \rightarrow (CH_2)_6N_4 + 6H_2O + 4H^+ \qquad (14\text{-}6)$$

One hydrogen ion is produced for each ammonium ion reacting. The procedure, known as the formol titration, takes the following course. The sample solution containing ammonium ion and only the anions of strong acids is adjusted to the transition color of methyl red. Formaldehyde, which has been neutralized to phenolphthalein, is now added in excess and then a few drops of this indicator. After several minutes, the solution is titrated with a standard base to the alkaline color of phenolphthalein (here screened by the alkaline color of methyl red to orange, that is, pink + yellow). If anions of weak acids are present, the initial pH adjustment is made with phenol red.

14.11 Errors in Acid-Base Titrations

The correctness of the result of a titration is influenced by many factors and to varying degrees, including uncertainties in the weighing of the sample, delivery of a solution by a pipet, the reading and nonreproducible drainage of a buret, impurities in the sample solution, changes in temperature, failure to titrate always to the same indicator hue (visual discrimination error), and others. All these factors produce small deviations, so that the individual results in a series of titrations performed under "identical" conditions and with "identical" samples fluctuate around their average. These deviations are known as random errors. Depending on the size of these errors, the repeatability, that is, the precision of a titration, is high or low. Although random errors can never be completely avoided, they can be reduced by careful operation.

Another kind of error, known as a systematic error, is inherent in the method. Such an error causes a departure of the result from the correct value in a single direction only, and its size is a measure of the accuracy of a titration result. This classification of errors is of theoretical interest and of importance in treating experimental data (see Chapter 2).

In practice, however, it is difficult to separate random and systematic errors, because every systematic error is accompanied by a random component. This is especially true of the drop error in titrimetry. When the end point is near, one more drop of titrant is added and in the visual titration the color change is seen. The closeness of the approach to the end point before addition of the final drop will vary in a series of titrations, and hence the small amount of titrant added in excess varies from titration to titration. This error may be negligible if a large volume of titrant is required but significant with a small volume of titrant. The drop error usually includes the visual discrimination error. Careful work, the use of a color comparison solution (Sections 12.4 and 14.5), if necessary, and completion of the titration with fractions of a drop will reduce the drop error.

The indicator error is a further, largely systematic error. The indicator as a weak acid or base consumes some material to be titrated or some titrant. Since only a minute amount of indicator is added, this error is usually negligible. With small samples and dilute titrant solutions, however, it may become significant and require correction through an indicator blank or exclusion by other means (see Section 12.2). Further systematic errors may be introduced, for example, by inaccurate standardization of the titrant solution and incorrect calibration of the volumetric glassware.

With all these possibilities for error, you might think that a correct titration result is hard to reach. However, the random errors from various sources vary in magnitude and sign—one may be large when another is small; one may be positive when another is negative. Hence, there is partial compensation, and the net random error may be small. With good equipment and appropriate care, the net random error, that is, the precision of the titration, may readily be within one or two drops of titrant solution, that is, within ±0.05 to ±0.10 mL.

Especially important is the titration error, or chemical error, which corresponds to the difference between the observed end point and the equivalence point. Calculating the size of this error will be considered for the titration of strong monofunctional acids and bases and will give a further insight into titration processes. The calculation for titrations involving weak acids or bases is possible, but involved.

Example 14-1 A volume of exactly 50 mL of exactly 0.1F HCl is titrated with exactly 0.1F NaOH to a certain hue of methyl orange corresponding to exactly pH 4. What is the titration error?

As in the calculation of titration curves [Section 11.2, equation (11-1)], the molar concentration of hydrogen is computed by subtracting the millimoles of base added from the millimoles of acid initially present and dividing by the total solution volume in milliliters. The volume of base added may be denoted by V_b. The expression becomes

$$10^{-4} = \frac{50 \times 0.1 - 0.1V_b}{50 + V_b}$$

$$50 \times 10^{-4} + 1 \times 10^{-4} V_b = 5 - 0.1 V_b$$

$$0.1001 V_b = 4.995$$

$$V_b = 49.90 \text{ mL}$$

The equivalence point corresponds to exactly 50 mL of base; hence, the titration error ϵ expressed in percentage is

$$\epsilon = \frac{(49.90 - 50) \times 100}{50} = -0.2\%$$

Since in conventional titrimetry, a result accurate to 0.5 to 1% is acceptable, it may be concluded that methyl orange is a suitable indicator for this titration. Titration to a more yellow hue of the indicator might be considered, thereby decreasing the titration error.

Example 14-2 A volume of exactly 50 mL of exactly 0.01F NaOH is titrated with exactly 0.01F HCl to exactly pH 4 (methyl orange). What is the titration error?

The solution at the end point has a pH below 7 and hence is overtitrated; that is, a small amount of acid in excess is added. The calculation is analogous to Example 14-1, and V_a denotes the volume of acid added.

$$10^{-4} = \frac{0.01V_a - 50 \times 0.01}{V_a + 50}$$

$$10^{-4}V_a + 50 \times 10^{-4} = 0.01V_a - 0.5$$

$$0.0099V_a = 0.505$$

$$V_a = 51.01 \text{ mL}$$

The titration error in percentage is therefore

$$\epsilon = \frac{(51.01 - 50) \times 100}{50} = +2\%$$

Comparing the data in Examples 14-1 and 14-2 shows, as expected, that when a 10 times more dilute titrant solution is used, a 10 times greater error results. The overtitration corresponds to a positive sign for the error. For this titration, methyl orange is not a suitable indicator, and an indicator that changes color at a higher pH should be used (say, methyl red).

Example 14-3 A volume of exactly 25 mL of exactly 0.1F HCl is titrated with exactly 0.05F NaOH to exactly pH 9 (phenolphthalein). What is the titration error?

Since an overtitration obviously occurs, let V_b denote the volume of base added.

$$\text{pH } 9 = \text{pOH } 5 \quad \text{and} \quad [\text{OH}^-] = 10^{-5}M$$

Hence,

$$10^{-5} = \frac{0.05\,V_b - 25 \times 0.1}{25 + V_b}$$

$$25 \times 10^{-5} + 10^{-5}\,V_b = 0.05\,V_b - 2.5$$

$$0.04999\,V_b = 2.50025$$

$$V_b = 50.015 \text{ mL}$$

The percentage of titration error is therefore

$$\epsilon = \frac{(50.015 - 50) \times 100}{50} = +0.03\%$$

These few examples show the importance of selecting an appropriate indicator to obtain a correct result. In the calculations, the amount of hydrogen or hydroxide ion furnished by water itself has been neglected and permittedly so, since these pH values are far from 7. When the end point is close to pH 7, this source of ions can no longer be neglected, and a more involved formula must be used. If, however, in a strong acid–strong base titration, the end point is close to pH 7 (that is, close to the equivalence point), the titration error is obviously negligible and no calculation is needed.

14.12 Questions

14–1 What is a primary standard? Describe its use and requirements.

14–2 What are the difficulties found in using primary standards for standardizing acid solutions?

14–3 How would you analyze a mixture containing hydrochloric and acetic acids? What indicators would you select?

14–4 Why is a back-titration recommended in the acid-base titration of calcium carbonate?

14–5 How could you determine zinc oxide by an acid-base titration?

14–6 Discuss in your own words the terms *temporary hardness* and *permanent hardness.*

14–7 Explain why phenolphthalein is not used in determining temporary hardness. What would be the result if this indicator were used?

14–8 How can nitrate be determined through an acid-base titration?

14–9 Elaborate on the possibility of titrimetrically determining in a mixture both hydrochloric acid and nitric acid, using silver acetate as an auxiliary reagent.

14-10 What chemical processes are involved in the distillation method for determining borate?

14-11 What chemical reactions are involved in the formol method of ammonia?

14-12 What difficulties will be encountered if a sample of hydrochloric acid containing some iron(III) chloride is titrated with sodium hydroxide to determine the free acid?

14-13 What is the purpose of adding a calcium chloride solution when phosphoric acid is titrated as a triprotic acid?

14-14 A certain amount of sulfur is burned under conditions yielding both SO_2 and SO_3. The gases are absorbed in water, yielding a mixture of H_2SO_3 and H_2SO_4. Would it be possible to differentiate between the two acids on the basis of a simple acid-base titration?

14-15 Sodium dihydrogen phosphate, NaH_2PO_4, can be titrated with a sodium hydroxide solution. Write the equation for the titration reaction. What indicator should be used? What is the equivalent weight of the salt in this titration?

14-16 What kinds of errors can be differentiated?

14-17 Describe the indicator error and discuss ways of avoiding or reducing it.

14-18 Name some sources of systematic errors in acid-base titrimetry.

14-19 What percentage error in the result is caused by a drop error of 0.05 mL of titrant when the volume of titrant required is (a) 5.00 mL and (b) 50.00 ml?

14-20 Name some sources of random errors found in titrimetry.

14-21 How would you reduce the visual discrimination error in a titration?

14-22 Using an inaccurately standardized titrant solution introduces what kind of error – random or systematic?

14-23 A general formula for the titration error in the strong acid–strong base titration can be derived in which the volume of the solution being titrated and the volume of the titrant solution are *not* factors; only the concentrations of the two solutions affect the size of the titration error. Verify this fact by repeating the calculations of Examples 14–1 and 14–2, using other volumes besides exactly 50 mL.

14-24 Define the following terms: *titration, acid-base titration, titrant, titrant solution, titrand* (that is, species titrated), *back-titrant, theoretical titration curve, experimental titration curve,* and *titer.*

14-25 Define the following terms: *standard solution, standardized solution,* and *primary standard.*

14-26 Highlight the use of the various terms defined in Questions 14–24 and 14–25 in the description of various acid-base titrations.

14.13 *Problems*

14–1 *A volume of 100 mL of 0.1*F* HCl is titrated with 0.05*F* NaOH to pH 8.5 (phenolphthalein). What is the titration error? *Ans.:* +0.01%

14-2 Calculate the titration error that arises when a volume of 20 mL of 0.01*F* HCl is titrated with 0.001*F* NaOH to an end point at pH 3.3. *Ans.:* –37%

14–3 Calculate the titration error that arises when a volume of 50 mL of 0.01*F* HCl is titrated with 0.01*F* KOH to an end point at pH 5.3 (Methyl red). *Ans.:* –0.1%

---

*In solving these problems, assume that all volumes, formalities, and pH values are exactly known.

15

Calculating Results in Acid-Base Titrations

15.1 Basis of Calculations

Regardless of the acids or bases involved, the principal titration equation is always the transfer of a proton: HA + B → A + HB (charges are omitted for simplicity). Consequently, the question as to what amount of an acid is equivalent to what amount of a base reduces to how many protons are donated per molecule or ion of the acid and accepted per molecule or ion of the base. Therefore, one equivalent of an acid is defined as the amount of acid that donates under the actual titration conditions one mole of protons. Analogously, one equivalent of a base is defined as the amount of base that under the actual titration condition accepts one mole of protons. Then, with regard to titrations, the equivalent weight EW of an acid is obtained by dividing the formula weight FW by the equivalence number f:

$$EW = \frac{FW}{f} \tag{15-1}$$

The equivalence number denotes the moles of protons donated per mole of acid or accepted per mole of base and *actually transferred under the conditions of the titration.*

The italicized part of the preceding sentence is extremely important. It is *not* the number of protons that have the *potential* for being donated or accepted, but the number that *are* actually donated or accepted. This definition of equivalence may be different from what you have learned in previous courses, which might have been more convenient for the purposes then at hand. For titrimetry, however, the definitions given above are the most suitable. The equivalence number will be an integer, or in rare cases, a ratio of integers. An equivalence number, in the sense here used, is not an experimentally measured value, and hence it will have no influence on how many significant figures should be retained in the presentation of the results. The equivalence number is dimensionless, but depending on whether the reference is to an acid or a base, it has the units "moles of protons donated per mole of acid" or "moles of protons accepted per mole of base."

As previously shown, the amount of a substance in moles is obtained by dividing the grams of substance, g, by the formula weight:

$$\text{moles} = \frac{\text{g}}{\text{FW}} \tag{15-2}$$

Likewise, the amount of a substance in equivalents is obtained by dividing the grams by the equivalent weight:

$$\text{eq} = \frac{\text{g}}{\text{EW}} \tag{15-3}$$

Combining equations (15-1) and (15-3) yields

$$\text{eq} = \frac{\text{g}}{\text{FW}/f} \tag{15-4}$$

The definition of formality is

$$F = \frac{\text{mol}}{\text{L}} \tag{15-5}$$

Combining this equation with equation (15-2) gives

$$F = \frac{\text{g}}{\text{FW} \times \text{L}} \tag{15-6}$$

Substituting FW from equation (15-4) yields

$$F = \frac{\text{eq}}{f \times \text{L}} \tag{15-7}$$

Rearranging the last equation gives

$$\text{L} \times F \times f = \text{eq} \tag{15-8}$$

The equivalent is usually too large a unit, and one-thousandth of it, that is, the milliequivalent, meq, is commonly used. Dividing both sides of equation (15-8) by 1000 yields

$$\text{mL} \times F \times f = \text{meq} \tag{15-9}$$

The last two equations show the relation between the volume and the formality of a solution and the equivalents or milliequivalents of reacting substance provided.

The concentration of a solution can also be expressed in equivalents per liter of solution, termed the normality of the solution, and denoted N. Hence, $F \times f = N$. Although the concept of normality has merit in certain cases of practical titrimetry, it is not stressed in this textbook because normality in general does not have a fixed value, depending instead on the conditions under which the titration is performed. For example, phosphoric acid can react with hydroxide ion (added as NaOH) in the following three ways:

$$H_3PO_4 + OH^- \rightarrow H_2PO_4^- + H_2O \qquad N = F \times 1$$

$$H_3PO_4 + 2OH^- \rightarrow HPO_4^{2-} + 2H_2O \qquad N = F \times 2$$

$$H_3PO_4 + 3OH^- \rightarrow PO_4^{3-} + 3H_2O \qquad N = F \times 3$$

As indicated at the right of the reaction equations, the normality of a phosphoric acid solution of given formality can have three different values. The most common titrants in acid-base titrimetry—HCl, $HClO_4$, HNO_3, NaOH, and KOH—are monofunctional, and hence their equivalence number is always unity. Only two common acid-base titrants, H_2SO_4 and $Ba(OH)_2$, have an equivalence number of 2 and *always* 2, since for aqueous media their two constants (respectively, acidity and basicity) are not sufficiently separated to allow a titration to two distinct end points.

At the equivalence point, the equivalents of titrant added exactly equal the equivalents of substance titrated. If the subscripts t and s are used to refer to titrant and substance titrated, this condition can be expressed as

$$\left.\begin{aligned} \text{eq}_t &= \text{eq}_s \qquad (15\text{-}10a) \\ \text{meq}_t &= \text{meq}_s \qquad (15\text{-}10b) \end{aligned}\right\} \text{ at the equivalence point only}$$

Combining (15-10b) with (15-9) yields the first fundamental equation for titrimetry:

$$\text{mL}_t \times F_t \times f_t = \text{mL}_s \times F_s \times f_s \qquad (15\text{-}11)$$

The right-hand side of equation (15-11) is the milliequivalents of substance titrated, and can be expressed by

$$\text{meq}_s = \frac{\text{mg}_s}{\text{FW}_s/f_s}$$

Substituting this term yields

$$\text{mL}_t \times F_t \times f_t = \frac{\text{mg}_s}{\text{FW}_s/f_s} \qquad (15\text{-}12)$$

On rearrangement, the second fundamental equation is obtained:

$$\text{mL}_t \times F_t \times f_t \times \frac{\text{FW}_s}{f_s} = \text{mg}_s \qquad (15\text{-}13)$$

This expression relates volume and concentration of the titrant solution with the formula weight, equivalence number, and amount of substance titrated. Again, the relation holds *only* at the equivalence point. A dimensional analysis applied to equation (15-13) is of interest:

$$\mathrm{mL}_t \times F_t \times f_t \times \frac{\mathrm{FW}_s}{f_s} = \mathrm{mg}_s$$

$$\text{Volume} \times \frac{\text{mass}}{\text{volume}} \times \text{none} \times \frac{\text{none}}{\text{none}} = \text{mass}$$

Since "volume" cancels, the identity "mass = mass" shows dimensional satisfaction for the equation.

More practical is an analysis of the units of this equation:

$$\mathrm{mL} \times \frac{\mathrm{mol}_t}{\mathrm{L}} \times \overbrace{10^{-3}\,\frac{\mathrm{L}}{\mathrm{mL}}}^{\text{to convert liter to milliliter}}$$

$$\times \frac{\text{moles, protons donated (accepted)}}{\mathrm{mol}_t}$$

$$\times \frac{\mathrm{g}_s/\mathrm{mol}_s}{\text{moles, protons accepted (donated)}/\mathrm{mol}_s}$$

$$= 10^{-3} \times \mathrm{g}_s = \mathrm{mg}_s$$

Only the term $10^{-3} \times \mathrm{g}_s$ remains on the left, which corresponds to mg_s, proving the units to be correct. All other terms cancel, including the terms referring to the moles of protons accepted or donated. The last-mentioned two terms cancel, because the number of protons donated (or accepted) by the titrant *must* exactly equal the number of protons accepted (or donated) by the substance titrated.

Often it is not the amount of substances titrated that is to be reported, but its content in the sample material in percent. Obviously, this percentage is determined as follows, with W denoting the sample weight:

$$\frac{\mathrm{mg}_s \times 100}{W\,(\text{in mg})} = \%_s \tag{15-14}$$

Combining this expression with (15-13) yields

$$\frac{\mathrm{mL}_t \times F_t \times f_t \times \mathrm{FW}_s \times 100}{W\,(\text{in mg}) \times f_s} = \%_s \tag{15-15}$$

It is very important to realize that the sample weight must be expressed in milligrams.

Often when a substance cannot be determined by a direct titration, a back-titration can be successfully applied. Calculating the results is then based on the following formula:

$$(\text{mL}_t \times F_t \times f_t - \text{mL}_r \times F_r \times f_r) \times \frac{\text{FW}_s}{f_s} = \text{mg}_s \qquad (15\text{-}16)$$

where the subscript *r* refers to the back-titrant. Here the expression within parentheses corresponds to the milliequivalents of the titrant actually caused to react with the substance titrated, and is the difference between the milliequivalents of titrant added and the milliequivalents of back-titrant needed in the back-titration.

The analog to equation (15-15) is

$$\frac{(\text{mL}_t \times F_t \times f_t - \text{mL}_r \times F_r \times f_r) \times \text{FW}_s \times 100}{W\,(\text{in mg}) \times f_s} = \%_s \qquad (15\text{-}17)$$

Where the term FW_s appears in a formula, it refers to the substance considered to be titrated. Frequently, however, the result of the titration is expressed in terms of another substance. For example, phosphoric acid may be titrated, but the result is to be expressed in terms of P_2O_5. In such cases, an additional chemical factor is applied, or more than one.

Example 15-1 A standard solution of HCl, 0.1000*F*, is used to standardize a NaOH solution. A volume of exactly 25 mL of the NaOH solution is transferred to a beaker, some phenolphthalein indicator added, and the titration performed. For the end point to be reached, a volume of 26.03 mL of the acid is required. What is the formality of the NaOH solution? Here HCl is the titrant and NaOH the substance titrated. Applying equation (15-11) yields

$$\underbrace{\text{mL}_{\text{HCl}} \times \text{F}_{\text{HCl}} \times f_{\text{HCl}}}_{\text{meq}_{\text{HCl}}} = \underbrace{\text{mL}_{\text{NaOH}} \times \text{F}_{\text{NaOH}} \times f_{\text{NaOH}}}_{\text{meq}_{\text{NaOH}}}$$

Substituting the numerical values given above, recognizing that the equivalence number for both the acid and base is unity, yields

$$26.03 \times 0.1000 \times 1 = 25.00 \times F_{\text{NaOH}} \times 1$$

and

$$F_{\text{NaOH}} = \frac{26.03 \times 0.1000}{25.00} = 0.1041F$$

Since $f_{\text{NaOH}} = 1$, the normality of the base is 0.1041*N*.

Example 15-2 On the titration of 20.00 mL of a H_2SO_4 solution with 0.1030*F* NaOH to the phenolphthalein end point, a volume of 38.30 mL of the base is required. What is the formality of the H_2SO_4 solution?

The reasoning used in Example 15-1 is also relevant here and equation

(15–11) is applicable. The equivalence number of H_2SO_4, whatever the indicator used, is 2; hence

$$38.30 \times 0.1030 \times 1 = 20.00 \times F_s \times 2$$

and

$$F_{H_2SO_4} = \frac{38.30 \times 0.1030}{20.00 \times 2} = 0.0986F$$

Since $f_{H_2SO_4} = 2$, the acid is also $2 \times 0.0986 = 0.1972N$.

Example 15–3 A NaOH solution is to be standardized using potassium hydrogen phthalate (*204.14*), KHP, which is a monoprotic acid. This primary standard, in amount 0.8632 g, is dissolved in water and titrated with the NaOH solution. To reach the phenolphthalein end point, a volume of 38.64 mL of the base is required. What is the formality of the base?

Applying equation (15–13) is appropriate, with the subscripts t and s referring to NaOH and KHP, as the titrant and substance titrated. The equivalence number of both NaOH and KHP is unity. Substituting the numerical data yields

$$38.64 \times F_t \times 1 \times \frac{204.14}{1} = 863.2$$

Solving for the formality yields

$$F_t = 0.1094F \text{ NaOH}$$

Since $f_{NaOH} = 1$, the base is also $0.1094N$.

Example 15–4 A nitric acid solution is to be standardized using Na_2CO_3 (*105.99*). A 0.1562-g sample of this standard is dissolved in water and titrated with the HNO_3 solution. To reach the methyl orange end point, a volume of 31.22 mL of the acid is needed. What is the formality of the HNO_3 solution?

Again equation (15–13) can be applied. For HNO_3, the equivalence number is unity; for Na_2CO_3, it is 2, since the ion CO_3^{2-} accepts two protons on titration to the methyl orange end point. Inserting the numerical data yields

$$31.22 \times F_t \times 1 \times \frac{105.99}{2} = 156.2$$

Solving for the formality yields

$$F_t = 0.0944F \text{ HNO}_3$$

Since $f_{HNO_3} = 1$, the acid is also $0.0944N$.

Example 15-5 A sample, weighing 0.1935 g, of a weak organic acid is titrated with 0.1034*F* NaOH to the phenolphthalein end point. The volume of base required is 29.44 mL. What is the equivalent weight of the acid? Applying equation (15-13)—with the recognition that FW_s/f_s is the equivalent weight of the substance titrated, here the acid—yields

$$29.44 \times 0.1024 \times 1 \times EW_s = 193.5$$

Solving gives

$$EW_s = 64.19$$

Example 15-6 A 0.5000-g sample of a material containing $CaCO_3$ (*100.09*) as the only active substance is dissolved in 50.00 mL of 0.1080*F* HCl, CO_2 is boiled off, and the excess of acid is back-titrated with 0.0904*F* NaOH, of which a volume of 4.22 mL is required. What is the percentage of $CaCO_3$ in the material?

To the sample was added 50.00 X 0.1080 X 1 = 5.400 meq of HCl. The excess of acid required 4.22 X 0.0904 X 1 = 0.381_5 meq of NaOH. The number of milliequivalents of HCl consumed by the $CaCO_3$ is 5.400 - 0.381_5 = 5.018_5 and that is also how many milliequivalents of $CaCO_3$ present. The equivalence number of $CaCO_3$ is 2, since two protons are accepted by one molecule of $CaCO_3$. (Write the reaction equation.) Consequently, the amount of $CaCO_3$ present is

$$5.018_5 \times \frac{100.09}{2} = 251.2 \text{ mg of } CaCO_3 \text{ in the sample weight taken}$$

The percentage is then given by

$$\frac{251.2 \times 100}{500.0} = 50.23\% \ CaCO_3$$

Note that the sample weight has to be expressed in milligrams because the weight of the $CaCO_3$ is in milligrams. The entire calculation can be presented in an integrated way according to equation (15-17):

$$\frac{(\overbrace{\overbrace{50.00 \times 0.1080 \times 1}^{\text{meq acid added}} - \overbrace{4.22 \times 0.0904 \times 1}^{\text{meq base for back-titration}}}^{\text{meq acid reacting with } CaCO_3 = \text{meq of } CaCO_3}) \times \overset{\text{formula wt of } CaCO_3}{\overset{\downarrow}{100.09}} \times \overset{\text{conversion factor for \%}}{\overset{\downarrow}{100}}}{\underset{\text{sample wt in mg}}{\underset{\uparrow}{500.0}} \times \underset{f_{CaCO_3}}{\underset{\uparrow}{2}}}$$

$$= 50.23\% \ CaCO_3$$

Example 15-7 A 2.52-g sample of a mixture containing Na_2CO_3 (*105.99*), $NaHCO_3$ (*84.01*), and inert material is dissolved in water and diluted with water to exactly 250 mL. A 50.00-mL aliquot of this solution is titrated with 0.1020*F* HCl to the phenolphthalein end point; a volume of 16.44 mL is required. Next a 25.00-mL aliquot is titrated with the same acid to the methyl orange end point; a volume of 22.46 mL is required. Calculate the percentage Na_2CO_3 and the percentage $NaHCO_3$ in the mixture.

The principles involved in this determination of hydrogen carbonate and carbonate, when present together, have been considered in Section 14.7. In the first titration, the carbonate present is titrated to hydrogen carbonate ion and $f_{Na_2CO_3} = 1$. The calculation takes the following form:

$$\frac{\overbrace{16.44 \times 0.1020}^{\text{meq acid added}} \times 1 \times \underset{\substack{\text{formula wt of} \\ Na_2CO_3}}{105.99} \times \underset{\substack{\text{conversion} \\ \text{factor for \%}}}{100} \times \underset{\substack{\text{only } \frac{1}{5} \text{ of total} \\ \text{sample titrated}}}{5}}{\underset{\text{sample wt in mg}}{2520.3} \times \underset{f_{Na_2CO_3}}{1}} = 35.26\%\ Na_2CO_3$$

For the second titration, the calculation for the hydrogen carbonate percentage is based on the difference in the titration to the methyl orange end point and to the phenolphthalein end point. The calculation takes the following form:

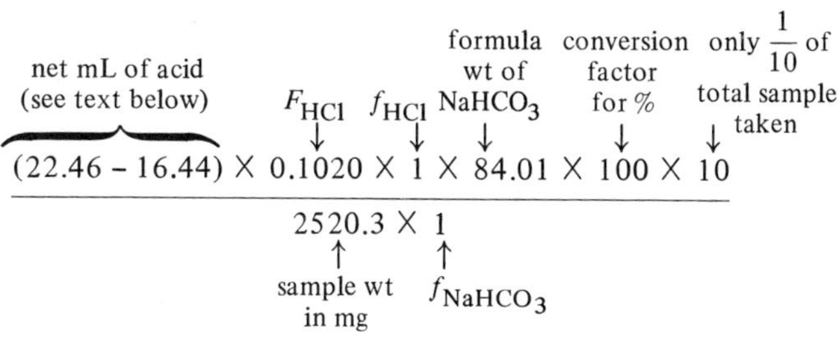

$$= 20.5\%\ NaHCO_3$$

The acid required for reaching the methyl orange end point in the second titration is consumed in three processes:

1. The acid converts all carbonate present to hydrogen carbonate ion. In the first titration, it was established that for a 50.00-mL aliquot a volume of 16.44 mL of acid is required for this purpose; hence, for a 25.00-mL aliquot, a volume of $0.5 \times 16.44 = 8.22$ mL.

2. The acid converts the hydrogen carbonate ion thus formed to carbonic acid. In the 25-mL aliquot, obviously a volume of 8.22 mL of acid is required for this conversion of the hydrogen carbonate that was originally present as carbonate. In other words, to completely convert the carbonate present in a 25.00-mL aliquot to carbonic acid, a volume of $2 \times 8.22 = 16.44$ mL is required.

3. The acid converts the hydrogen carbonate initially present to carbonic acid, and for this purpose a volume of 22.46 - 16.44 mL of acid is required for the 25.00-mL aliquot. This reasoning explains the factor labeled "net mL of acid" in the above calculation.

15.2 Indirect Analysis and Acid-Base Titrations

Indirect analysis, discussed in Section 7.3 for gravimetric methods, is also possible with titrimetric methods. The principles of calculation are like those previously outlined and are readily understood from an example.

Example 15-8 A mixture is known to contain only calcium carbonate (*100.09*) and magnesium carbonate (*84.32*). A 0.4000-g sample of this mixture is treated with 50.00 mL of 0.2000*F* HCl, and the solution is boiled to expel CO_2; then the excess of acid is back-titrated with 0.1000*F* NaOH to the phenolphthalein end point, requiring 12.00 mL of base. Calculate the percentage $CaCO_3$ and the percentage $MgCO_3$ in the mixture.

Let x and y represent the % $CaCO_3$ and % $MgCO_3$ in the mixture. The first equation of the required pair follows from the description of the mixture:

$$x + y = 100.00$$

The result of the titration of total carbonate content in the sample can be expressed in terms of $CaCO_3$:

$$\frac{(50.00 \times 0.2000 - 12.00 \times 0.1000) \times 100.09 \times 100}{400.0 \times 2}$$

$$= 110.10\% \text{ total carbonate expressed as } CaCO_3$$

Finding a percentage over 100% is associated with expressing the result in terms of the substance with the *higher* equivalent weight, $CaCO_3$. (The example can be solved alternatively by expressing the result as $MgCO_3$, an exercise you should undertake yourself.)

The second equation of the required pair is then based on expressing $MgCO_3$ in terms of $CaCO_3$ and applying the titration result:

$$x + y\,\frac{FW_{CaCO_3}}{FW_{MgCO_3}} = \%\text{ total carbonate expressed as } CaCO_3$$

$$x + 1.1870y = 110.10$$

Solving the pair of simultaneous equations by subtracting one from the other yields

$$0.1870y = 10.10$$

$$y = 54.01 = 54.0\%\ MgCO_3$$

$$x = 100.00 - 54.0 = 46.0\%\ CaCO_3$$

Because of the limited accuracy of indirect analysis, it is best to report 54% $MgCO_3$ and 46% $CaCO_3$.

The discussion of the advantages and disadvantages of indirect analysis given in Section 7.3 is also relevant to titrations and needs no further elaboration. As an exercise and for understanding better the limitations of indirect analysis, you can recalculate Example 15-8 with the assumption that an error of +0.10 mL was made in establishing the volume of acid needed and an error of -0.10 mL in the base required.

15.3 Problems

15-1 A 0.8326-g sample of potassium hydrogen phthalate (*204.23*), KHP, is titrated with an NaOH solution, of which a volume of 33.33 mL is required. Calculate the formality of the base. *Ans.:* 0.1223*F*

15-2 To standardize an HCl solution, a volume of 25.00 mL of the acid is titrated with 0.1035*F* NaOH, of which a volume of 24.30 mL is required. What is the formality of the acid? *Ans.:* 0.1006*F*

15-3 A 50.00-mL volume of a solution of Na_2CO_3 (*105.99*) is titrated to the methyl orange end point with 0.1000*F* HCl. A volume of 8.20 mL of this acid is required. Calculate the concentration of the Na_2CO_3 solution expressed as milligrams per liter. *Ans.:* 869 mg/L

15-4 A 10.00-mL sample of aqueous ammonia is titrated with 0.0488*F* H_2SO_4, of which a volume of 7.35 mL is required. What amount, in milligrams, of nitrogen (*14.01*) is present in the volume of the base taken? *Ans.:* 10.0_5 mg of nitrogen

15-5 A 0.5000-g sample of a material containing $CaCO_3$ (*100.09*) as the only active ingredient is treated with 50.00 mL of 0.0936*F* HNO_3. The excess acid is then back-titrated with 0.1045*F* NaOH, of which a volume of 4.25 mL is required. What is the percentage $CaCO_3$ in the material? *Ans.:* 42.40%

15-6 A solution was prepared by mixing 25.00 mL of 0.0882*F* H_3PO_4 and 20.00 mL of 0.1030*F* NaH_2PO_4. What volume of 0.1130*F* NaOH is required to titrate this solution (a) to the methyl orange end point and (b) to the phenolphthalein end point? *Ans.:* (a) 19.51 mL; (b) 57.26 mL

15-7 To 50.00 mL of a HCl solution, an excess of $AgNO_3$ solution is added. The AgCl (*143.32*) formed is separated, dried, and found to weigh 0.1205 g. What is the formality of the acid? *Ans.:* 0.01682*F*

15-8 An 0.6030-g sample of a material containing Na_3PO_4 as the only active component is titrated to the methyl orange end point with 0.1035*F* $HClO_4$, of which a volume of 12.24 mL is required. (a) What is the percentage P_2O_5 (*141.94*) in the material? (b) What volume of an H_2SO_4 solution of the same formality as the $HClO_4$ solution would be needed for the same titration? *Ans.:* (a) 7.46%; (b) 6.12 mL

15-9 Pure $BaCO_3$ (*197.35*) is used for standardizing an HCl solution in the following way. A 0.8320-g sample of the carbonate is dissolved in 50.00 mL of the acid. The solution is boiled and cooled. Then methyl orange indicator is added and the excess of acid back-titrated with a NaOH solution, of which a volume of 5.25 mL is needed. Then a 10.00-mL portion of the acid is added to the solution and titrated with the same base, of which now a volume of 12.05 mL is required. Calculate the formality of the HCl solution.
Ans.: 0.1847*F*

15-10 A 0.3500-g sample of a material containing Na_2CO_3 (*105.99*) and $NaHCO_3$ (*84.01*) as the only reactive substances is dissolved in water, phenolphthalein is added, and a titration performed with 0.0955*F* HCl, of which a volume of 25.40 mL is required. Methyl orange indicator is now added, the buret is refilled, and a further titration performed to the methyl orange end point, now requiring 34.60 mL of the acid. Calculate the percentage Na_2CO_3 and the percentage $NaHCO_3$ in the material.
Ans.: 73.46% Na_2CO_3, 21.09% $NaHCO_3$

15-11 A material is known to contain only $CaCO_3$ (*100.1*) and $MgCO_3$ (*84.3*). A 0.500-g sample of this material is dissolved in 60.00 mL of 0.200*F* HCl; then CO_2 is expelled by boiling, and the excess of acid back-titrated with 0.100*F* NaOH to the phenolphthalein end point, requiring 10.48 mL of the base. Calculate the percentage $CaCO_3$ and the percentage $MgCO_3$ in the material. *Ans.:* 49% $CaCO_3$, 51% $MgCO_3$

15-12 The weights of potassium hydrogen phthalate (*204.23*), KHP, given below required the listed volumes of NaOH solutions to reach the phenolphthalein end point. Calculate the formality of each of the NaOH solutions.

	Grams, KHP taken	mL, NaOH solution required
(a)	1.0000	45.64
(b)	1.1256	39.25
(c)	0.8567	28.75
(d)	0.8169	40.00
(e)	0.0679	8.96
(f)	1.1212	35.62

15-13 The monoprotic weak base 2-amino-2-(hydroxymethyl)-1,3-propanediol (*121.14*), $(HOCH_2)_3 \cdot NH_2$, sometimes termed *tris*, is used as a standard for acid solutions (see Section 14.2). Calculate the formalities of the HCl solutions that are standardized by titration of the amounts of tris given below.

	Grams, tris	mL, HCl solution required
(a)	0.4592	32.67
(b)	0.5115	40.11
(c)	0.4179	38.67
(d)	0.3992	45.95

15-14 A 1.0000-g sample of Na_2CO_3 (*105.99*) is caused to react with 65.00 mL of HCl solution, and the excess of HCl is back-titrated with a NaOH solution, of which 24.62 mL is required. In a separate titration, 25.00 mL of the

HCl solution required 27.85 mL of the NaOH solution. Calculate the formalities of both the solutions, HCl and NaOH.

15–15 An amount of 0.1523 g of calcium carbonate (*100.09*) of 99.50% w/w purity is treated with 50.00 mL of a nitric acid solution and the excess acid is back-titrated; 18.03 mL of a barium hydroxide solution is required. A volume of 24.34 mL of the nitric acid is equivalent to 29.94 mL of the base. Calculate the formalities of the acid and the base.

15–16 What volume of 0.1000*F* HCl is required to titrate 0.2000 g of a material that contains 34.2% w/w $Ba(OH)_2$ (*171.35*) and 65.8% w/w NaOH (*40.00*)?

15–17 What volume of 0.2000*F* HCl is required to titrate 0.4000 g of a mixture containing 30.0% w/w NaOH (*40.00*) and 53.0% w/w Na_2CO_3 (*106.0*) to the phenolphthalein end point?

15–18 A volume of 20.00 mL of a solution containing 1.600 g of NaOH (*40.00*) per 100 mL and a volume of 50.00 mL of a solution containing 0.530 g of Na_2CO_3 (*106.0*) per 100 mL are mixed in a 250-mL volumetric flask and diluted to the mark with pure water. A one-fifth aliquot of the solution is titrated to the methyl orange end point with 0.1000*F* HCl. What volume of HCl solution is required?

15–19 What volume of 0.1000*F* HCl will be required for titrating 200.0 mg of a material, containing 60.0% w/w $NaHCO_3$ (*84.01*) and 40.0% w/w Na_2CO_3 (*106.0*), to the phenolphthalein end point?

15–20 A material contains 35.0% w/w $NaHCO_3$ (*84.01*), 45.0% w/w Na_2CO_3 (*106.0*), and inert material. What volume of 0.1500*F* HCl will be required to titrate a 1.000-g sample to the methyl orange end point?

15–21 A mixture contains 60.0% w/w $NaHCO_3$ (*84.01*) and 40.0% w/w Na_2CO_3 (*106.0*). What volume of 0.1200*F* HCl will be required to titrate a 1.000-g sample to the methyl orange end point?

15–22 A material contains only NaOH (*40.00*) and Na_2CO_3 (*106.0*). An amount of 0.4240 g of the material requires 93.00 mL of 0.1000*F* HCl for the methyl red end point to be reached. Calculate the % w/w composition of the material.

15–23 A mixture is known to contain only NaOH (*40.00*), Na_2CO_3 (*106.0*), and inert material. An amount of 0.5000 g is titrated with 0.1230*F* HCl. A volume of 28.32 mL is required to reach the phenolphthalein end point and an additional volume of 22.04 mL to reach the methyl orange end point. What is the % w/w composition of the mixture?

15–24 A mixture contains only $CaCO_3$ (*100.09*) and CaO (*56.08*). A 0.5000-g sample is dissolved in 118.52 mL of 0.1000*F* HCl and the excess of acid back-titrated to the methyl orange end point; 4.26 mL of 0.2000*F* NaOH is required. Calculate the % w/w composition of the mixture.

15–25 A material in the amount of 0.5000 g, containing only $CaCO_3$ (*100.1*), $MgCO_3$ (*84.32*), and some acid-insoluble material, is dissolved in 80.00 mL of 0.1000*F* HCl. The excess of the acid is back-titrated, and 7.50 mL of 0.2000*F* NaOH is required. The insoluble material is separated by filtration, washed, dried, and found to weigh 195.0 mg. Calculate the % w/w composition of material.

15-26 An amount of 1.000 g of a mixture containing Na_2CO_3 (*106.0*), $NaHCO_3$ (*84.01*), and inert material requires 74.12 mL of 0.1500*F* HCl to reach the methyl orange end point. A second 1.000-g sample is ignited to 550°C, and a loss of 110.7 mg in weight is observed. Calculate the % w/w composition of the mixture.

15-27 A solution contains both H_3PO_4 (*98.00*) and H_2SO_4 (*98.08*). When a 25.00-mL portion of the solution is titrated to the phenolphthalein end point, a volume of 38.32 mL of 0.1000*F* NaOH is required. A 10.00-mL portion requires 12.21 mL of the same NaOH solution to reach the end point. Calculate the grams per liter of each of orange methyl the two acids present in the solution.

15-28 In forensic analysis, the carbon monoxide content of blood can be determined as follows. In a dry test tube are mixed 2.00 mL of blood (having a few crystals of sodium fluoride added to prevent coagulation) and 1.00 mL of water. Exactly 2 mL of this mixture is pipetted into a Conway chamber. This small device has two separate compartments and when it is closed, a gaseous constituent can diffuse from one compartment to the other. The blood preparation and 0.25 mL of 2*F* sulfuric acid are placed in one compartment; in the other is placed 1.00 mL of 0.01*F* palladium chloride. The chamber is closed and allowed to stand in the dark for 3 hr. The carbon monoxide diffuses into the latter solution and the following reaction takes place

$$Pd^{2+} + H_2O + CO \rightarrow Pd + 2H^+ + CO_2$$

The chamber is then opened and the acid in the palladium solution is titrated with 0.0200*F* sodium hydroxide. A blank is performed under identical conditions but with 2.00 mL of water placed into the Conway chamber instead of the blood preparation. The following formula is used to calculate the result:

$$(\text{mL}_{\text{titrant in titration}} - \text{mL}_{\text{titrant in blank}}) \times (\text{factor})$$
$$= \text{vol. \% CO} \qquad \text{(gas volume at standard conditions)}$$

Calculate the numerical value of this *factor* for the determination performed exactly as described above. Carbon monoxide has a formula weight of 28.0 and is assumed to behave as an ideal gas with a mole occupying 22.4 L at exactly 0°C and 1.013×10^5 Pa (=1 atm) pressure. *Ans.:* 16.8

15-29 A 0.2500-g sample of a material containing calcium carbonate (*100.1*) as the only active ingredient is treated with 50.00 mL of 0.1000*F* HCl. After the reaction has been completed, the excess of acid is back-titrated with 0.0800*F* NaOH, of which 35.00 mL is needed. What is the percentage of $CaCO_3$ in the material. *Ans.:* 44.0%

15-30 The sulfur dioxide content of air can be determined as follows: Air is bubbled though a solution of H_2O_2, which traps the SO_2 and converts it to H_2SO_4; the solution is then titrated with a standard NaOH solution. Calculate the ppm w/w of SO_2 *(64.1)* in the air from the following data: Air at the rate of 5.0 L per min is passed through the H_2O_2 trap for 5.0 hr and the titration of the total solution requires 1.72 mL of 0.0100*F* NaOH. (A liter of air weighs 1.06 g under the experimental conditions.)

Ans.: 0.35 ppm w/w

16

Acid-Base Titrations in Nonaqueous Media

16.1 Introduction

So far our study of acid-base phenomena has been essentially restricted to aqueous media. Acid-base reactions, that is, proton-transfer reactions, take place however, in other solvents also. Liquid ammonia is an amphiprotic solvent, that is, it can function as either an acid or a base, and undergoes the following autoprotolytic reaction:

$$\underset{\text{acid}_1}{NH_3} + \underset{\text{base}_2}{NH_3} \rightleftharpoons \underset{\text{base}_1}{NH_2^-} + \underset{\text{acid}_2}{NH_4^+} \tag{16-1}$$

This scheme is completely analogous to the autoprotolysis of water. The ions formed are the amide ion and the ammonium ion. If an acid, HA, is dissolved in liquid ammonia, the protolysis reaction taking place is

$$HA + NH_3 \rightleftharpoons A^- + NH_4^+ \tag{16-2}$$

Ammonium ion is the strongest acid that can exist in liquid ammonia. It thus parallels the hydronium ion, that is, the hydrated hydrogen ion, in water. Because of the experimental difficulties of working in liquid ammonia, the system has not found application in analytical practice. But this does not detract from the importance of the analogy with water. Organic derivatives of ammonia, that is, amines, such as butylamine, $C_4H_9NH_2$, do find such application.

Another system, which has great significance, is water-free acetic acid. Its autoprotolysis proceeds according to

$$CH_3COOH + CH_3COOH \rightleftharpoons CH_3COO^- + CH_3COOH_2^+ \tag{16-3}$$

The acetate and acetonium ions formed are fully analogous to the hydroxide and the hydronium ions in water. If an acid, for example, perchloric acid, is dissolved in water-free acetic acid, the following protolytic reaction occurs:

$$HClO_4 + CH_3COOH \rightleftharpoons ClO_4^- + CH_3COOH_2^+ \qquad (16\text{-}4)$$

Note that in each of the three amphiprotic solvents just considered, the "active" acid species is the solvated proton.

The "dissociation" of an acid does not necessarily require an amphiprotic solvent. Such a reaction can take place in so-called aprotic or nonprotonic solvents that offer a pair of nonbonding electrons. The solvent *p*-dioxane is one:

$$HClO_4 + :O\!\!<\!\!\begin{matrix}CH_2-CH_2\\CH_2-CH_2\end{matrix}\!\!>\!\!O: \rightleftharpoons \left[:O\!\!<\!\!\begin{matrix}CH_2-CH_2\\CH_2-CH_2\end{matrix}\!\!>\!\!O{:}H^+, ClO_4^-\right] \qquad (16\text{-}5)$$

Dioxane has no proton to donate, but a pair of nonbonding electrons can accept a proton. Dioxane has quite a low dielectric constant and therefore does not support the separation of opposite charges. As a consequence, the dioxonium cation and the perchlorate anion by electrostatic attraction remain close together as an entity, known as an ion pair. In the ion pair, the transfer of the proton is complete, and the cation and anion with their separate charges persist as such. An ion pair also can form in amphiprotic solvents, such as acetic acid, but to a lesser degree.

16.2 *Leveling and Differentiating Effects*

Two factors govern the extent to which the protolysis of an acid proceeds (Section 8.1). The first is the "intrinsic" strength of the acid, that is, its inherent tendency to donate a proton. The second is the basicity of the solvent, that is, its tendency to accept a proton. If the basicity of the solvent is high, then acids with even relatively low intrinsic strengths are able to donate their protons fully to the solvent and consequently to behave as strong acids in that solvent. Water as a solvent has relatively high basicity, and perchloric acid, sulfuric acid (first step), hydrochloric acid, and nitric acid all have enough intrinsic strength to donate their protons completely to the water. Consequently, these four acids behave like strong acids in water and their strength is leveled to the strength of the hydronium ion, which is the strongest acid that can exist in water. This is known as the leveling effect, and water may be described as a leveling solvent for these acids.

Water-free acetic acid is a far less basic solvent than water; that is, its tendency to accept a proton is smaller. Nitric acid has much less intrinsic strength than perchloric acid, and its protolysis is not complete in acetic acid. Consequently, nitric acid in this solvent does not behave like a strong acid, but perchloric acid does. In acetic acid, the strength of the acids listed above decreases in the order in which they are named. Because acetic acid shows this differentiating effect, it can be described as a differentiating solvent for these acids.

Analogous reasoning holds for other solvents and for bases. It is the leveling effect of water that prevents reaction (16-3) from taking place in aqueous media. Acetonium ion is a very strong acid, but it is leveled in water to the strength of the hydronium ion. Likewise, amide ion is a very strong base, which in water is leveled to the strength of the hydroxide ion.

16.3 Nonaqueous Acid-Base Titrimetry

From the preceding discussion, you can appreciate that proton-transfer reactions in nonaqueous media can be made the basis of titrimetric processes, and that such an expansion of acid-base titrimetry not only should supplement titrations in aqueous media but would provide distinctive advantages in some cases, including the following:

1. Acids or bases too weak to be titrable in water may be stronger in an appropriate nonaqueous solvent and may become titrable in that medium. Acetate, for example, has too low an intrinsic basicity for its protolysis with hydronium ion to go to completion. Therefore, in water acetate is too weak a base to be readily titrated with acid. In contrast, in the more acidic solvent water-free acetic acid, acetate accepts protons more completely from the much stronger acid acetonium ion, consequently acts as a stronger base, and can be titrated to a sharp end point, for example, with a solution of perchloric acid in acetic acid. Many other bases (per se or as the salts of weak acids) that cannot be titrated in water can be determined in acetic acid or other acidic solvents.
2. Acids or bases identical in strength in water can often be differentiated in a nonaqueous solvent of appropriate acidity or basicity. For example, perchloric acid and hydrochloric acid, due to the leveling effect of water, are both strong acids in that medium; thus the titration curve for a mixture of these two acids shows only a single break. However, in water-free acetic acid, which acts as a differentiating solvent, the mixture can be titrated with two breaks, using a strong base, for example, acetate ion (as sodium acetate) as the titrant.
3. Substances that are insoluble in water may be titrated in a solvent in which

they are soluble. For example, many water-insoluble organic compounds can thereby be subjected to simple titrimetric procedures.

Nonaqueous titrimetry has found special interest for determining pharmaceuticals. With the selection of suitable differentiating solvents, mixtures of several compounds can often be analyzed titrimetrically, avoiding involved and tedious separations.

Common solvents in nonaqueous acid-base titrimetry, besides acetic acid and dioxane, including acetonitrile, dimethylformamide, and the so-called G-H solvent mixtures. These mixtures contain a glycol (usually ethylene glycol) and a hydrocarbon solvent (such as a hydrocarbon, a halocarbon, or an alcohol), commonly in a 1:1 proportion.

The titration process proper is not significantly different in nonaqueous media and in aqueous media. The end point can be detected visually in exactly the same way—by using organic dyes that show acid-base properties and that change color when transformed from the acid form to the base form. The dyes methyl violet and crystal violet find frequent use as indicators in nonaqueous media. Potentiometric end-point detection with a glass electrode is possible in solvent systems having a high enough dielectric constant (see Section 27.3).

The exclusion of water is in many cases a critical factor for the success of a nonaqueous titration. Water, depending on the solvent, may act as an acid or a base, and then, of course, can upset the situation. Removing water from acetic acid media is readily achieved by adding acetic anhydride, which undergoes the following reaction:

$$(CH_3CO)_2O + H_2O \rightarrow 2CH_3COOH \qquad (16\text{-}6)$$

Special precautions are often necessary to avoid errors due to the high volatility and the large thermal expansion coefficient of the organic solvent. The standard solution may increase significantly in its concentration if losses of solvent by evaporation are not prevented. Temperature changes may cause significant changes in weight/volume concentration.

16.4 *Lewis Acid-Base Concepts*

Many organic reactions that are catalyzed by Brønsted acids are also catalyzed by species that have no protons that can be donated. Thus it seems that acid-base properties are not restricted to the transfer of protons. Lewis, in 1923, proposed a concept of acids and bases that was more general than Brønsted's. Consider as an example the base ammonia. Its acceptance of a proton in terms of electronic structure can be written

$$\begin{matrix} & \text{H} & \\ \text{H}: & \ddot{\underset{\cdot\cdot}{\text{N}}} & : \\ & \text{H} & \end{matrix} + H^+ \rightarrow \left[\begin{matrix} & \text{H} & \\ \text{H}: & \ddot{\underset{\cdot\cdot}{\text{N}}} & :\text{H} \\ & \text{H} & \end{matrix} \right]^+ \qquad (16\text{-}7)$$

The essential aspect is that ammonia, or any Brønsted base, shares its nonbonding pair of electrons with the proton it accepts [compare the analogous situation with dioxane, equation (16-5)]. The extension of the concept of a base effected by Lewis consists in removing the restriction that the sharing of a pair of electrons must be with a proton. The sharing may be with a proton but equally well with another species. An analog to the above reaction can be written with the copper(II) ion replacing the proton:

$$\begin{array}{c} \mathrm{H} \\ \mathrm{H{:}\ddot{\underset{\cdot\cdot}{N}}{:}} \\ \mathrm{H} \end{array} + \mathrm{Cu^{2+}} \rightarrow \left[\begin{array}{c} \mathrm{H} \\ \mathrm{H{:}\ddot{\underset{\cdot\cdot}{N}}{:}Cu} \\ \mathrm{H} \end{array}\right]^{2+} \tag{16-8}$$

This reaction is commonly termed complex formation, but by the Lewis theory is viewed only as one type of acid-base reaction. According to this theory, the formation of any coordinate covalent bond is an acid-base reaction, a neutralization. Since those species that have an unshared pair of electrons can combine with a proton and also with other electron acceptors, the number of bases according to Lewis is not greatly altered. However, the number of substances that are considered to be acids is greatly increased. Any substance that can accept a pair of electrons is termed a Lewis acid. Thus, for example, boron trifluoride is a Lewis acid that reacts with ammonia as follows:

$$\begin{array}{ccccccc} & \mathrm{F} & & \mathrm{H} & & \mathrm{F} & \mathrm{H} \\ & | & & | & & | & | \\ \mathrm{F{-}} & \mathrm{B} & + & \mathrm{{:}N{-}H} & \rightarrow \mathrm{F{-}} & \mathrm{B{:}} & \mathrm{N{-}H} \\ & | & & | & & | & | \\ & \mathrm{F} & & \mathrm{H} & & \mathrm{F} & \mathrm{H} \end{array} \tag{16-9}$$

Lewis concepts are rarely used in theoretical considerations of inorganic analysis because almost all relevant phenomena can be described satisfactorily either with the Brønsted concepts or as complex formation (Chapter 20). Of the few exceptions, acid-base phenomena in fused media may be mentioned; these are briefly considered in Chapter 51.

16.5 Questions

16-1 Speculate on the reasons why perchloric acid in a suitable solvent is the most widely applied titrant solution in nonaqueous titrimetry.

16-2 Differentiating solvents are also called nonleveling solvents. Is this term appropriate? Explain.

16-3 What is meant by the statement: Water levels $HClO_4$ and HCl?

16-4 What are the properties required by a solvent to be designated as differentiating?

16–5 Why is it necessary with some solvents to exclude water strictly when titrations are performed?

16–6 What is the strongest acid that can exist in liquid ammonia?

16–7 What is a G-H solvent?

16–8 Explain why many organic substances that are excellent indicators in aqueous media are unsuitable as indicators for titrations in organic solvent.

16–9 Substances A, B, and C are assumed to behave in water and the nonaqueous solvents 1 and 2 as indicated by the table below.

	Water	Solvent 1	Solvent 2
A	$K_b = 10^{-12}$	Medium strong base	Very weak base
B	$K_b = 10^{-4}$	Very weak base	Medium-strong base
C	$K_b = 10^{-10}$	Very weak base	Strong base

Elaborate on a scheme to determine these substances in their admixture solely by titrimetry.

16–10 To titrate a weak base successfully, the solvent should, at most, have only weakly basic properties. Explain.

17

Review Questions and Problems on Acid-Base Titrimetry

17.1 *Questions*

17-1 Is an aqueous solution that is "neutral" to methyl orange actually neutral in the sense of having equal concentrations of hydrogen ion and hydroxide ion?

17-2 Elaborate on the statement: The equivalent weight of a substance is independent of the titration reaction.

17-3 How could you standardize a hydrochloric acid solution gravimetrically?

17-4 What would be the effect on the result if pH 7.0 were taken as the end point in the titration of a weak base with a strong acid?

17-5 What anionic species will predominate in a solution obtained by dissolving 5 mmol of Na_2HPO_4 and 3 mmol of Na_3PO_4 in water?

17-6 What points on an acid-base titration curve should be calculated using approximate formulas to permit a rough, rapid judgment of the feasibility of the titration?

17-7 Could the sodium salt of a very weak acid ($K_a = 10^{-11}$) be used reasonably well as a titrant in titrating a strong acid?

17-8 What is stated by the electroneutrality rule and what are the possible applications of the "charge balance"?

17-9 A very weak monoprotic organic acid ($K_a = 1 \times 10^{-12}$) forms a water-insoluble silver salt. How might an acid-base titration of this acid be achieved?

17-10 Discuss the interrelation of formality, molarity, and normality of a titrant. Elaborate on the meaning of *equivalence* and *equivalent weight* in acid-base titrimetry.

17-11 Differentiate between the concepts of the equivalence point and the end point in broad terms and in relation to acid-base titrimetry. What is the most desirable relation between these two points?

17–12 Why are weak acids and bases hardly ever used as titrants in acid-base titrimetry?

17–13 Using your knowledge of the color transition interval of an acid-base indicator, explain why a factor of at least 10^3 must separate the dissociation constants of a diprotic acid so that its visual titration of two distinct end points is possible.

17–14 How many useful visual end points can be secured in titrating a triprotic acid with the successive acidity constants 1×10^{-4}, 1×10^{-6}, and 1×10^{-12}? In practice, to what pH should the titration be made in an aqueous medium? *Ans.:* pH 9.0. Explain.

17–15 Give reasons why calcium carbonate, although readily obtainable in a highly pure form, is not a very suitable standard for standardizing an acid.

17–16 What is temporary water hardness? How is it determined? What is its significance for industry?

17–17 Explain why perchloric acid is preferred as the titrant in many acid-base titrations in nonaqueous media.

17–18 List some errors that may be encountered in an acid-base titration. Which errors may be related to the selection of the indicator and its functioning?

17–19 Discuss the functioning of an acid-base indicator.

17–20 A carbonate-containing sodium hydroxide solution was standardized using phenolphthalein as the indicator. Explain why incorrect results are obtained if this solution is then used in a titration in which methyl orange is the indicator. Would applying an indicator blank remedy the situation?

17–21 What can be concluded from the presence in a titration curve of an acid or a base of (a) a nearly horizontal portion and (b) a nearly vertical portion?

17–22 Chloride ion is a base. Why does this base added in the form of NaCl, for example, not interfere with an acid-base titration?

17–23 A weighed sample containing only two components is given. Compare the indirect analysis of the mixture by gravimetry and by acid-base titrimetry. (For example, what parameters must be considered and what are the best conditions?)

17–24 Discuss the statement: The titration of substance A can never yield a correct result since the pH at the end point does not coincide with the pH at the equivalence point.

17–25 Predict how an aqueous solution of sodium carbonate will react to phenolphthalein. When this solution is treated with an excess of barium chloride and the precipitate separated by filtration, how will the filtrate react toward the same indicator? What will be the result if a solution of sodium carbonate is treated with excess barium chloride and then titrated with hydrochloric acid to the methyl orange end point (a) with and (b) without removal of the precipitate? Write equations for all the reactions taking place.

17-26 The equivalence point in an otherwise satisfactory titration of an acid falls within the range of the high buffering action of another system that is present in the aqueous solution. Consider the feasibility of the titration under this condition.

17-27 Explain why, in determining the carbonate content of calcium carbonate by back-titration of an added excess amount of standard acid, it is advisable to boil the solution to expel carbon dioxide even when methyl orange is used as the indicator. Could phenolphthalein be substituted as the indicator in this back-titration? State how the titration results with each indicator would be influenced if the carbon dioxide were not expelled fully.

17-28 In an aqueous solution of H_2SO_4 also containing $NaNO_3$ and KCl, the sulfate content is determined both gravimetrically and by an acid-base titration. Compare and discuss the two methods with regard to (a) speed, (b) the possible influence of the neutral salts, and (c) how using a balance enters each determination. State what reagents are needed for each of the methods and suggest how they might be prepared. Which set of reagents can be prepared more rapidly? List the laboratory equipment required for the two methods.

17.2 Problems

17-1 The monoequivalent base 2-amino-2-(hydroxymethyl)-1,3-propanediol (that is, tris) (*121.14*) is a primary standard for acids. In a standardization, 306.2 mg of this base required 25.32 mL of an H_2SO_4 solution. What is the formality of this solution? *Ans.:* 0.04991*F*

17-2 In the titration of 50.00 mL of a solution of H_3PO_4 (*98.00*) with 0.1026*F* NaOH to the "second" end point (thymol blue), a volume of 48.30 mL of the base is required. (a) If the titration of the same volume of the H_3PO_4 solution were performed to the "first" end point, what volume of base would be required? (b) What is the formality of the H_3PO_4 solution? (c) What weight, in grams, of H_3PO_4 is present in 50.00 mL of the solution? (d) What weight, in grams, of silver ion (*107.87*) is equivalent to the phosphate present in 50.00 mL of the solution if silver is precipitated as Ag_3PO_4?

Ans.: (a) 24.15 mL; (b) 0.04956*F*; (c) 0.2428 g; (d) 0.8018 g

17-3 In water, the pK_a of HCl has a value about –3. Calculate K_a for HCl and K_b for its conjugate base, chloride ion. *Ans.:* $K_a \cong 10^3, K_b \cong 10^{-17}$

17-4 A volume of 50.00 mL of an NH_4NO_3 (*80.04*) solution is treated with concentrated NaOH solution, and the ammonia is distilled into 20.00 mL of an HCl solution. The excess of HCl is back-titrated with 0.1010*F* NaOH, of which a volume of 6.22 mL is required. In a separate titration, a volume of 28.66 mL of the HCl solution is found to be equivalent to

24.50 mL of the standard base. What weight of NH_4NO_3 is present in the sample taken? *Ans.:* 87.9 mg

17-5 In Problem 17–4, what volume of the acid would be needed to exactly neutralize the total ammonia distilled if the nitrate ion in another 50.00-mL portion of the solution were reduced to ammonia by the action of Devarda's alloy? *Ans.:* 25.4 mL

17-6 A 1.000-g sample of a uniform mixture known to contain only Na_2CO_3 (*106.0*) and $NaHCO_3$ (*84.0*) is titrated with 0.2000*F* HCl to the methyl orange end point, requiring 73.45 mL of the acid. Calculate the percentage composition of the mixture. (*Hint:* When formulating the simultaneous equations, remember that two HCO_3^- are equivalent to one CO_3^{2-}.)
Ans.: 40% Na_2CO_3, 60% $NaHCO_3$

17-7 To a 0.1232-g sample of MgO (*40.32*) known to be of 95.66% purity (the remainder is inert material) is added 50.00 mL of a HNO_3 solution. The excess of the acid is back-titrated with 0.0987*F* NaOH, of which a volume of 14.40 mL is required. Calculate the formality of the HNO_3 solution. *Ans.:* 0.1453*F*

17-8 A volume of 50.00 mL of a HCl solution is treated with $AgNO_3$ in excess and the precipitate is separated by filtration and dried. The weight of the dried AgCl (*143.32*) produced is 0.7225 g. Calculate the formality of the HCl solution. *Ans.:* 0.1008*F*

17-9 A material contains 65.0% K_2HPO_4 (*174.2*) and 35.0% Na_3PO_4 (*164.0*). How many milliliters of 0.1000*F* HCl will be needed for titrating 0.7000 g of this material to the methyl orange end point? *Ans.:* 56.0 mL

17-10 A volume of 100.0 mL of a 0.1000*F* solution of a triprotic acid ($K_{a,1} = 1.0 \times 10^{-4}$, $K_{a,2} = 1.0 \times 10^{-6}$, $K_{a,3} = 3.0 \times 10^{-8}$) is titrated with 0.1000*F* NaOH. (a) Calculate the pH value of the three equivalence points (use the proper solution volume in calculating C_{salt}). (b) Which of these equivalence points would be used in practice and what volume of base would be needed for reaching that point? (c) When the solution is titrated as in (b) and then 150 mL of 0.100*F* HCl is added, what will be the pH of the resulting solution?
Ans.: (a) pH 5.00, 6.76, 9.96; (b) the last one, 300 mL; (c) pH 6.00

17-11 A solution is made by mixing 20.00 mL of 0.1000*F* NaH_2PO_4 and 30.00 mL of 0.0500*F* Na_2HPO_4. (a) How many milliliters of 0.1000*F* NaOH will be required to titrate this solution to the second equivalence point of H_3PO_4? (b) How many milliliters of 0.1000*F* HCl will be required to bring the solution resulting from this titration to the pH of the first equivalence point of H_3PO_4? *Ans.:* (a) 20.00 mL; (b) 35.00 mL

17-12 How many milliliters of 0.10*F* $Ba(OH)_2$ must be added to 50 mL of 0.10*F* formic acid ($K_a = 1.8 \times 10^{-4}$) to attain pH 3.30? Would the resulting solution exert a reasonable buffering action? *Ans.:* 6.6 mL; yes

17-13 A volume of 20.00 mL of a solution containing 1.500 g of NaOH (*40.00*) per 100 mL and a volume of 50.00 mL of a solution containing 0.500 g of

Na_2CO_3 (*106.0*) per 100 mL are mixed in a 250-mL volumetric flask and diluted to mark with pure water. A one-fifth aliquot of the solution is titrated to the methyl orange end point with 0.1000*F* HCl. What volume of HCl solution is needed?

17-14 Calcium carbonate (*100.09*) of 99.50% w/w purity in the amount of 0.1523 g is treated with 50.00 mL of a nitric acid solution and the excess of acid is back-titrated; 18.03 mL of a barium hydroxide solution is needed. A volume of 24.34 mL of the nitric acid is equivalent to 29.94 mL of the base. Calculate the formalities of the acid and the base.

17-15 When Rochelle salt, $KNaC_4H_4O_6 \cdot 4H_2O$, is ignited, $KNaCO_3$ is formed. An 0.8050-g sample of Rochelle salt is ignited, and the residue dissolved in pure water; the resulting solution is titrated with 0.0635*F* H_2SO_4 to the methyl orange end point. What volume of titrant solution will be required if the original material was 98.55% w/w pure?

17-16 A solution is known to have been prepared by dissolving H_3PO_4 (*98.00*) and Na_3PO_4 (*163.94*) in pure water. This solution requires 40.00 mL of 0.100*F* HCl in the titration to the methyl orange end point. Then a neutral solution of $CaCl_2$ is added and the solution titrated to the phenolphthalein end point; 100.0 mL of 0.100*F* NaOH is required. Calculate the milligrams of H_3PO_4 and Na_3PO_4 present in the original solution.

17-17 A solution contains both H_3PO_4 (*98.00*) and H_2SO_4 (*98.08*). When a 25.00-mL portion of the solution is titrated to the methyl orange end point, a volume of 38.32 mL of 0.1000*F* NaOH is required. A 10.00-mL portion requires 21.21 mL of the same NaOH solution to reach the phenolphthalein end point. Calculate the grams per liter of each of the two acids present in the solution.

17-18 An amount of 0.0821 g of a material is dissolved, the solution made slightly acidic, and potassium precipitated as the tetraphenylborate salt. The precipitate is separated by filtration and dissolved on the filter by acetone, and the resulting solution collected in a platinum crucible. After evaporation of the acetone, the residue is ignited to potassium metaborate, KBO_2. The KBO_2 is dissolved in water and titrated with 0.01122*F* HCl (HBO_2 is a very weak acid), of which a volume of 24.07 mL is needed for reaching the methyl red end point. Calculate the % w/w of K_2O (*94.20*) in the sample.

17-19 A material contains as the only active component about 25% of a solid monoprotic acid of formula weight 220. The content of this acid is determined by titration with 0.100*F* NaOH. What minimum sample weight must be taken so that an error in reading the buret of ±0.06 mL will not cause more than a 0.5% relative error in the final result? *Ans.*: $2._1$ g

17-20 A volume of 30.00 mL of 0.2000*F* HCl is diluted with water to 250.0 mL. A 50.00-mL portion of this solution is titrated with a $Ba(OH)_2$ solution, and a volume of 60.00 mL of that solution is consumed. Calculate the formality of the base. *Ans.*: 0.01000*F*

17-21 The ammonia content of river water is to be determined. A 400.0-mL sample is adjusted to pH 9.5 with 1*F* NaOH and transferred to a Kjeldahl flask, and 25 mL of borate buffer is added. Then about 300 mL of this solution is distilled into a 500-mL conical flask containing 50 mL of 2% boric acid solution. Then 3 drops of an indicator solution of methyl red screened with methylene blue are added and the distillate is titrated with 0.0201*N* (= 0.01005*F*) H_2SO_4; a volume of 18.25 mL is required to reach the end point. Calculate the mg/L of NH_3 (*17.03*) in the sample.

17-22 *Total alkalinity* of surface waters is determined by titration with a standard solution of acid. A 50.0-mL sample of lake water is titrated, using bromocresol green–methyl red mixed indicator, with 0.0214*F* HCl; a volume of 21.22 mL is required. Calculate the total alkalinity expressed as milligrams of $CaCO_3$ (*100.09*) per liter of water.

17-23 *Acidity* of surface waters is determined by titration with a standard, carbon-dioxide-free NaOH solution. A 100.0-mL sample of river water is titrated to the phenolphthalein end point with 0.0197*F* NaOH; a volume of 32.18 mL is required. Calculate the acidity expressed as mg/L $CaCO_3$ (*100.09*). (*Note:* In this context it is understood that the value is the milligrams of $CaCO_3$ required to neutralize the acid present in 1 L of the water. In this way, the same expression is used both for acidity and total alkalinity (cf. Problem 17–22).

17-24 *Free* ammonia in urine can be determined by a formol titration (see Section 14.10). (A Kjeldahl determination fails, since in the alkaline digestion urea would be hydrolyzed to ammonia.) The specimen is initially boiled and filtered to coagulate and remove proteins, which would interfere. To a portion of the filtrate, some phenolphthalein is added, and then 0.1*F* NaOH until a light red color develops. In a second vessel, some phenolphthalein is added to 2 to 5 mL of 40% formaldehyde solution, and then 0.1*F* NaOH to the appearance of a pink color. The two solutions are mixed and a titration is performed with 0.1*F* NaOH to the reappearance of a pink color. In an actual determination, a 10.00-mL sample of urine required 8.75 mL of 0.1008*F* NaOH. Calculate the free ammonia (*17.0*) content in milligrams per 100 mL. *Ans.:* 150 mg/100 mL

17-25 The acidity of gastric juice is determined by titrating with carbonate-free 0.1*F* NaOH to the dimethyl yellow end point and then continuing to the phenolphthalein end point. The first end point corresponds to unbound strong acids (*free acid*), and the second additionally to weak acids and strong acids bound to proteins (*total acid*). The results are expressed as the milliliters of 0.1000*F* NaOH needed to neutralize 100 mL of filtered gastric contents (they are often reported without units, or as "degrees"). To a 3.00-mL centrifuged portion of an actual gastric specimen, dimethyl yellow and phenolphthalein were added and a titration was performed with 0.1000*F* NaOH. To reach the slight change from red to yellow, 5.20 mL was required; then on delivery of a total of 6.80 mL, the yellow solution developed a red tint. Calculate the free and total acids.

Ans.: Free acid, 173; total acid, 227

17-26 Fertilizers are described by three numbers that are, in order, the percentages of total nitrogen (as N), total "phosphoric acid" (as P_2O_5), and total "potash" (as K_2O). A premium lawn fertilizer, for example, might be labeled 23-7-7. For a water-soluble fertilizer with all nitrogen present as ammonium or nitrate, nitrogen was determined as follows. A 1.000-g sample was dissolved and diluted with water to 250 mL. A 50-mL aliquot with added water was treated with Devarda's alloy and sodium hydroxide solution. The ammonia formed was distilled and collected; and for its titration 28.6 mL of 0.1000*F* HCl was needed. (a) Calculate the percentage of nitrogen (*14.0*). (b) Assuming that the nitrogen comes from *equimolar* amounts of ammonium nitrate and potassium nitrate (*111.1*), calculate the potash content. *Ans.:* (a) 20% N; (b) 22% K_2O

17-27 An ester (R_aCOOR_b) forms from an acid (R_aCOOH) and an alcohol (R_bOH) with the loss of a molecule of water for each carboxylic acid (—COOH) and hydroxyl (—OH) group. Fats (and oils) are esters of the trihydric alcohol glycerol, $HOCH_2CHOHCH_2OH$, and have the general formula $RCOOCH(OCOR')CH_2COOR''$. Esters can be hydrolyzed to the corresponding acids and alcohols. For fats, this hydrolysis in the presence of "alkalis" (NaOH or KOH) leads to glycerol and the alkali metal salts of the higher fatty acids, that is, soaps. This classical process for soapmaking is called saponification (Latin, *sapo, -onis,* "soap"). The saponification value, a valuable measure of the molecular weight of the contained acids, is defined as the milligrams of KOH needed to saponify 1 g of the fat. (The value is often loosely reported without units.) The weighed sample is placed in a flask with a known amount of KOH, as an ethanolic solution, a condenser is fitted, and the mixture heated under reflux until saponification has been completed. The condenser is rinsed down with water, and the combined washings and solution mixture is titrated with a standard acid solution to the decolorization of phenolphthalein. In the control of a processed coconut oil, 1.250 g of the oil was heated under reflux with 25.00 mL of approximately 0.5*F* ethanolic KOH (*56.1*) and the cooled solution titrated with 0.4980*F* HCl, of which 12.45 mL was required. In the titration of 25.00 mL of the base, 23.86 mL of the HCl solution was needed. Calculate the saponification value. *Ans.:* 255 (mg of KOH/g of oil)

17-28 The saponification value of an oil or a fat is the number of milligrams of KOH (*56.1*) consumed per gram of sample in its saponification. (See Problem 17-27.) A 2.600-g sample of a fat is heated under reflux with 50.00 mL of a 0.320*F* alcoholic KOH solution. After the reaction is complete, the excess of base in the cooled solution is titrated with 0.4200*F* HCl; 15.20 mL is required. Calculate the saponification value of the fat.

18

Precipitation Titrations

18.1 General

In a precipitation titration, the species to be determined forms a sparingly soluble compound with the titrant. Although it might appear that countless precipitation reactions could be made the basis of a titration, requirements must be met that seriously limit the number. The reaction must be sufficiently rapid and complete and lead to a product of reproducible composition and of low solubility. A method must exist to locate the end point. However, meeting these requirements can present difficulties. Precipitation frequently proceeds slowly or starts only after some time. Many precipitates tend to absorb and thereby coprecipitate the species being titrated or the titrant. The indicator may also be adsorbed on the precipitate formed during the titration and thereby be unable to function appropriately at the end point. If a precipitate is highly colored, visual detection of a color change at the end point may be impossible; even when a white precipitate is formed, the milky solution may still make locating the end point troublesome.

Because of these difficulties, precipitation titrations receive only limited attention in present-day routine analysis. The chief areas of application include the determination of silver or a halide (or pseudohalide) by precipitating the silver salt (that is, so-called argentometric titrations) and the determination of sulfate by its precipitation as barium sulfate. Although the considerations here are limited largely to these applications, the principles apply to all precipitation titrations.

18.2 Titration Curve for Precipitation Titration

The curve for a precipitation titration is like the curve for the titration of a strong acid with a strong base. As you can see from the following example, the reasoning involved and the calculations necessary are analogous and straightforward, pH is replaced by pX and K_w by K_{sp}, where X is the ion titrated and K_{sp} is the solubility product of the compound precipitated.

Assume that a 100.0-mL portion of a 0.100*F* sodium chloride solution is titrated with 0.100*F* silver nitrate solution. During the titration, silver chloride is precipitated, the solubility product of which for simplicity is taken as $K_{sp} = [Ag^-][Cl^-] = 1.0 \times 10^{-10}$. The titration curve, defined in Section 10.1, consists of a plot of the negative logarithm of the molar concentration of the species being titrated, Cl^-, versus the milliliters of the titrant solution added.

The value of pCl at the start of the titration is readily calculable: pCl $= -\log[Cl^-] = -\log(1.0 \times 10^{-1}) = 1.00$. After the addition of, for example, 5.00 mL of the silver nitrate solution, the calculation of the pCl value takes the following form. Initially the chloride ion in solution amounted to $100 \times 0.100 = 10.0$ mmol. The 5.00-mL portion of the titrant solution introduces $5.00 \times 0.100 = 0.5$ mmol of silver ion and removes an equivalent amount of chloride ion from the solution as insoluble silver chloride. Consequently, the number of millimoles of chloride ion remaining in solution is 10.0 - 0.5 = 9.5. The total solution volume is 100 + 5 = 105 mL. Thus, the molar concentration of chloride ion is 9.5/105, and pCl = -log (9.5/105) = log 105 - log 9.5 = 2.02 - 0.98 = 1.04.

Further points of the titration curve to the vicinity of the equivalence point can be calculated likewise. At the equivalence point, this mode of calculation fails since it yields the value $[Cl^-] = 0$, which is impossible. However, at this point silver and chloride ions are present in exactly equivalent amounts, and the solution is a saturated solution of silver chloride. From the solubility product of silver chloride, it follows that

$$[Cl^-] = [Ag^+] = \sqrt{1.0 \times 10^{-10}} = 1.0 \times 10^{-5}$$

and

$$pCl = 5.00.$$

Beyond the vicinity of the equivalence point, the following reasoning is applied. The silver ion contributed by the saturated solution of silver chloride is neglected and the concentration of silver ion is assumed to depend solely on the amount of silver nitrate added in excess. Then the chloride concentration is obtained through the solubility-product expression. Assume that a total of 110 mL of silver nitrate solution is added. A volume of 100 mL is required for the equivalence point to be reached. The excess of 10 mL corresponds to $10 \times 0.100 = 1.0$ mmol of silver, and this is present in a volume of 210 mL. Hence, the molar concentration of silver ion, $[Ag^+]$, is 1.00/210. The molar concentration of chloride is then calculated from the solubility-product relation.

$$[Cl^-] = \frac{1.0 \times 10^{-10}}{[Ag^+]} = \frac{1.0 \times 10^{-10}}{1.00/210}$$

$$= 2.10 \times 10^{-8} M$$

and

$$pCl = -\log(2.1 \times 10^{-8}) = 7.68$$

The curve, with further additions of silver nitrate, approaches asymptotically a pCl value of 9.00, corresponding to saturating the 0.1*F* titrant solution with silver chloride.

The complete titration curve, thus calculated, is shown in Fig. 18-1 as curve

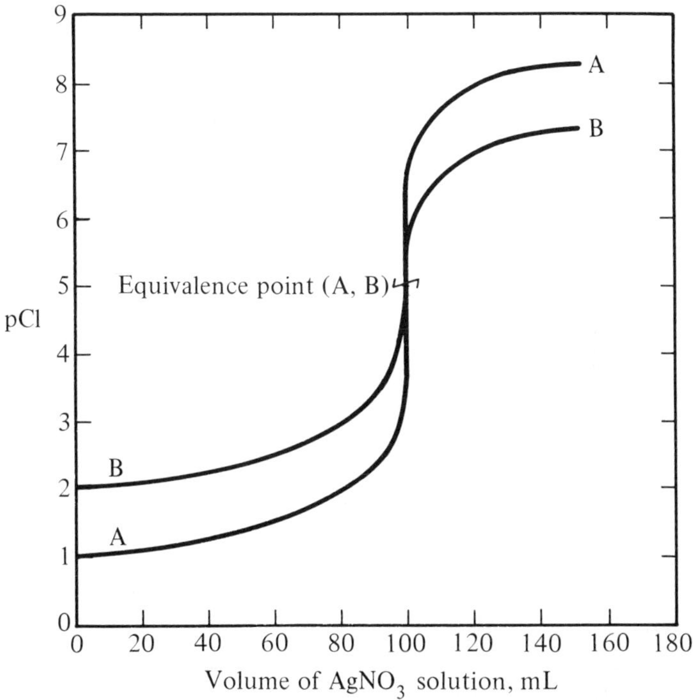

Fig. 18-1. Theoretical titration curves for the titration of NaCl with $AgNO_3$. Curve A: 100 mL of 0.100*F* NaCl titrated with 0.100*F* $AgNO_3$. Curve B: 100 mL of 0.0100*F* NaCl titrated with 0.0100*F* $AgNO_3$.

A. The figure also shows (curve B) the curve for the titration of 100 mL of a 0.010*F* sodium chloride solution with a 0.010*F* silver nitrate solution. This curve starts at a pCl value of 2.00 and approaches asymptotically a pCl value of 8.00. This curve, and also the curves for titrations involving the chloride-silver system at any other concentrations, pass at the equivalence point through the value pCl = 5.00. The reverse titration, that is, the titration of a silver ion solution with chloride ion as the titrant, is also practical. The plot of the titration curve will be identical except that pAg is plotted on the ordinate. The shape of the titration curve is not peculiar to the example selected but characteristic of all precipitation titrations in which the reactants combine in a 1:1 ratio. The extent of the "break" around the equivalence point increases as the concentrations of the titrated species and of the titrant increase. Further, the break is greater the lower the solubility of the compound precipitated. Thus, in the respective titrations of bromide and iodide ion with silver nitrate to form silver bromide ($K_{sp} = 1 \times 10^{-12}$) and silver iodide ($K_{sp} = 1 \times 10^{-16}$), the equivalence points are at pBr = 6.0 and pI = 8.0. If the titrant is 0.1*F*, the asymptote approached is at pBr = 11.0 and pI = 15.0. Curves for the two titrations are shown in Fig. 18-2. Your attention is again called to the analogy of a precipitation titration with the titration of a strong acid with a strong base. In the

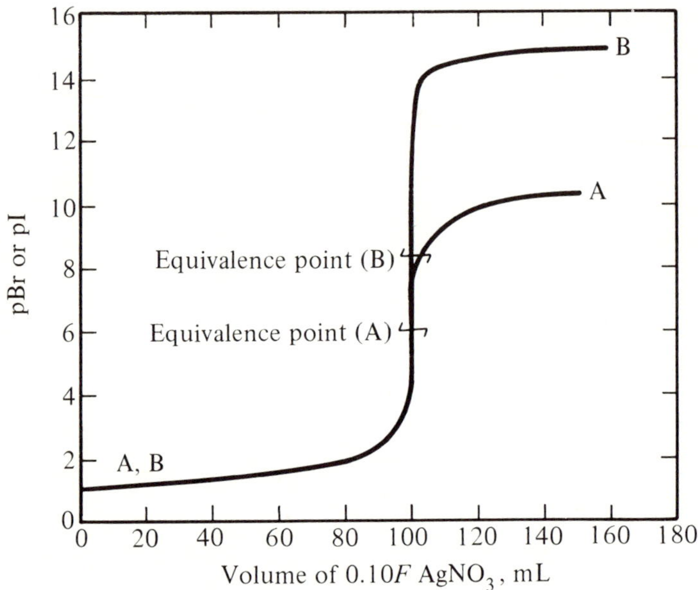

Fig. 18-2. Theoretical titration curves for the titration of NaBr and NaI with $AgNO_3$. Curve A: 100 mL of 0.10*F* NaBr titrated with 0.10*F* $AgNO_3$. Curve B: 100 mL of 0.10*F* NaI titrated with 0.10*F* $AgNO_3$.

second titration, a single titration reaction, $H^+ + OH^- \rightleftharpoons H_2O$, and a single constant K_w are involved. Hence, for given formal concentrations, a single titration curve is obtained regardless of the acid titrated. In contrast in precipitation titrations for given formal concentrations, different titration curves are obtained, depending on the particular titration equation and the solubility product of the substance precipitated.

18.3 *Visual End-Point Detection in Argentometry*

In neutralization titrations, the visual indication functions as follows. The indicator is a weak acid (or base) and undergoes an acid-base reaction with the titrant. This reaction is accompanied by a color change and takes place at a pH close to that of the equivalence point. By analogy, it may be surmised that in a precipitation titration the indicator would be a substance that gives with the titrant a precipitation reaction that is accompanied by a color change (that is, a colored precipitate is formed) and that occurs at a pX value close to that of the equivalence point. Indeed, this approach to indication is used in the so-called Mohr titration (Section 18.4). However, there exist several additional possibilities for indicating a precipitation titration, and the most common ones will be considered subsequently.

18.4 Method of Mohr

Mohr in 1856 introduced chromate ion as the indicator in the titration of chloride with silver. If some potassium chromate is addded to the neutral chloride solution and the titration with silver ion is started, a reddish color appears due to the precipitation of silver chromate. However, the color disappears rapidly as the solution is stirred because the silver is incorporated in the less soluble silver chloride. During the titration, the red disappears more and more slowly. Finally, when all the chloride has been precipitated as silver chloride, the next drop of titrant solution imparts a permanent red-brown color to the liquid, thereby signaling the end point. Ideally the chromate concentration should be such that the amount of silver chromate precipitated is just enough to yield a visible color change exactly at the equivalence point. However, the chromate concentration necessary to attain this would impart to the solution a yellow color so intense as to obscure the end point. In pràctice, a chromate concentration of about 0.005*M* is used; this causes an end point that is somewhat late. Consequently, the use of an indicator blank is advisable and is mandatory when very dilute solutions are involved.

Bromide, iodide, and thiocyanate, on the basis of their solubility products, should be titrable by this method; however, in practice, the titration can be extended only to bromide, since silver iodide and silver thiocyanate strongly adsorb the chromate ion. The reverse titration of silver ion with a standard chloride solution is not feasible when chromate is used as the indicator. Should chromate be added to the silver salt solution, a precipitate of silver chromate would form; this precipitate would react only very slowly at the end point.

Since the solubility of silver chromate increases rapidly with temperature, the titration must be performed at room temperature. Silver chromate is readily soluble in acidic solutions since chromate is converted to dichromate ion:

$$2CrO_4^{2-} + 2H \rightleftharpoons Cr_2O_7^{2-} + H_2O \qquad (18\text{-}1)$$

As a consequence, the titration cannot be performed below about pH 6.5. On the other hand, the pH cannot be higher than about 10, since in too alkaline solutions brown hydrous silver oxide precipitates. In practice the appropriate pH is commonly maintained by buffering with hydrogen carbonate or with borate.

Interferences in the Mohr titration of chloride are many. Any anion that forms a silver salt more insoluble than the chloride (for example, I^-, Br^-, SCN^-) will be precipitated first. Cations that form insoluble chromates (for example, Ba^{2+}) also interfere by rendering the indicator inactive. The relatively high pH at which the Mohr titration must be performed imposes a serious limitation, because difficulties occur if ions are present that form insoluble hydroxides or oxides, including iron and aluminum. Reducing substances may interfere by converting chromate ion to chromium(III). Finally, no substance can be tolerated that forms strong complexes with chloride ion (for example, Hg^{2+}) or with silver ion (for example, CN^-, and above pH 7, NH_3).

18.5 Method of Fajans

For the precipitation titration of chloride and certain other ions, Fajans and his co-workers introduced indicators that function through being adsorbed on the precipitate. For an understanding of the action of adsorption indicators, the adsorptive behavior of a precipitate must be appreciated. When silver nitrate is added to a dilute sodium chloride solution, turbidity develops, but if other electrolytes are absent, coagulation is not immediate. The silver chloride particles adsorb chloride ions and these attract counterions. The situation can be represented by $AgCl{\cdot}Cl^{-}{:}Na^{+}$. Consequently, these colloidal particles are electrically charged and repel each other, and coagulation is prevented. As the titration proceeds, the chloride ion concentration decreases and consequently also the extent of surface charge. This charge is so diminished shortly before the equivalence point that coagulation is initiated and can visually signal that the equivalence point is close. Abrupt coagulation occurs only in more concentrated solutions, however; in dilute solution, the turbidity persists. When an excess of silver nitrate is added, silver ions are adsorbed on the precipitate and attract counterions. The situation can now be represented by $AgCl{\cdot}Ag^{+}{:}NO_3^{-}$. The point at which equal quantities of positive and negative charge are adsorbed is called the isoelectric point. This point may not coincide with the equivalence point.

The adsorption indicators of Fajans interact with this adsorption process. Such indicators are colored, weakly acidic or basic organic compounds. Their action can be exemplified by fluorescein. This compound is a weak acid, denoted by HFl, and its aqueous solution is yellow green. When a small amount of this indicator is present during the titration, shortly after the isoelectric point is past, the indicator anion, Fl^{-}, is attracted to the positively charged particles; it replaces nitrate ion as the counterion, imparting a pink color to the particles. The process can be represented by

$$\underset{\text{white}}{AgCl{\cdot}Ag^{+}{:}NO_3^{-}} + \underset{\text{yellow}}{HFl} \rightleftharpoons \underset{\text{pink}}{AgCl{\cdot}Ag^{+}{:}Fl^{-}} + NO_3^{-} + H^{+} \qquad (18\text{-}2)$$

The titration of chloride using fluorescein as the adsorption indicator takes the following course. The chloride sample solution containing a small amount of fluorescein is initially clear and yellow green. When titrant is added, the white silver chloride precipitate forms, imparting a milky look to the colored solution. Shortly before the end point, coagulation occurs, and the end point is signaled by rapid adsorption of the dye on the precipitate; a pink color is the result. The solution remains yellow green, since only a small portion of the indicator is adsorbed. On addition of more titrant, the color of the particles does not deepen. It is therefore difficult to judge whether or not the end point has been exceeded, and to what extent.

In the use of fluorescein, if coagulation is avoided, the color change to pink is not confined to the bulky and rapidly settling precipitate but extends throughout the entire volume. This advantageous situation can be achieved by keeping the concentration of inert electrolytes low, by operating in an adequately dilute solution, and by adding a "protective" colloid, such as dextrin.

Since the indicator is adsorbed only in its anionic form, the acidity constant of the dye defines the lowest pH value at which a satisfactory end point can be secured. With fluorescein, the titration of chloride can be performed in the pH range 6.5 to 10.3. The upper pH limit for the titration is again controlled by the precipitation of hydrous silver oxide. Fluorescein can also serve in the titration of bromide and iodide. Dichlorofluorescein is a stronger acid and can be used in the titration of chloride down to pH 4. Eosin, a tetrabromofluorescein, can be used in titrating bromide, iodide, or thiocyanate down to pH 2, but not in titrating chloride, since this dye is adsorbed on the silver chloride precipitate well before the equivalence point.

The dyes sensitize silver halides toward photochemical reduction, manifested by blackening of the precipitate. Much of the particle surface is blocked by the silver metal formed, and the end point becomes less distinct. Consequently, the titration using an adsorption indicator should be conducted not in direct sunlight but in diffuse daylight or under subdued illumination from an incandescent (not fluorescent) light source.

In the titration of chloride with fluorescein as the indicator, the required pH range is usually maintained by buffering with hydrogen carbonate or with borate. As in the titration after Mohr (Section 18.4), among others, cations that precipitate under the prevailing pH conditions are potential interferences.

18.6 *Method of Volhard and Modifications*

Silver thiocyanate, AgSCN, is only slightly soluble in water ($K_{sp} = 10^{-12}$), and hence a very accurate precipitation titration of silver can be effected with thiocyanate ion as the titrant. Iron(III), added as iron(III) nitrate or iron(III) ammonium sulfate, serves as the indicator; the first excess of thiocyanate ion imparts a reddish color in the supernatant liquid due to the formation of the soluble iron(III)-thiocyanate complex, $FeSCN^{2+}$. Since the reverse titration of thiocyanate ion with silver ion is also possible, the Volhard method can be applied to the determination of halides by a back-titration procedure. The halide solution is treated with an excess of standard silver ion solution, and the excess silver is back-titrated with a standard thiocyanate solution. In this way, chloride, bromide, and iodide can be determined. With iodide, the iron(III) must be added *after* the silver iodide has been precipitated; otherwise iodide would be oxidized to iodine by the iron(III):

$$Fe^{3+} + I^- \rightleftharpoons \tfrac{1}{2} I_2 + Fe^{2+} \tag{18-3}$$

A significant advantage of the Volhard method is that the titrations can be performed in fairly acidic media; indeed the method fails at higher pH values because hydrous iron(III) oxide precipitates. At the low pH employed, interference from hydrolyzable ions is excluded; the Volhard method, consequently, has a much wider range of application than the methods of Mohr and Fajans.

In the determination of chloride, the end point fades after a few seconds and more titrant must be added to obtain a persistent red coloration. One explanation is that silver thiocyanate is less soluble than silver chloride, causing the metathetical reaction $AgCl + SCN^- \rightarrow AgSCN + Cl^-$ to proceed.

Division of the solubility-constant expressions of the two precipitates involved yields

$$\frac{K_{sp,AgCl}}{K_{sp,AgSCN}} = \frac{[Ag^+][Cl^-]}{[Ag^+][SCN^-]} = \frac{[Cl^-]}{[SCN^-]} = \frac{2 \times 10^{-10}}{1 \times 10^{-12}} = 2 \times 10^2$$

The chloride ion must have a molar concentration about 200 times that of thiocyanate ion, or else the metathetical reaction occurs. A fading end point can be prevented by removing the silver chloride by filtration and performing the back-titration in the filtrate. The results are then satisfactory, but the filtration is tedious. A second approach is to inactivate the silver chloride precipitate by coating it with an immiscible organic solvent. Commonly the suspension of silver chloride is shaken vigorously with nitrobenzene, which is adsorbed on the surface of the precipitate; then in the back-titration, the metathesis proceeds so slowly that a stable end point is secured.

Schulek has suggested that metathesis is not the main cause of the fading end point, but rather that silver ions are adsorbed on the surface of the precipitate and the adsorbed silver reacts with the titrant only slowly. He has proposed that after silver chloride has been precipitated, potassium nitrate should be added and the solution boiled; thereby coagulation and desorption of the silver ion are promoted. The end point in the subsequent back-titration with thiocyanate is then sharp and stable.

18.7 Method of Swift

If a high concentration of iron(III) is provided in the chloride solution and only very little thiocyanate is added, the chloride can be titrated directly. When silver nitrate is added, silver chloride is precipitated, *not* the less soluble silver thiocyanate. At an iron(III) concentration about 0.2*M*, the equilibrium

$$Fe^{3+} + SCN^- \rightleftharpoons FeSCN^{2+} \qquad (18\text{-}4)$$

is shifted far to the right, and the concentration of "free" thiocyanate ion becomes very small. The value of the ion product $[Ag^+][SCN^-]$ does not reach the value of the solubility product of silver thiocyanate, and consequently the thiocyanate does not precipitate. Only when all the chloride is precipitated does the silver ion concentration become high enough for silver thiocyanate to precipitate. Then the above equilibrium is shifted to the left and the end point is indicated by the disappearance of the red solution color.

18.8 Precipitation Titration of Sulfate

Barium sulfate has about the same solubility product as silver chloride, and hence the precipitation titration of sulfate with barium might be expected to be as feasible as the titration of chloride. Unfortunately, barium sulfate shows such pronounced tendencies to coprecipitation phenomena that accurate results are not readily attained. This titration finds application in rapid control analysis, however, where an error of up to 3% can be tolerated. Various indicators have been used, probably the most explored being tetrahydroxy-*p*-benzoquinone. The titration can sometimes be improved by making the titration medium 30 to 80% in ethanol. In such a medium and also water, Thorin is often used as the indicator (see Section 46.6).

18.9 Other Precipitation Titrations

Of other precipitation titrations receiving attention in routine analysis, two are noteworthy. Fluoride can be titrated with thorium ion to form sparingly soluble thorium fluoride, ThF_4. The end point can be detected when the insoluble red-colored precipitate of thorium forms with the dye Alizarin Red S. This titration is extremely sensitive to foreign ions and hence is most frequently applied after the fluoride is separated, commonly by being distilled as hexafluorosilicic acid from solutions containing nonvolatile mineral acids in the presence of silica.

Zinc can be titrated with hexacyanoferrate(II) ion according to the reaction

$$3Zn^{2+} + 2K^+ + 2[Fe(CN)_6]^{4-} \rightleftharpoons \underline{K_2Zn_3(Fe(CN)_6]_2} \qquad (18\text{-}5)$$

The end point can be detected in various ways, including the use of redox indicators (Section 24.9).

18.10 Questions

18-1 What are the requirements for a reaction so that it can be made the basis for a precipitation titration?

18-2 What methods of end-point detection can be used in precipitimetry?

18-3 What are the pH limits in the titration of chloride according to the methods of Mohr, Fajans, Volhard, and Swift?

18-4 Discuss the advantages and disadvantages of the four titration methods named in Question 18-3.

18-5 For the four titration methods named in Question 18-3, name the indicators used and describe the end-point color changes and the indicator mechanism.

18-6 Compare the interferences in the four methods named in Question 18-3.

18-7 Explain why the titration of iodide according to Mohr's method fails.

18-8 What is to be understood by the terms *coagulation* and *isoelectric point?*

18-9 What is the principle of the Swift method?

18-10 Explain why it is impossible to titrate silver ion with chloride ion using chromate ion as the indicator.

18-11 Explain why a pH of about 10 is the upper limit for argentometric titrations.

18-12 How will strong light affect argentometric titrations? Which of the modifications is most affected? Explain.

18-13 Explain how the acidity constant of the adsorption indicator is related to the lower pH limit at which the halide titration after Fajans can be performed.

18-14 Would barium ion interfere in the titration of chloride after Mohr? Explain your answer.

18-15 Explain why the results of titrating sulfate with barium ion are limited in accuracy.

18-16 Contrast the titration curve for a precipitation titration with the curve of an acid-base titration. Compare the shape of the two curves. What is plotted on the ordinate and the abscissa? What factors govern the position with respect to the ordinate of the equivalence point of the two titrations?

18-17 At the concentrations of sample and titrant solutions most often used, the pH changes by about 10 to 12 units during an acid-base titration in aqueous medium. (a) Within what approximate range might the pX change during precipitation titrations? (b) What would be the range in acid-base titrations if nonaqueous solvents were included in the considerations? Elaborate your answers.

18-18 Discuss the requirements for the precipitate in a precipitation titration and in a gravimetric determination. Which requirements are identical and which are different?

18-19 Suggest an example in which a solution to be used as a titrant in a precipitation titration could be standardized (a) by an acid-base titration; (b) by a gravimetric determination.

18.11 Problems

18-1 The bromide present in 80.0 mL of 0.050*F* KBr is titrated with 0.080*F* $AgNO_3$. For simplicity, take the solubility product of AgBr to be 1.0 × 10^{-12}. Calculate the molar concentration of silver ion and bromide ion for the following points of the titration curve:

(a) At the start of the titration. *Ans.:* $[Ag^+] = 0, [Br^-] = 5 \times 10^{-2}M$

(b) After adding 20.0 mL of silver solution.
Ans.: $[Ag^+] = 4.2 \times 10^{-11}M$, $[Br^+] = 2.4 \times 10^{-2}M$

(c) At the equivalence point. *Ans.:* $[Ag^+] = [Br^-] = 1.0 \times 10^{-6}M$

(d) After adding 10.00 mL of silver solution in excess.
Ans.: $[Ag^+] = 5.7 \times 10^{-3}M$, $[Br^-] = 1.8 \times 10^{-10}M$

18-2 Calculate what the pAg values are at the starting point, halfway to the equivalence point, 10.0% before the equivalence point, at the equivalence point, and 10.0% beyond the equivalence point for the following titrations of silver ion. The solubility-product constants for AgCl, AgBr, AgI, and AgSCN are 1.8×10^{-10}, 5.2×10^{-13}, 8.3×10^{-17}, and 1.0×10^{-12}.

	$AgNO_3$,mL	$AgNO_3$,F	Titrant
(a)	25.00	0.108	0.150F KCl
(b)	50.00	0.162	0.110F NaCl
(c)	20.00	0.120	0.100F $BaCl_2$
(d)	30.00	0.110	0.100F KBr
(e)	25.00	0.101	0.132F KBr
(f)	10.00	0.098	0.109F KI
(g)	20.00	0.132	0.092F KI
(h)	25.00	0.119	0.096F KSCN

18-3 Calculate the pBa values after the addition of 0.0, 5.0, 10.0, 15.0, 19.0, 20.0, 21.0, and 25.0 mL of 0.150F Na_2SO_4 to 100.00 mL of a solution containing 0.6248 g of $BaCl_2$ (*208.25*). The K_{sp} value of $BaSO_4$ is 1.3×10^{-10}.

18-4 A solution made by mixing 10.0 mL of 0.100F KCl, 10.0 mL of 0.200F KBr, 10.0 mL of 0.150F KI, and 70.0 mL of pure water is titrated with 0.200F $AgNO_3$. The solubility-product constants of AgCl, AgBr, and AgI are 1.8×10^{-10}, 5.2×10^{-13}, and 8.3×10^{-17}. Calculate the pAg values after the addition of 1.0, 4.0, 7.5, 10.0, 14.0, 17.5, 20.0, and 22.5 mL of titrant.

19

Calculating Results of Precipitation Titrations

19.1 General Considerations

The result for a precipitation titration can be calculated in a way that in principle is analogous to that for acid-base titrations (Chapter 15). There is no need to repeat the reasoning and derivations leading to the final formulas; you may refer to them as given in Chapter 15. Slight modifications will become necessary in some cases, however, for the following reasons. In an acid-base titration, regardless of the acid and base involved, the titration proper is a transfer of protons; and at least in aqueous media, a single principal titration reactions exists, $H^+ + OH^- \rightarrow H_2O$. Also, the relevant equivalence numbers are simply the number of moles of hydrogen ion or hydroxide ion provided per mole of acid or base and actually reacting according to this titration equation. In precipitation titrations, however, different reacting species are provided by the various titrants and substances titrated, in different molar amounts. In addition, these reacting species may not combine in a 1:1 molar ratio, as hydrogen and hydroxide ions do. This possibility sometimes calls for the additional consideration of the combining ratio; in other words, it then becomes necessary to base the calculation on the number of moles of sought-for substance corresponding to one mole of the titrant.

The best way of introducing the novice to this approach is to illustrate the calculations by examples, starting with simple cases and passing to more involved ones.

Example 19-1 The chloride in an NaCl solution is titrated with 0.1000*F* $AgNO_3$, of which a volume of 25.30 mL is required (a) What amount of Cl^- (*35.45*) in milligrams is present? (b) What amount of NaCl (*58.44*) in milligrams is present? The titration reaction is

$$Cl^- + Ag^+ \rightarrow \underline{AgCl}$$

The equation previously derived and extensively used in Chapter 15 can be applied.

$$\text{mL}_t \times F_t \times f_t \times \frac{\text{FW}_s}{f_s} = \text{mg}_s$$

Since one mole of $AgNO_3$ provides one mole of Ag^+, that is, of reacting species, the equivalence number of the titrant f_t is unity.

(a) The sought-for substance in this part of the problem is Cl^-, which is also the reacting species; consequently, f_s is also unity. Substituting the numerical values gives

$$25.30 \times 0.1000 \times 1 \times \frac{35.45}{1} = 89.69 \text{ mg of Cl}^-$$

(b) Getting the answer to this part of the problem requires merely applying a chemical factor that converts the result for Cl^- to a result for NaCl; this factor is $\text{FW}_{\text{NaCl}}/\text{FW}_{\text{Cl}}$. Then

$$\text{mL}_t \times F_t \times 1 \times \frac{\text{FW}_{\text{Cl}}}{1} \times \frac{\text{FW}_{\text{NaCl}}}{\text{FW}_{\text{Cl}}} = \text{mg of NaCl}$$

As you can see, the term FW_{Cl} cancels, and the following is obtained:

$$\text{mL}_t \times F_t \times 1 \times \frac{\text{FW}_{\text{NaCl}}}{1} = \text{mg of NaCl}$$

This equation could have been reached directly by the following reasoning. One mole of Cl^- corresponds to one mole of NaCl, and consequently the formula weight of NaCl and its equivalence number of 1 could have been substituted in the equation from the very beginning.

Substituting the numerical values gives

$$25.30 \times 0.1000 \times 1 \times \frac{58.44}{1} = 147.9 \text{ mg of NaCl}$$

Example 19-2 The chloride in a solution of $BaCl_2$ is titrated with 0.0800*F* $AgNO_3$, of which a volume of 18.34 mL is required. (a) What amount of Cl^- (*35.45*) in milligrams is present? (b) What amount of $BaCl_2$ (*208.25*) in milligrams is present?

The titration reaction is the same as in Example 19-1. For part (a) of the problem, the reasoning is the same as in Example 19-1 and the calculation takes the form

$$18.34 \times 0.0800 \times 1 \times \frac{35.45}{1} = 52.0 \text{ mg of Cl}^-$$

We get the answer to part (b) as we did the answer of Example 19-1; but here the conversion factor is $FW_{BaCl_2}/2FW_{Cl}$.

Substitution yields

$$mL_t \times F_t \times 1 \times \frac{FW_{Cl}}{1} \times \frac{FW_{BaCl_2}}{2FW_{Cl}} = \text{mg of } BaCl_2$$

Again FW_{Cl} cancels.

Also here the formula can be derived directly by the following reasoning. Two moles of reacting species, that is, of Cl^-, are provided by one mole of the sought-for substance, that is, $BaCl_2$, consequently the equivalence number of $BaCl_2$ is 2.

$$mL_t \times F_t \times 1 \times \frac{FW_{BaCl_2}}{2} = \text{mg of } BaCl_2$$

Substituting the numerical values gives

$$18.34 \times 0.0800 \times 1 \times \frac{208.25}{2} = 152.8 \text{ mg of } BaCl_2$$

Example 19-3 What is the formality of a solution of $BaCl_2$, of which a volume of 25.00 mL requires 16.00 mL of 0.07500*F* $AgNO_3$?

The relevant formula previously derived applies:

$$mL_t \times F_t \times f_t = mL_s \times F_s \times f_s$$

According to the reasoning in Examples 19-1 and 19-2, the equivalence number of the titrant is 1 and of the substance titrated is 2. Substituting the numerical data yields

$$16.00 \times 0.07500 \times 1 = 25.00 \times F_s \times 2$$

And solving gives

$$F_s = \frac{16.00 \times 0.07500 \times 1}{25.00 \times 2} = 0.02400F$$

Example 19-4 The Ag^+ in a solution of Ag_3PO_4 is titrated with 0.0400*F* $BaCl_2$, of which a volume of 27.20 mL is required. (a) What amount of Ag (*107.87*) in milligrams is present? (b) What amount of Ag_3PO_4 (*418.58*) in milligrams is present?

The titration reaction is the same as in Example 19-1. With $BaCl_2$ the titrant, two moles of reacting species, that is, Cl^-, are provided per one mole. Consequently, we have as the equivalence number of the titrant $f_t = 2$.

For the answer to part (a), Ag^+ is the reacting species and also the sought-for substance; consequently, $f_s = 1$. Substituting the numerical values yields

$$27.20 \times 0.0400 \times 2 \times \frac{107.87}{1} = 235 \text{ mg of Ag}$$

In solving part (b), it must be recognized that three moles of the reacting species, that is, of Ag^+, are furnished by one mole of the sought-for substance, that is, by Ag_3PO_4, consequently, $f_s = 3$. Hence

$$\text{mL}_t \times F_t \times 2 \times \frac{FW_{Ag_3PO_4}}{3} = \text{mg of } Ag_3PO_4$$

You should derive this formula by applying factors as shown in part (b) of both Examples 19–1 and 19–2.

Substituting the numerical values gives

$$27.20 \times 0.0400 \times 2 \times \frac{418.58}{3} = 304 \text{ mg of } Ag_3PO_4$$

Example 19–5 The Zn^{2+} in a $ZnSO_4$ solution is titrated with $0.1020F$ $K_4[Fe(CN)_6]$, of which a volume of 20.40 mL is required. (a) What amount of Zn (*65.37*) in milligrams is present? (b) What amount of $ZnSO_4$ (*161.43*) in milligrams is present?

The titration reaction can be written

$$3Zn^{2+} + 2K^+ + 2[Fe(CN)_4)]^{4+} \rightarrow \underline{Zn_3K_2[Fe(CN_6]_2}$$

For part (a) of the example, since Zn^{2+} is both the reacting species and the sought-for substance, its equivalence number is unity. So is the equivalence number of $[Fe(CN)_6]^{4+}$. However, the molar combining ratio is not 1:1 as in the previous examples, but $[Fe(CN)_6]^{4+}:[Zn^{2+}] = 2:3$, or referred to one mole of sinc ion, 1:3/2. Consequently, this fraction 3/2 must be incorporated into the formula:

$$\text{mL}_t \times F_t \times 1 \times \frac{FW_{Zn}}{1} \times \frac{3}{2} = \text{mg of Zn}$$

Substituting the numerical data yields

$$20.40 \times 0.1020 \times 1 \times \frac{65.37}{1} \times \frac{3}{2} = 204.0 \text{ mg of Zn}$$

Part (b) is simply solved. Since one mole of Zn corresponds to one mole of $ZnSO_4$, the equivalence number of the sulfate is also unity. The calculation takes the form

$$20.40 \times 0.1020 \times 1 \times \frac{161.43}{1} \times \frac{3}{2} = 503.9 \text{ mg of } ZnSO_4$$

Example 19-6 A solution of $K_4[Fe(CN)_6]$ is standardized by using it in the titration of 625.5 mg of $Zn_2P_2O_7$ (*277.68*) dissolved in acid. What is the formality of this titrant solution if a volume of 28.32 mL is required? The titration reaction is the same as in Example 19-4. The equivalence number of the titrant is unity. The equivalence number of the substance titrated, however, is $f_t = 2$, since one mole of $Zn_2P_2O_7$ provides two moles of reacting species, that is, of Zn^{2+}. Again the combining ratio must be taken care of by incorporating the factor 3/2 as in Example 19-5.

The calculation takes the form

$$28.32 \times F_t \times 1 \times \frac{277.68}{2} \times \frac{3}{2} = 625.5$$

Solving yields

$$F_t = \frac{625.5 \times 2 \times 2}{28.32 \times 277.68 \times 3} = 0.1061F$$

Example 19-7 A 0.5324-g sample of $BaCl_2$ is dissolved in water and the chloride (*35.45*) titrated with 0.0924*F* $AgNO_3$, of which a volume of 27.67 mL is needed. Calculate the percentage of chloride in $BaCl_2$.

The calculation, consolidated, is

$$\frac{27.67 \times 0.0924 \times 1 \times 35.45 \times 100}{532.4 \times 1} = 17.02\%\ \text{Cl}$$

To the novice, the problem may seem like part (b) of Example 19-2, and the temptation may be to use the formula weight of $BaCl_2$ and its equivalence number of 2. The situation is different, however. The present problem asks for Cl^- to be determined and to be expressed as its percentage in $BaCl_2$.

19.2 Use of Titer in Precipitation Titrations

The concept of a titer was briefly considered in Chapter 3. You are now better prepared, having encountered both acid-base and precipitation titrimetry, to understand the use and significance of titer values. The concentration of a titrant can be directly expressed in terms of the milligrams of the sought-for substance corresponding to 1 mL of the titrant solution. Such an expression is called a titer.

Using a titer has special practical value when a given titration must be performed many times with the same kind of sample. Indeed, the concentration of the titrant solution can be adjusted in such a way that 1 mL corresponds to a certain number of milligrams of the sought-for substance, or for a specified sample weight, to 0.1%, 1%, and so on, as desired. The calculation of results is thereby simplified and arithmetic errors reduced. When the titration reaction proceeds stoichiometrically, the concentration of the titrant solution that is needed for such simplification can be readily calculated.

A titer value is of special interest either if the titration reaction does not proceed stoichiometrically or go to completion, or if other difficulties are met. For example, in a precipitation titration, the end point signaled by the indicator used may be too early or too late, the precipitate formed during the titration may show an unduly high solubility, the titrant may be absorbed on the precipitate, and impurities may exert an unfavorable influence. Even in such cases, a practical titration can be effected and a correct value determined provided that the titration is performed under reproducible conditions. The titrant solution can then be standardized under these conditions against a known amount of the substance to be determined. The concentration is then best expressed as a titer.

A titer value expressed for one substance can be used analogously to a formality value in calculating the results for the titration of other substances with this titrant. All that is needed is to combine the established titer value with appropriate chemical factors.

Titer values and the relevant calculations can be appreciated from illustrative examples.

Example 19-8 If a volume of 35.34 mL of a $AgNO_3$ solution is needed for the titration of 0.2461 g of NaCl, what is the NaCl titer of the $AgNO_3$ solution?

$$\text{NaCl titer} = \frac{246.1}{35.34} = 6.964 \text{ mg NaCl/mL}$$

Example 19-9 How much KBr (*119.01*) in milligrams is present in a sample if the titration of its bromide content requires 28.30 mL of an $AgNO_3$ solution having an NaCl (*58.44*) titer of 1.205 mg of NaCl per milliliter?

The result can be first expressed in terms of NaCl by applying the given titer value:

$$28.30 \times 1.205 = 34.10 \text{ mg of NaCl}$$

The result can then be converted to terms of KBr by applying the relevant chemical factors:

$$34.10 \times \frac{FW_{KBr}}{FW_{NaCl}} = 34.10 \times \frac{119.01}{58.44} = 69.44 \text{ mg of KBr in sample}$$

Example 19-10 Zinc (*65.37*) is to be determined routinely in an alloy by titration with potassium hexacyanoferrate(II). What must be the formality of the $K_4[Fe(CN)_6]$ titrant solution so that its zinc titer will be exactly 0.1% Zn per milliliter if a sample weight of 0.5000 g is always taken?

Extension of the reasoning of Example 19-5 leads to

$$\frac{mL_t \times F_t \times 1 \times \dfrac{FW_{Zn}}{1} \times \dfrac{3}{2} \times 100}{\text{sample wt in mg}} = \%Zn$$

Introducing the numerical data yields

$$\frac{1.000 \times F_t \times 65.37 \times 3 \times 100}{500.0 \times 2} = 0.1\%$$

$$F_t = 0.00509F\ K_4[Fe(CN)_6]$$

19.3 Indirect Analysis and Precipitation Titrations

Indirect analysis (discussed in Section 7.4 for gravimetric methods and in Section 15.3 for acid-base titrations) is possible with precipitation titrations. The calculation procedure is like what has already been outlined, and a single example will illustrate the approach.

Example 19–11 A mixture is known to contain only NaCl (*58.44*) and $BaCl_2$ (*208.25*). A 0.5000-g sample of this mixture is dissolved and the chloride titrated with 0.1000*F* $AgNO_3$, of which 63.00 mL is required. Calculate the percent NaCl and percent $BaCl_2$ in the mixture.

Let x represent the percentage NaCl in the mixture and y the percent $BaCl_2$. From the description of the mixture, it follows that

$$x + y = 100.00$$

This is the first equation of the required pair. The result for the titration of chloride can be expressed either as % NaCl or % $BaCl_2$. Selecting the first yields

$$\frac{63.00 \times 0.1000 \times 58.44 \times 100}{500} = 73.63\% \text{ total chloride expressed as NaCl}$$

The second equation of the required pair is obtained as follows:

$$x + y \times \frac{2 \times FW_{NaCl}}{FW_{BaCl_2}} = \% \text{ total chloride expressed as NaCl}$$

$$x + y \times \frac{2 \times 58.44}{208.25} = \% \text{ total chloride expressed as NaCl}$$

$$x + 0.56125y = 73.63$$

Solving the simultaneous equations for y yields y = 60% $BaCl_2$ and $x = 100.00 - y$ = 40% NaCl.

19.4 Problems

19–1 A silver nitrate solution is standardized against KCl (*74.56*) of 99.60% purity. A 0.3254-g sample of this salt is titrated according to Mohr and

44.22 mL of the $AgNO_3$ solution is needed. Calculate the formality of this solution. *Ans.:* 0.09830*F*

19-2 A coin weighing 12.52 g is analyzed for silver (*107.87*). The coin is dissolved in nitric acid and the solution diluted with water to exactly 250 mL in a volumetric flask. A 25.00-mL portion of the solution is titrated with 0.1050*F* KCl, of which a volume of 44.22 mL is needed. What is the percentage of silver in the coin? *Ans.:* 40.00%

19-3 A mixture of $BaCl_2$ and Na_2CO_3 is to be analyzed for its barium chloride content by the Volhard method. The following data are given. The $AgNO_3$ titrant solution is prepared by dissolving 4.983 g of silver metal of 99.10% purity in nitric acid and diluting with water to exactly 500 mL in a volumetric flask. In the standardization of the KSCN solution, it is found that a volume of 25.00 mL of the $AgNO_3$ solution requires 22.00 mL of the KSCN solution. A 0.5000-g sample is dissolved and acidified, and a volume of 40.30 mL of $AgNO_3$ solution is added; the excess of silver is then back-titrated, requiring 6.22 mL of the KSCN solution.

(a) Calculate the formality of the silver (*107.87*) solution. *Ans.:* 0.0915_6F

(b) Calculate the formality of the KSCN solution. *Ans.:* 0.1040_4F

(c) Calculate the percent of $BaCl_2$ (*208.25*) in the mixture. *Ans.:* $63.3_7\%$

19-4 Sulfur (*32.1*) in coal is determined by dry combustion of the sample, absorption of the SO_3 formed in water, and then titration of the sulfate thus formed with a $BaCl_2$ solution. The titer of the titrant solution is established by titrating a freshly dried sample of Na_2SO_4 (*142.0*). In a particular determination, a 0.2500-g sample of coal is subjected to combustion. Titration of the sulfate obtained requires 4.80 mL of a $BaCl_2$ solution, of which a volume of 17.22 mL is needed in the titration of 0.0188 g of Na_2SO_4. (a) What is the titer of the $BaCl_2$ solution in milligrams of sulfur per milliliter? (b) What is the percent of sulfur in the coal?

Ans.: (a) 0.2468 mg S/mL, (b) 0.474% S

19-5 A mixture is known to contain only NaCl (*58.44*) and KBr (*119.0*). In the argentimetric titration of the total halide in a 0.2500-g sample of this mixture, a volume of 27.64 mL of 0.1000*F* $AgNO_3$ is required. Calculate the composition of the mixture. *Ans.:* 30.5% NaCl, 69.5% KBr

19-6 One approach to determining fluoride is as follows. Fluoride is precipitated quantitatively as PbClF. The isolated precipitate is either dried and weighed or dissolved in acid and the chloride in the solution is titrated by the Volhard method. A 0.8400-g sample of a material was dissolved and fluoride precipitated as PbClF. The precipitate was separated and dissolved. To the resulting solution a volume of 35.00 mL of 0.09530*F* $AgNO_3$ was added, and the titration with 0.1025*F* KSCN required 2.16 mL. Calculate the percent of fluorine (*19.0*) present in the material.

19-7 The production of a preparation containing barium chloride (*208.25*) as the active ingredient is to be checked routinely by a Mohr titration. How many grams of silver (*107.87*) must be dissolved to 2 L to get a solution of which each milliliter is equivalent to 1.00% $BaCl_2$ if a 1.00-g sample of the preparation is to be taken?

19-8 The capacity of an irregularly shaped vessel (a boiler, for example) cannot be computed from geometric data; it can be established from analytical data, however, as follows. The vessel is filled to its capacity mark with chloride-free water. A known amount of sodium chloride is introduced, and a homogeneous solution obtained by stirring. A measured portion of the solution is withdrawn and titrated with silver nitrate.

For an actual case, 1.32 kg of NaCl (*58.44*) was dissolved, a 0.500-L portion of the solution taken, and 46.5 mL of 0.502*F* $AgNO_3$ consumed in a Mohr titration. What is the capacity of the vessel in liters? What error in capacity would result from an error in the titration amounting to 0.1 mL of $AgNO_3$ solution? *Ans.:* 484 L; about 1 L

20

Complexation Equilibria and Compleximetric Titrations

Compleximetric titrations are based on the formation of a soluble complex when the species titrated reacts with the titrant. Many reactions lead to a soluble complex but only a few can be used for a titration. Besides the general requirements for titration reactions, such as reasonable rate, stoichiometry, and the like, (see Chapter 10), special requirements must be met as to the stability of the complex formed and the number of steps involved in its formation. Insoluble complexes may be formed, and titrations can be based on such reactions. However, such titrations are advantageously considered as precipitation titrations. Formally, protonation reactions can be described as complexation reactions. When this view is adopted, an acid is considered the "proton complex" of its conjugate base. However, titrations based on protonation reactions are usually better classified as acid-base rather than as compleximetric titrations. You should compare the calculation of the titration curve and the treatment of metal indicators given below with the relevant sections on acid-base titrations and note the analogies.

Almost all compleximetric titrations of practical significance entail the formation of a complex involving the cation of a metal; consequently, we must first examine the concept of metal complexes.

20.1 Nature of Metal Complexes

A metal complex involves the coordination of a metal ion with a species that has one or more pairs of nonbonding electrons available for sharing. The electron-donating species is known as a ligand (Latin, *ligare,* "to bind"), or a complexing agent. The metal ion in the complex is known as the central ion. The ligand may have one or more electron-donating groups (or atoms) in each molecule and is then said to uni-, bi-, tri-, quadri-, . . ., multidentate (Latin, *dentatus,* "toothed"). The greatest number of unidentate ligands known to coordinate with a given central ion is termed (maximum) coordination number of that ion. The charge of a complex is the algebraic sum of the charges of the central ion and of the ligand species. Thus a metal complex can be neutral, or either positively or negatively charged.

When a ligand is multidentate, that is, has two or more donor groups as part of a single molecule, there is a possibility of forming a ring structure containing the central ion. Such rings are known as chelate rings. The term is apt, as the central ion is held in each ring as in the claws of a crab (Greek, *chele,* "crab's claw"). The entire multidentate complex is known as a chelate complex, or simply a chelate. The stability of a chelate is usually the greatest if the rings are five-membered or six-membered, that is, if each ring involves the linking of five or six atoms.

All metal ions in aqueous solution are present in the form of complexes, at least as aquo complexes. The oxygen atom of the water molecule is an electron donor, H:$\ddot{\underset{\cdot\cdot}{O}}$:H. For example, the copper (II) ion in water persists as $Cu(H_2O)_4^{2+}$, but the unsolvated ion Cu^{2+} is usually written for simplicity, unless the solvation must be explicitly shown.

20.2 Nomenclature of Metal Complexes

Complexes are named systematically in the following way. First, the number of ligand species bonded to a central ion is indicated by a Greek prefix: *mono-, di-, tri-, tetra-, penta-, hexa-,* and so on. Next, the ligand is named, and if anionic is usually terminated by *-o.* Water and ammonia, as ligands, are named *aquo* and *ammine.* Finally, the name of the central ion is given. In some cases, the root of the Latin name of the element is preferred, for example, *argent-* for silver, *plumb-* for lead, and *ferr-* for iron. If the complex is anionic, the name is terminated by *-ate.* The oxidation state of the central ion is indicated, when desirable, by a Roman numeral within parentheses. (For a discussion of the determination of the oxidation state, see Section 21.5.) You can best appreciate the rules by some simple examples:

$Cu(NH_3)_4^{2+}$	tetraamminecopper(II) ion
$Ag(NH_3)_2^{+}$	diamminesilver ion
HgI_4^{2-}	tetraiodomercurate(II)
$Fe(CN)_6^{4-}$	hexacyanoferrate(II)
$Fe(CN)_6^{3-}$	hexacyanoferrate(III)
$[CrCl_4(H_2O)_2]^-$	tetrachlorodiaquochromate(III)
$Ag(CN)_2^-$	dicyanoargentate(I)

It is often convenient to use the following simpler description of a complex. The name of the metal is separated from the name of the ligand (and its prefix, if any) by a hyphen; for example, copper(II)-tetraammine complex, mercury(II)-tetraiodo complex, metal-chloro complexes, and the like. Some complex species have traditional names that are commonly used, for example, ferrocyanide for hexacyanoferrate(II) and ferricyanide for hexacyanoferrate(III).

20.3 Complexation Equilibria

The formation of a soluble metal complex is a reversible process. The equilibrium is dynamic, and the law of mass action applies. For the formation of the copper(II)-tetraammine complex, the reaction and the expression for the equilibrium constant can be written

$$Cu^{2+} + 4NH_3 \rightleftharpoons Cu(NH_3)_4^{2+} \tag{20-1}$$

$$K_{st} = \frac{[Cu(NH_3)_4^{2+}]}{[Cu^{2+}][NH_3]^4} \tag{20-2}$$

The equilibrium constant corresponding to the reaction leading to the formation of the complex is known as the stability constant, or the formation constant, of the complex. The equilibrium constant for the reverse reaction is known as the instability constant, or dissociation constant, for the complex. Obviously the two constants are reciprocals:

$$K_{st} = \frac{1}{K_{inst}}$$

Complexes involving more than one ligand molecule or one ion form in a stepwise fashion, and an equilibrium expression can be written for each step. Silver, for example, has a coordination number of 2, and hence the silver ion combines in two steps with ammonia to form the diamminesilver(I) ion:

$$Ag^{+} + NH_3 \rightleftharpoons Ag(NH_3)^{+} \tag{20-3}$$

$$k_{st,1} = \frac{[Ag(NH_3)^{+}]}{[Ag^{+}][NH_3]} \tag{20-4}$$

$$Ag(NH_3)^{+} + NH_3 \rightleftharpoons Ag(NH_3)_2^{+} \tag{20-5}$$

$$k_{st,2} = \frac{[Ag(NH_3)_2^{+}]}{[Ag(NH_3)^{+}][NH_3]} \tag{20-6}$$

The overall stability constant K_{st} is the product of the two stepwise stability constants:

$$K_{st} = k_{st,1} \times k_{st,2} = \frac{\cancel{[Ag(NH_3)^{+}]}}{[Ag^{+}][NH_3]} \times \frac{[Ag(NH_3)_2^{+}]}{\cancel{[Ag(NH_3)^{+}]}[NH_3]}$$

$$= \frac{[Ag(NH_3)_2^{+}]}{[Ag^{+}][NH_3]^2} \tag{20-7}$$

Stability constants have the advantage over dissociation constants in having positive exponents. The reasoning and the method of calculation, however, are fully analogous to those encountered with dissociation constants as used, for example, in the consideration of acid-base equilibria.

Example 20–1 The complex MY, formed from a metal ion M and a complexing agent Y (charges omitted since they are not relevant here) has a stability constant of 4.0×10^8. What is the concentration of free, that is, uncomplexed metal ion, in a solution $1.0 \times 10^{-2} F$ in the complex?

The expression for the stability constant of the complex is

$$K_{st} = \frac{[MY]}{[M][Y]}$$

According to the material balance, the total concentration of the metal ion, C_M, which equals $1.0 \times 10^{-2} F$, is given by

$$C_M = [MY] + [M]$$

Since on dissociation of the complex, one free ion of the complexing agent is formed for each free metal ion, the relation

$$[M] = [Y]$$

holds. Substituting the last two equations into the expression for the stability constant yields

$$K_{st} = \frac{C_M - [M]}{[M]^2}$$

Since the complex is stable, neglecting [M] relative to C_M is a reasonable approximation, leading to

$$K_{st} \cong \frac{C_M}{[M]^2}$$

Solving for [M] and substituting the numerical data gives

$$[M] = \sqrt{\frac{C_{MY}}{K_{st}}} = \sqrt{\frac{1.0 \times 10^{-2}}{4.0 \times 10^8}} = 5.0 \times 10^{-6} M$$

The concentration of free metal ion is much smaller than the concentration of the complex and consequently the approximation was allowed.

Example 20–2 A complex MY has the stability constant 2.0×10^3. What is the concentration of free metal ion in a solution $2.0 \times 10^{-3} F$ in the complex? Solving as we did for Example 20–1 leads to

$$[M] = \sqrt{\frac{2.0 \times 10^{-3}}{2.0 \times 10^3}} = 1.0 \times 10^{-3} M$$

A check of the approximation reveals that the concentration of the metal ion is half the concentration of the complex; the 5% rule adopted previously is therefore violated, and the approximation is not permitted. Consequently, the solution must involve the full quadratic equation, as follows:

$$2.0 \times 10^3 = \frac{2.0 \times 10^{-3} - [M]}{[M]^2}$$

$$2.0 \times 10^3 [M]^2 + [M] - 2.0 \times 10^3 = 0$$

$$[M] = \frac{-1.0 \pm \sqrt{(1.0)^2 + 4 \times 2.0 \times 10^3 \times 2.0 \times 10^{-3}}}{2 \times 2.0 \times 10^3}$$

$$= \frac{-1.0 \pm \sqrt{1.0 + 16}}{4.0 \times 10^3}$$

Only the positive root has physical meaning; hence

$$[M] = \frac{-1.0 + \sqrt{17}}{4.0 \times 10^3} = 7.8 \times 10^{-4} M$$

Example 20–3 A complex MY has a stability constant of 5.0×10^6. What is the concentration of free metal ion in a solution $2.0 \times 10^{-3} F$ in the complex and made additionally $2.0 \times 10^{-1} M$ in the free complexing agent? The material balance for the metal, as in Example 20–1, is

$$C_{MY} = [MY] + [M] \cong [MY]$$

However, the equality $[M] = [Y]$ no longer applies. Free complexing agent stems from two sources: the dissociation of the complex, and the addition. The complex is stable and the contribution from the dissociation will be small, especially since the high concentration of the added agent will repress the dissociation. Consequently, the concentration of the free complexing agent [Y] can be taken as equal to $2.0 \times 10^{-1} M$. Substituting the numerical values in the expression for the stability constant yields

$$5.0 \times 10^6 = \frac{2.0 \times 10^{-3}}{[M] \times 2.0 \times 10^{-1}}$$

and

$$[M] = 2.0 \times 10^{-9} M$$

Checking the validity of the approximation expressed in the material balance is left to the student.

Example 20–4 Into 20.0 mL of a $5.0 \times 10^{-2} F$ solution of ammonia is pipetted 5.0 mL of a $1.0 \times 10^{-3} F$ solution of copper sulfate. Calculate the concentration of free copper(II) ion.

The solution contains the various ammine complexes of copper(II) ion, and an exact treatment would require the consideration of this fact. The calculation would then involve equations of higher order, and the mathematical effort would be great. The treatment, however, can be simplified. The ammonia concentration is considerably greater than the copper ion concentration. As an approximation, the complex $Cu(NH_3)_4^{2+}$ can therefore be considered to be the only one present; that is, complexes with fewer than four ammonia molecules coordinated to copper are assumed to be present in negligible quantities. Then the overall stability constant (2.0×10^{12}) can be used. Applying this and the approximations made in the preceding examples leads to

$$2.0 \times 10^{12} = \frac{(5.0 \times 1.0 \times 10^{-3})/25}{[Cu^{2+}] \times [(20.0 \times 5.0 \times 10^{-2})/25]^4}$$

and

$$[Cu^{2+}] = 3.9 \times 10^{-11} M$$

20.4 EDTA Titrations, General Considerations

Most compleximetric titrations involve the titrant *e*thylen*d*iamine*t*etra*a*cetic acid, also known as (*e*thylene*d*initrilo)*t*etra*a*cetic acid. The acronym EDTA applies to either name (as indicated here by the italic letters in the name), and also to the anionic form of the compound (that is, the A may represent *a*cetate).

EDTA, introduced as a complex-forming titrant by G. Schwarzenbach shortly after World War II, forms stable, water-soluble, 1:1 chelates with many polyvalent metal ions. It can be obtained in high purity, as the free acid and as the disodium salt dihydrate. Either compound has a favorably high equivalent weight. The salt can be dissolved directly in water; the acid requires sodium hydroxide to be added or another base for dissolution. A titrant solution prepared in either way is stable indefinitely.

In Fig. 20-1, the structural formula of EDTA, as the free acid, is shown and also a representation of the "normal" 1:1 chelate of a dispositive metal ion, M^{2+}. A count of atoms will reveal that all the chelate rings are five-membered and that there is the potential for six rings. The high stability of EDTA complexes can thereby be appreciated.

EDTA is a tetraprotic acid. The values of the four acid dissociation constants are 1×10^{-2}, 2.1×10^{-3}, 6.9×10^{-7}, and 7.4×10^{-11}. From these values it is evident that two of the protons are strongly acidic and two only weakly acidic. From the values of these constants, it can be concluded that in a solution of about pH 5 (closely attained in a solution of the disodium salt of EDTA) the predominant species is H_2Y^{2-} if EDTA is formally represented by H_4Y. At that pH, the formation of a complex between EDTA and a metal ion, M^{n+}, can be presented as

$$M^{n+} + H_2Y^{2-} \rightleftharpoons MY^{n-4} + 2H^+ \qquad (20\text{–}8)$$

As you can see, two hydrogen ions are freed. This reaction is reversible, and consequently the complex is more extensively dissociated the higher the acidity of the solution. In other words, the proton competes with the metal ion for complexation

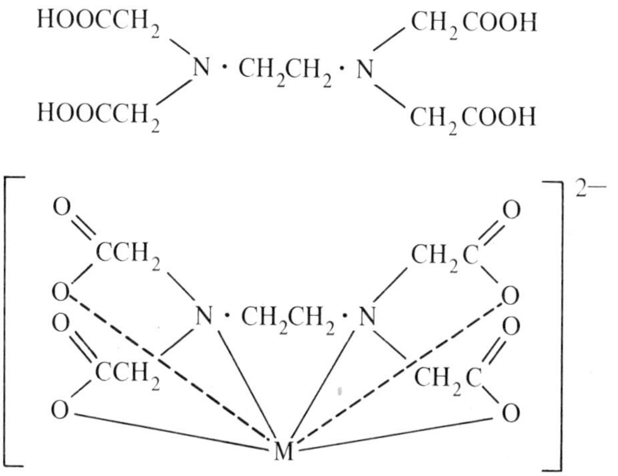

$HOOCCH_2$ CH_2COOH
$N \cdot CH_2CH_2 \cdot N$
$HOOCCH_2$ CH_2COOH

O O 2−
CCH_2 CH_2C
O $N \cdot CH_2CH_2 \cdot N$ O
O O
CCH_2 CH_2C
O O
M

Fig. 20–1. Structural formulas of EDTA, as free acid, and of a 1:1 chelate of a dipositive metal ion and EDTA.

with the EDTA. This is an important fact because it delineates the necessity of performing any EDTA titration above a certain limiting pH value; otherwise, the complex dissociates to too great an extent and the titration fails. The limiting pH value depends on the stability of the complex formed during the titration. The lower the stability of the relevant metal-EDTA complex is, the higher the pH value maintained during the titration must be.

Consequently, evaluation of the feasibility of a compleximetric titration cannot be based simply on the value of the stability constant as such, but the pH, among other factors, must be taken into account. The influence of the pH can be assessed by an α factor in the same way that coexistent solubility and acid-base equilibria were treated (see Section 8.13). Consideration of the α factor leads to the conditional stability constant.

20.5 *Conditional Stability Constants*

The expression of the stability constant of a complex formed between a metal ion M^{n+} and EDTA is based on the reaction

$$M^{n+} + Y^{4-} \rightleftharpoons MY^{n-4} \tag{20-9}$$

and is then written as

$$K_{st} = \frac{[MY^{n-4}]}{[M^{n+}][Y^{4-}]} \tag{20-10}$$

In this expression, only the fully dissociated form of EDTA, that is, the anion Y^{4-}, occurs. By M^{n+} is implied the free metal ion, that is, the aquo complex.

However, only at high pH values (above about pH 10) is the portion of the EDTA not bound to the metal ion present in the form Y^{4-}. The lower the pH, the larger the portion of the uncomplexed EDTA present in various protonated forms. It is possible to write the expression for the conditional stability constant K_{st}^{H} that takes care of the influence of the pH as follows:

$$K_{st}^{H} = \frac{[MY^{n-4}]}{[M^{n+}]\,[Y]^{*}} \tag{20-11}$$

Here $[Y]^{*}$ is the molar concentration of the EDTA uncombined with the metal ion and in whatever stage of protonation it may be present. The following relation obviously exists:

$$[Y]^{*} = [Y^{4-}] + [HY^{3-}] + [H_2Y^{2-}] + [H_3Y^{-}] + [H_4Y] \tag{20-12}$$

The function α_H is defined as the ratio of the concentration of total uncomplexed EDTA to the concentration of the EDTA present in the form Y^{4-}. Hence

$$[Y]^{*} = [Y^{4-}]\alpha_H \tag{20-13}$$

Combining equations (20-11) and (20-13) yields

$$K_{st}^{H} = \frac{[MY^{n-4}]}{[M^{n+}]\,[Y^{4-}]\alpha_H} = \frac{K_{st}}{\alpha_H} \tag{20-14}$$

Consequently, if α_H is known, the value of the conditional stability constant can be calculated from the stability constant.

With the acid dissociation constants of EDTA known, the factor α_H can be calculated for any pH. Deriving the required formula proceeds by successively replacing all concentration terms except $[Y^{4-}]$ in equation (20-12) by expressions from the equations for the acid dissociation constants:

$$[H_3Y^{-}] = \frac{[Y^{4-}]\,[H^{+}]}{K_{a,4}}$$

$$[H_2Y^{2-}] = \frac{[H_3Y^{-}]\,[H^{+}]}{K_{a,3}} = \frac{[Y^{4-}]\,[H^{+}]^{2}}{K_{a,3}K_{a,4}} = \ldots$$

The result is

$$[Y]^{*} = [Y^{4-}]\underbrace{\left(1 + \frac{[H^{+}]}{K_{a,4}} + \frac{[H^{+}]^{2}}{K_{a,3}K_{a,4}} + \frac{[H^{+}]^{3}}{K_{a,2}K_{a,3}K_{a,4}} + \frac{[H^{+}]^{4}}{K_{a,1}K_{a,2}K_{a,3}K_{a,4}}\right)}_{\parallel \atop \alpha_H} \tag{20-15}$$

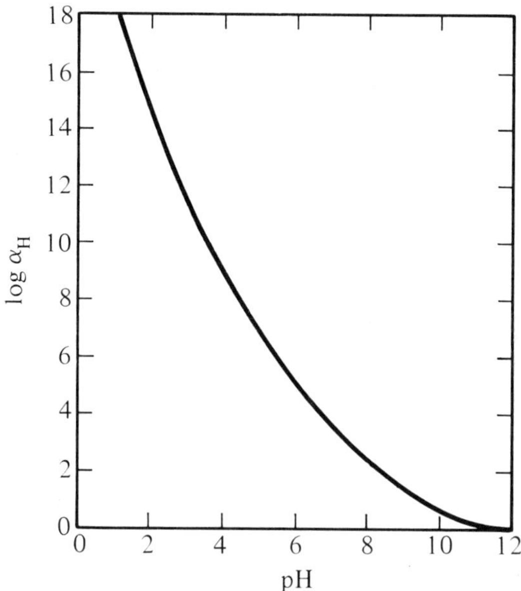

Fig. 20-2. Plot of log α_H versus pH for EDTA.

If the relevant numerical data are substituted in the parenthetical portion of equation (20-15), one can secure the values for plotting the log α_H versus pH curve for EDTA shown in Fig. 20-2. From this curve, values for the log α_H, and consequently α_H, can be obtained for any pH within the limits at which titrations are performed.

An α_H factor is independent of the metal ion involved and solely related to the ligand. For a given ligand, the factor depends only on the pH of the solution.

It is also possible for the metal ion to undergo a side reaction. For example, if some ammonia is added to a solution containing the copper(II)-EDTA complex, ammonia will compete with the EDTA for the copper in the way that the proton competes with the metal for reaction with EDTA. It is possible to derive a factor α_{NH_3}, to take care of this influence. The method of calculation is like that for α_H, and a formula can be derived that contains the concentration of the additional complexing agent (in the example, ammonia) and the stability constants of the various complexes formed by it with the metal.

Such factors have an advantage because the stability constants can be tabulated and the various α factors can be either tabulated or plotted; with these data at hand, the value of the conditional constant for any actual solution conditions can be calculated rapidly, and thereby an evaluation of the situation is expedited.

20.6 Compleximetric Titration Curves

The understanding of the functioning and requirements of a compleximetric titration can be greatly advanced from a discussion of a titration curve. The curve for such a titration involves the plot of the negative logarithm of the molar concentration of the uncomplexed metal ion, that is, of pM, versus the volume of titrant solution added. Only an example involving the formation of a simple 1:1 complex, as in an EDTA titration, will be considered. The titration reaction can be written

$$M + Y \rightleftharpoons MY \tag{20-16}$$

where M represents the metal ion and Y the EDTA (or any other ligand yielding a 1:1 complex). Charges are omitted because they are not relevant here.

The stability constant of the complex is given by the following expression:

$$K_{st} = \frac{[MY]}{[M]\,[Y]} \tag{20-17}$$

Actually the conditional constant should be used. However, for simplicity we assume that the titration is performed under conditions in which neither the hydrogen ion nor any other constituent of the solution can significantly influence the stability of the complex; then the constant as written in equation (20-17) can be used.

Assume that exactly 100 mL of a 0.100*F* solution of metal M is titrated with a 0.100*F* solution of EDTA, and that the stability constant has a value of 1.0×10^{11}. (The case is close to the titration of calcium at pH 10 or higher.)

For the start of the titration, the pM value of the original sample solution is

$$\text{pM} = -\log 0.100 = 1.00$$

For 10.0 mL of titrant solution added, the pM value is calculated as follows. Initially the metal ion present amounted to $100 \times 0.100 = 10.0$ mmol. The ligand added amounts to $10.0 \times 0.100 = 1.00$ mmol. Since a 1:1 complex is formed, this quantity of titrant reacts with an equimolar amount of the metal ion. As a consequence, 10.0 - 1.0 = 9.0 mmol of "free" metal ion remains in a total solution volume of 100 + 10 = 110 mL. Hence

$$[M] = \frac{9.0}{110}$$

and

$$\text{pM} = -\log\left(\frac{9.0}{110}\right) = \log 110 - \log 9.0$$

$$= 2.04 - 0.95 = 1.09$$

Since the complex is stable, shown by the large value of the stability constant, essentially all the titrant added combines with the metal ion; in other words, the dissociation of the complex can be neglected. This approximation also applies to further additions of titrant solution.

For 20.0 mL of titrant added, the calculation is analogous and can be consolidated:

$$[M] = \frac{100.0 \times 0.100 - 20.0 \times 0.100}{100.0 + 20.0} = \frac{8.0}{120} = \frac{1.0}{15}$$

and $$pM = -\log (1.0/15) = \log 15 = 1.18$$

When 99.0 mL of the complexing agent is added, the calculation gives

$$[M] = \frac{100.0 \times 0.100 - 99.0 \times 0.100}{100.0 + 99.0} = \frac{0.10}{199} \cong \frac{1.0}{2000}$$

and $$pM = -\log\left(\frac{1}{2000}\right) = 3.30$$

At the equivalence point, this method of calculation results in [M] = 0, an unacceptable result. At this point, dissociation of the complex can no longer be neglected. As a matter of fact, dissociation is the only process by which free M is provided. Here the calculation is based on the expression for the stability constant (20-17) and takes the form already discussed in Example 20-1. However, C_M now is the total concentration of metal at the equivalence point. Its numerical value can be calculated from the initial concentration and the dilution that has taken place:

$$C_{M,eq\,pt} = \frac{C_{M,initial} \times 100}{200} = 5.0 \times 10^{-2} F$$

Then

$$[M] = \sqrt{\frac{C_{M,eq\,pt}}{K_{st}}} = \sqrt{\frac{5.0 \times 10^{-2}}{1.0 \times 10^{11}}} = \sqrt{5.0 \times 10^{-13}}$$

and $$pM = \tfrac{1}{2}(13 - \log 5.0) = 6.15$$

For the calculation of pM values beyond the equivalent point, the approach shown in Example 20-3 is applied. On the addition of 110.0 mL of titrant solution, that is, 10.0 mL beyond the equivalence point, the molar concentration of the free ligand is

$$[Y] = \frac{10.0 \times 0.100}{100 + 110} = \frac{1.00}{210}$$

The molar concentration of the complex is given by

$$[MY] = C_M - [M] \cong C_M$$

The value of C_M at this point is given by

$$C_M = \frac{100 \times 0.100}{100 + 110} = \frac{10.0}{210}$$

Substituting the numerical values into the expression for the stability constant gives

$$1.00 \times 10^{11} = \frac{10.0/210}{[M] \times 1.00/210}$$

and

$$[M] = \frac{10.0}{1.00 \times 10^{11}} = 1.00 \times 10^{-10} M$$

$$pM = 10.00$$

The value [M] is very small compared with the value C_M; consequently, the above approximation $[MY] \cong C_M$ is permissible. Additional points beyond the equivalence point are calculated similarly.

The entire curve is shown in Fig. 20-3. The shape is again the typical one met in acid-base and precipitation titrations; the calculation methods and the steps in reasoning should be compared. In analogy to these other titration types, one can appreciate that the size of the break for a given system will depend on the concentration of both the titrant and the metal ion; the lower the concentrations, the smaller the break. Further, for different systems at given concentrations of titrant and metal ion, the extent of the break is larger as the stability of the complex formed is greater. An increase in the stability constant for a 1:1 complex MY by a factor of 10^2 shifts the equivalence point up by 1 pM unit.*

---

*Here as in other cases, the equivalence point is taken as the inflection point. Correctly, for the equivalence point, the condition is $C_M = C_Y$; and for the inflection point, it is $C_M = C_Y - 1/K_{st,MY}$. For a useful titration, the value of the stability constant must be 10^5 or larger (see Section 20.7); consequently, $1/K_{st,MY}$ is so small that the inflection and equivalence points are almost identical. The student of calculus will understand the following sketch of how the relation for the inflection point is derived:

1. Combine the relations $C_M = [M] + [MY]$, $C_Y = [Y] + [MY]$, and $K_{st,MY} = [MY]/([M][Y])$ to eliminate [MY] and [Y].
2. Solve the resulting quadratic equation $K_{st,MY}[M]^2 + (K_{st,MY}C_Y - K_{st,MY}C_M + 1)[M] - C_M = 0$ and express the result as $-\log [M]$.
3. Differentiate the new equation twice with regard to C_Y and set $d^2(-\log [M])/dC_Y^2$ equal to zero (a condition for an inflection point).
4. Substitute the expression for $-\log [M]$, to get $C_M = C_Y - 1/K_{st,MY}$.

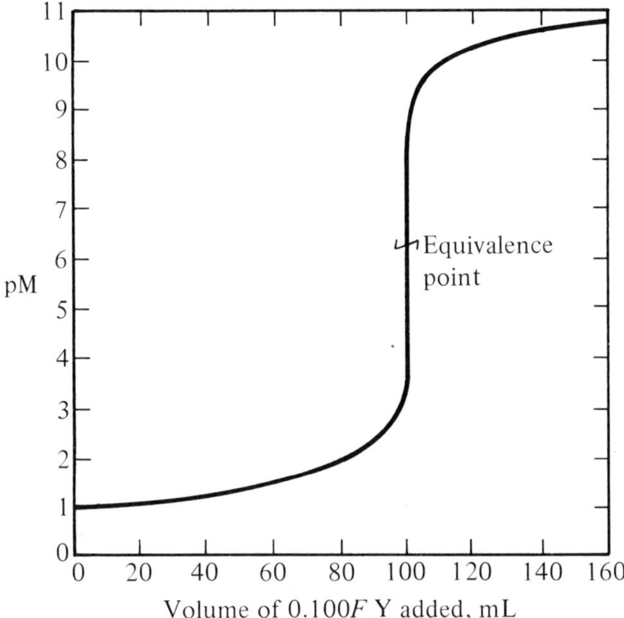

Fig. 20–3. Calculated titration curve for titration of 100 mL of a 0.100F solution of metal ion M with a solution 0.100F in complexing agent Y to form MY ($K_{st,MY} = 1.0 \times 10^{11}$).

20.7 *Requirements for a Compleximetric Titration*

For a compleximetric titration, the general requirements of any titration process (Section 10.1) must be met, including an essentially complete and relative rapid titration reaction and a method for detecting the end point. How far the titration reaction goes toward completion is reflected in the stability of the complex formed. It is not the stability as such but the product $K_{st}C_M$ that essentially determines the feasibility. As a rule of thumb, this product should have at least a value about 10^7 for the titration to yield acceptable results. In addition, however, the number of steps involved in the complex formation is important. This number should be as low as possible, preferably 1. The above data and considerations consequently hold only for titrations involving the formation of a 1:1 complex. The influence of the number of steps needs elaboration.

Suppose the complexation takes place in two steps. Then two end-point breaks are possible; but they will be fully developed only if the two stepwise stability constants differ sufficiently in their values. Actually to give two well-defined breaks, the stepwise constants must be separated by a factor of 10^4 or 10^5. (Compare the analogous situation in the titration of a diprotic acid, or the consecutive titration of two substances in a precipitation titration.) In practice, such a large factor is seldom found, at least not for systems that fulfill the other requirements for a titration. The two titration steps usually overlap, and one poorly defined break results. Even a high value for the overall stability constant will not necessarily assure a satisfactory titration, since this constant is "split" into two "step" constants. The situation becomes even worse if more than two steps are involved.

Most inorganic ligands are unidentate; therefore, compleximetric titrations involving such ligands are restricted to only a few metal ions, notably silver(I), which has a coordination number of 2, and mercury(II), which, although it has a maximum coordination number of 4, effectively coordinates with only two particles of many unidentate ligands (see Section 20.11). Compleximetric titrations for numerous metals are therefore based on the use of organic multidentate ligands that permit the complexation to proceed in a single step. Since chelate complexes are formed, this kind of titration may be assigned the more restrictive term *chelatometric titration,* a useful distinction. EDTA and some of its analogs are the most used representatives of this class of chelate-forming titrants.

So far the discussion has centered on titrating of a metal ion with a complexing agent. Obviously the possibility exists for the reverse titration, that is, the titration of a complexing agent with a suitable metal ion. The treatment is analogous and needs no special consideration.

20.8 Indicators in Compleximetric Titrations

There are a number of possibilities for indicating the end point in compleximetric titrations, including the formation of a precipitate (see Section 20.11) and the use of redox indicators or instrumental methods. However, the technique most often employed is visual indication with a complex-forming indicator. Metallochromic indicators, or simply metal indicators, form colored complexes with a metal ion, and the "metallized" indicator has a color different from that of the free, "unmetallized" form. The value of the stability constant of the metal-indicator complex must be sufficiently high to prevent extensive dissociation of this complex; such dissociation would cause a sluggish end point. The value must, however, be sufficiently smaller than the value of the metal-titrant complex to permit complete removal of the metal from the indicator complex at the end point.

The indicator equilibrium and the expression of the stability constant of the metal-indicator complex take the following form, where charges are again omitted since they are not relevant to the discussion:

$$\mathrm{M} + In \rightleftharpoons \mathrm{M}In \qquad (20\text{-}18)$$

$$K_{\mathrm{st,M}In} = \frac{[\mathrm{M}In]}{[\mathrm{M}]\,[In]} \qquad (20\text{-}19)$$

Rearranging of equation (20-19) yields

$$[M] = \frac{1}{K_{st,MIn}} \times \frac{[MIn]}{[In]} \tag{20-20}$$

which, written in logarithmic form, is

$$pM = \log K_{st,MIn} - \log \frac{[MIn]}{[In]} \tag{20-21}$$

The analogy of equation (20-21) with equation (12-4) for acid-base indicators is obvious, and all the considerations advanced in Chapter 12 concerning visibility, transition range, and the like apply fully here. The ideal case would be for the pM value of the midpoint of the color transition range to be identical with the pM value of the equivalence point in the compleximetric titration. In practice, however, because of the large breaks usually encountered, it is enough to have the transition range located on the steep portion of the titration curve, and then to perform the titration to the disappearance of one limiting color. Metal indicators exist that also form other complexes besides 1:1 complexes. The theoretical treatment is then more involved, but for practical application the considerations are the same as those given above.

Metal indicators complex not only with metal ions but also with protons and thus show acid-base indicator properties. Consequently, in the functioning of a metal indicator, the pH is important not only for its effect on the conditional stability constant but also for its effect on the color change.

Eriochrome Black T is an example of an important metal indicator for EDTA titrations. This indicator is supplied as the sodium salt of the compound having

HO

—OH

HO_3S— —N═N—

NO_2

the accompanying structural formula. The compound is a triprotic acid, H_3D. The sulfonic acid group, $-SO_3H$, is a relatively strong acid and the two phenolic groups, —OH, are weak ones. Consequently the indicator can be denoted by H_2D^-.

Eriochrome Black T can act as an acid-base indicator, and the color changes and respective pH values involved are as follows:

$$H_2D^- \underset{+H^+}{\overset{-H^+}{\rightleftarrows}} HD^{2-} \underset{+H^+}{\overset{-H^+}{\rightleftarrows}} D^{3-}$$

$$\begin{array}{ccccccccc} \text{red} & \leftarrow & \text{pH 6–7} & \rightarrow & \text{blue} & \leftarrow & \text{pH 11–12} & \rightarrow & \text{orange} \\ & & (\text{p}K_{a,1}, 6.3) & & & & (\text{p}K_{a,2}, 11.6) & & \text{(decomposition)} \end{array} \tag{20-22}$$

Eriochrome Black T forms 1:1 metal complexes with many metal ions and the chelate rings involve the phenolic oxygen atoms and the nitrogen atoms of the indicator molecule. All the metal complexes are red, but the shade differs for different metals. As you can see from equation (20–22), for this dye to be a metal indicator that gives a distinctive color change, the pH of the solution must be in the range 7 to 11. The color change is then *red* (metal complex) to *blue* ("free" indicator).

Such a pH range is not the only consideration in choosing the conditions for a titration. The indicator as a weak acid has a high affinity for the protons. Consequently, at too low a pH, even when within the range 7 to 11, the hydrogen ion concentration may become so high that the indicator tends to combine with protons instead of with the particular metal; the compound will then not function to indicate the end point. Thus it becomes obvious that using metal indicators calls for adjusting the pH appropriately for the titration and maintaining it within reasonable limits during the titration. For this purpose, it is common to use appropriate buffers.

Some metals form more stable complexes with a given indicator than with the titrant. For those metals, the given indicator is inappropriate. Moreover, in the presence of even traces of such metals the indicator will react with them and not function. The indicator is then said to be "blocked" (see Section 20.10).

20.9 Masking

Masking is defined as excluding the action of an interfering substance by adding an appropriate reagent and not removing the substance. Masking can be effected by precipitation, oxidation, reduction, complexation, or a combination of more than one of these processes. Masking is widely used as a powerful tool in many branches of analytical chemistry. However, because masking is often applied in EDTA titrations and because most masking techniques involve complex formation, it is appropriate to consider it here. Some examples randomly drawn from practical analysis illustrate various masking approaches.

The EDTA titration of magnesium is performed at pH 10. If barium (among various other metals) is present, it reacts with the titrant. If, however, sodium sulfate is added, barium precipitates as barium sulfate and then no longer interferes with the magnesium determination. The titration is performed in the presence of the precipitate. If the barium sulfate were removed, separation would be involved instead of masking.

Precipitation masking frequently suffers many of the adverse phenomena discussed for precipitation in connection with gravimetry. Coprecipitation of the substance to be determined or the reagent used in the determination has detrimental effects. The presence of the precipitate may obscure other phenomena, for example, the color change at the end point in a titration. Consequently, only a few cases of precipitation masking exist that have practical value.

The chemical behavior of a metal ion often changes drastically when the oxidation state is changed. Chromium(III), for example, precipitates readily as the hydrous oxide and thereby interferes with the gravimetric determination of, say, iron(III). If the chromium is oxidized to chromium(VI), it is present in anionic form as chromate or dichromate ion and as such does not precipitate together with iron when ammonia is added. Another example of oxidation masking involves arsenic(III), which interferes with the color reactions of some other elements; if it is oxidized to arsenic(V), the interference is often obviated.

Reduction masking is effective when iron(III) causes disturbances. Converting it to iron(II), often by reduction with ascorbic acid, is in many analytical procedures the choice for eliminating the interference. Another example is the reduction of mercury(II) to mercury(I), which at the same time is precipitated as the insoluble chloride; here there is a mixed masking involving both reduction *and* precipitation. The number of metals that can readily change their oxidation state is limited, however, and, therefore, so is the application of reduction masking.

In contrast, the number of complexation reactions is enormous, and such reactions are less prone to difficulties than, for example, precipitation reactions. Accordingly, complexation is most often the choice for masking metal ions.

A very versatile and effective complexing agent in masking is the cyanide ion. It forms very stable complexes with a variety of metal ions including iron, copper, mercury, cobalt, nickel, cadmium, zinc, and the noble metals. The masking of iron(II) and of iron(III) leads to the formation of the respective hexacyanoferrates. The masking is so effective that no chemical reaction known for either iron(II) or iron(III) gives a positive test. This "vanishing" of iron was very puzzling to early chemists. The cyano complexes of zinc and cadmium are relatively weak and the cyanide is able to react with formaldehyde; as a consequence zinc and cadmium can be freed, that is, "demasked." By this reaction, zinc or cadmium can be selectively detected and determined in the presence of the other elements named above, whose cyano complexes are not affected by formaldehyde.

Iron(III) in chloride-containing solutions produces an intensely yellow color due to the formation of chloro complexes. Although the iron(III) itself may undergo no interfering reaction, the intense color can cause difficulties in determining (or detecting) other substances. If phosphoric acid is added, a colorless iron(III)-phosphato complex forms and the yellow solution color disappears. Silver forms an ammine complex that is stable enough to prevent silver chloride from precipitating, but not silver iodide; consequently, ammonia masks silver against chloride and allows the selective detection or determination of iodide. The stability of the complexes formed by many metals with EDTA is the reason that EDTA has become one of the most versatile masking agents. An example of mixed redox-complexation masking may be cited. Cobalt(II) is oxidized to cobalt(III) when ammonia and hydrogen peroxide are added and is then strongly bound in an ammine complex. The ammine complexes of other metals are weak enough to permit reaction of these metals with other reagents. This masking of cobalt is especially valuable when nickel is to be determined.

Additional examples of masking will be found below.

20.10 Application of EDTA Titrations

Probably the most important application of an EDTA titration is the determination of the *total* hardness of water. (The concept of water hardness has already been discussed in Section 14.8.) Since both calcium and magnesium complex strongly with EDTA, the sum of these metals is readily determined. The determination can be effected by adding pH 10 ammoniacal buffer and some Eriochrome Black T to the water sample, and then titrating with a standard EDTA solution to an end point marked by a color change of red to blue. Usually some potassium cyanide is also added to mask traces of heavy metals that would block the indicator. EDTA titrations also permit the rapid determination of both the calcium and the magnesium hardness separately. In one aliquot of the water sample, the *sum* of calcium and magnesium is titrated as described above. To a second aliquot, sodium hydroxide is added to reach a pH above 12. Magnesium is precipitated as the hydroxide and is thereby masked in the titration of calcium with EDTA. Murexide is an indicator often used in this titration. The magnesium hardness is found by difference.

The principles for determining calcium and magnesium in water described above underlie the determination of these two elements in a large variety of materials, ranging from blood serum and pharmaceuticals to rocks and minerals.

Alloys can frequently be analyzed rapidly and simply by EDTA titrations. For example, a bismuth-lead alloy can be analyzed as follows. A sample of the material is dissolved in nitric acid and then the bismuth is titrated with EDTA at a pH 1 to 1.5, with Pyrocatechol Violet as the indicator. When the end point is reached (blue to yellow), the volume of EDTA needed is recorded and the pH is increased to about 5. Then lead is titrated with EDTA in the same solution using the same indicator (again a blue to yellow color change).

A mixture of zinc and lead can be analyzed in the following manner. To the weakly acidic sample solution some tartaric acid is added; the tartrate masks the lead against precipitation as the hydroxide (but does not impair its reaction with EDTA) when ammoniacal buffer of pH 10 is added in the next step. (The ammonia complexes with the zinc and thus masks it against hydroxide precipitation.) To the buffered solution potassium cyanide is added next. This agent masks the zinc against EDTA but does not affect the lead, which is then titrated with EDTA using Eriochrome Black T as the indicator. The volume of EDTA is recorded and formaldehyde is added to demask the zinc (see Section 20.9), which is then titrated using the same titrant and indicator.

20.11 Other Compleximetric Titrations

Of the limited examples of complexation titrations not involving a chelating titrant, two are of practical interest.

The titration of cyanide with silver was developed as early as 1851 by Liebig. The titration reaction is

$$2CN^- + Ag^+ \rightleftharpoons Ag(CN)_2^- \tag{20-23}$$

This actually is a two-step reaction, but the formation constant of the complex has an exceptionally high value (1.1×10^{21}). The titration is self-indicating, and no added indicator is necessary. Near the equivalence point, all the cyanide ion is transferred into the dicyanoargentate(I) ion. When more titrant solution (that is, silver nitrate) is added, insoluble silver dicyanoargentate(I) is formed. The equation for the indicator reaction is

$$Ag(CN)_2^- + Ag^+ \rightleftharpoons \underline{Ag[Ag(CN)_2]} \tag{20-24}$$

The appearance of a permanent turbidity signals the end point. Good stirring and slow addition of titrant solution are essential in this titration. Because of local overconcentration of titrant, a transient turbidity may become evident during the titration. The permanent turbidity indicating the end point develops slowly. The cyanide titration is routinely applied to the analysis of cyanide-containing plating baths.

The mercurimetric determination of chloride offers another example of a compleximetric titration of practical significance. The titration involves simply adding a titrant solution of mercury(II) nitrate or perchlorate to the acidic sample solution until an appropriate indicator signals the end point. The combining ratio between mercury(II) and chloride ion in this titration is 1:2.

Although mercury(II) ion has a coordination number of 4, it often acts with an "effective" coordination number of 2. This is evident, for example, from the logarithmic values of the stability constants of the chloro complexes of mercury(II) –$HgCl^+$, $HgCl_2$, $HgCl_3^-$, and $HgCl_4^{2-}$; these are, in turn, 6.75, 6.48, 0.95, and 1.05. The constants for the first and second steps are close in value. In a simple way, it may be reasoned that the coordination of the two chloride ions takes place simultaneously in a quasi-single step. In addition, the overall stability constant for the two steps is high (1.7×10^{13}). Mercury(II) chloride, as a complex, is a weak electrolyte and is little dissociated in aqueous solution. In contrast, mercury nitrate and perchlorate are strong electrolytes and are common titrants. A stable, mercury(II) standard solution is prepared by dissolving a weighed amount of mercury(II) oxide in either nitric or perchloric acid and diluting to known volume with water; mercury(II) oxide is an excellent primary standard due to its high equivalent weight, exact stoichiometric composition, and ready availability in high purity.

In mercurimetric titrations, *sym*-diphenylcarbazide [that is, 1,5-diphenylcarbohydrazide, $(C_6H_5NHNH)_2CO$] or its oxidation product *sym*-diphenylcarbazone is often used as a metallochromic indicator. These compounds are yellow in solution and form blue-violet complexes with mercury(II). With the first indicator, the optimum range for the titration is pH 1.5 to 2.0, and with the second, pH 3.0 to 3.5.

Since the compleximetric titration of chloride with mercury(II) ion proceeds in acidic solution, this method is often used in place of the Mohr titration with silver ion, especially in determining chloride in blood serum and other biological fluids.

20.12 Calculating Results in Compleximetric Titrations

Calculations for compleximetric titrations are based on the reaction describing the complex formation. This reaction can be written in the general form

$$nS + mT \rightleftharpoons S_nT_m \qquad (20\text{–}25)$$

where S is the species titrated and T the titrant. From this equation it follows that one mole of the titrant is equivalent to n/m moles of the species titrated. If the species titrated and the sought-for substance are not identical, an appropriate conversion factor can be additionally applied. The calculations are especially simple for EDTA titrations, because here 1:1 complexes are formed almost exclusively, yielding $n = m = 1$. (Only molybdenum and tungsten form 2:1 complexes.)

Example 20–5 How much $CaCl_2$ (*110.99*) in milligrams is present in a sample that on titration requires 25.22 mL of 0.01350*F* EDTA?

Since calcium and EDTA react in a 1:1 ratio, the calculation takes the form

$$mL_t \times F_t \times FW_s \times 1 = mg_s$$

$$25.22 \times 0.01350 \times 110.99 \times 1 = 37.79 \text{ mg of } CaCl_2 \text{ in sample}$$

Example 20–6 What amount of KCN (65.12) in milligrams is present in a sample if 18.35 mL of 0.1050*F* $AgNO_3$ is needed for its compleximetric titration?

From the titration equation, (20–23), silver and cyanide ions react in a 1:2 ratio; hence

$$mL_t \times F_t \times FW_s \times 2 = mg_s$$

Substituting the numerical data yields

$$18.35 \times 0.1050 \times 65.12 \times 2 = 250.9 \text{ mg of KCN in sample}$$

Example 20–7 A solution containing nickel ion is titrated in an ammoniacal medium with 0.1160*F* KCN, of which a volume of 20.15 mL is required. Express the titration result in milligrams of nickel pyrophosphate, $Ni_2P_2O_7$ (*291.36*).

The titration reaction is

$$Ni^{2+} + 4CN^- \rightarrow Ni(CN)_4^{2-}$$

Thus, one-fourth mole of nickel is equivalent to one mole of cyanide ion or of potassium cyanide; however, two moles of nickel ion are furnished by one mole of nickel pyrophosphate. Consequently, one-eighth mole of nickel pyrophosphate is equivalent to one mole of potassium cyanide and the required formula becomes

$$\text{mL}_t \times F_t \times \text{FW}_s \times \frac{1}{8} = \text{mg}_s$$

Here the factor 1/8 contains both the combining ratio according to the complexation reaction (1/4) and the equivalence factor for the conversion of nickel to nickel pyrophophate (1/2).

Alternatively, the overall stoichiometry can be shown by writing the reaction of interest as

$$Ni_2P_2O_7 + 8KCN \rightarrow 2K_2[Ni(CN)_4] + K_4P_2O_7$$

From the equation it is immediately evident that one mole of potassium cyanide is equivalent to one-eighth mole of nickel pyrophosphate.

Substituting the numerical data yields

$$20.15 \times 0.1160 \times 291.36 \times \frac{1}{8} = 85.13 \text{ mg of } Ni_2P_4O_7$$

20.13 *Questions*

20-1 What is the relation between the dissociation constant and the stability constant of a complex?

20-2 Define in your own words the terms *chelate ring, chelate complex, central ion, coordination number, masking,* and *blocking* (of an indicator).

20-3 Explain why a low number of steps is essential if a complexation process is to be made the basis of a titration.

20-4 Compare the titration curves and parameters influencing their shape for acid-base, precipitation, and compleximetric titrations. What influences the size of the "break" and the position of the equivalence point? What changes take place in the shape of the titration curves and in the location of the equivalent point if changes are made in concentrations of the species titrated and of the titrant?

20-5 When zinc is titrated with EDTA using Eriochrome Black T, an ammoniacal buffer solution is added. What function does this solution have besides buffering the solution? Explain.

20–6 Explain how a metallochromic indicator functions. What is required of a substance for it to be a useful indicator for an EDTA titration?

20–7 For a single pH value and without resort to masking, discuss the possibility for the stepwise titration of two metal ions with a single titrant that complexes with both ions. Parallel your reasoning with the reasoning for the stepwise titration of a diprotic acid with a base.

20–8 In the EDTA titration of magnesium, any iron(III) present in the original sample solution precipitates as hydrous oxide when ammoniacal buffer of pH 10 is added. The hydrous oxide does not react with the indicator and the titrant. Critically discuss the significance of these facts for masking iron.

20–9 Would the total hardness of a water sample change if it were heated before analysis?

20–10 Could ordinary tap water be used in preparing a standard EDTA solution? Elaborate your answer.

20–11 Explain why adequate buffering is important in EDTA titrations.

20–12 Give several examples for each type of masking technique. (Draw the examples from your own knowledge and experience and do not cite those discussed in Section 20.9.)

20–13 What is the meaning of the statement, EDTA can function as a hexadentate ligand?

20–14 What do you understand by the expression "'free' metal ion concentration in a solution"?

20–15 Tetracycline hydrochloride is an antibiotic with nitrogen and oxygen atoms so placed in the structure that stable five-membered rings with metal ions, including aluminum, calcium, and magnesium, can be formed. Since this is so, elaborate on the warning on a tetracycline label that the intake of dairy products should be reduced and that if antacids are to be used in cotherapy, those containing the above-named metals should be avoided. Suggest a compound that could be substituted as the antacid.

20.14 Problems

20–1 A 1:1 complex MZ, having a stability constant of 1.00×10^{10}, is present as a $2.50 \times 10^{-3}F$ solution. Calculate the free metal ion concentration.

Ans.: $5.00 \times 10^{-7}M$

20–2 The solution of the complex MZ of Problem 20–1 is made $0.100M$ in Z. What is the free metal ion concentration? *Ans.:* $2.5 \times 10^{-12}M$

20–3 Exactly 10 mL of a $0.0100F$ solution of M is titrated compleximetrically with $0.00500F$ Z. The reactants form a 1:1 complex MZ ($K_{st} = 1.00 \times 10^{11}$). Calculate what the pM value is at (a) the start of the titration,

(b) after exactly 5 mL of titrant is added, (c) after exactly 20 mL of titrant is added, and (d) after exactly 30 mL of titrant is added.

Ans.: (a) 2.00; (b) 2.30; (c) 6.74; (d) 10.70

20–4 What is the concentration of free metal ion in a $0.0010F$ solution of the complex MZ ($K_{st} = 5.0 \times 10^2$)? (*Hint:* Since the complex is weak, C_{MZ} cannot be taken as equal to [MZ].) *Ans.:* $7.3 \times 10^{-4}M$

20–5 A metal ion forms a complex MZ ($K_{st} = 1.0 \times 10^8$) and a slightly soluble compound MP_2 ($K_{sp} = 1.0 \times 10^{-10}$). Can a $0.100F$ solution of MZ be made $0.010M$ in P without precipitation of MP_2? Calculate the free metal ion concentration according to the relevant solubility and complexation equilibria and compare the figures in deciding.

Ans.: No; $[M]_{complex} = 3.2 \times 10^{-5}M$, $[M]_{solubility} = 1.0 \times 10^{-6}M$

20–6 A $0.200F$ EDTA solution is used to titrate exactly 100 mL of a solution of $ZnCl_2$ (*136.3*). A volume of 50.0 mL of EDTA was needed to reach the Eriochrome Black T end point. What is the % w/v concentration of the zinc salt in the solution? *Ans.:* 1.36% w/v $ZnCl_2$

20–7 What volume, in milliliters, of $0.0500F$ $MgSO_4$ is needed for determining the substances listed below by a back-titration? In each case, a volume of 30.00 mL of $0.0500F$ EDTA has been added to a solution containing 100.0 mg of the substance indicated.

	Substance titrated	
(a)	$CaCO_3$	(*100.1*)
(b)	$MgSO_4$	(*120.4*)
(c)	$Pb(NO_3)_2$	(*331.2*)
(d)	$Al_2(SO_4)_3$	(*342.1*)

20–8 A mixture contains KCl (*74.56*), LiCl (*42.39*), and $MgSO_4 \cdot 7H_2O$ (*246.48*). An amount of 1.8650 g of the mixture is dissolved in pure water and diluted to mark with water in a 250-mL volumetric flask. A one-fifth aliquot of the solution is treated with 100.0 mL of $0.1000F$ $AgNO_3$. The excess silver is back-titrated with $0.1885F$ KSCN, 20.00 mL being required. Another one-fifth aliquot of the sample solution needs 11.50 mL of $0.0100F$ EDTA. Calculate the % w/w composition of the sample.

20–9 Calculate the molar concentrations of uncomplexed metal ion M and ligand Z in a solution $1.0 \times 10^{-3}F$ in complex MZ, from the assumed values of the stability constant.

	Stability constant of MZ
(a)	1.0×10^9
(b)	5.0×10^{10}
(c)	7.5×10^8
(d)	6.2×10^6
(e)	8.4×10^9
(f)	4.0×10^5
(g)	3.2×10^{14}

20–10 What are the molar concentrations of uncomplexed metal ion M and ligand Z when the concentrations of the complex MZ ($K_{st} = 2.0 \times 10^8$) and ligand Z in the solution have the specified values?

	Formality in MZ	Molarity in excess Z
(a)	1.0×10^{-2}	1.0
(b)	1.0×10^{-3}	5.0×10^{-1}
(c)	1.0×10^{-1}	1.0×10^{-3}
(d)	5.0×10^{-2}	2.0×10^{-2}
(e)	5.0×10^{-1}	1.0×10^{-3}
(f)	5.0×10^{-3}	8.0×10^{-1}

20–11 A standard mercury(II) solution is prepared by dissolving exactly 1 g of mercury (*200.59*) of 99.71% purity in nitric acid and diluting to exactly 1 L with pure water. A sample solution is titrated with the mercury(II) solution, 24.85 mL being required. Calculate the amount of NaCl, in milligrams, in the sample.

20–12 A solution of $FeCl_3$ is titrated with 0.04930*F* mercury(II) nitrate, 35.21 mL being required. What volume, in milliliters, of 0.0832*F* EDTA will be required to titrate the iron in an identical aliquot of that solution?

20–13 The chloride content of blood serum is often determined by a mercurimetric titration. In a centrifuge tube are mixed 2.00 mL of water, 0.50 mL of blood serum, and 0.50 mL of trichloroacetic acid. Protein is thereby precipitated and is removed by centrifugation. Exactly 2 mL of the clear supernatant liquid is transferred to a vessel and titrated with 0.0100*F* mercury(II) nitrate using *sym*-diphenylcarbazide as the indicator. A blank is performed in exactly the same manner, but an equal volume of water is substituted for the serum. The formula used to calculate the result is

$$(\text{mL}_{\text{titrant in titration}} - \text{mL}_{\text{titrant in blank}}) \times (\text{factor}) = \text{mg \% Cl}$$

By mg % Cl in the jargon of clinical analysis is meant "the milligrams of chloride in 100 mL of the serum." If the determination is performed as described, the *factor* has a certain numerical value. Calculate this value. Chloride has a formula weight of 34.45. Take into account that one mercury(II) ion in the titration complexes with two chloride ions.

Ans.: 213

20–14 The normal concentration of calcium (*40.1*) in blood serum is 9 to 11 mg % (that is, 9 to 11 mg of calcium in 100 mL of serum). The value 7 mg % or lower or 13 mg % or higher is considered a sign of an abnormal condition. A 1-mL sample of serum in a compleximetric titration of calcium was found to require 2.85 mL of 1.560×10^{-3}*F* EDTA.

(a) Express the result of the titration in mg % Ca.

(b) Does the result suggest an abnormal condition?

(c) The concentration is often given in milliequivalents of calcium per liter of serum, with one equivalent here taken as one-half mole. Express the titration result in this way.

(d) Express the titration result in millimoles of calcium per liter.
Ans.: (a) 17.8 mg % Ca; (b) yes; (c) 8.9_1 meq Ca/L; (d) 4.4_5 mmol Ca/L

20–15 The normal level of magnesium (*24.3*) in blood serum is 2 to 3 mg % (that is, 2 to 3 mg of magnesium in 100 mL of serum).
(a) How many grams of EDTA as the disodium salt dihydrate (*372.2*) must be dissolved in water and be diluted to a volume of 100.0 mL to get a solution of which 0.100 mL is equivalent to 0.20 mg % Mg when a 1-mL sample of serum is taken and titrated after calcium is removed by an oxalate precipitation?
(b) What is the formality of this EDTA solution?
(c) What mg % Ca of calcium (*40.1*) is equivalent to 0.100 mL of this EDTA solution when 1 mL of serum is taken for the titration?
(d) Describe the conditions, including the indicator used, for the titration of the magnesium. What should be the titration result of the magnesium titration if calcium were not removed before this titration? Explain.
(e) What is the normal serum concentration range for magnesium expressed in millimoles per liter?
Ans.: (a) 0.0306_3 g; (b) $8.23 \times 10^{-4}F$; (c) 0.330 mg % Ca; (d) the result would be high; (e) 0.8 to 1 mmol/L

20–16 A metal ion M^+ forms a 1:1 complex with the anion of the weak acid HA ($K_a = 1.00 \times 10^{-8}$). The complex has a stability constant of 1.00×10^{10}.
(a) Calculate the conditional stability constant for pH 2.0.
(b) What is the molar concentration of uncomplexed metal ion at pH 2.0 when the concentration of the complex is $1.00 \times 10^{-4}F$?
Ans.: (a) 1.0×10^4; (b) $6.2 \times 10^{-5}M$

20–17 The concept of indirect analysis, exemplified for acid-base and precipitation titrations (Sections 15.2 and 19.3), can be applied with compleximetric titrations. For a *binary* alloy of magnesium (*24.3*) and zinc (*65.4*), 0.1500 g is dissolved in acid. The solution is neutralized and ammoniacally buffered to pH 10; the solution is then titrated to the Eriochrome Black T end point with 0.1000*F* EDTA, of which 37.22 mL is needed. Calculate the percentage composition of this alloy. *Ans.:* 36.8% Mg; 63.2% Zn

20–18 Chloride can be determined in a variety of samples by titrating the acidified solution with mercury(II) nitrate using diphenylcarbazone indicator. To a 100.0-mL sample of drinking water, 1 mL of acidified diphenylcarbazone indicator solution is added. The solution is titrated to a purple end point with 0.0141*N* (0.007050*F*) $Hg(NO_3)_2$; a volume of 28.72 mL is needed. For a blank of 100.0 mL of distilled water with 10 mg of $NaHCO_3$, 0.24 mL is needed. Calculate that the chloride (*35.45*) concentration, expressed in milligrams per liter.

20–19 The *total hardness* of river waters can be determined by an EDTA titration. A 25.0-mL sample is diluted to 50 mL with distilled water and ammoniacally buffered to pH 10.1. Then 2 drops of an Eriochrome Black T indicator solution are added. The solution is titrated with 0.0098*F* EDTA; a volume of 12.37 mL is required. Calculate the total hardness of the river water, expressed as mg/L $CaCO_3$ (*100.09*).

21

Oxidation-Reduction Equations (Redox Equations)

21.1 Introduction

Originally, the term *oxidation* was used to designate a reaction in which a substance was caused to react or to combine with oxygen. The term *reduction* indicated a reaction in which oxygen was removed from a substance. These concepts have undergone generalization, and in terms of the electron concepts of matter are applied to reactions in which electrons are lost or gained. If a substance loses one or more electrons, it is said to undergo oxidation; if it gains one or more electrons, it is said to undergo reduction. The combined *red*uction-*ox*idation process, abbreviated *redox* process, involves the transfer of one or more electrons from the reducing agent (also known as the reductant or reducer) to the oxidizing agent (the oxidant or oxidizer). Since free electrons do not accumulate in chemical reactions, the number of electrons lost must exactly equal the number of electrons gained.

A redox reaction, involving the transfer of one or more electrons, is formally analogous to an acid-base reaction, in which one or more protons are transferred (see Chapter 16).

Because electrons do not appear as reactants or products in the complete redox equation, it is often profitable to split the reaction into two partial reactions, known as half-reactions—one related to the oxidation and having one or more electrons appearing as a product; the other related to the reduction and having one or more electrons appearing as a reactant. The correctly balanced half-reaction equations can be multiplied by appropriate factors so that when the equations are added, the electrons cancel and the balanced complete redox equation is obtained. The details of this balancing process for redox equations are considered below.

The concept of the half-reaction has an experimental basis since many such reactions can occur in electrochemical half-cells (Chapter 22). When two such half-cells are suitably connected, the redox process proceeds without physical mixing of the components of the two half-cells. Half-reaction equations, however—like most chemical equations—do not delineate the actual mechanism of the process. The considerations in this textbook are confined to half-reactions that take place in aqueous medium; consequently, when necessary, water and hydrogen ion or hydroxide ion can be considered participants in half-reactions, to achieve a balance of material in the equation.

A half-reaction equation must be balanced in material (that is, the same number of atoms of each element must appear on both sides of the equation) and also in electrical charge. The preferred approach, delineated below, is balancing material first, then adding water, hydrogen ion, or hydroxide ion as a reactant or a product when necessary. Charge can then be balanced by simply adding the arithmetically necessary number of electrons as either a product or a reactant. An alternative approach (considered in Section 21.4) involves establishing initially the number of electrons involved in a half-reaction equation; this calls for assigning oxidation states to the atoms involved in the reduction and oxidation processes. Such assignments must often be arbitrary for composite ions and covalent compounds, and may then be a source of confusion. Either method, correctly followed, leads to the same redox equation.

21.2 Balancing Redox Equations: Electron-Change Method

The first step in balancing a redox equation is to write down the reactants and products, which must either be established experimentally or be inferred from chemical knowledge. Once the reactants and products are written, the balancing of the redox equation involves merely balancing material and charge.

In our study of balancing, redox equations and half-reaction equations are written with equals signs; when the reactions are known to be fully reversible, the double arrow, $\rightleftharpoons$, could be appropriately substituted. The balancing, however, is independent of the reversibility or nonreversibility of the redox process.

Example 21-1 Write the redox equation for the reaction of iron(III) with tin(II) to yield iron(II) and tin(IV).

The half-reaction equations are

$$Sn^{2+} = Sn^{4+} + 2e \qquad \text{(oxidation, balanced)}$$

$$Fe^{3+} + e = Fe^{2+} \qquad \text{(reduction, balanced)}$$

The second half-reaction equation is then multiplied by 2 and the two equations added:

$$\begin{array}{r} Sn^{2+} = Sn^{4+} + 2e \\ 2Fe^{3+} + 2e = 2Fe^{2+} \\ \hline 2Fe^{3+} + Sn^{2+} = 2Fe^{2+} + Sn^{4+} \end{array} \qquad \text{(material and charge balanced)}$$

In this example, the number of electrons involved in each half-reaction equation is obvious since simple ions are involved. With composite ions or covalent compounds, however, determining the number of electrons becomes more difficult.

Example 21-2 Iron(II) is oxidized in *acidic* medium by permanganate ion, MnO_4^-, to yield iron (III) and manganese(II). Balance the redox equation.

$$Fe^{2+} + MnO_4^- = Fe^{3+} + Mn^{2+} \quad \text{(unbalanced)}$$

The half-reactions are

$$Fe^{2+} = Fe^{3+} \quad \text{(unbalanced)}$$

$$MnO_4^- = Mn^{2+} \quad \text{(unbalanced)}$$

Notice that at this point it is unnecessary to recognize explicitly which half-reaction represents oxidation and which reduction.

The first half-reaction equation is already balanced for material, and to achieve a balance of charge, one electron is added to the right side:

$$Fe^{2+} = Fe^{3+} + e \quad \text{(balanced)}$$

In the second half-reaction equation, manganese is already balanced, since one atom appears on both sides. To balance oxygen, four molecules of water are added to the right side; and then to balance the added hydrogen, eight hydrogen ions are added to the left side. Hydrogen ion and water are used, instead of hydroxide ion and water, because the reaction occurs in acidic medium. The result thus obtained is

$$MnO_4^- + 8H^+ = Mn^{2+} + 4H_2O \quad \text{(material balanced, charge unbalanced)}$$

To balance charge, it is now only necessary to note the net electric charge for each side of the equation, +7 on the left and +2 on the right, and to add electrons so that the net charges become equal. Consequently, five electrons are added to the left side. The fully balanced half-reaction equation is therefore

$$MnO_4^- + 8H^+ + 5e = Mn^{2+} + 4H_2O \quad \text{(balanced)}$$

To secure the complete redox equation, the first half-reaction equation, which involves a single electron, is multiplied by 5 and the two half-reaction equations are added. The final result is

$$5Fe^{2+} + MnO_4^- + 8H^+ = 5Fe^{3+} + Mn^{2+} + 4H_2O$$

Example 21-3 The following oxidation of the hexanitritocobaltate(III) ion occurs in acidic solution:

$$Co(NO_2)_6^{3-} + MnO_4^- = Mn^{2+} + Co^{2+} + NO_3^- \quad \text{(unbalanced)}$$

The half-reactions can be written

$$MnO_4^- = Mn^{2+} \qquad \text{(unbalanced)}$$

$$Co(NO_2)_6^{3-} = Co^{2+} + NO_3^- \qquad \text{(unbalanced)}$$

The first half-reaction equation has already been considered in Example 21–2:

$$MnO_4^- + 8H^+ + 5e = Mn^{2+} + 4H_2O \qquad \text{(balanced)}$$

In the second half-reaction equation, six nitrate ions must appear on the right side to balance the six atoms of nitrogen on the left:

$$Co(NO_2)_6^{3-} = Co^{2+} + 6NO_3^- \qquad \text{(unbalanced)}$$

Twelve oxygen atoms appear on the left side and eighteen on the right; therefore to balance oxygen, six molecules of water are added to the left side:

$$Co(NO_2)_6^{3-} + 6H_2O = Co^{2+} + 6NO_3^- \qquad \text{(unbalanced)}$$

To balance hydrogen, twelve hydrogen ions must be added to the right side. (Water and hydrogen ion are added because the reaction occurs in acidic medium.) The balancing of material is thereby completed:

$$Co(NO_2)_6^{3-} + 6H_2O = Co^{2+} + 6NO_3^- + 12H^+$$

(material balanced, charge unbalanced)

The net charge on the left side is –3 and on the right side +8; to balance charge, we must add eleven electrons to the right side to yield the fully balanced half-reaction equation:

$$Co(NO_2)_6^{3-} + 6H_2O = Co^{2+} + 6NO_3^- + 12H^+ + 11e \qquad \text{(balanced)}$$

The first half-reaction equation is multiplied by 11 and the second by 5, and the two equations are added:

$$11MnO_4^- + 88H^+ + 55e = 11Mn^{2+} + 44H_2O$$

$$5Co(NO_2)_6^{3-} + 30H_2O = 5Co^{2+} + 30NO_3^- + 60H^+ + 55e$$

$$5Co(NO_2)_6^{3-} + 11MnO_4^- + 28H^+ = 5Co^{2+} + 30NO_3^- + 11Mn^{2+} + 14H_2O$$

Example 21–4 In an alkaline aqueous medium, the citrate anion, $C_6H_5O_7^{3-}$, is completely oxidized by permanganate ion according to the equation

$$C_6H_5O_7^{3-} + MnO_4^- = CO_2 + MnO_2 \qquad \text{(unbalanced)}$$

Notice that in alkaline media the reduction product of the permanganate ion is insoluble manganese(IV) oxide, and not manganese(II), which is obtained in acidic media. Carbon dioxide is written here as a product. Depending on solution conditions, however, the carbon dioxide may persist in the solution as hydrogen carbonate ion or carbonate ion. When the exact conditions are known, one of these ions could be written as the product and the redox equation be balanced accordingly. The half-reactions are

$$MnO_4^- = MnO_2 \qquad \text{(unbalanced)}$$

$$C_6H_5O_7^{3-} = CO_2 \qquad \text{(unbalanced)}$$

The first half-reaction equation is already balanced for manganese; to reach a material balance, we must consider oxygen. Since an alkaline medium is involved, it is desirable to operate with water and hydroxide ion. In such a case, add the difference in oxygen in the form of water to the side with the excess in oxygen and then add twice that amount of hydroxide ion to the other side. In the reaction of immediate interest, this results in

$$MnO_4^- + 2H_2O = MnO_2 + 4OH^- \qquad \text{(material balanced, charge unbalanced)}$$

Since the net charge on the left side is −1 and on the right side −4, charge is balanced by adding three electrons to the left side to yield the fully balanced half-reaction equation:

$$MnO_4^- + 2H_2O + 3e = MnO_2 + 4OH^- \qquad \text{(balanced)}$$

In the second half-reaction equation, citrate contains both hydrogen and oxygen. Whenever such a situation exists, the measures suggested for balancing the permanganate half-reaction in alkaline medium may be inadequate. However, the material balance can be readily achieved as follows.

First, balance all elements except hydrogen and oxygen. For the second half-reaction, this gives

$$C_6H_5O_7^{3-} = 6CO_2 \qquad \text{(unbalanced)}$$

Then balance oxygen by adding water when needed. Thus

$$C_6H_5O_7^{3-} + 5H_2O = 6CO_2 \qquad \text{(unbalanced)}$$

Next balance hydrogen. For an alkaline medium, as in this example, add water to the side deficient in hydrogen and an equal number of hydroxide ions to the other side. (For an acidic medium, add the appropriate number of hydrogen ions to the side deficient in hydrogen.) Thus,

$$C_6H_5O_7^{3-} + 5H_2O + 15OH^- = 6CO_2 + 15H_2O$$

(material balanced; charge unbalanced)

Often, as in this example, water molecules will appear on both sides of the equation, and some can be canceled.

$$C_6H_5O_7^{3-} + 15OH^- = 6CO_2 + 10H_2O$$

(material balanced; charge unbalanced)

Because the net charge on the left side is −18 and on the right side is zero, charge is balanced by adding eighteen electrons to the right side. The fully balanced half-reaction equation is

$$C_6H_5O_7^{3-} + 15OH^- = 6CO_2 + 10H_2O + 18e \qquad \text{(balanced)}$$

Since the first half-reaction involves three electrons and the second eighteen, the first equation is multiplied by 6 and the two half-reactions equations are added, to yield the complete redox equation:

$$\begin{array}{r} 6MnO_4^- + 12H_2O + 18e = 6MnO_2 + 24OH^- \\ C_6H_5O_7^{3-} + 15OH^- = 10H_2O + 6CO_2 + 18e \\ \hline C_6H_5O_7^{3-} + 2H_2O + 6MnO_4^- = 6CO_2 + 6MnO_2 + 9OH^- \end{array}$$

It is always advisable to inspect the final equation to ensure that a balance of both material and charge has been attained, in other words, to rule out any arithmetical error. The coefficients of the reactants and products in the final equation should be examined to see whether a common denominator is present; if so, the coefficients should be divided by it, so that the smallest integers possible appear as the final coefficients.

Redox reactions may also take place in a slightly acidic, slightly alkaline, or neutral medium. The question then arises whether to use hydrogen ion or hydroxide ion in securing a material balance. You recall that both hydrogen ion and hydroxide ion are present in an aqueous solution and the designation *acid* or *alkaline* refers only to the predominance of one of these species, not to the absence of the other. As noted in Section 21.1, neither the complete redox equation nor the half-reaction equations represent the actual reaction mechanism; the writing of hydrogen ion rather than hydroxide ion or the reverse is therefore to some degree arbitrary. For example, the reduction of permanganate to manganese(IV) oxide can also occur in slightly acidic medium; since this process is the general one that takes place in alkaline solution, the redox equation may be appropriately written with hydroxide ion as a product, as in Example 21-4.

21.3 Determination of Oxidation State

In applying the electron-change method in balancing redox equations, as presented in Section 21.2, it is not necessary to determine the oxidation state of the various atoms involved, since the number of electrons in each half-reaction equation is obtained, in one sense, automatically. It is sometimes of interest to determine the oxidation state and the changes in oxidation states involved in a reduction or an oxidation without the necessity of balancing an equation. Further, the calculation of an oxidation state is necessary in writing the systematic names for those substances for which the oxidation state is given in parenthetical Roman numerals (a positive sign is understood). It is also possible to balance a half-reaction through the change in oxidation states (Section 21.4).

The oxidation state is determined by applying the following rules:

1. The oxidation state of a simple (monoatomic) ion equals its ionic charge.
2. The sum of the oxidation states of the atoms in a molecule or an ion must equal the total charge of that molecule or ion.
3. The oxidation state of oxygen is -2 except in peroxy compounds, oxygen gas, and substances in which oxygen is bonded to fluorine, where it is -1, 0, and +2, respectively.
4. The oxidation state of hydrogen is +1 except in hydrogen gas, where it is 0, and in hydrides, where it is -1.

These rules can be exemplified.

In the ion Fe^{2+} by rule 1, the oxidation state is +2; hence the ion is termed the iron(II) ion.

In the permanganate ion, MnO_4^-, the oxidation state of manganese is determined by rules 2 and 3. The sum of the oxidation states of manganese and oxygen must equal -1; four oxygens equals a total of $-2 \times 4 = -8$; hence manganese is present in the oxidation state +7. Or in other terms, manganese(VII) is present in the permanganate ion. When this ion is reduced in acidic solution to manganese(II), it follows that $7 - 2 = 5$ electrons are involved.

The oxidation state of manganese in MnO_2, by rules 2 and 3, is +4, and the compound can be named manganese(IV) oxide. In the reduction of permanganate in alkaline medium to this compound, it follows that $7 - 4 = 3$ electrons are involved.

The oxidation states of cobalt and nitrogen in the ion $Co(NO_2)_6^{3-}$ are calculated as follows. In the nitrite ion, NO_2^-, nitrogen has an oxidation state of +3, since the charge on the ion is -1. Six nitrite ions are present in the complex ion $Co(NO_2)_6^{3-}$, and consequently the oxidation state of cobalt must be +3. The systematic name of this ion is hexanitritocobaltate(III); the ion was known classically as the cobaltinitrite ion. If this ion is oxidized to the products Co^{2+} and NO_3^- (Example 21-3), the oxidation state of cobalt is +2, and indeed cobalt is reduced. The oxidation state of nitrogen in the nitrate ion, NO_3^-, is +5, and hence in the oxidation each nitrogen has lost two electrons. Hence in the complete oxidation of the $Co(NO_2)_6^{3-}$ ion, $2 \times 6 - 1 = 11$ electrons are involved.

21.4 Balancing Redox Equations: Oxidation-State Method

A half-reaction can also be balanced by considering the oxidation states involved. In this procedure, the number of electrons involved in the half-reaction is determined from the difference in the oxidation state between the reactant and its corresponding product. Consider, for example, the half-reaction for the reduction of permanganate ion in acidic medium:

$$MnO_4^- = Mn^{2+} \qquad \text{(unbalanced)}$$

The oxidation states of manganese, calculated as considered above, are +7 in the reactant and +2 in the product. The half-reaction therefore involves 7 - 2 = 5 electrons:

$$MnO_4^- + 5e = Mn^{2+} \quad \text{(unbalanced)}$$

Material is then balanced as in the previously described method. The four atoms of oxygen of the reactant will combine with eight hydrogen ions to yield four molecules of water:

$$MnO_4^- + 8H^+ + 5e = Mn^{2+} + 4H_2O \quad \text{(balanced)}$$

As a check, charge balance is calculated for both sides of the equation; the net charge on each side is found to be +2.

21.5 *Questions*

21-1 Establish the oxidation state of chromium in the dichromate ion, $Cr_2O_7^{2-}$, and explain each step in the reasoning.

21-2 What are the possible oxidation states of oxygen in compounds?

21-3 Define the terms *half-reaction, half-reaction equation, oxidation state,* and *reductant.*

21-4 In a fully balanced reaction equation, is it allowed that electrons appear on either side? Explain.

21-5 As the final step in securing a balanced redox equation, what check should be made on the coefficients?

21.6 *Problems*

21-1 The reaction of iodate ion with iodide ion in acidic solution yields elemental iodine. Write the balanced redox equation.

Ans.: $IO_3^- + 5I^- + 6H^+ = 3I_2 + 3H_2O$

21-2 The complete oxidation of methanol, CH_4O, to carbon dioxide in alkaline solution by the action of permanganate ion yields manganese(IV) oxide. Write the balanced redox equation.

Ans.: $CH_4O + 2MnO_4^- = CO_2 + 2MnO_2 + H_2O + 2OH^-$

21-3 In the oxidation of iron(II) ion in acidic solution, dichromate ion is reduced to chromium(III) ion. Write the balanced overall equation.

Ans.: $6Fe^{2+} + Cr_2O_7^{2-} + 14H^+ = 6Fe^{3+} + Cr^{3+} + 7H_2O$

21-4 Complete and balance the following redox equations. (Check your results to ensure that material and charge are fully balanced. The description in parentheses refers to the solution.)

(a) $IO_4^- + I^- = I_2$ (acidic)
(b) $IO_4^- + Mn^{2+} = MnO_4^- + IO_3^-$ (acidic)

(c) $MnO_4^- + Mn^{2+} = MnO_2$ (alkaline)
(d) $MnO_4^- + Mn = Mn^{2+}$ (acidic)
(e) $PbO_2 + Cr^{3+} = Pb^{2+} + Cr_2O_7^{2-}$ (acidic)
(f) $AsO_3^{3-} + I_2 = AsO_4^{3-} + I^-$ (alkaline)
(g) $ZnS + NO_3^- = Zn^{2+} + S + NO\uparrow$ (acidic)
(h) $BrO_3^- + Br^- = Br_2$ (acidic)
(i) $Hg^{2+} + Cl^- + Sn^{2+} = Hg_2Cl_2 + Sn^{4+}$ (acidic)
(j) $Bi(OH)_3 + Na_2SnO_2 = Bi + Na_2SnO_3$ (alkaline)
(k) $FeCl_3 + H_2S = FeCl_2 + HCl + S$ (acidic)
(l) $As_2O_3 + Br_2 = H_3AsO_4 + Br^-$ (acidic)
(m) $S_2O_3^{2-} + I_2 = SO_4^{2-} + I^-$ (alkaline)
(n) $Ag + ClO_3^- + Cl^- = AgCl$ (acidic)
(o) $H_3PO_3 + Ce^{4+} = H_3PO_4 + Ce^{3+}$ (acidic)
(p) $Cu(NH_3)_4^{2+} + CN^- = Cu(CN)_3^{2-} + NCO^- + NH_3$ (alkaline)
(q) $CuSCN + IO_3^- + Cl^- = Cu^{2+} + HCN + SO_4^{2-} + ICl_2^-$ (acidic)
(r) $Ce(IO_3)_4 + H_2C_2O_4 = I_3^- + Ce_2(C_2O_4)_3 + CO_2\uparrow$ (acidic)
(s) $NH_3 + H_2O_2 = N_2\uparrow$ (alkaline)
(t) $PbS + H_2O_2 = PbSO_4$ (acidic)

21-5 Determine the oxidation state of the *metal* in the following compounds and anions: Sb_2O_5, BiOCl, CrO_2Cl_2, $Fe_3(PO_4)_2 \cdot 5H_2O$, Fe_3O_4, $Pb(CN)_2$, WO_4^{2-}, $UO_2(NO_3)_2$, OsO_4, $B_4O_7^{2-}$.

22

Electrochemical Principles

Electrical conductance is the process by which electrical charges pass through a conductor. Two kinds of electrical conductors can be distinguished, metallic and electrolytic. The former kind, usually a metal, conducts electricity by the passage of electrons; the latter, by the movement of ions (cations and anions). If a current flows across the junction between a metallic conductor and an electrolytic one, an electrochemical reaction occurs. The nature of this reaction is the central topic of electrochemistry. Electrochemical phenomena underlie some of the most important methods of quantitative analysis.

22.1 *Definition of Electrical Units*

The SI units of measurement are used in this textbook (see Section 3.1 and Table K of the Appendix). The definitions of the relevant electrical units are summarized here.

The *ampere* (A) is the unit of strength of a current and is an SI base unit. One ampere is the constant current that produces a force of 2×10^{-7} $N \cdot m^{-1}$ when it is maintained between two straight, parallel conductors of infinite length and negligible cross section. This somewhat abstract definition can be related more closely to electrochemistry by recalling the earlier (now invalid) definition of one ampere as the strength of an unvarying current that under specified conditions deposits 1.1180 mg of silver from a silver nitrate solution in one second.

The *volt* (V) is the unit of electrical potential, potential difference, or electromotive force. One volt is the potential difference between two points of a conducting wire carrying a constant current of one ampere when the power dissipated between these points is one watt (see below).

The *ohm* (Ω) is the unit of electrical resistance. One ohm is the resistance between two points of a conductor when a constant potential difference of one volt applied to these points causes a current of one ampere to flow.

The *watt* (W) is the unit of electrical power. One watt is that power that in one second gives rise to energy of one joule (J). In other terms, one watt is the product of one volt and one ampere.

The *coulomb* (C) is the unit of electrical charge or the quantity of electricity. One coulomb is the quantity of electricity passed by a constant current of one ampere in one second.

Ohm's law describes the relation of the electromotive force E, the current I, and the resistance R, and can be expressed $E = IR$, or substituting the relevant units, volts = amperes × ohms.

22.2 Electrochemical Cells

If a piece of zinc metal is placed in an aqueous solution containing copper(II) ion, the zinc can be seen to dissolve, forming zinc ion and copper plates on the surface of the zinc. The redox reaction can be written*

$$Zn + Cu^{2+} \rightarrow Zn^{2+} + Cu \qquad (22\text{-}1)$$

Examining the electrical phenomena involved and also allowing practical use of the changes in electrical energy associated with the reaction call for physically separating the two redox systems, although they must still be connected by an electric conductor. Such an arrangement is shown in Fig. 22-1 and is called a cell. It consists of the two corresponding half-cells, which are connected by a U-tube filled with a salt solution. So that this solution can be "solidified," gelatin or agar-agar is added. This salt bridge effects a conducting connection without allowing the two solutions to mix. An electrolytically conducting connection must be used instead of a simple wire loop for the following reason. In a metallic conductor, an electric current is the flow of electrons; in an electrolytic conductor, a current is sustained by the translatory movement of ions. For the current to pass the phase boundary between a metallic and an electrolytic conductor in either direction, an electrochemical reaction must take place as this phase boundary. If a metal wire were used to connect the two half-cells, an additional half-cell would be present in each of the two vessels, and instead of the desired single cell, two cells connected by the wire would exist.

A cell can be described schematically. The zinc-copper cell of Figure 22-1 can be represented as

$$Zn|ZnSO_4\|CuSO_4|Cu$$

A single vertical line here denotes a phase boundary between a metallic conductor and an electrolyte solution (often a solid-liquid interface); a double line, a liquid-liquid junction, that is, denotes an electrolytically conducting connection (for example, a salt bridge or a porous, sintered-glass disk) between two solutions. Unless it has been otherwise stated, the electrolytes, zinc(II) sulfate and copper(II) sulfate in the example, will be assumed to be present as aqueous solutions. When the

---

*Zinc and copper metals, as solids, might be underlined in equation (22-1), however, in considering electrochemistry, we shall often omit this practice as an unnecessary complication.

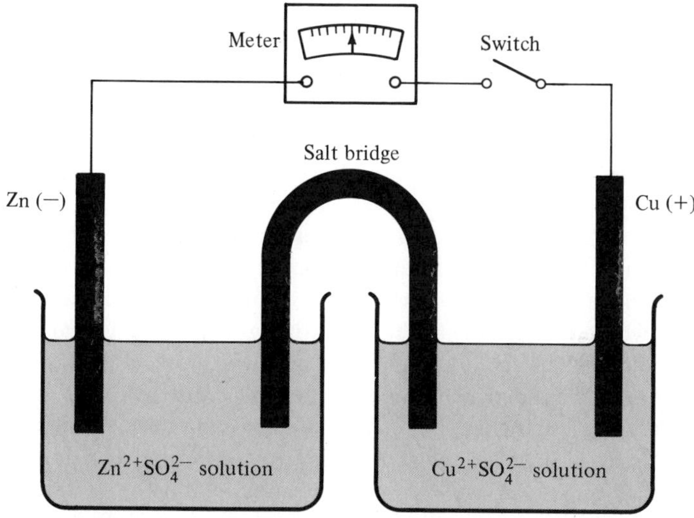

Fig. 22-1. Zinc-copper voltaic cell.

state of a substance must be stated explicitly, the letters s and g are used parenthetically to designate solid and gas, and concentrations are given parenthetically, for example, AgCl(s), H_2(g), and $ZnSO_4$ (0.1*F*). The use of these conventions will become clearer as further examples are considered.

22.3 *Voltaic Cells and Electrolysis Cells*

When the pieces of zinc and copper metal of the zinc-copper cell are connected by a metallic wire, the zinc dissolves progressively and copper plates on the copper piece. In addition, electrons flow through the wire from the zinc to the copper, and the zinc piece is negatively charged in relation to the copper piece. From these findings, you can infer that in the right and left half-cells, respectively, the following reactions are occurring:

$$Cu^{2+} + 2e \rightarrow Cu \quad \text{(reduction)}$$

$$Zn \rightarrow Zn^{2+} + 2e \quad \text{(oxidation)}$$

The complete reaction equation of the whole cell is

$$Zn + Cu^{2+} \rightleftharpoons Zn^{2+} + Cu \tag{22-2}$$

The double arrow is used since the reaction is reversible.

In this cell, a chemical reaction is used to generate electrical energy. Such a cell is known as a voltaic cell, or a galvanic cell.

If an outside source of electrical energy is used to force electrons on to the zinc metal and off at the piece of copper, under appropriate conditions zinc ion will be reduced to zinc metal and copper metal will be oxidized to copper(II) ion; that is, the direction of the cell reaction will be reversed. Here, then, electrical energy is used to cause a chemical reaction. This process is known as electrolysis, and the cell is called an electrolysis cell, or an electrolytic cell.

It is important to appreciate that the piece of zinc is negative and the copper is positive regardless of whether the cell acts as a voltaic cell or as an electrolysis cell. In the voltaic cell, oxidation proceeds at the negative electrode; in the electrolytic cell, reduction occurs at this electrode. Instead of the terms *positive electrode* and *negative electrode,* the terms *cathode* and *anode* are often used. Unfortunately there is no uniform agreement on the definition of these terms. Some workers define the anode and cathode as the positive and negative electrodes. Others define the anode as the electrode at which oxidation occurs and the cathode as the electrode at which reduction occurs. From the above considerations, you should realize that these two definitions are equivalent for an electrolysis cell but are directly opposed for a voltaic cell. To avoid confusion, it is best to limit the use of the terms *cathode* and *anode* to the consideration of electrolysis cells and never use them for voltaic cells. We follow this practice in this textbook.

22.4 *Faraday's Laws*

Often we can treat the electrons in a balanced half-reaction equation as a reactant or a product just like any other participating species. Then the principle of Le Châtelier can be applied. For example, for the equilibrium

$$Cu^{2+} + 2e \rightleftharpoons Cu \qquad (22\text{-}3)$$

it can be concluded that adding electrons will shift the position of the equilibrium toward the right. This treatment, however, is strictly formal; electrons cannot be added from a reagent bottle in the form of a solution. In a solution, electrons never occur as free, isolated entities but are provided or consumed either by another redox system or by an electrode.

Equation (22-3) implies that two electrons are needed to convert one copper(II) ion into one copper atom. Consequently, for the conversion of one mole of copper(II) ion, two moles of electrons are necessary. As careful measurements have established, one mole of electrons corresponds to 9.649×10^4 C. Since 1 C equals 1 A • s, it can be established that converting one mole of copper(II) ion to one mole of copper metal requires $2 \times 9.649 \times 10^4$ A • s. Consequently, the quantity of material reacting at an electrode is related both to the strength of the direct current flowing and to the time that it flows.

A half-cell can never operate alone; there must always be two electrodes in one cell, even when the two half-cells forming the cell are physically separated but joined by an electrolytic conductor. When the circuit has been completed, the electrode reactions proceed, and if no branches exist, the same number of amperes flow in any element of the circuit. Consequently, the number of electrons involved in the reaction at one electrode must exactly equal the number of electrons involved in the reaction taking place at the other electrode; in chemical terms, the quantity of substance reacting at one electrode must be *equivalent* to the quantity of substance reacting at the other electrode. Thus, if one mole of copper(II) reacts at one electrode in the cell under consideration, one mole of zinc metal reacts at the other electrode. If a cell consists of a copper(II)-copper half-cell and a silver(I)-silver half-cell, then for one mole of copper(II) that is reacting, two moles of silver should and will react, since only one electron is involved in the silver(I)-silver half-reaction ($Ag^+ + e \rightarrow Ag$). It is therefore possible to define the equivalent weight of a substance in an electrochemical reaction as the formula weight divided by the number of electrons released or consumed per reacting molecule, ion, or atom of that substance. The quantity 9.649×10^4 C·mol^{-1} is termed the Faraday constant (symbolized by F) and is the quantity of electricity, in coulombs, corresponding to one mole of electrons, that is, to 6.023×10^{23} electrons (Avogadro's number). In other terms, "one Faraday" (1 F) or 9.649×10^4 coulombs is needed for the electrochemical reaction of one equivalent of a substance.

The considerations of the preceding paragraphs are implicit in what Michael Faraday discovered about 1833. Those findings can be expressed in two laws:

1. The quantity of a substance reacting at an electrode of a cell is directly proportional to the quantity of electricity passed. Since the quantity of electricity is given by current multiplied by time (ampere × second = coulombs), the quantity of a substance reacting at the electrode in unit time is directly proportional to the current.
2. The quantities of different substances reacting during the passage of equal quantities of electricity are proportional to the chemical equivalent weights of the substances. In other words, for a given quantity of electricity, the number of chemical equivalents reacting at an electrode is identical for different substances.

With this information, it is possible to derive an equation for the relation among time, current, and the quantity of a reacting substance.

The chemical equivalents of the reacting substance are also given by g/EW, where g is the grams of the substance. The equivalent weight, EW, is given by FW/f, where f is the equivalence number, that is, here, the number of electrons reacting per ion, atom, or molecule. Thus the equivalents are g × f/FW. Since F coulombs are required for the reaction of one equivalent of substance, the number of coulombs required for g grams of substance is F × g × f/FW. This number of coulombs is the number of ampere-seconds necessary:

$$\frac{\mathrm{F} \times \mathrm{g} \times f}{\mathrm{FW}} = It \qquad (22\text{-}4)$$

where I is the current in amperes and t the time in seconds.

Rearranging, and substituting the numerical value for the Faraday constant, yield

$$\frac{I \times t \times \mathrm{FW}}{9.649 \times 10^4 \times f} = \mathrm{g} \qquad (22\text{-}5)$$

A few examples show how to apply the relevant facts and how to use these equations.

Example 22–1 How many grams of copper (*63.5*) will be deposited on the cathode if a constant current of 0.50 A is passed for 5.0 min?

Applying equation (22–5) yields

$$\frac{\overset{\text{amp}}{0.50} \times \overbrace{5.0 \times 60}^{\text{sec}} \times \overset{\text{FW}}{63.5}}{9.65 \times 10^4 \times \underset{f}{2}} = 0.049 \text{ g of Cu}$$

Example 22–2 Permanganate can be cathodically reduced according to the reaction

$$MnO_4^- + 8H^+ + 5e \rightleftharpoons Mn^{2+} + 4H_2O$$

For what time must a constant current of 0.0250 A be applied to reduce 5.2 mg of $KMnO_4$ (*158.0*)?

$$\frac{\overbrace{2.50 \times 10^{-2}}^{\text{amp}} \times \overset{\text{sec}}{t} \times \overset{\text{FW}}{158.0}}{9.65 \times 10^4 \times \underset{f}{5}} = 5.20 \times 10^{-3} \text{ g}$$

Solving for time, we obtain

$$t = \frac{5.20 \times 5 \times 9.65 \times 10^4}{25 \times 158} = 0.0635 \times 10^4 = 635 \text{ s},$$

or 10 min 35 s

Example 22–3 A cell in which silver is deposited and a cell in which dichromate is reduced ($Cr_2O_7^{2-} + 14H^+ + 6e \rightleftharpoons 2Cr^{3+} + 7H_2O$) are connected in series, and an unvarying current is passed for a certain time. After that time, it is found that 125.0 mg of Ag (*107.9*) has been plated. How much Cr^{3+} (*52.0*) in grams has formed?

According to Faraday's second law, the quantities of Ag deposited and of Cr^{3+} formed must be equivalent. The deposition of one atom of Ag involves one electron and the formation of two ions of Cr^{3+} involves six electrons; that is, for each ion Cr^{3+} formed, three electrons have reacted, or in turn, three Ag atoms have plated. Consequently the relation

$$(\text{moles of Ag deposited}) = (\text{moles of } Cr^{3+} \text{ formed}) \times 3$$

exists. Then

$$\frac{125 \times 10^{-3}}{3 \times 107.9} = \frac{x}{52.0}$$

and
$$x = \frac{52.0 \times 125 \times 10^{-3}}{3 \times 107.9} = 0.0201 \text{ g of } Cr^{3+}$$

22.5 *Half-Cell Potential; Standard Hydrogen Electrode*

It is impossible to obtain the *absolute* potential of a single half-cell by measuring cell voltages. This difficulty arises because reducing a certain amount of one substance calls for an exactly equivalent amount of another substance in the system to be oxidized, and vice versa. Although it is possible to separate the oxidation and reduction half-reactions so that they take place in two separate half-cells, only the combined voltage can be measured, since the half-cell reactions occur simultaneously and interdependently. For quantitative comparisons between various half-cells, it is necessary to assign arbitrarily a potential value to one selected, closely defined half-cell and to relate the potentials of all other half-cells to it.

By international agreement, the standard hydrogen electrode is assigned the exact value zero volt at all temperatures.* This electrode consists of a piece of platinum metal (coated with finely divided platinum) dipping into an aqueous solution that has hydrogen ion at unit activity and that is saturated with hydrogen gas at a pressure of 1 atm. The approximation is often made that molar concentration equals activity and this practice is followed in this textbook (see Section 4.3 for the definition of *activity*). Thus, the hydrogen electrode can be represented by

$$\text{Pt}, H_2 \text{ (g, 1 atm)} | H^+ \text{ (1}M\text{)}$$

or

$$\text{Pt}, H_2 \text{ (g, } 1.01325 \times 10^5 \text{ Pa)} | H^+ \text{ (1}M\text{)}$$

The first of these representations expresses the pressure of the hydrogen in atmospheres and is the traditional one. The standard-state condition for a gas is still taken as exactly 1 atm. With SI units, however, the atmosphere is only sanctioned

---

*The term *electrode* is used in two senses relative to electrochemical cells. On one hand, the piece of metal that dips into the electrolyte solution is often called the electrode. This metal may participate in the cell reaction, as in the zinc-copper cell considered above. Alternatively, the metal may simply provide a means of electrical connection or a site for an electron transfer. In these cases, the metal does not participate in the cell reaction and is called an inert electrode (or indicator electrode). Platinum often serves for this purpose. On the other hand, the term *electrode* is applied to an entire half-cell. The two usages can be seen in a single context. The hydrogen electrode (that is, half-cell) consists of an inert platinum electrode (a rod, a wire, or a sheet of platinum metal) dipped into an electrolyte solution, at which the reaction $H^+ + e \rightleftharpoons \frac{1}{2} H_2$ occurs. When an inert metal is used, the meaning of the term *electrode* is usually obvious from the context. When the metal participates in the cell reaction, however, the term is ambiguous.

for "temporary use" and pressure is to be given in pascals (Pa) (see Table K in the Appendix) (1 atm = 1.01325×10^5 Pa = 101.325 kPa). The second of the above representations, with the pressure in pascals, is the form used in this textbook when gases are involved. Pressures in pascals can be converted to atmospheres (1 Pa = 9.8692×10^{-6} atm; see also Example 22-7).

As a consequence of the assignment of zero volt to the standard hydrogen electrode, the total voltage of any cell in which one half-cell is this standard hydrogen electrode can be ascribable to the other half-cell. Thus, the cell

$$\text{Pt}, \text{H}_2(\text{g}, 1.01325 \times 10^5 \text{ Pa})|\text{H}^+ (1M)||\text{Zn}^{2+} (1M)|\text{Zn}$$

has experimentally a total voltage of 0.76 V and the zinc electrode is found to be negative. The electrode potential (relative) of the zinc(II)-zinc electrode is then stated to be -0.76 V. Because in the above cell zinc metal and zinc ion ($1M \cong$ unit activity) are at "standard state," the value of -0.76 V represents the "standard potential," designated by the superscript zero. Thus, we may write $E^0_{Zn^{2+}/Zn}$ = -0.76V.

Under the conventions adopted, the term *potential* (that is, *electrode potential* or *half-cell potential*), without any further qualification, should be used to describe the electrical potential (including sign) of the half-cell relative to the standard hydrogen electrode only under the condition that the reaction equation of the half-cell to be specified is written as a reduction.* In reference to the electromotive force of whole cells, batteries, circuitry, and so on, the term *voltage* will be used and not *potential.*

A cell consisting of the standard hydrogen electrode and the standard copper(II)-copper electrode

$$\text{Pt}, \text{H}_2 (\text{g}, 1.01325 \times 10^5 \text{ Pa})|\text{H}^+ (1M)||\text{Cu}^{2+} (1M)|\text{Cu}$$

has experimentally a voltage of 0.34 V and the copper electrode is positive. The standard electrode potential of the copper(II)-copper half-cell $E^0_{Cu^{2+}/Cu}$ therefore has the value +0.34 V.

22.6 *Nernst Equation*

In the zinc-copper voltaic cell, if both copper(II) and zinc ions are present at a concentration of one mole per liter, the cell at 25 °C will be found experimentally

*In practice, the hydrogen electrode has some disadvantages, and other electrodes of constant and well-defined electrode potential are substituted (see Section 22.12). In such a case, the electrode potential of a half-cell can be numerically expressed in reference to the substitute electrode. Expressing an electrode potential in this way has the further qualification that the exact nature of the substitute (reference) electrode must be stated.

to have the voltage 1.10 V, which corresponds to the difference in the electrode potentials, 0.34 - (-0.76). If the electrodes are connected externally and the cell is allowed to discharge, progressively the copper(II) ion concentration decreases, the zinc ion concentration increases, and the cell voltage decreases. Thus, it is evident that the voltage of a voltaic cell and the potential of a half-cell will depend on the concentration of the species involved in the cell reaction.

By the convention expressed above, a half-cell reaction is written as a reduction and in general terms takes the form

$$\mathrm{Ox} + ne \rightleftharpoons \mathrm{Red} \tag{22-6}$$

where Ox and Red denote the oxidized and reduced species forming the so-called redox pair or redox couple. The quantitative relation of the potential of the half-cell to concentration, termed the Nernst equation, after the electrochemist who derived it, is given by

$$E = E^0 + \frac{RT}{n\mathrm{F}} \ln \frac{[\mathrm{Ox}]}{[\mathrm{Red}]} \tag{22-7}$$

In the Nernst equation, which is applicable only if the half-cell reaction is reversible, the parameters have the following meaning.* This value E is the potential of the half-cell relative to the standard hydrogen electrode. The value E^0, the standard electrode potential, is the potential of the half-cell relevant to the standard hydrogen half-cell with each of the species involved in the cell reaction in its standard state: the pure liquid, the pure solid, the gas at a pressure of 1 atm (1.01325×10^5 Pa), and the dissolved species at unit concentration (strictly unit activity but 1*M* in the approximation used here). The value R is the gas constant of the ideal gas law and has the value $8.314\ \mathrm{J \cdot K^{-1} \cdot mol^{-1}}$. The temperature in degrees kelvin is represented by T. The symbol F is the Faraday constant, $9.649 \times 10^4\ \mathrm{C \cdot mol^{-1}}$; and n is the number of electrons transferred in the half-cell reaction as written. The terms [Ox] and [Red] stand for the products of the concentrations of all species occurring on the oxidation and reduction sides of the half-reaction. Concentrations are molarities except for gases; their concentrations are expressed as partial pressures in pascals (or traditionally in atmospheres). Solids and water, however, are at constant concentration and do not occur in the expression; in other words, they are at standard state and thus their activity is unity. As in expressions of the law of mass action, a coefficient in the half-cell reaction equation appears as the power of the concentration of the relevant species in the Nernst equation.

On the conversion from natural logarithms to Briggsian logarithms (that is,

---

*Some irreversible reactions, however, follow the Nernst equation reasonably well, like the reduction of permanganate. This and other reactions (for instance, reduction of dichromate) will be used in some of the examples and problems.

to base 10) and insertion of the values of the constants with T equal to 298.1°K (that is, 25.0°C), the result is*

$$E = E^0 + \frac{0.059}{n} \log \frac{[\text{Ox}]}{[\text{Red}]} \quad \text{(at 25°C)} \qquad (22\text{–}8)$$

Here n is dimensionless and 0.059 has the unit of volt, as you can verify by inserting the units for RT/F, given above, and recalling that J = V·C (see Table K in the Appendix).

Example 22–4 What is the potential of the electrode Pt|Fe^{2+} ($1.0 \times 10^{-1} M$), Fe^{3+} ($1.0 \times 10^{-2} M$)? The value of $E^0_{Fe^{3+}/Fe^{2+}}$ is +0.771 V at 25°C.

The half-cell reaction is

$$Fe^{3+} + e \rightleftharpoons Fe^{2+}$$

and the potential after equation (22–8) is given by

$$E = E^0 + \frac{0.059}{1} \log \frac{[Fe^{3+}]}{[Fe^{2+}]}$$

Substituting the numerical data yields

$$E = +0.771 + 0.059 \log \frac{1.0 \times 10^{-2}}{1.0 \times 10^{-1}}$$

$$= +0.771 + 0.059(-1.00) = +0.71 \text{ V}$$

This implies that a cell consisting of the standard hydrogen electrode and this particular iron(III)-iron(II) electrode has an experimental voltage of 0.712 V and that the iron electrode is positive.

Often the reduced species is the metal of the electrode (as in the zinc or copper electrode), and since the metal is in its standard state (mentioned above), its con-

*The value of the factor in the Nernst equation varies, as follows, with the temperature:

Temperature, °C	10	15	20	25	30	35	40
Value of factor, V	0.05618	0.05717	0.05817	0.05916	0.06015	0.06114	0.06213

In many calculations, especially when formal or molar concentrations are used instead of activities, the value 0.059 is precise enough for the temperature range 20 to 30°C.

centration term is not written. In such cases the Nernst equation simplifies to

$$E = E^0 + \frac{0.059}{n} \log [\text{Ox}] \qquad (\text{at } 25^\circ\text{C}) \qquad (22\text{-}9)$$

Example 22–5 What is the potential of the half-cell $Ni|Ni^{2+}$ $(1.0 \times 10^{-3}M)$ at 25°C? $E^0_{Ni^{2+}/Ni} = -0.25$ V.

The potential after equation (22–9) is given by

$$E = E^0 + \frac{0.059}{2} \log [Ni^{2+}]$$

Hence

$$E = -0.25 + \frac{0.059}{2} \log(1.0 \times 10^{-3}) = -0.25 - 0.09 = -0.34 \text{ V}$$

This finding implies that a cell consisting of the standard hydrogen half-cell and the given nickel electrode would have a voltage of 0.34 V, with the nickel electrode negative.

Example 22–6 What is the potential of the following half-cell?

$$Pt|Cr_2O_7^{2-}\ (1.0 \times 10^{-2}M),\ Cr^{3+}\ (2.0 \times 10^{-3}M),\ H^+\ (1.0 \times 10^{-2}M)$$

The standard potential for the reaction

$$Cr_2O_7^{2-} + 14H^+ + 6e \rightleftharpoons 2Cr^{2+} + 7H_2O$$

is $E^0_{Cr_2O_7^{2-}/Cr^{3+}} = +1.33$ V. The Nernst equation for the given half-cell is

$$E = E^0 + \frac{0.059}{6} \log \frac{[Cr_2O_7^{2-}][H^+]^{14}}{[Cr^{3+}]^2}$$

Inserting the relevant numerical data and solving yields

$$E = +1.33 + \frac{0.059}{6} \log \frac{(1.0 \times 10^{-2})(1.0 \times 10^{-2})^{14}}{(2.0 \times 10^{-3})^2} = +1.0_9 \text{ V}$$

Example 22–7 What is the change in the potential of the *standard* hydrogen electrode (a) if the partial pressure of hydrogen is decreased from 1.01325×10^5 Pa (1.000 atm) to 9.737×10^4 Pa (0.961 atm or (b) if the concentration of hydrogen ion is increased from 1*M* to 2*M*?

(a) For the reaction $H^+ + e \rightleftharpoons \frac{1}{2}H_2$, the Nernst equation takes the form

$$E = 0.059 \log \frac{[H^+]}{\sqrt{p_{H_2} \times 9.8692 \times 10^{-6}}}$$

where the pressure of hydrogen p_{H_2} is in pascals (1 atm = 9.8692×10^{-6} Pa) (see Section 22.5). Substituting the value for the reduced pressure yields

$$E = 0.059 \log \frac{1}{\sqrt{9.737 \times 10^{4} \times 9.869 \times 10^{-6}}}$$

$$= 0.059 \log \frac{1}{0.9803} = -5.1 \times 10^{-4} \text{ V}$$

Since for the standard hydrogen electrode a potential of zero volt is assigned, the change in potential is $E = -5.1 \times 10^{-4}$ V.

(b) Since for the standard conditions $p_{H_2} \times 9.8692 \times 10^{-6}$ is unity, the Nernst equation for the hydrogen electrode reduces to

$$E = 0.059 \log [H^+]$$

$$\Delta E = E - E^0$$

$$= 0.059 (\log 2 - \log 1)$$

$$= 0.059 \log 2$$

$$= 0.059 \times 0.301 = +0.018 \text{ V}$$

22.7 *Cell Voltage and Electrode Polarity*

If the necessary data are available, the Nernst equation can be used to calculate half-cell potentials. From these potentials cell voltages can be computed, and a scheme developed for establishing the polarity of the electrodes and the direction in which the cell reaction proceeds spontaneously. Several different procedures are possible, all of which are satisfactory if followed consistently. The procedure to be described is concordant with the conventions recommended by the International Union of Pure and Applied Chemistry, most of which have already been introduced in this chapter. The procedure is best explained through examples.

Example 22–8 For the following cell calculate the cell voltage, indicate the polarity of the electrodes, write the cell reaction, and indicate in which direction it will proceed spontaneously.

$$Zn|Zn^{2+}\ (1.0 \times 10^{-2} M)\|Cu^{2+}\ (1.0 \times 10^{-1} M)|Cu$$

At 25°C $\quad E^0_{Cu^{2+}/Cu} = +0.34$ V $\quad$ and $\quad E^0_{Zn^{2+}/Zn} = -0.76$ V

First, write the reaction equations as reductions with that for the right electrode above that for the left electrode. Do the same with the Nernst equations for the two electrodes.

$$Cu^{2+} + 2e \rightleftharpoons Cu$$

$$E_r = E^0 + \frac{0.059}{n} \log [Cu^{2+}] = +0.34 + \frac{0.059}{2} \log(1.0 \times 10^{-1})$$

$$Zn^{2+} + 2e \rightleftharpoons Zn$$

$$E_l = E^0 + \frac{0.059}{n} \log [Zn^{2+}] = -0.76 + \frac{0.059}{2} \log(1.0 \times 10^{-2})$$

Then subtract the reaction equation for the left electrode from the equation for the right electrode, and subtract the corresponding Nernst equations similarly.

$$Cu^{2+} + Zn \rightleftharpoons Cu + Zn^{2+}$$

$$E_{cell} = E_r - E_l = +0.34 - (-0.76) + \frac{0.059}{2} \log \frac{1.0 \times 10^{-1}}{1.0 \times 10^{-2}}$$

$$= +0.34 + 0.76 + \frac{0.059}{2} \log(1.0 \times 10) = (+)1.13 \text{ V}$$

The voltage of the cell is thus 1.13 V. The positive sign indicates that the right electrode is positive and that the cell reaction, as written, proceeds spontaneously from left to right, that is, copper(II) ion is reduced and zinc metal oxidized.

It should be stressed that a cell has a certain voltage and that this voltage has no sign. The plus or minus sign for the cell voltage is an algebraic consequence and is placed in parentheses. This sign is useful, however, since it indicates the polarity of the right half-cell and the direction of the spontaneous reaction. Example 22-9 shows how to proceed if other species besides those involved directly in the redox process participate in the half-reaction.

Example 22-9 What is the voltage and polarity of the cell given below? What is the cell reaction and in which direction will it proceed spontaneously?

$Zn|Zn^{2+}$ ($1.0 \times 10^{-4} M$)||H^+ ($1.0 \times 10^{-2} M$),

V^{3+} ($1.0 \times 10^{-1} M$), VO^{2+} ($1.0 \times 10^{-3} M$)|Pt

$$E^0_{Zn^{2+}/Zn} = -0.763 \text{ V} \quad \text{and} \quad E^0_{VO^{2+}/V^{3+}} = +0.337 \text{ V}$$

First, write the reaction equations and Nernst equations as described in Example 22-7.

$$VO^{2+} + 2H^+ + e \rightleftharpoons V^{3+} + H_2O$$

$$E_r = +0.337 + 0.059 \log \frac{[VO^{2+}][H^+]^2}{[V^{3+}]}$$

$$= +0.337 + 0.059 \log \frac{(1.0 \times 10^{-3})(1.0 \times 10^{-2})^2}{1.0 \times 10^{-1}}$$

$$Zn^{2+} + 2e \rightleftharpoons Zn$$

$$E_l = -0.763 + \frac{0.059}{2} \log [Zn^{2+}]$$

$$= -0.763 + \frac{0.059}{2} \log (1.0 \times 10^{-4})$$

Before the reaction equations can be subtracted, the one for the right electrode must be multiplied by 2 so that the number of electrons involved will cancel on subtraction. The question arises whether the terms in the Nernst equation for this electrode should also be multiplied by 2. The answer is no! In an analogy to a water-distribution system, the electrode potential is like the water pressure, and this is independent of the volume of the system. It is true that when the reaction equation is multiplied throughout by 2, in the Nernst equation the n will be multiplied by that factor. However, the concentration terms will also be raised to a power equal to this factor, so that the increase in n is nullified. Thus in the example

$$E_r = E_r^0 + \frac{0.059}{1} \log \frac{[VO^{2+}][H^+]^2}{[V^{3+}]}$$

$$= E_r^0 + \frac{0.059}{2} \log \frac{[VO^{2+}]^2[H^+]^4}{[V^{3+}]^2}$$

Multiplication by 2, as indicated above, and subtraction yields the balanced complete reaction equation for the cell, and mere subtraction of the Nernst equations yields the expression for the cell voltage.

$$2VO^{2+} + 4H^+ + Zn \rightleftharpoons 2V^{3+} + Zn^{2+} + 2H_2O$$

$$E_{cell} = E_r - E_l$$

$$= +0.337 - (-0.763) + 0.059 \times (-6.00) - \frac{0.059}{2} \times (-4.00)$$

$$= (+)0.864 \text{ V}$$

The voltage of the cell is therefore 0.86V, the vanadyl-vanadium(III) electrode is positive, and the reaction goes spontaneously from left to right, as the equation is written.

The procedure for calculating the voltage and the polarity of a voltaic cell can be summarized:

1. Write each of the electrode reactions as a reduction and each of the Nernst equations for the potential of the electrodes with those for the right electrode above those for the left one.*

---

*In Chapter 21, in the balancing of redox reactions, one half-reaction equation is written as an oxidation and one as a reduction. This can be done, since

2. Multiply the reaction equations by the factors needed to make the number of electrons involved in each identical. (Do not multiply the Nernst equations by these factors.)
3. Subtract the equations for the left electrode from the corresponding equations for the right one, thus securing the balanced equation for the cell reaction and an expression for the voltage of the cell, which is solved.
4. The sign associated with the cell voltage is the polarity of the right electrode. A positive sign indicates that the cell reaction will proceed spontaneously from left to right, as written in step 3. A negative sign indicates that this reaction proceeds spontaneously from right to left. If the cell voltage is zero, the system is at equilibrium and there will be no reaction.

22.8 Formal Potentials

In analytical chemistry, we seldom find solutions in which the species involved in a cell or a half-cell reaction are present alone. Foreign substances often influence the voltage either by changing activities or by forming soluble complexes. It is difficult to evaluate the extent of these effects, and even when calculations are possible, they are tedious. For this reason, electrode potentials are often determined and tabulated for solutions in which the given redox couple is frequently required in practice. These are called formal electrode potentials and may be designated E^f. When a formal electrode potential is used in the Nernst equation, the concentration terms are expressed in the *formalities* of the relevant species in the solution of interest rather than in *molarities*. Thus, in general terms

$$E = E^f + \frac{0.059}{n} \log \frac{C_{\mathrm{Ox}}}{C_{\mathrm{Red}}} \quad \text{(at 25 °C)} \qquad (22\text{-}10)$$

For example, the standard potential of the iron(III)-iron(II) redox couple is +0.771 V. In 0.1F, 1F, and 5F hydrochloric acid, the formal potential E^f is +0.73, +0.70, and +0.64 V. The change in potential with increasing concentration of hydrochloric acid is due to the formation of chloro complexes such as $FeCl_4^-$. The concentration C_{Ox} used in this Nernst equation should be the formal concentration of iron(III) and not the molar concentration of the uncomplexed iron(III).

---

by the statement of the problem the species that are oxidized and reduced are known. In the present considerations, this information is in general to be ascertained by following a definite scheme, and this scheme requires all half-reactions to be written as reductions. Furthermore, electrode potentials are involved and these by definition are related to half-reactions written as reductions.

22.9 Measuring Cell Voltage: The Potentiometer

Measuring the voltage of a voltaic cell is complicated by the necessity of avoiding, as much as possible, the passage of current through the cell. If current flows, the voltage does not correspond to the "spontaneous" cell voltage sought. Therefore, a simple voltmeter operating on the principle of a wire coil and magnet cannot be used because it requires the passage of a current. But an electronic voltmeter (see Section 23.4) or a potentiometer circuit can be used. Such a circuit is not as convenient and may not be preferred in practice, but its consideration will afford a deeper insight and exemplify an important kind of circuitry involved in "null" measurements (see also Section 32.4).

Potentiometer* Circuitry. The essential parts of a potentiometer circuit are shown schematically in Fig. 22-2. The working battery WB is connected across a uniform slide-wire resistance SW in series with a variable resistance VR. The total voltage drop E_{ac} across the slide wire is controlled by the setting of VR. The voltage drop E_{ab} from point a to the movable slide-wire contact b is related to the total voltage drop by $E_{ab} = E_{ac} \times (\overline{ab}/\overline{ac})$, where the symbols $\overline{ab}$ and $\overline{ac}$ are the length of the slide wire between points a and b and points a and c. Thus, if $\overline{ac}$ is 100 cm and $\overline{ab}$ is 75 cm and the total voltage drop across the slide wire is 1.00 V, the voltage drop between points a and b, E_{ab}, is $1.00 \times (75/100) = 0.75$ V.

The remainder of the circuit is a means of momentarily applying the voltage between points a and b to a cell and observing whether the applied voltage is greater or smaller than the cell voltage. The cell is momentarily connected to the potentiometer circuit by taping the key S, and if a current flows, its size and direction are

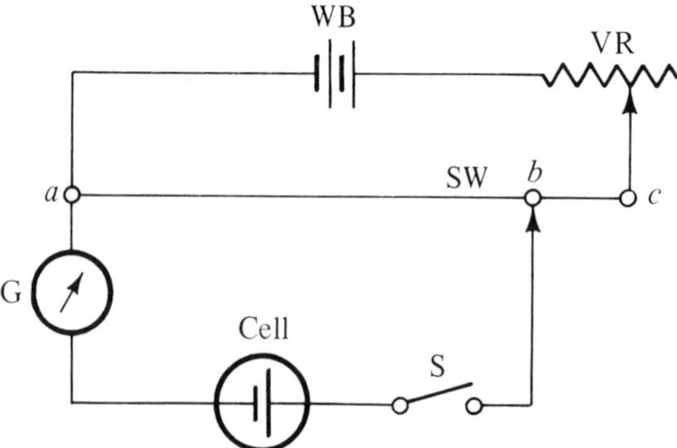

Fig. 22-2. Potentiometer circuit.

*The potentiometer here referred to should not be confused with a device used in electric and electronic circuitry that is simply a variable resistor with three terminals.

revealed by the displacement of the needle of the galvanometer G. The slide-wire contact is moved until a point is found at which no current flows when the key is closed momentarily. At this point the cell voltage is then equal to the voltage E_{ab}.

22.10 *Determining Half-Cell Potentials*

In determining the potential of a half-cell, coupling it experimentally with the standard hydrogen electrode as the reference is not mandatory. Any other half-cell can be used provided that its potential remains constant during the measurement and is exactly known relative to the standard hydrogen electrode. The voltage of the cell thus formed is the difference of the two half-cell potentials. Consequently, the unknown potential is obtained by adding the potential of the reference half-cell to the measured cell voltage. Full recognition of signs, including the formal sign of the cell voltage, is mandatory. The procedure is as follows. Let E_c, E_x, and E_{ref} stand for the voltage of the cell, the unknown potential, and the reference potential. Then

$$E_c = E_x - E_{\text{ref}} \tag{22-11}$$

$$E_x = E_c + E_{\text{ref}} \tag{22-12}$$

and for the formal computation, E_c must have a sign attached. The equation is analogous to $E_c = E_r - E_l$ used in the previously established computation scheme, in which the sign of E_c indicated the polarity of the right half-cell. Some examples show the procedure, and reference to the accompanying "scale" will help your understanding.

Volts:	-0.76	-0.52	-0.32	0.00	+0.34	+0.49
Electrode:	Zn^{2+}/Zn	X	Z	H^+/H_2	Cu^{2+}/Cu	Y

Example 22-10 A cell consists of a half-cell of unknown potential X and a zinc(II)-zinc electrode of potential −0.76 V. The cell voltage is found to be 0.24 V with the unknown electrode the positive electrode of the cell. What is the electrode potential of the unknown electrode?

$$X = (+)0.24 + (-0.76) = -0.52 \text{ V}$$

Example 22-11 A cell consists of an electrode of unknown potential Y and a copper(II)-copper electrode of potential +0.34 V. The cell voltage is found to be 0.15 V and the unknown electrode to be the positive electrode. What is the electrode potential of the unknown electrode?

$$Y = (+)0.15 + 0.34 = +0.49 \text{ V}$$

Example 22-12 A cell consists of an electrode of unknown potential Z and a copper(II)-copper electrode of potential +0.34 V. The cell voltage is found to be 0.66 V and the unknown electrode to be the negative electrode. What is the electrode potential of the unknown electrode?

$$Z = (-)0.66 + 0.34 = -0.32 \text{ V}$$

22.11 Secondary Reference Electrodes

The standard hydrogen electrode is troublesome when used in practice. This electrode is easily "poisoned" by impurities in either the solution or the hydrogen gas. The electrode reaches equilibrium slowly, and consequently a stable voltage is not attained promptly. Also the adjustment of the appropriate partial pressure of hydrogen gas, although not overly critical (see Problem 22-5), is tedious and the excess of gas must be carried away safely to avoid the formation of an explosive hydrogen–air mixture. For these reasons, secondary reference electrodes that are free of most of these shortcomings are usually used, most frequently calomel electrodes.

One form of the calomel electrode is shown in Fig. 22-3. A pool of mercury is covered with a thin layer of solid calomel, Hg_2Cl_2. The solution is saturated with calomel and also contains potassium chloride. A platinum wire sealed in glass tubing makes electrical contact with the mercury. The side arm is filled with gelatin con-

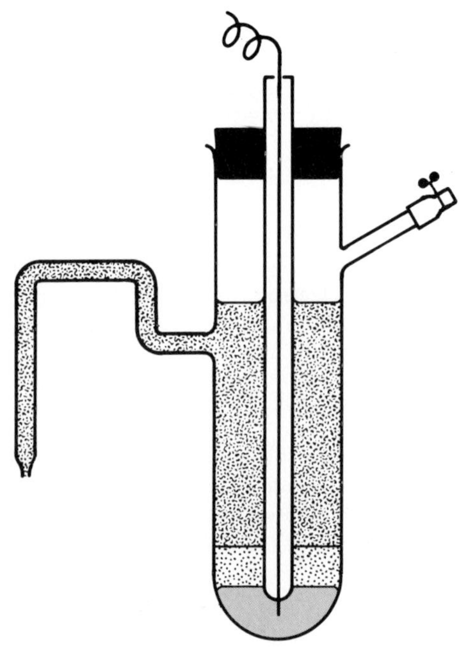

Fig. 22-3. Calomel reference electrode.

taining potassium chloride and functions as the salt bridge. The potential of a calomel electrode depends on the chloride concentration and is given for 25°C by the Nernst equation $E = E^0 - 0.059 \log [Cl^-]$. The saturated calomel electrode (assigned the acronym SCE), containing a solution saturated with potassium chloride, is preferred for routine analytical applications, since on any evaporation of the solution the potential remains unchanged. This particular calomel electrode has a potential of +0.241 V at 25°C.

Another reference electrode often used in practice is the silver chloride-silver electrode. It consists of a silver wire coated with silver chloride that dips into a solution $1M$ in chloride ion (usually as potassium chloride) saturated with silver chloride. This electrode at 25°C has a potential of +0.222 V.

Example 22-13 The potential of a half-cell A is found to be −0.362 V versus SCE. (a) What is the electrode potential (that is, versus the standard hydrogen electrode) of A? (b) What is the potential of A versus the silver chloride–silver ($1F$ in KCl) electrode? The electrode potential of the SCE is +0.241 V; the potential of the silver chloride–silver ($1F$ in KCl) electrode, +0.222 V.

(a) Since A is the negative electrode of the cell, the calculation takes the form $E = -0.362 + 0.241 = -0.121$ V (versus standard hydrogen electrode).

(b) With the value calculated in part (a), the calculation takes the form $E = -0.121 - 0.222 = -0.343$ V versus the silver chloride–silver ($1F$ in KCl) electrode.

You can understand the calculation scheme better by examining the scale given below.

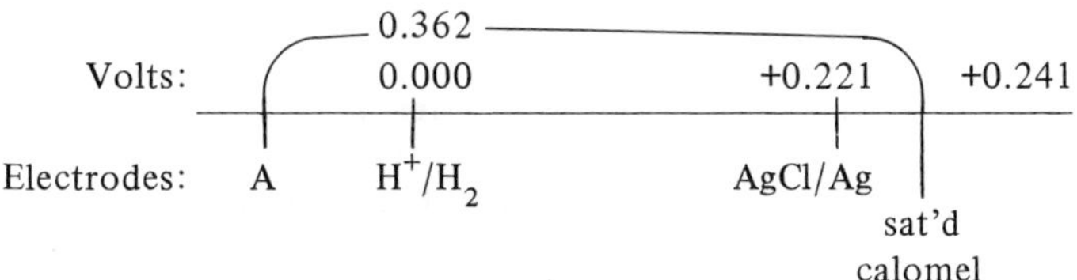

22.12 Questions

22-1 What is the basic distinction between voltaic and electrolytic cells?

22-2 What is the function of a salt bridge?

22-3 In your own words, explain the impossibility of measuring the *absolute* potential of a single half-cell.

22-4 At what temperature is the standard hydrogen electrode defined to have a potential of zero volt?

22-5 Discuss the various meanings of the term *electrode* in the consideration of electrochemical cells.

22-6 What does the sign of a calculated cell voltage mean?

22-7 What does a cell voltage of zero imply about the cell reaction?

22-8 Explain why the term for the metal concentration is omitted from the Nernst equation when the electrode material is the metal itself.

22-9 Elaborate on the distinction between the terms *voltage* and *potential.*

22.13 Problems

For each of the following voltaic cells, calculate the cell voltage for 25°C, give the polarity of the right-hand electrode, write a balanced equation for the complete cell reaction, and indicate by a single arrow the direction of the spontaneous reaction.

22-1 Pt|VO_2^+ (1.0 × $10^{-2}M$), VO^{2+} (1.0 × $10^{-3}M$), H^+ (1.0 × $10^{-4}M$)‖Fe^{2+} (1.0 × $10^{-1}M$), Fe^{3+} (1.0 × $10^{-3}M$)|Pt. $E^0_{VO_2^+/VO^{2+}}$ = +1.00 V, $E^0_{Fe^{3+}/Fe^{2+}}$ = +0.77 V.
Ans.: 0.07 V, positive,
$Fe^{3+} + VO^{2+} + H_2O \rightarrow Fe^{2+} + VO_2^+ + 2H^+$

22-2 Pt|$Cr_2O_7^{2-}$ (1.0 × $10^{-1}M$), Cr^{3+} (1.0 × $10^{-3}M$), H^+ (1.0 × $10^{-2}M$)‖Fe^{2+} (1.0 × $10^{-2}M$), Fe^{3+} (1.0 × $10^{-3}M$)|Pt. $E^0_{Cr_2O_7^{2-}/Cr^{3+}}$ = +1.33 V, $E^0_{Fe^{3+}/Fe^{2+}}$ = +0.77 V.
Ans.: 0.39 V, negative,
$Cr_2O_7^{2-} + 6Fe^{2+} + 14H^+ \rightarrow 2Cr^{3+} + 6Fe^{3+} + 7H_2O$

22-3 Pt|Ti^{3+} (1.0 × $10^{-1}M$), TiO^{2+} (1.0 × $10^{-3}M$), H^+ (1.0 × $10^{-1}M$) ‖ V^{2+} (1.0 × $10^{-2}M$), V^{3+} (1.0 × $10^{-1}M$)|Pt. $E^0_{V^{3+}/V^{2+}}$ = −0.26 V, $E^0_{TiO^{2+}/Ti^{3+}}$ = +0.10 V.
Ans.: 0.06 V, negative,
$V^{2+} + TiO^{2+} + 2H^+ \rightarrow V^{3+} + Ti^{3+} + H_2O$

22-4 The standard electrode potential E^0 of the copper(II)-copper half-cell has the value +0.337 V. What is the potential of this half-cell (a) versus a silver chloride–silver (1*F* in KCl) reference electrode (E^0 = +0.222 V) and (b) versus the saturated calomel electrode (E^0 = +0.241 V)?

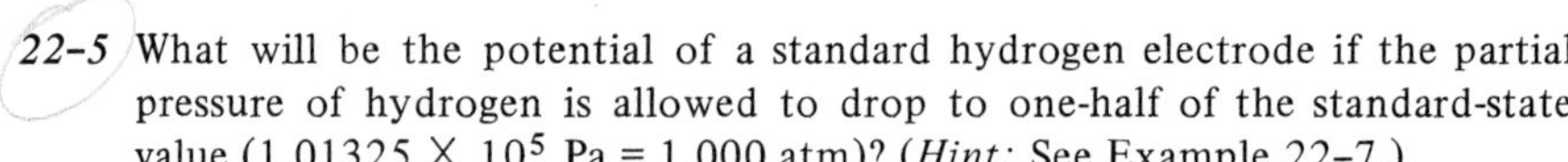

Ans.: (a) +0.115 V versus AgCl/Ag (1*F* in KCl) electrode; (b) +0.096 V versus SCE

22-5 What will be the potential of a standard hydrogen electrode if the partial pressure of hydrogen is allowed to drop to one-half of the standard-state value (1.01325 × 10^5 Pa = 1.000 atm)? (*Hint:* See Example 22-7.)

22-6 How much time, in hours, is needed to deposit all the copper from 100.0 mL of 0.10*F* $CuSO_4$ when the current is 0.10 A? Assume 100% efficiency for the electrolysis.

22-7 A solution containing Cu^{2+} and Ag^+ is electrolyzed for 32 min and 10 s with a current of 200.0 mA, thereby depositing copper and silver completely. The total weight of deposited copper (*63.54*) and silver (*107.87*) is 279.3 mg. Calculate the milligrams of copper and of silver present in the solution assuming that only the ions of these two metals reacted at the cathode.

22-8 Pure crystals of $CuSO_4 \cdot 5H_2O$ are dissolved in 100.0 mL of pure water and the solution is electrolyzed. The cathode gains 0.4280 g. What is the pH of the resulting solution?

22-9 A volume of 200.0 mL of 0.10*F* $NiSO_4$ is electrolyzed with a current of 6.0×10^{-2} A for 1 hr and 20 min. Oxygen is evolved at the anode. Calculate the pH of the solution after electrolysis for an initial pH of 4.00 due to the presence of a small amount of H_2SO_4. Assume 100% efficiency for the electrolysis.

22-10 Calculate the cell voltage, indicate the polarity of the right electrode, and write a balanced equation for the complete cell reaction, indicating the direction of the spontaneous reaction by a single arrow, for each of the following cells at 25 °C.

(a) $Pt|Tl^{3+}$ $(1.0 \times 10^{-2}M)$,
Tl^{+} $(0.10M)||Cu^{2+}$ $(1.0 \times 10^{-3}M)|Cu$
$E^0_{Tl^{3+}/Tl^+} = +0.13$ V, $E^0_{Cu^{2+}/Cu} = +0.34$ V

(b) $Sb|SbO^{+}$ $(5.0 \times 10^{-2}M)$,
H^{+} $(1.0 \times 10^{-2}M)||Pb^{2+}$ $(1.0 \times 10^{-4}M)|Pb$
$E^0_{SbO^+/Sb} = +0.2$ V, $E_{Pb^{2+}/Pb} = -0.13$ V

(c) $Ag|Ag^{+}$ $(6.0 \times 10^{-4}M)||Fe^{3+}$ $(1.0 \times 10^{-3}M)$,
Fe^{2+} $(5.0 \times 10^{-1}M)|Pt$
$E^0_{Ag^+/Ag} = +0.80$ V, $E^0_{Fe^{3+}/Fe^{2+}} = +0.77$ V

22-11 How many milliliters of a solution 0.10*F* in iron(II) and 1.0*F* in HCl must be added to 50.0 mL of a solution that is $1.0 \times 10^{-2}F$ in iron(III), $1.0 \times 10^{-3}F$ in iron(II), and 1.0*F* in HCl, so that a platinum indicator electrode in this solution will have a potential of +0.440 V versus SCE? $E^f_{Fe^{3+}/Fe^{2+}} = 0.700$ V, $E_{SCE} = +0.241$ V.

22-12 What is the potential of a half-cell obtained by mixing 30.0 mL of $1.0 \times 10^{-1}M$ Fe^{2+} with 60.0 mL of $1.0 \times 10^{-3}M$ Fe^{3+}? $E^0_{Fe^{3+}/Fe^{2+}} = +0.77$ V.

23 Electrometric Determination of pH and Ionic Concentrations

In many cases it is possible to compute (by means of the Nernst equation) the concentration of an ion from the measured potential of a half-cell involving this ion, and the known standard potential of the half-cell. The most common analytical application of this technique is the determination of the hydrogen ion concentration of solutions and consequently the pH. This determination, under certain circumstances, can be effected with a simple potentiometer circuit of the type described in Section 22.10. Usually, however, a more elaborate assembly is used, the pH meter, which you may know from previous courses. The assembly consists of the measuring device proper (Section 23.4) and the sensing device, the pH-sensitive electrode (Section 23.3). In recent years, progress has been made in developing sensing devices that respond selectively to an ion other than hydrogen ion. An ion-selective electrode can be connected with the meter proper and thereby allow the electrometric determination of the relevant ion.

23.1 Measuring pH by Means of Voltaic Cells

Consider the following cell:

$$\text{Pt}, H_2(\text{g}, 1.01325 \times 10^5 \text{ Pa}) \mid H^+ (xM) \parallel \text{KCl}, Hg_2Cl_2(\text{sat'd soln}) \mid Hg_2Cl_2(\text{s}), \text{Hg}$$

The right half-cell is a saturated calomel electrode, which at 25°C has a potential of +0.241 V. The left half-cell is a hydrogen electrode, the potential of which depends on the concentration of the hydrogen ion x, and at 25°C follows the relation

$$E = 0.0592 \log [H^+]$$

Note that this equation does not include a factor for the partial pressure of hydrogen gas. That factor has a value of unity since hydrogen is at standard pressure (see Example 22-7).

The voltage of the cell is given by

$$E_{\text{cell}} = E_r - E_l = +0.241 - 0.0592 \log [\text{H}^+]$$

Rearrangement yields

$$\frac{E_{\text{cell}} - 0.241}{0.0592} = -\log [\text{H}^+] = \text{pH} \qquad (23\text{-}1)$$

Thus the pH of the solution in the left half-cell can, at least theoretically, be computed from the measured cell voltage.

23.2 *Operational Definition of pH*

In trying to determine the concentration of hydrogen ion as just described, problems arise, both theoretical and practical. First, the Nernst equation and the definition of pH, on a rigorous basis, involve activities; the use of molar concentrations is an approximation. Second, the measured value of the cell voltage contains, besides the potentials of the two half-cells, a small, unknown "junction potential" of the order of a few millivolts. Such a potential occurs at liquid-liquid junctions, for example, at the ends of the salt bridge, and has as its source the difference in the rates of diffusion of anions and cations across the junction.

To eliminate the difficulties in practical pH measurements, an operational pH scale is established, in the following way. A standard buffer solution of assigned pH value, pH_s, is made the electrolyte in the left half-cell and the cell voltage E_s is measured. Then the buffer is removed and replaced by the solution of unknown pH value, pH_x, and under otherwise identical conditions, the cell voltage E_x is obtained. For the pH standard and the sample solution, respectively, the following analogs to equation (23-1) can be written:

$$\frac{E_s - k}{0.0592} = \text{pH}_s \qquad (23\text{-}2)$$

$$\frac{E_x - k}{0.0592} = \text{pH}_x \qquad (23\text{-}3)$$

The term k includes the electrode potential of the reference half-cell and any junction potentials, and its value is assumed to remain constant if the measurements of E_s and E_x are made consecutively and with the same experimental arrangement. Then k can be eliminated by combining equations (23-2) and (23-3), to give the following equation, which allows the pH of the sample solution to be calculated:

$$\mathrm{pH}_x = \mathrm{pH}_s + \frac{E_x - E_s}{0.0592} \qquad (23\text{-}4)$$

Note that the operational pH scale does not solve the theoretical problems mentioned above. It merely allows the pH measurements to be made comparably and on a useful basis.

Through extensive studies, pH_s values have been assigned to six solutions as primary pH standards. The pH_s values for these solutions at various temperatures are given in Table E in the Appendix.

23.3 The Glass Electrode

In Section 22.12 various experimental difficulties experienced with the use of the hydrogen electrode are mentioned. Owing to these shortcomings, much effort has been expended in developing other pH-sensitive electrodes. Most of these electrodes, including the antimony(III)-antimony and the quinhydrone electrodes, have been largely replaced by the glass electrode in association with the required electronic circuitry of the "pH meter."

One of the many existing forms of a glass electrode is shown schematically in Fig. 23-1. A very thin membrane of a special glass in the form of a small bulb B is sealed onto a tube T of thicker ordinary glass. Within the bulb is placed a solution S of definite pH. This internal solution usually has a substantial buffer capacity so that its hydrogen ion concentration remains constant. Into this internal solution dips a metal wire W. The interior of the glass electrode is thereby made a half-cell, in practice frequently a silver chloride-silver electrode. The internal electrode has a shielded wire lead, L, which allows connection to a voltage-measuring device.

For the pH to be determined, the glass electrode and a salt bridge leading to an external reference electrode are immersed in the sample solution and the voltage developed across the cell thus formed is measured. To simplify the discussion, we shall assume that the internal and external reference electrodes are identical silver chloride-silver electrodes, although in practice the saturated calomel electrode is usually the external reference. The complete cell can then be represented as follows:

internal electrode — glass membrane — external (reference) electrode

$$\mathrm{Ag, AgCl(s)} \mid \mathrm{Cl^-}(aM), \mathrm{H^+}\,(iM) \;\vdots\vdots\vdots\; \mathrm{H^+}\,(xM) \parallel \mathrm{Cl^-}(aM) \mid \mathrm{AgCl(s), Ag}$$

glass electrode

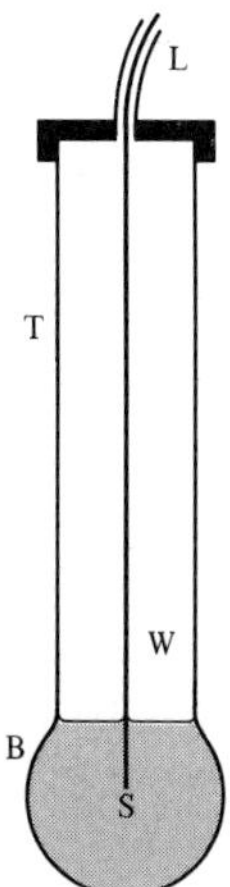

Fig. 23-1. Glass electrode.

Since the chloride concentration is the same in the two electrodes, no net cell voltage results from these electrodes. If the concentration of hydrogen ion in the sample solution, *x,* differs from the hydrogen ion concentration in the internal solution, *i,* a potential develops across the glass membrane and therefore across the cell. The origin of this potential is not the same as for the voltaic cells previously discussed. The phenomena underlying this so-called membrane potential have not been fully elucidated, but the following nonrigorous description should give you some insight into how the glass electrode functions.

The unique feature of the special glass used in the glass bulb is its pronounced selectivity for protons. Protons, unlike most other ions, can accumulate on the membrane surface. This accumulation is by adsorption, or according to another interpretation, proceeds by exchange with other cations present in the outer layer of the silicate structure of the membrane. Whatever the mechanism, the number of protons accumulated depends on the hydrogen ion concentration of the solution. When the concentrations of hydrogen ion in the internal and the sample solution differ, more protons accumulate at one side of the membrane than at the other and a membrane potential develops.

The relation between this potential and the hydrogen ion concentration of the external solution follows the equation*

$$E = k + 0.0592 \log[H^+] \qquad (23\text{-}5)$$

*The equation is an approximation, since the factor of the logarithmic term is not 0.0592, but depending on the glass, has a value close to it. This has no bearing on actual pH measurements, however, since the formula is not used as such; instead, the pH meter is calibrated using pH standards.

or when it is rewritten in terms of pH, it becomes

$$E = k - 0.0592 \text{ pH} \tag{23-6}$$

Thus the potential is a linear function of the pH of the test solution. The value k depends on the composition of the glass, the temperature, and the hydrogen ion concentration of the internal solution.

Equations (23-5) and (23-6) apply even if the internal and external electrodes are not identical. The potential difference between these two electrodes is constant, can be incorporated into k, and consequently merely changes the numerical value of this term. For practical determinations, k is not evaluated as such; a calibration with a pH standard is used instead, and the reasoning given in Section 23.2 applies.

Even when the internal and external electrodes and the hydrogen ion concentrations of the internal and the sample solutions are identical, a small voltage usually develops across the cell. This potential, known as the asymmetry potential, is the result of strain in the glass membrane, unequal conditions of the inner and outer membrane surfaces, and other causes not fully understood. This potential is also of no practical consequence, since during the calibration, it is incorporated into k. However, because the value of the asymmetry potential undergoes changes with time, frequent recalibration of a glass electrode is warranted.

23.4 *The pH Meter*

The voltage of a cell containing a glass electrode cannot be measured by a potentiometer with a galvanometer as the null detector (Section 22.9). The resistance of a glass electrode is of the order of 100 megohms ($1 \text{ M}\Omega = 1 \times 10^6 \ \Omega$). With so great a resistance in the circuit, the current flowing is far too small to move the needle of the galvanometer. Consequently it is necessary to resort to more elaborate instruments that can measure potentials essentially without the need for a current to flow. They involve electronic circuits that need not be described. We shall examine only how they are applied in pH meters.

One type of pH meter uses a conventional potentiometer circuit (Fig. 22-2), in which the galvanometer is replaced by a high resistance. Any imbalance between the potentiometer and cell circuits will result in a voltage (IR) drop across the high resistance. This voltage drop is amplified by an electronic circuit so that it can be presented by even a relatively insensitive meter. When the potentiometer and cell voltages are exactly balanced, no current flows, no voltage drop occurs across the resistance, and the meter reads zero. Here the amplifier circuit aids only in the detection of the balance point of the potentiometer (null detector). Consequently, such pH meters are described as the potentiometer, or null, type. The slide-wire dial of such instruments is usually calibrated both in pH units and volts.

The second type of pH meter is direct-reading, and simpler to operate, and it has largely replaced pH meters of the first type. The voltmeter contains electronic circuits, and the cell voltage is applied across a high resistance at the input of the instrument, and the pH (or voltage) is read directly from a meter that is driven by the output of the amplifier circuit or from a digital display. The electronic requirements are more rigorous for such a pH meter than for the potentiometer type because a strictly linear relation between input voltage and the reading is mandatory. This kind of instrument is usually called a direct-reading pH meter. It has another practical advantage over the potentiometer type in that the electrical output can be led directly to a chart recorder, permitting a graphical presentation of pH changes as a function of time or of any parameter proportional to time. Such an instrument coupled with a recorder, for example, can be used to plot the titration curve during an acid-base titration if the titrant is added from a buret of constant flow rate. The electrical output of a direct-reading pH meter can also be fed to a servomechanism to control an industrial process.

23.5 Alkali and Acid Errors

With common glass electrodes, there is a sizable error in the pH measurement if the sample solution contains sodium ion and has a pH value greater than about 9.5. This error imposes a serious limitation, since sodium is often a primary constituent of alkaline solutions. For solutions 1*M* in sodium ion, the sodium error is negligible below pH 9 but may reach -0.2 pH unit at pH 10 and -1 pH unit at pH 12. Manufacturers provide a nomograph, which allows a correction to be read for the sodium error. The magnitude of the correction depends on the temperature, the pH value, and the sodium ion concentration. Other alkali metal cations cause an analogous, but smaller error. Thus the error is sometimes called the alkali-metal error, or simply the alkali error.

Consideration of the processes at and in the glass electrode allows derivation of an equation relating the effect of a monovalent ion, M^+, other than hydrogen ion, to the electrode potential. This equation is an expansion of equation (23-5).

$$E = k + 0.0592 \log \left(a_{H^+} + \frac{f_{H^+}}{f_{M^+}} a_{M^+}\right) \qquad (23\text{-}7)$$

Here, correctly, are written the activities, instead of a molar concentration as in equation (23-5). The values f_{H^+} and f_{M^+} are parameters that depend on the activities of the hydrogen ion and the other monovalent ion, and on the composition of the glass. It is possible to manufacture glasses having compositions such that the second term within the parentheses is small compared with the first one. In other words, glass electrodes can be made that show no alkali error or only a very small one (up to about pH 12).

A typical glass electrode also shows an error in acidic solutions, below about pH 1. This error is opposite to the alkali error; that is, the pH readings are somewhat greater than the true pH. There is evidence that this departure is due to anion penetration of the pH-sensitive surface layer of the glass. Contrary to the alkali error, the acid error is not reproducible and thereby is more difficult to compensate for. This situation is not too serious, however, since in practice, pH measurements in this region are seldom needed.

23.6 Comparison of Glass and Hydrogen Electrodes

The glass electrode has some advantages over the hydrogen electrode: (1) It is not readily poisoned; (2) it does not change the composition of the sample solution; (3) it functions properly in solutions containing oxidants or reductants; and (4) it comes rapidly to equilibrium. The glass electrode has the following shortcomings: (1) It is fragile and cannot readily be improvised; (2) its surface can become fouled with large organic molecules such as proteins; (3) its high electrical resistance requires special circuitry; (4) it is subject to an alkali error unless it has been fashioned from special glass; and (5) it must be stored in an aqueous solution, or if dried, be reconditioned by prolonged soaking in water before use.

23.7 Ion-Selective Electrodes

The potential of a half-cell can be calculated from the concentration of the metal (or other) ion by the Nernst equation (Chapter 22). Inversely, in principle, the concentration of the ion can be calculated from the potential. The voltage can, for example, be measured across a silver wire and a reference cell dipping into a solution containing silver ion. From that voltage and the potential of the reference cell, the potential of the silver ion–silver half-cell can be found by difference. From that potential value, in turn the silver ion concentration can be calculated. Such an electrode assembly can also be used to measure indirectly the concentration of certain other ions. If a solution, for example, containing chloride ion is saturated with silver chloride, a silver ion concentration is established that is related to the chloride concentration by the solubility product. By use of the abovementioned cell assembly, the silver ion concentration can be determined and from it by calculation, or better by standardization, the chloride concentration. In these two examples, the silver wire, it is said, acts as an electrode of the *first order* and the *second order,* respectively. The values obtained, however, are for various reasons not reliable, and absolute data are not often developed in this way in practice. In contrast, when potential *differences* are involved, such simple "indicator electrodes" are of great service (see Chapter 27, Potentiometric Titrations).

In a more restrictive sense, the term *ion-selective electrode* usually refers to devices analogous to the glass electrode used for measuring pH. A variety of such electrodes is available and additional ones are being developed.

These electrodes actually measure the activity of the ion involved. Because, however, their application is predominantly to measurements for dilute solutions, replacing activity by concentration does not introduce a serious error. When these conditions are not met, standardization is a partial remedy. The response function of an ion-selective electrode is usually of the Nernstian type; that is, a plot of the logarithm of the concentration versus potential gives a straight line with a slope of 2.303 $RT/n\text{F}$. At both high and extremely low concentrations, deviations from this relation are usually observed. Such deviations, however, do not prevent analytical use of the electrode if appropriate calibration is used.

Some of the early research toward ion-selective electrodes led to glass-bulb electrodes much like those for pH measurement. The response of a glass electrode to hydrogen ion and metal ions when present together in the solution is roughly described by equation (23–7). As mentioned in Section 23.6, with changes in the composition of the glass, the values of the f terms change and thereby the response characteristic of the electrode.

For example, a glass of composition 11% Na_2O, 18% Al_2O_3, and 71% SiO_2 responds more strongly to sodium than to either potassium or hydrogen ion. A Nernstian relation is maintained down to about $10^{-5}M$ sodium ion. Useful measurements can be made one order of magnitude smaller, but below that, the electrode fails to provide an adequate signal.

The selectivity of an ion-selective electrode can be expressed by the potentiometric selectivity coefficient. For the difference in the response to sodium and potassium, this constant is written as $k^{\text{pot}}_{\text{Na}^+,\text{K}^+}$ and for the glass mentioned has at pH 11 a value of about 2800. This value implies that at equal concentrations (activities) of the two ions, the response to sodium is about 2800 times the response to potassium. Consequently the measurement of the sodium activity is unaffected by a large relative excess of potassium. In neutral solution, the situation is less favorable, since the value of the constant is then 300. A glass with about twice as much sodium and less than one-third the aluminum content of the glass mentioned above has a greater response to potassium than to sodium; for this second glass, the potentiometric selectivity constant, which then takes the form $k^{\text{pot}}_{\text{K}^+,\text{Na}^+}$, is only about 20. Glasses selective to monovalent cations, including the alkali metal, ammonium, and silver ions, are available.

Solid-state and precipitate electrodes are analogous to glass electrodes. A fluoride-selective electrode is shown schematically in Fig. 23–2 and can be discussed as an example of this type of device. You should compare this electrode and its various components and their function with the pH electrode, as shown in Fig. 23-1. At the end of a supporting plastic tube is inserted a single crystal of sparingly soluble lanthanum(III) fluoride, LaF_3, that is doped with europium(II) (that is, the crystal contains a deliberately added minute amount of that rare earth ion) to give it some degree of electrical conductance.

The internal electrolyte consists of a solution of sodium chloride and sodium fluoride, each typically at a concentration of 0.1F. The internal conductor is a silver wire that becomes coated with silver chloride and forms with the chloride ion the internal reference electrode. When such an electrode and an external reference electrode (commonly a calomel one) are dipped into a sample solution of

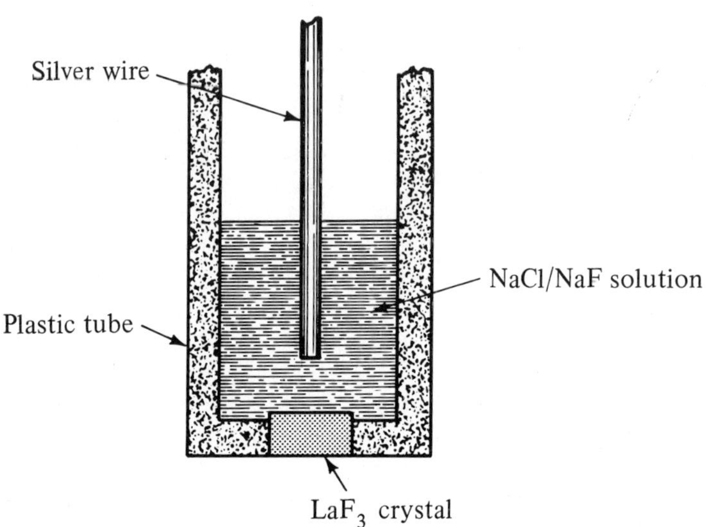

Fig. 23-2. Schematic representation of fluoride-selective electrode (see text).

unknown fluoride concentration, a cell is formed that can be described by the following scheme:

$$\overbrace{\underbrace{\text{Ag, AgCl(s)}}_{\text{internal reference electrode}} \mid \text{Cl}^-(0.1M),\ \text{F}^-\ (0.1M)}^{\text{internal system}} \mid \overbrace{\text{LaF}_3}^{\text{“membrane”}} \mid \overbrace{\underbrace{\text{F}^-\ (xM)}_{\text{sample solution}} \parallel \underbrace{\text{calomel}}_{\text{external reference electrode}}}^{\text{external system}}$$

This cell follows a Nernstian relation

$$E = \text{constant} + 0.059 \log \frac{[\text{F}^-]_{\text{int}}}{[\text{F}^-]_{\text{ext}}} \quad (\text{at } 25\,^\circ\text{C}) \tag{23-8}$$

Because $[\text{F}^-]_{\text{int}}$ the concentration of fluoride ion in the internal electrolyte, is fixed, its value can be incorporated into the constant and the relation simplifies to

$$E = \text{constant} + 0.059\,\text{pF}_{\text{ext}} \tag{23-9}$$

where pF_{ext} represents the negative logarithm of the fluoride ion concentration (or rather its activity) in the sample solution. For causes similar to those given for pH electrodes, actual computation is not used in practice; instead, fluoride concentrations are obtained by the use of a calibration curve that has been established by carrying solutions of known concentration and composition through the procedure.

Ion-selective electrodes are also available with a membrane consisting of a pellet cast, for example, from a silver halide a silver sulfide. The resulting electrode is selective to the halide ion or to the sulfide ion, accordingly. For some materials, the pellets do not have the desired properties and cannot be cast; then other means must be used to obtain a membrane. A salt can be incorporated into a membrane formed from paraffin or polymerized silicone rubber. The membrane is then cemented to a glass tube or a plastic tube. When the salt used, for example, is silver chloride or barium sulfate, the electrodes obtained are selective to chloride and sulfate ion, respectively.

Another kind of ion-selective electrode involves a membrane consisting of a liquid ion exchanger. For the present purpose, the function of such a material will become sufficiently clear from considering the calcium-selective electrode shown schematically in Fig. 23-3. (For information on ion exchange, see Chapter 45.) The ion exchanger involved here is a liquid organic compound containing phosphate groups that "bind" calcium preferentially over other cations. The ion exchanger is contained in a reservoir and seeps through a porous disk. By means of the mem-

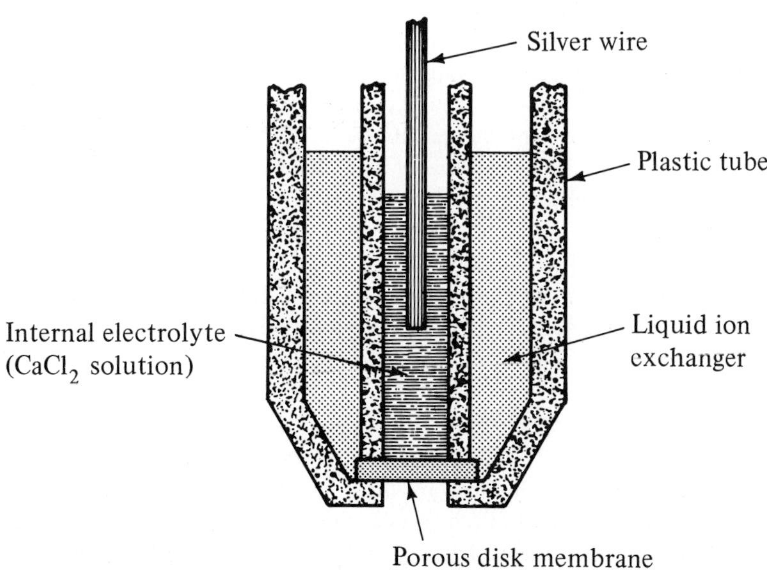

Fig. 23-3. Schematic representation of a calcium-selective electrode (see text).

brane thus formed, electrical connection is established between the internal electrolyte and the sample solution. A silver wire dipping into the internal calcium chloride solution completes the internal reference electrode. Into the sample solution also dips an external reference electrode (commonly a saturated calomel electrode). The cell scheme thus formed is analogous to that presented for the fluoride selective electrode. When the calcium concentrations at the inner and outer surfaces of the membrane are different, a potential develops that can be measured and related to the calcium concentration of the sample solution in the manner described above. The selectivity constant $k^{\text{pot}}_{\text{Ca}^{2+},\text{Mg}^{2+}}$ for the calcium electrode has a value of about 50, and Nernstian response is observed for solutions of pH 6 to 11. At lower pH values, calcium ion is exchanged for hydrogen ion and the working of the electrode is impaired. Shortcomings of this kind of ion-selective electrode, at least at the present state of development, are the slow response (about 30 s), and the need for renewing the ion exchanger from time to time—and in some cases, also the membrane.

Electrodes with membranes that are permeable to gases have been designed. For example, an ammonia-selective electrode allows ammonium concentrations to be measured. Electrodes have been devised that respond selectively to certain enzymes or to enzymatic substrates (that is, substances involved in an enzyme-catalyzed reaction).

Measurements with ion-selective electrodes can be made continuously and yield an electrical signal. Consequently these devices can monitor and control processes. A more subtle but important feature is that the measurements commonly do not disturb an equilibrium in which the species to be measured is involved. Consider a situation of physiological interest. Calcium is an important constituent of blood and is important in the clotting process. Calcium is present partly as the free, hydrated ion and partly loosely bound to protein. It is of interest to know to what extent the calcium is in the "free" form. In older methods for determining free calcium, a reagent was added. This reagent reacted with the free calcium and thereby the position of the equilibrium was shifted so that some bound calcium was released. This calcium then reacted with the reagent, and so forth. Clearly the final analytical result did not relate fully to the true concentration of free calcium. In contrast, when the calcium-selective electrode is used, the equilibrium does not shift and the free calcium level can be established.

23.8 *Limitations of Direct Potentiometric Determinations*

Although determinations by direct potentiometric measurements are simple and practical, their application faces limitations. First, it must be realized that such measurements yield the activity of the ion of interest and not its concentration, which is what is wanted in most practical analyses. Whereas this error may be negligible at low concentrations, it cannot be neglected at higher concentrations, and a correction factor must be introduced. Unfortunately, the mathematical conversion of activity to concentration is not always simple and reliable. Second, an error is inherent in potentiometric determinations that cannot be eliminated even by most careful measurement of the potential. The implications of this statement will become clear from the following derivation. The measurement generally involves comparing the potential of the indicator electrode, E_{in}, with the potential of the reference electrode, E_{ref}. Within the connections between the system, there is also a salt bridge or equivalent, and consequently, a junction potential E_j must be considered (see Section 23.2). The observed potential E_{obs} can be expressed by

$$E_{obs} = E_{ref} - E_{in} + E_j \qquad (23\text{–}10)$$

For an ion M^{m+}, the potential of the indicator electrode will usually be given by the Nernst equation:

$$E_{in} = K' + \frac{0.0592}{n} \log a_{M^{m+}} \qquad \text{(at 25°C)} \qquad (23\text{–}11)$$

Here, $a_{M^{m+}}$ is the activity of the ion M^{m+}. Combining these two equations yields

$$\text{pM} = -\log a_{M^{m+}} = \frac{E_{obs} - (E_{ref} + E_j - K')}{0.0592/n} \qquad (23\text{–}12)$$

Replacing the parenthetical term by K gives

$$\text{pM} = -\log a_{M^{m+}} = \frac{E_{obs} - K}{0.0592/n} \qquad (23\text{–}13)$$

The problem is that K incorporates the junction potential, the value of which can be neither calculated nor determined exactly by experiment. The way out of this dilemma is to establish a value for K by using one or more standards. To do so, however, involves assuming that the value of K is unchanged when the sample solution is substituted for a standard solution. Under the best conditions, it is estimated, this assumption is valid within an absolute uncertainty of typically ±1 mV.

Regardless of how carefully the potential measurements are made, the relative error in the activity obtained is, because of this uncertainty, about ±4*n* percent, where *n* is the number of electrons involved in the electrode reaction.*

---

*The student familiar with calculus will understand the following derivation. Take equation (23–13), assume that E_{obs} is constant, and convert to differentials. The result is

$$d(-\log a_{M^{m+}}) = d\frac{E_{obs} - K}{0.0592/n}$$

$$0.434\frac{da_{M^{m+}}}{a_{M^{m+}}} = \frac{dK}{0.0592/n}$$

By rearrangement, replacement of d by Δ, and multiplication by 100, the

23.9 Questions

23–1 What is the purpose of adopting an operational definition of pH?

23–2 Explain why frequent standardization of a pH meter is necessary.

23–3 Explain why the potential of a glass electrode cannot be measured with an ordinary potentiometer.

23–4 Describe the difference between the two types of pH meters mentioned and discuss the advantages and disadvantages of each.

23–5 What difficulty may be experienced in using a glass electrode to measure the pH of a sodium hydroxide solution?

23–6 Reasoning from a knowledge of how to use a hydrogen-ion-sensitive electrode and a pH meter, indicate the steps that might be involved in establishing the chloride activity of a biological fluid using a chloride-selective electrode. If the chloride content of the fluid were determined by titration with silver nitrate, how might the chloride-selective electrode be used?

23–7 What is understood by the term *Nernstian response*?

23–8 An ion-selective electrode does not show Nernstian response. Does this prevent its useful, practical application? Explain.

23–9 What are electrodes of the first order and of the second order?

23–10 What limits the accuracy and precision in the potentiometric determination of the concentration of a metal ion?

23–11 Elaborate on the rationale and necessity for standardization in the measurement of pH.

23.10 Problems

23–1 A hydrogen electrode in combination with a saturated calomel electrode is used at 25°C to determine the pH of a solution. The measured cell voltage is 0.321 V with the hydrogen electrode negative in polarity. Assuming the partial pressure of hydrogen gas to be at the standard value of 1.01325 × 10^5 Pa (1 atm), calculate the pH of the solution. *Ans.:* pH 1.3_6

---

percentage error is obtained:

$$\% \text{ error} = \frac{\Delta a_{\mathrm{M}^{m+}}}{a_{\mathrm{M}^{m+}}} \times 100 = 3900n\ \Delta K$$

or with

$$\Delta K = \pm 1 \text{ mV},$$

$$\% \text{ error} = \pm 3900n \times 10^{-3} = \sim \pm 4n\%$$

23-2 A hydrogen electrode–calomel electrode pair is used to measure the pH of a solution. Neither the pressure of the hydrogen gas nor the chloride ion concentration is known exactly. However, the measured cell voltage is 0.561 V, with the hydrogen electrode having negative polarity, when the cell contains a phthalate solution of pH 4.01. When the cell is filled with the unknown solution, other conditions remaining constant, the voltage is found to be 0.828 V. Calculate the pH of the solution. *Ans.:* pH 8.54

23-3 A hydrogen electrode–saturated calomel electrode pair is used to measure the pH of solutions. Calculate the voltage of this cell for each of the pH values given, assuming that the hydrogen partial pressure is at the standard value of 1.01325×10^5 Pa (1 atm), and indicate the polarity of the SCE. $E_{SCE} = +0.241$ V. *Ans.:* 0.300 V with SCE positive
(a) 1.00
(b) 3.75
(c) 8.26
(d) 9.07
(e) 5.22
(f) 6.98

23-4 The voltage of the following cell is found to be 0.118 V. Calculate the dissociation constant of the weak acid, HA.

$$^{(+)}\mathrm{Pt} \mid H_2(1.01325 \times 10^5\ \mathrm{Pa}),\ H^+(0.20M) \parallel \mathrm{HA}(0.040F),$$
$$H_2(1.01325 \times 10^5\ \mathrm{Pa}) \mid \mathrm{Pt}^{(-)}$$

Ans.: $K_a = 1.0_5 \times 10^{-4}$

23-5 The voltage of the following cell is found to be 0.477 V. Calculate the dissociation constant of the weak acid. $E_{SCE} = +0.241$ V.

$$^{(+)}\mathrm{Hg} \mid Hg_2Cl_2(s),\ \mathrm{KCl(sat'd)} \parallel \mathrm{HA}(1.0 \times 10^{-2}F),$$
$$H_2(1.01325 \times 10^5\ \mathrm{Pa}) \mid \mathrm{Pt}^{(-)}$$

23-6 Calculate the voltage of the following cell.

$$\mathrm{Hg} \mid Hg_2Cl_2(s),\ \mathrm{KCl(sat'd)} \parallel NH_4Cl(0.18F),\ H_2(1.01325 \times 10^5\ \mathrm{Pa}) \mid \mathrm{Pt}$$
$$K_{b,NH_3} = 1.8 \times 10^{-5},\ E_{SCE} = +0.241\ \mathrm{V}$$

23-7 A saturated solution of a sparingly soluble silver salt AgX was placed in a beaker. A silver wire was inserted as the indicator electrode and a saturated calomel electrode for reference. The voltage of the cell thus formed was found to be zero. Calculate the solubility product of AgX.

$$E^0_{Ag^+/Ag} = +0.810\ \mathrm{V},\ E_{SCE} = +0.241\ \mathrm{V}$$

Ans.: 10^{-19}

23-8 A fluoride-selective electrode is used with a reference electrode for the electrometric determination of the fluoride concentration of municipal waters. Each standard solution and sample is brought to the same ionic strength by adding a noninterfering electrolyte. Solutions of sodium fluoride in deionized water were used to establish calibration data:

Cell voltage, V	Fluoride molarity
+0.122	1.00×10^{-5}
+0.062	1.00×10^{-4}
+0.002	1.00×10^{-3}
-0.058	1.00×10^{-2}

(a) Plot the data (voltage versus logarithm of the fluoride molarity) and draw the best-fitting line through the points. Is the electrode response Nernstian in relation to linearity and slope?

(b) For a given sample, the cell response was found to be +0.043 V. What is the concentration of fluoride (*19.0*) in the sample, expressed in milligrams per liter (that is, in parts per million)?

Ans.: (b) 3.6 mg/L

24

Theory of Redox Titrations

A redox reaction, to be made the basis of a titration, must fulfill the general requirements for any titration reaction: The reaction should proceed rapidly and to completion; a definite, known equivalence should exist between the oxidant and reductant, and a suitable technique should be available for detecting the end point (cf. Chapter 10). Since many of the elements can exist in more than one oxidation state, applications for redox titrations are numerous (Chapter 25). We now examine the feasibility of such titrations, using the Nernst equation to calculate titration curves.

24.1 *Titration Curves*

For redox titrations, as for titrations previously considered, a titration curve can be obtained by plotting the negative logarithm of the concentration of the species being titrated as a function of the volume of titrant solution added. However, plotting half-cell potentials is preferable for several reasons. First, from the Nernst equation, such potentials are a logarithmic function of the concentration of the reacting species. Second, this way makes it easier to choose visual redox indicators, since these are redox couples for which electrode potentials are available from tables (Section 24.8). Finally, the potentiometric method of end-point detection, widely used in redox titrimetry, is directly related to the half-cell potential (Chapter 27).

The model used for calculating a redox titration curve is a cell consisting of the half-cell containing the solution that is to be titrated and the standard hydrogen electrode, connected by a salt bridge. The cell potential is calculated after the assumed addition of a known volume of the titrant solution. It will become evident in the subsequent consideration of methods of end-point detection that such a physical arrangement may not be used in the actual redox titration. Such a model, however, helps in understanding the titration.

Although the same calculation approach applies to all redox titrations, it is convenient to distinguish several cases: (1) titrations in which the same number of electrons appear in each of the balanced half-reaction equations (Section 24.2);

(2) titrations in which different numbers of electrons appear in the two balanced half-reaction equations (Section 24.3); (3) titrations in which a half-reaction equation involves the solvent and its dissociation products explicitly (Section 24.4); and (4) titrations in which one molecule or one ion of a reactant yields two or more molecules or ions of product, or vice versa (Section 24.5).

24.2 *Titrations with Same Number of Electrons in Each Half-Reaction Equation*

Consider the titration of exactly 50 mL of a solution 0.10F in iron(II) sulfate with 0.10F cerium(IV) sulfate solution. The titration equation is

$$Fe^{2+} + Ce^{4+} \rightarrow Fe^{3+} + Ce^{3+} \qquad (24\text{–}1)$$

The half-reaction equations and Nernst equation expressions take the form*

$$Fe^{3+} + e \rightarrow Fe^{2+} \qquad E = +0.68 + 0.059 \log \frac{[Fe^{3+}]}{[Fe^{2+}]} \qquad (24\text{–}2)$$

$$Ce^{4+} + e \rightarrow Ce^{3+} \qquad E = +1.44 + 0.059 \log \frac{[Ce^{4+}]}{[Ce^{3+}]} \qquad (24\text{–}3)$$

At the start of the titration, the solution as a first approximation might be considered to contain only iron(II) ion. This assumption implies that the iron(III) concentration is zero, and hence from the Nernst equation, that the potential is minus infinity. This conclusion is unreasonable and conflicts with experimental observation. There is an extremely small, but unknown, concentration of Fe^{3+} present; the potential, although finite, cannot be calculated for this point.

When a small volume of the titrant is added, the situation changes. Since the reaction goes essentially to completion, it can be seen from the titration equation, (24–1), that the number of millimoles of iron(III) ion formed equals the number of millimoles of cerium(IV) ion added. The number of millimoles of iron(II) remaining

*This titration is carried out in a sulfuric acid solution, and cerium(IV) is largely present as an anionic sulfato complex, $Ce(SO_4)_3^{2-}$. To simplify the treatment of this example, we shall assume that the cerium(IV) ion is present as the simple ion and that its concentration can be shown as $[Ce^{4+}]$. However, the formal potential of the couple, $E^f_{Ce^{4+}/Ce^{3+}} = +1.44$ V in 1F H_2SO_4, will be used rather than the standard potential. The formal potential of the iron couple, $E^f_{Fe^{3+}/Fe^{2+}} = +0.68$ in 1F H_2SO_4, will also be used. The concept of formal potential was introduced in Section 22.8 and is further treated in Section 28.8. Throughout Chapter 24, wherever formal potentials are used with the Nernst equation, concentrations will be shown as molarities, although formalities should properly be used for the relevant species.

unreacted equals the original number of millimoles of iron(II) minus the number of millimoles of iron(III) formed.

After 1.0 mL of the 0.10*M* Ce^{4+} solution is added, the amount of iron(III) ion formed equals 1.0 × 0.1 = 0.10 mmol. The amount of iron(II) remaining is 50 × 0.10 - 0.10 = 4.9 mmol. The half-cell potential can be calculated from the Nernst equation for the iron couple:

$$E = +0.68 + 0.059 \log \frac{0.10}{4.9}$$

$$= +0.68 - 0.059 \times 1.69 = +0.58 \text{ V}$$

In this calculation, millimoles can be substituted for molarities since the solution volume cancels. (You recall that a similar simplification was possible in calculating the titration curve for a weak acid, Section 11.4.) This possibility implies that for this and analogous examples, the potential is independent of the dilution of the solution.

In principle, the potential could also be calculated from the Nernst equation for the cerium couple. However, the reduction goes essentially to completion and only an extremely small concentration of cerium(IV) ion remains. This cerium(IV) concentration could be calculated from the value of the equilibrium constant if this were known. It is easier, however, to calculate the half-cell potential from the expression for the iron couple. (The value of an equilibrium constant can be calculated from potentiometric data; see Section 24.7.)

When a total volume of 10 mL of the Ce^{4+} solution has been added, the amount of iron(III) ion formed is 10 × 0.10 mmol, and the amount of iron(II) ion unreacted corresponds to 5.0 - 1.0 = 4.0 mmol. At this point, the half-cell potential is given by

$$E = +0.68 + 0.059 \log \frac{1.0}{4.0}$$

$$= +0.68 - 0.059 \times 0.60 = +0.65 \text{ V}$$

At the midpoint of the titration, that is, when a total volume of exactly 25 mL of the Ce^{4+} solution has been added, the amount of iron(III) formed as 25 × 0.10 = 2.5 mmol and the amount of iron(II) remaining is 5.0 - 2.5 = 2.5 mmol. Hence, the half-cell potential is given by

$$E = +0.68 + 0.059 \log \frac{2.5}{2.5} = +0.68 \text{ V}$$

Note that at the midpoint of this titration, the potential equals the formal potential (or standard electrode potential) of the system being titrated, $E^f_{Fe^{3+}/Fe^{3+}}$. This equality offers an analogy to the titration of a weak acid with a strong base, where at the midpoint pH = pK_a; the analogy is more obvious when the Nernst equation is contrasted with the "buffer" equation, pH = pK_a + log([salt]/[acid]) (see Sections 11.4 and 13.3).

The equivalence point is reached when a total volume of exactly 50 mL of the Ce^{4+} solution has been added. At this point, following the same mode of calculation, the amount of iron(III) formed is 50 × 0.10 = 5.0 mmol, and the amount of iron(II) unreacted is 5.0 - 5.0 = 0.0 mmol. Substituting these values into the Nernst equation yields an electrode potential of infinity, which is obviously incorrect. (Calculating the equivalence point for the titration of a strong acid with a strong base is analogous; see Section 11.2.) Actually at equilibrium, small amounts of iron(II) and cerium(IV) species must be present. The potential at this point can be calculated by expressing the equivalence-point potential in terms of the Nernst equation for each of the two couples involved:

$$E_{\text{eq pt}} = +0.68 + 0.059 \log \frac{[Fe^{3+}]}{[Fe^{2+}]}$$

$$E_{\text{eq pt}} = +1.44 + 0.059 \log \frac{[Ce^{4+}]}{[Ce^{3+}]}$$

Then, adding these two equations yields

$$2E_{\text{eq pt}} = +0.068 + 1.44 + 0.059 \log \frac{[Fe^{3+}][Ce^{4+}]}{[Fe^{2+}][Ce^{3+}]}$$

At the equivalence point, the total millimoles of the iron species exactly equals the total millimoles of the cerium species. Looking at the titration equation reveals that the change of one ion of Fe^{2+} to Fe^{3+} requires the conversion of one ion of Ce^{4+} to Ce^{3+}. In other words, at the equivalence point, for each Fe^{3+} ion present one Ce^{3+} ion has formed, and for each Fe^{2+} ion remaining unoxidized one Ce^{4+} ion remains unreduced. Consequently,

$$[Fe^{3+}] = [Ce^{3+}]$$

$$[Fe^{2+}] = [Ce^{4+}]$$

Dividing one of these equations by the other yields

$$\frac{[Fe^{3+}]}{[Fe^{2+}]} = \frac{[Ce^{3+}]}{[Ce^{4+}]}$$

Substituting in the above equation yields

$$2E_{\text{eq pt}} = +0.68 + 1.44 + 0.059 \log \frac{\cancel{[Ce^{3+}]}\,\cancel{[Ce^{4+}]}}{\cancel{[Ce^{4+}]}\,\cancel{[Ce^{3+}]}}$$

Since log 1 = 0,

$$2E_{\text{eq pt}} = +0.68 + 1.44$$

Hence,

$$E_{\text{eq pt}} = \frac{+0.68 + 1.44}{2} = +1.06 \text{ V}$$

Thus, in this example, the potential of the half-cell at the equivalence point is the average of the formal electrode potentials of the two redox couples. (When standard electrode potentials are applicable, the equivalence-point potential will be the average of these two values.)

Beyond the equivalence point, the half-cell potential can be calculated from the Nernst equation for the cerium couple. In principle, the potential could be calculated from the Nernst equation for the iron couple; however, the concentration of the Fe^{2+} ion is not readily available. The reasoning is like that mentioned for not using the cerium couple before the equivalence point.

When a total volume of 60 mL of the Ce^{4+} solution has been added, 5 mmol of cerium(III) and 1 mmol of cerium(IV) are present in the solution. Here

$$E = +1.44 + 0.059 \log \frac{1.0}{5.0} = +1.40 \text{ V}$$

After a total of 100 mL of the Ce^{4+} solution has been added,

$$E = +1.44 + 0.059 \log \frac{5.0}{5.0} = +1.44 \text{ V}$$

Thus, after addition of twice the volume of titrant solution needed to reach the equivalence point, the half-cell potential equals the formal potential (or standard electrode potential) of the redox system of the titrant, $E^f_{Ce^{4+}/Ce^{3+}}$. The complete titration curve, calculated in this way, is given in Fig. 24-1.

24.3 *Titrations with Different Numbers of Electrons in the Half-Reaction Equations*

When the redox titration involves a different number of electrons in each of the half-reaction equations, the shape of the titration curve differs from what is relevant when the same number of electrons is involved.

Consider the titration of exactly 50 mL of a solution 0.050*F* in tin(IV) chloride with a solution 0.10*F* in chromium(II) chloride. Since tin(IV) is present as a chloro complex, the formal potential in 1*F* hydrochloric acid will be used for the tin couple, but to simplify the notation used, the symbol $[Sn^{4+}]$ will be allowed to denote the formal, instead of the molar, concentration of tin(IV). Similar considerations apply to the chromium couple, for which the formal potential in 1*F* hydrochloric acid will also be used.

The titration reaction is

$$Sn^{4+} + 2Cr^{2+} \rightleftharpoons Sn^{2+} + 2Cr^{3+} \qquad (24\text{-}4)$$

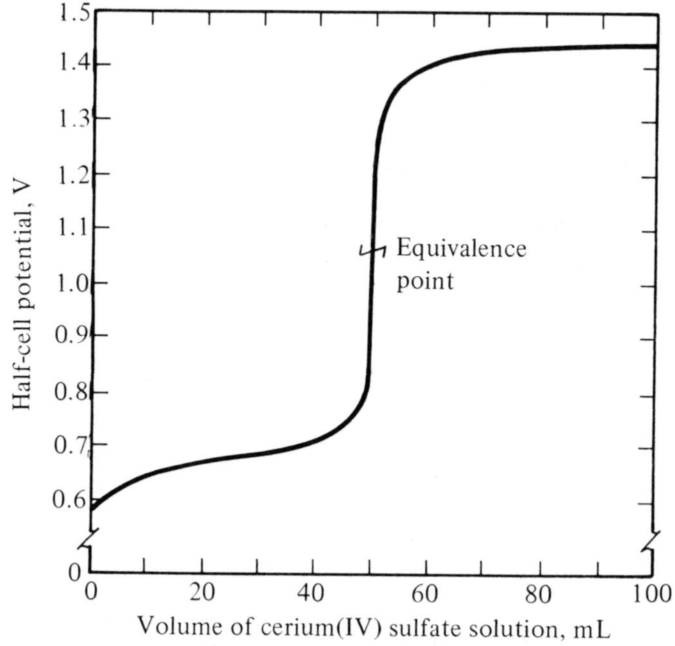

Fig. 24-1. Calculated titration curve for the titration of 50 mL of 0.10*F* $FeSO_4$ with 0.10*F* cerium(IV) solution (in a titration medium 1*F* in H_2SO_4).

The half-reaction equations and Nernst equation expressions take the form

$$Sn^{4+} + 2e \rightleftharpoons Sn^{2+} \qquad E = +0.14 + \frac{0.059}{2} \log \frac{[Sn^{4+}]}{[Sn^{2+}]} \tag{24-5}$$

$$Cr^{3+} + e \rightleftharpoons Cr^{2+} \qquad E = -0.38 + 0.059 \log \frac{[Cr^{3+}]}{[Cr^{2+}]} \tag{24-6}$$

At the start of the titration, the potential cannot be calculated, for reasons noted in the case considered in Section 24.2.

When 10 mL of the Cr^{2+} solution is added, the amount of chromium(III) formed is $10 \times 0.10 = 1.0$ mmol. Initially, the amount of tin(IV) present was $50 \times 0.050 = 2.5$ mmol. Since from the titration equation two ions of Cr^{3+} are produced by the reaction of one Sn^{4+} ion, the amount of tin(IV) that reacts is $1.0/2 = 0.50$ mmol. The amount of tin(IV) remaining is therefore $2.5 - 0.5 = 2.0$ mmol. The half-cell potential can be calculated from the Nernst equation for the tin couple, (24-5):

$$E = +0.14 + \frac{0.059}{2} \log \frac{2.0}{0.5} = +0.16 \text{ V}$$

In this calculation, millimoles can be substituted for molarities since the solution volume cancels; this implies that for this example and others like it, the potential is independent of the dilution of the solution.

When a total volume of 25 mL of the Cr^{2+} solution has been added, the amount of chromium(III) formed is 25 × 0.10 = 2.5 mmol. The amount of tin(IV) that reacts is 2.5/2 = 1.25 mmol. The amount of tin(IV) remaining is 2.5 - 1.25 = 1.25 mmol. At this point the potential is given by

$$E = +0.14 + \frac{0.059}{2} \log \frac{1.25}{1.25} = +0.14 \text{ V} = E^f_{Sn^{4+}/Sn^{2+}}$$

Here, as in the case considered in Section 24.2, the potential at the midpoint of the titration equals the formal potential (or standard electrode potential) of the system being titrated.

The equivalence point is reached when exactly 50 mL of the Cr^{2+} solution is added. Here the total number of millimoles of the chromium species equals twice the total number of millimoles of the tin species present. At equilibrium, not all the chromium is present as chromium(III) and not all the tin is present as tin(II). Inspecting the titration equation reveals that changing one ion of Sn^{4+} to Sn^{2+} calls for converting two ions of Cr^{2+} to Cr^{3+}. As a consequence, at the equivalence point

$$2[Sn^{4+}] = [Cr^{2+}] \quad \text{and} \quad 2[Sn^{2+}] = [Cr^{3+}]$$

Dividing the first of these equations by the second yields

$$\frac{[Sn^{4+}]}{[Sn^{2+}]} = \frac{[Cr^{2+}]}{[Cr^{3+}]}$$

Substituting this equation in the Nernst equation for the tin couple, (24-5), yields for the half-cell potential at the equivalence point

$$E_{\text{eq pt}} = +0.14 + \frac{0.059}{2} \log \frac{[Cr^{2+}]}{[Cr^{3+}]} \tag{24-7}$$

To eliminate the logarithmic term when this equation is added to the Nernst equation for the chromium couple, (24-6), it is necessary to multiply (24-7) by 2:

$$2E_{\text{eq pt}} = 2 \times 0.14 + 0.059 \log \frac{[Cr^{2+}]}{[Cr^{3+}]} \tag{24-8}$$

For the chromium couple, from (24-6)

$$E_{\text{eq pt}} = -0.38 + 0.059 \log \frac{[Cr^{3+}]}{[Cr^{2+}]} \tag{24-9}$$

Adding equations (24-8) and (24-9) yields

$$3E_{\text{eq pt}} = 2 \times 0.14 - 0.38$$

and

$$E_{\text{eq pt}} = \frac{2 \times 0.14 - 0.38}{3} = -0.03\text{ V}$$

We see that the value of the potential of the equivalence point does not fall halfway between the values of the formal potentials (or standard electrode potentials) of the two redox couples, as in the case considered in Section 24.2. Rather, the value is two-thirds of the way and closer to the potential involving the two-electron change.

Beyond the equivalence point, the potential can be calculated from the Nernst equation for the chromium couple. Thus, the potential after the addition of a total of 60 mL of the Cr^{2+} solution is given by

$$E = -0.38 + 0.059 \log \frac{5.0}{1.0} = -0.34\text{ V}$$

After a total volume of 100 mL of the Cr^{2+} solution is added, the potential is

$$E = -0.38 + 0.059 \log \frac{5.0}{5.0} = -0.38\text{ V} = E^f_{Cr^{3+}/Cr^{2+}}$$

Here, as in the case previously considered, the potential after addition of twice the volume of titrant needed for reaching the equivalence point corresponds to the formal potential (or standard electrode potential) represented by the titrant. The complete titration curve for this example is given in Fig. 24-2.

Generally, when the titration equation can be written in the form $n_1\text{Ox}_1 + n_2\text{Red}_2 \rightleftharpoons n_1\text{Red}_1 + n_2\text{Ox}_2$, the half-cell potential at the equivalence point is given by

$$E_{\text{eq pt}} = \frac{n_1 E^f_{\text{Ox}_1/\text{Red}_1} + n_2 E^f_{\text{Ox}_2/\text{Red}_2}}{n_1 + n_2} \tag{24-10}$$

In the two cases considered so far, $n_1 = n_2$ and $2n_1 = n_2$, respectively. While formal potentials appear in equation (24-9), standard electrode potentials can be substituted when applicable.

24.4 *Titrations in Which a Half-Reaction Equation Involves the Solvent and Its Dissociation Products*

Calculating the titration curve for titrations in which a half-reaction equation involves the solvent and its dissociation products is more complicated than the cases previously considered. The difficulty may arise when a different number of combined oxygen atoms appears in a reactant and its product.

Consider the titration of exactly 50 mL of a solution 0.10*M* in Fe^{2+} with a

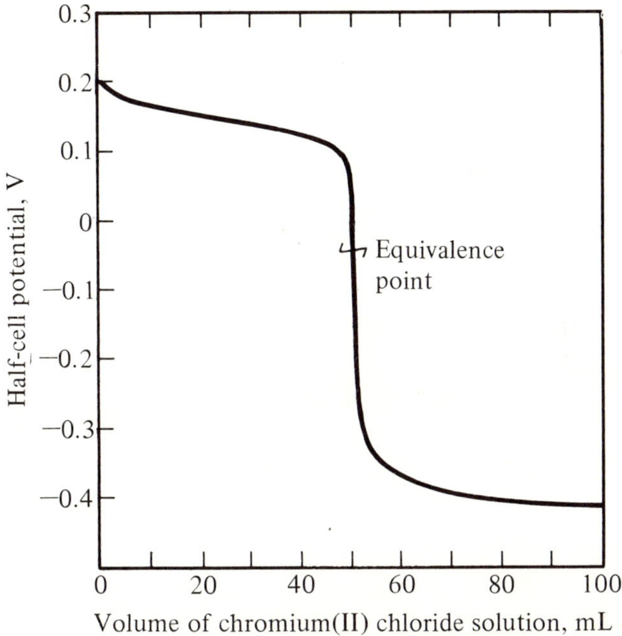

Fig. 24-2. Calculated titration curve for the titration of 50 mL of 0.050*F* $SnCl_4$ with 0.10*F* $CrCl_2$ solution (in a titration medium 1*F* in HCl).

solution 0.02*M* in MnO_4^-, under conditions in which the titration solution is acidic. The titration reaction is

$$5Fe^{2+} + MnO_4^- + 8H^+ \rightleftharpoons 5Fe^{3+} + Mn^{2+} + 4H_2O \tag{24-11}$$

The reduction of permanganate ion to manganese(II) ion is not fully reversible and the Nernst equation, therefore, does not strictly apply. However, the agreement between calculated and experimental titration curves is close enough to make this a worthwhile example, especially because permanganate titrations have practical importance.

Since the half-cell potential before the equivalence point is reached is determined by the ratio $[Fe^{3+}]/[Fe^{2+}]$ and the concentration ratios are identical to those in the example considered in Section 24.3, the titration curve for both examples will be identical almost to the vicinity of the equivalence point. At the equivalence point, the following equalities exist:

$$[Fe^{3+}] = 5[Mn^{2+}] \qquad \text{and} \qquad [Fe^{2+}] = 5[MnO_4^-]$$

Hence,

$$\frac{[Fe^{3+}]}{[Fe^{2+}]} = \frac{[Mn^{2+}]}{[MnO_4^-]}$$

Substituting this equation in the Nernst equation for the iron couple, (24–2), yields

$$E_{\text{eq pt}} = +0.68 + 0.059 \log \frac{[\text{Mn}^{2+}]}{[\text{MnO}_4^-]} \tag{24–12}$$

The Nernst equation for the permanganate-manganese(II) couple takes the form

$$E_{\text{eq pt}} = +1.51 + \frac{0.059}{5} \log \frac{[\text{MnO}_4^-]\,[\text{H}^+]^8}{[\text{Mn}^{2+}]} \tag{24–13}$$

Multiplying equation (24–12) by 5 and adding to (24–11) yields

$$6E_{\text{eq pt}} = +0.68 + 5 \times 1.51 + 0.059 \log \frac{\cancel{[\text{Mn}^{2+}]}\,\cancel{[\text{MnO}_4^-]}\,[\text{H}^+]^8}{\cancel{[\text{MnO}_4^-]}\,\cancel{[\text{Mn}^{2+}]}}$$

On rearrangement,

$$\begin{aligned} E_{\text{eq pt}} &= \frac{+0.68 + 5 \times 1.51}{6} + 0.0098 \log [\text{H}^+]^8 \\ &= +1.37 + 0.0098 \log [\text{H}^+]^8 \\ &= +1.37 - 0.078\cdot\text{pH} \end{aligned} \tag{24–14}$$

We see that the half-cell potential at the equivalence point markedly depends on the hydrogen ion concentration. Only if $[\text{H}^+] = 1M$, that is, pH = 0, does equation (24–14) reduce to the form of equation (24–10), with $n_1 = 5n_2$, and $E_{\text{eq pt}} = +1.37$ V. If $[\text{H}^+] = 0.1M$, that is, pH = 1.0, then $E_{\text{eq pt}} = +1.29$ V. Points on the titration curve beyond the equivalence point can be calculated from the Nernst equation for the permanganate-manganese couple, (24–13), if the pH of the solution is known. The complete, calculated titration curve for a medium $1M$ in hydrogen ion is given in Fig. 24–3.

24.5 Titrations in Which One Particle of Reactant Yields Two or More Particles of Product, or Vice Versa

In the cases already described, the titration curve is independent of the total concentration of the redox couple. That is, the half-cell potential is essentially unchanged by dilution (or concentration) of the solution, except as this changes the pH in the systems considered in Section 24.4. Cases exist in which the species actually reduced (or oxidized), written on one side of the balanced equation, has a coefficient different from that of the species formed, written on the other side of the equation. Typical examples include

$$\text{S}_4\text{O}_6^{2-} + 2e \rightleftharpoons 2\text{S}_2\text{O}_3^{2-}$$

$$14\text{H}^+ + \text{Cr}_2\text{O}_7^{2-} + 6e \rightleftharpoons 2\text{Cr}^{3+} + 7\text{H}_2\text{O}$$

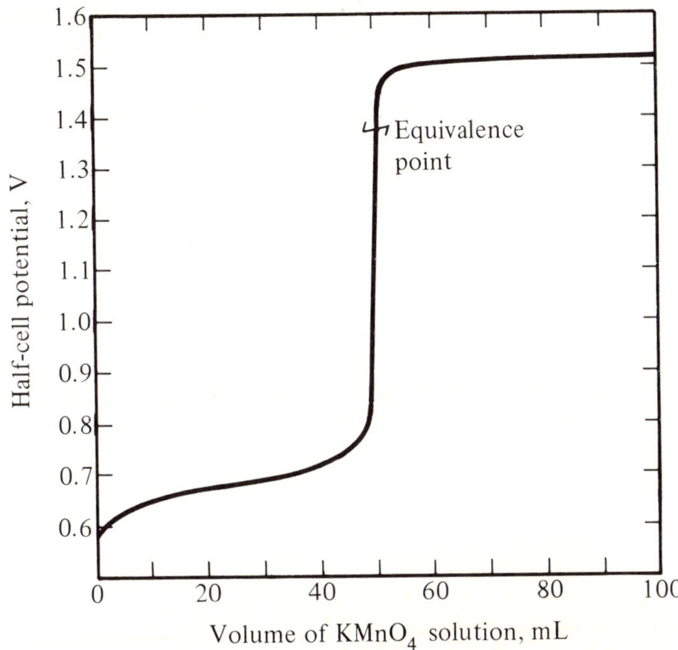

Fig. 24-3. Calculated titration curve for the titration of 50 mL of 0.10*F* $FeSO_4$ with 0.02*F* $KMnO_4$ (in a titration medium 1*M* in hydrogen ion).

Here one mole of tetrathionate or dichromate is equivalent to two moles of thiosulfate or chromium(III). An analogous situation exists, for example, in all titrations involving iodine, bromine, and oxalate. (Write the relevant half-reaction equations.)

In such cases, the concentration terms of the relevant species have different exponents in the Nernst equation. As a result, the solution volume will not cancel in the ratio of concentrations (if expressed as millimoles per milliliter), and the potential will vary with the dilution of the solution. Also, the logarithmic term will not vanish from the equation for the potential of the equivalence point, and consequently, the potential at this point will also be a function of the initial concentrations both of the species titrated and of the titrant.

24.6 Redox "Buffers"

Analogies between redox and acid-base titrations have already been stressed. One more is of interest. A redox titration may be compared with the titration of a weak acid with a strong base in the shape of their curves. Both curves show the following features: a sharp rise at the beginning, a section of low slope around the halfway point, a "jump" about the equivalence point, and a horizontal portion beyond that jump. Of special interest is the situation around the halfway point.

(The situation for an acid-base titration has been discussed in Section 13.3.) The portion of low slope indicates that adding oxidant or reductant to a solution having a composition corresponding to that region will not cause an appreciable change in potential. In this region, the redox system, it is said, is well "poised." The term *poised* is analogous to *buffered.*

24.7 Calculating Equilibrium Constants from Potentials

The equilibrium constant for the reaction equation in redox titrations (as well as for many other ionic reactions) can be calculated from potentiometric data. At any point in a redox titration when equilibrium has been established, the potential calculated from the Nernst equation of one of the redox couples present must equal the potential calculated from the Nernst equation of the second couple. Equating the two Nernst equations leads to a relation between formal potential (or standard electrode potential) data and the equilibrium constant that allows calculation of its value.

Example 24-1 What is the value of the equilibrium constant in $1F$ HCl for the reaction $Cr^{2+} + Fe^{3+} \rightleftharpoons Cr^{3+} + Fe^{2+}$ if $E^f_{Cr^{3+}/Cr^{2+}} = -0.38$ V and $E^f_{Fe^{3+}/Fe^{2+}} = +0.70$ V?

The expression for the equilibrium constant takes the form

$$K = \frac{[Cr^{3+}][Fe^{2+}]}{[Cr^{2+}][Fe^{3+}]}$$

The Nernst equations for the two redox couples, with the formal potentials in $1F$ HCl inserted, are

$$E = +0.70 + 0.059 \log \frac{[Fe^{3+}]}{[Fe^{2+}]}$$

$$E = -0.38 + 0.059 \log \frac{[Cr^{3+}]}{[Cr^{2+}]}$$

At equilibrium, the potentials of the two couples have the same value; hence

$$+0.70 + 0.059 \log \frac{[Fe^{3+}]}{[Fe^{2+}]} = -0.38 + 0.059 \log \frac{[Cr^{3+}]}{[Cr^{2+}]}$$

Rearrangement yields

$$0.70 + 0.38 = 0.059 \log \frac{[Cr^{3+}]}{[Cr^{2+}]} + 0.059 \log \frac{[Fe^{2+}]}{[Fe^{3+}]}$$

or

$$1.08 = 0.059 \log \frac{[Cr^{3+}][Fe^{2+}]}{[Cr^{2+}][Fe^{3+}]} = 0.059 \log K$$

Consequently,

$$\log K = \frac{1.08}{0.059} = 18.3$$

or

$$K = \sim 10^{18} \qquad \text{(in } 1F \text{ HCl)}$$

The large value of the equilibrium constant indicates that the reaction goes essentially to completion. As far as the criterion of the completeness of the reaction is concerned, the titration of chromium(II) with iron(III) or the reverse is feasible.

Example 24-2 What is the value of the equilibrium constant for the reaction $5Fe^{2+} + MnO_4^- + 8H^+ \rightleftharpoons 5Fe^{3+} + Mn^{2+} + 4H_2O$ in acidic solution if $E^0_{Fe^{3+}/Fe^{2+}} = +0.77$ V and $E^0_{MnO_4^-/Mn^{2+}} = +1.51$ V?

The expression for the equilibrium constant takes the following form:

$$K_{eq} = \frac{[Fe^{3+}]^5 [Mn^{2+}]}{[Fe^{2+}]^5 [MnO_4^-] [H^+]^8}$$

Equating the two Nernst equations yields

$$E = +0.77 + 0.059 \log \frac{[Fe^{3+}]}{[Fe^{2+}]}$$

$$= +1.51 + \frac{0.059}{5} \log \frac{[MnO_4^-][H^+]^8}{[Mn^{2+}]}$$

Next the concentration terms of the iron species are raised to the fifth power, and at the same time the factor in front of the logarithm is divided by 5. This does not change the numerical value of the term but makes it possible to combine the two logarithmic terms, to give

$$1.51 - 0.77 = \frac{0.059}{5} \log \frac{[Fe^{3+}]^5 [Mn^{2+}]}{[Fe^{2+}]^5 [MnO_4^-] [H^+]^8}$$

Hence

$$\log K = \frac{5 \times 0.74}{0.059} = 62.7$$

$$K = \sim 10^{63}$$

A solubility product is a special form of an equilibrium constant and can be calculated from suitable potentiometric data, notably from the known standard electrode potentials for the reduction of one of the ions involved from the simple (aquo) ion and from the slightly soluble strong electrolyte.

Example 24-3 Calculate the solubility product of mercury(I) bromide, Hg_2Br_2, given that

$$\underline{Hg_2Br_2} + 2e \rightleftharpoons 2Hg + 2Br^- \qquad E^0 = +0.14 \text{ V}$$

$$Hg_2^{2+} + 2e \rightleftharpoons 2Hg \qquad E^0 = +0.79 \text{ V}$$

If the second equation is subtracted from the first, the desired net reaction is obtained:

$$\underline{Hg_2Br_2} \rightleftharpoons Hg_2^{2+} + 2Br^-$$

If the two redox couples are in equilibrium, their potentials must be equal, and the two Nernst equations can be equated.

$$E = +0.79 + \frac{0.059}{2} \log [Hg_2^{2+}] = +0.14 + \frac{0.059}{2} \log \frac{1}{[Br^-]^2}$$

$$0.14 - 0.79 = \frac{0.059}{2} \log [Hg_2^{2+}][Br^-]^2$$

$$-0.65 = \frac{0.059}{2} \log K_{sp}$$

$$\log K_{sp} = \frac{2 \times (-0.65)}{0.059} = -22.0$$

$$K_{sp} = \sim 10^{-22}$$

24.8 Detecting End Points in Redox Titrations

The end point of a redox titration can be detected by measuring the half-cell potential by the potentiometer method described in Section 22.10. This may be described as a redox titration to a potentiometric end point or simply as a potentiometric redox titration. (Consideration of this technique is deferred until Chapter 27.)

The end point in a redox titration can also be detected visually if the titration system is self-indicating or if a suitable indicator is available. Permanganate ion is probably the most common example of a self-indicating oxidizing titrant; the end point corresponds to the appearance of a pink solution color on the addition of the first drop of permanganate solution in excess (Sections 25.5 and 25.6).

Redox indicators are organic substances that can be reduced or oxidized and that also show different colors in their oxidized and reduced forms. The color of at least one of the forms should be so intense that only a small amount of the indicator need be added, and hence the amount of titrant consumed in its oxidation or reduction can be neglected.

The color transition interval of a reversible, two-color redox indicator is centered around a definite potential, which is analogous to the transition interval of a two-color acid-base indicator. In Table H in the Appendix, some common redox indicators are listed with their formal potentials and color changes.

In a redox titration, the transition interval of the indicator should include the equivalence point, and ideally the titration should be stopped when the color is reached that corresponds exactly to the potential of the titration half-cell at that point. In practice, a titration is considered satisfactory when the color changes within the potential range corresponding to the "jump" in the titration curve. The end point is then close enough to the equivalence point and any difference between these two points becomes negligible.

A simple, reversible two-color redox indicator is treated like a two-color acid-base indicator (Chapter 12). The equation for the reaction is $In_{\text{ox}} + ne \rightleftharpoons In_{\text{red}}$. The corresponding Nernst equation is

$$E_{In} = E^f_{In} + \frac{0.059}{n} \log \frac{[In_{\text{ox}}]}{[In_{\text{red}}]} \tag{24-15}$$

where E^f_{In} is the formal potential of the indicator under the existing solution conditions.

For a one-color redox indicator, consider the case in which the oxidized form is colored. The formal (total) concentration of the indicator is C_{In}, hence, the Nernst equation takes the form

$$E_{In} = E^f_{In} + \frac{0.059}{n} \log \frac{[In_{\text{ox}}]}{C_{In} - [In_{\text{ox}}]} \tag{24-16}$$

Consequently for a one-color indicator, the end-point color (more correctly the color intensity) to which the titration should be performed depends on the concentration of the indicator. This parallels the behavior of a one-color, acid-base indicator (see Chapter 12).

24.9 *Complexing Agents in Redox Titrations*

Complexing agents are used in redox titrations for several purposes. If the product of a titration reaction is complexed more strongly than a corresponding reactant, the equilibrium position will be shifted toward the products and the reaction will be more complete. Thus, adding a suitable complexing agent may make feasible a titration that otherwise would not be so. For instance, the break around the equivalence point may be increased, or the curve may be shifted so that it passes through the transition interval of an available indicator. For example, phosphoric acid is often added in the titration of iron(II) ion with dichromate ion in acidic media. Iron(III) forms a stable, almost colorless soluble complex with the orthophosphate ion in acidic solution. As a consequence, the formal potential of the iron couple is lowered and the titration reaction goes further to completion. Diphenylaminesulfonic acid, that is, *N*-phenylanthranilic acid, as the sodium or barium salt, is usually used as the redox indicator. This indicator has a formal potential of about +0.84 V (see Table H in the Appendix). The dichromate-chromium(III) couple has a standard electrode potential of +1.36 V, and the iron(III)-iron(II) couple has one of +0.77 V. With this indicator, the end point would therefore come before the equivalence point. By formation of the iron(III)-phosphate complex, the electrode potential of the iron couple is lowered by about 0.12 V. Consequently, the formal potential of the indicator more nearly coincides with the potential of the equivalence point. The colorless nature of the iron(III) phosphate complex also makes the end-point color change more discernible. If permanganate is used in the analogous, self-indicating titration of iron(II), the decolorizing action of phosphate is even more important (Section 25.7).

24.10 Questions

24-1 At every point in a redox titration, the numerical values of the half-cell potentials of the two redox systems involved are identical. Explain.

24-2 What is the difficulty in calculating the half-cell potential at the start of a redox titration?

24-3 Is a redox titration possible if the formal potentials of the two redox couples involved are close in value? Discuss in terms of the shape of the titration curve.

24-4 Explain why a more precise result can be expected in a redox titration when the standard potentials of the two redox couples involved are widely separated in value.

24-5 Elaborate on a possible classification of redox titrants into groups wherein the titration curves are (a) independent of concentration and pH, (b) pH-dependent, (c) concentration-dependent, and (d) dependent on both concentration and pH.

24-6 Is a large difference in formal potentials sufficient to ensure a practical redox titration? What other requirements must be fulfilled?

24-7 Compare the action of a redox indicator with the action of an acid-base indicator.

24-8 Compare a redox titration curve (where the same number of electrons appears in each half-reaction equation) with the titration curves for (a) a strong monoprotic acid with a strong base and (b) a weak monoprotic acid with a weak base. Compare the methods of calculating the starting point, halfway point, equivalence point, and points beyond the equivalence point.

24-9 Elaborate on the possibility of titrating consecutively two or more redox couples in one solution. (*Hint:* The criteria can be deduced by analogy from the titration curve of a dibasic acid with a strong base.)

24-10 Values for standard and formal potentials are at hand (see Table F in the Appendix). Is it possible to use both types of values concurrently in trying to describe a particular equilibrium? Elaborate.

24.11 Problems

24–1 The titration of 50 mL of a 0.10M Fe^{2+} solution with 0.02F MnO_4^- solution is considered in Section 24.4. Calculate the points on the titration curve for the addition of 50, 60, 80, and 100 mL of the titrant, assuming that the hydrogen ion concentration is maintained (a) at 1.0M and (b) at $1.0 \times 10^{-2} M$.
Ans.: (a) +1.37, +1.50, +1.51, +1.51 V; (b) +1.21, +1.31, +1.32, +1.32 V

24–2 Calculate the solubility product of ZnS, given that

$$Zn^{2+} + 2e \rightleftharpoons \underline{Zn} \qquad E^0 = -0.76 \text{ V}$$

$$\underline{ZnS} + 2e \rightleftharpoons \underline{Zn} + S^{2-} \qquad E^0 = -1.44 \text{ V}$$

Ans.: $\sim 10^{-23}$

24–3 Uranium(VI) can be reduced to uranium(IV) by chromium(II). Derive the formula for calculating the potential at the equivalence point for the titration of uranyl ion, UO_2^{2+}, with a standard Cr^{2+} solution. What is the value of this potential if the pH of the solution is 2.00?

$$UO_2^{2+} + 4H^+ + 2e \rightleftharpoons U^{4+} + 2H_2O \qquad E^f = +0.31 \text{ V}$$

$$Cr^{3+} + e \rightleftharpoons Cr^{2+} \qquad E^f = -0.38 \text{ V}$$

Ans.: $E_{\text{eq pt}} = 0.08 - 0.08$ pH, $E_{\text{eq pt}}$ at pH 2.0 = −0.08 V

24–4 A volume of exactly 25 mL of a solution $1.0 \times 10^{-2} F$ in Cr^{2+} is titrated with a solution $5.0 \times 10^{-3} F$ cerium(IV) sulfate solution.

$$Cr^{3+} + e \rightleftharpoons Cr^{2+} \qquad E^f = -0.41 \text{ V}$$

$$Ce^{4+} + e \rightleftharpoons Ce^{3+} \qquad E^f = +1.44 \text{ V}$$

(a) Calculate what the values of the half-cell potential are after the addition of 10, 20, 30, 40, 49, 50, and 55 mL of the cerium(IV) solution.
(b) Estimate the value of the equilibrium constant for the titration reaction.
Ans.: (a) −0.45, −0.42, −0.40, −0.37, −0.31, +0.52, +1.38 V; (b) $\sim 10^{31}$

24–5 A volume of 50.0 mL of 0.0500F iron(II) nitrate is titrated with 0.100F cerium(IV) solution with both solutions 1F in HNO_3. Calculate what the potential of a platinum indicator electrode in the titration solution is after the addition of 5.0, 10.0, 12.5, 15.0, 20.0, 24.0, 25.0, 26.0, 30.0, 40.0, and 50.0 mL of titrant.

$$E^f_{Ce^{4+}/Ce^{3+}} = +1.60 \text{ V}, \; E^f_{Fe^{3+}/Fe^{2+}} = +0.68 \text{ V}$$

24–6 A volume of 10.0 mL of 0.0700F iron(II) perchlorate is titrated with 0.100F cerium(IV) solution with both solutions 1F in $HClO_4$. What is the potential of a platinum indicator electrode in the titration solution after the addition of 1.0, 2.0, 3.5, 5.0, 6.0, 6.9, 7.0, 7.1, 8.0, 10.0, 12.0, and 14.0 mL of titrant?

$$E^f_{Ce^{4+}/Ce^{3+}} = +1.70 \text{ V}, \; E^f_{Fe^{3+}/Fe^{2+}} = +0.75 \text{ V}$$

24-7 A volume of 30.0 mL of 0.100*F* iron(III) chloride is titrated with 0.0800*F* chromium(II) chloride with both solutions 1*F* in HCl. What is the potential of a platinum indicator electrode in the titration solution after the addition of 10.0, 20.0, 30.0, 37.0, 37.5, 38.0, 45.0, 60.0, and 75.0 mL of titrant?

$$E^f_{Fe^{3+}/Fe^{2+}} = +0.70\text{ V},\ E^f_{Cr^{3+}/Cr^{2+}} = -0.38\text{ V}$$

24-8 A volume of 20.0 mL of 0.0500*F* iron(III) chloride is titrated with 0.0200*F* tin(II) solution with both solutions 1*F* in HCl. What is the potential of a platinum indicator electrode in the titration solution after the addition of 5.0, 10.0, 15.0, 20.0, 24.0, 25.0, 26.0, 30.0, 40.0, and 50.0 mL of titrant?

$$E^f_{Fe^{3+}/Fe^{2+}} = +0.70\text{ V},\ E^f_{Sn^{4+}/Sn^{2+}} = +0.14\text{ V}$$

24-9 A volume of 25.00 mL of 0.020*F* uranium(IV) sulfate is titrated with 0.050*F* vanadium(V) solution. The titration solution is maintained 0.50*F* in H_2SO_4 throughout the titration. What is the potential of a platinum indicator electrode in the solution after the addition of 5.0, 10.0, 15.0, 19.0, 20.0, 21.0, 25.0, 30.0, and 40.0 mL of titrant? The relevant half-reaction equations and potentials are

$$VO_2^+ + 2H^+ + e \rightleftharpoons VO^{2+} + H_2O \qquad E^f = +1.02\text{ V}$$
$$UO_2^{2+} + 4H^+ + 2e \rightleftharpoons U^{4+} + 2H_2O \qquad E^f = +0.41\text{ V}$$

24-10 A volume of 30.0 mL of 0.0500*F* uranium(IV) sulfate is titrated with 0.0200*F* $KMnO_4$. The titration solution is kept 1*F* in H_2SO_4 throughout the titration. Calculate what the potential of a platinum indicator electrode in the solution is after the addition of 10.0, 15.0, 20.0, 29.0, 30.0, 31.0, 40.0, and 50.0 mL of titrant. The relevant half-reaction equations and potentials are

$$UO_2^{2+} + 4H^+ + 2e \rightleftharpoons U^{4+} + 2H_2O \qquad E^f = +0.41\text{ V}$$
$$MnO_4^- + 8H^+ + 5e \rightleftharpoons Mn^{2+} + 4H_2O \qquad E^0 = +1.51\text{ V}$$

24-11 Calculate the potential for the equivalence point in the titration of uranium (IV) with cerium(IV) in 0.5*F* H_2SO_4 solution.

$$UO_2^{2+} + 2H^+ + 2e \rightleftharpoons UO^{2+} + H_2O \qquad E^f = +0.41\text{ V}$$
$$Ce^{4+} + e \rightleftharpoons Ce^{3+} \qquad E^f = +1.44\text{ V}$$

24-12 Calculate the change in potential from 0.10% before the equivalence point to 0.10% after it in the titration of iron(III) with tin(II) in 1.0*F* HCl solution.

24-13 Calculate the change in electrode potential from 0.10% before the equivalence point to 0.10% after it in the titration of iron(II) with cerium(IV) in 1*F* $HClO_4$ solution.

24-14 A volume of 100.0 mL of a 0.100*F* solution of iron(III) is titrated with a 0.100*F* solution of titanium(III). Assuming that the solution is 2.00*F* in H_2SO_4, calculate the formal concentration of iron(III) remaining (a) at the equivalence point and (b) after the addition of one drop (0.050 mL) of titrant past the equivalence point.

$$Fe^{3+} + e \rightleftharpoons Fe^{2+} \qquad E^f = +0.68 \text{ V}$$

$$TiO^{2+} + 2H^+ + e \rightleftharpoons Ti^{3+} + H_2O \qquad E^f = +0.12 \text{ V}$$

24-15 Calculate the value of the solubility product of silver chloride from the following data.

$$Ag^+ + e \rightleftharpoons Ag \qquad E^0 = +0.80 \text{ V}$$

$$AgCl + e \rightleftharpoons Ag + Cl^- \qquad E^0 = +0.22 \text{ V}$$

24-16 Calculate the value of the solubility product of silver bromide from the following data.

$$Ag^+ + e \rightleftharpoons Ag \qquad E^0 = +0.80 \text{ V}$$

$$AgBr + e \rightleftharpoons Ag + Br^- \qquad E^0 = +0.07 \text{ V}$$

24-17 Calculate the value of the solubility product of lead(II) bromide from the following data.

$$Pb^{2+} + 2e \rightleftharpoons Pb \qquad E^0 = -0.13 \text{ V}$$

$$PbBr_2 + 2e \rightleftharpoons Pb + 2Br^- \qquad E^0 = -0.28 \text{ V}$$

24-18 Calculate the value of the solubility product of mercury(I) chloride from the following data.

$$Hg_2^{2+} + 2e \rightleftharpoons 2Hg \qquad E^0 = +0.79 \text{ V}$$

$$Hg_2Cl_2 + 2e \rightleftharpoons 2Hg + 2Cl^- \qquad E^0 = +0.27 \text{ V}$$

24-19 Calculate the value of the equilibrium constant for the reaction $Cu^{2+} + Zn \rightleftharpoons Cu + Zn^{2+}$ from the following data.

$$Cu^{2+} + 2e \rightleftharpoons Cu \qquad E^0 = +0.34 \text{ V}$$

$$Zn^{2+} + 2e \rightleftharpoons Zn \qquad E^0 = -0.76 \text{ V}$$

24-20 Calculate the value of the equilibrium constant for the reaction $Sn^{4+} + 2Cr^{2+} \rightleftharpoons Sn^{2+} + 2Cr^{3+}$ in 1.0*F* HCl.

$$Sn^{4+} + 2e \rightleftharpoons Sn^{2+} \qquad E^f = +0.14 \text{ V}$$

$$Cr^{3+} + e \rightleftharpoons Cr^{2+} \qquad E^f = -0.41 \text{ V}$$

24-21 Calculate the value of the equilibrium constant for the reaction $Tl^{3+} + 2Fe^{2+} \rightleftharpoons Tl^+ + 2Fe^{3+}$ in 1.0*F* HCl.

$$Tl^{3+} + 2e \rightleftharpoons Tl^+ \qquad E^f = +0.78 \text{ V}$$

$$Fe^{3+} + e \rightleftharpoons Fe^{2+} \qquad E^f = +0.70 \text{ V}$$

24–22 From relevant tables in the Appendix, select required data and calculate the solubility product for (a) AgBr, (b) CuI, (c) AgI, and (d) MnO_2. Compare the calculated values with those found in another table in the Appendix. If you are unable to do as asked, explain why.

24–23 For some two-color redox indicators, a color change can be seen when the ratio of reactants shifts by a factor of about 100; that is, say $[In_{ox}]/[In_{red}]$ changes from 10 to 1/10. Substitute these values in Equation (24–15) and establish the shift in potential, ΔE_{In}, corresponding to the color change. If $n = 2$, what is the shift in potential in millivolts? *Ans.*: 59 mV

25 Applied Redox Titrimetry

Oxidizing or reducing titrants are widely applied in either the direct or indirect titration of numerous substances. Some substances can be determined by either an oxidation or a reduction titration, depending on the oxidation state of the species present in the solution after preliminary treatment. Picking the best method must be based on various factors, including the stability of the oxidation states, the nature of accompanying substances, the presence of interferences, the convenience and rapidity of the method, the accuracy required, the number of samples to be analyzed, and the frequency of the determination. Selected examples of the use of the most common titrants will be described to show the principles and some of the possible practical approaches.

25.1 *Preliminary Treatment*

For the success of a redox titration, it is essential that the substance to be titrated should be present in a single oxidation state, a situation that is sometimes not attained by simple dissolution of the sample. In addition, it is frequently more convenient to titrate the substance in another oxidation state than the one existing after such dissolution. In these circumstances, a preliminary oxidation or reduction is necessary before the actual titration. Consider as an example the determination of iron, which is usually present as iron(III). This ion can be titrated with a strong reductant such as tin(II) or titanium(III), but the solutions of these agents require special storage to prevent their air oxidation (Section 25.11). An oxidimetric titration is therefore often preferred; for this purpose, iron(III) is first reduced to iron(II), which is more stable to air oxidation than tin(II) or titanium(III). The iron(II) is then titrated with an oxidizing titrant.

It is essential in any such pretreatment for the oxidation or reduction to go completely to a single oxidation state and for the excess of the oxidant or reductant to be able to be removed or destroyed without affecting the species to be titrated. Furthermore, it is often necessary for the pretreatment to be such that only the substance to be determined reacts. The preliminary treatments can be divided into reductions and oxidations.

25.2 Preliminary Reduction

Metals and Reductor Columns. Many metals are strong reductants and offer the advantage that their excess is readily separated. The metal, in the form of a sheet or a wire coil, may be immersed in the solution and after the reaction is complete, be simply removed with rinsing. Since the redox reaction goes on at the metal surface, the use of metal granules, shot, turnings, or powder is advantageous because they offer a larger surface area and therefore afford a more rapid reaction.

An especially convenient way to use a metal is to pack it into a vertically mounted tube to form a reductor column (known simply as a reductor). Thus a zinc reductor, also known as a Jones reductor, contains zinc granules. The tube is filled, and washed with dilute acid, and then the solution containing the species to be reduced is passed through the column. The liquid level must not be allowed to fall below the top of the zinc layer; this precaution avoids air bubbles. Such bubbles would decrease the available surface area, and also, oxygen might be reduced to hydrogen peroxide, which interferes in some titrations. The zinc(II) ion formed in the redox reaction does not interfere in the subsequent titration. Often the reduced species emerging from the column is susceptible to air oxidation. In such cases, the effluent should be collected under a protective atmosphere of nitrogen or carbon dioxide. Sometimes it is convenient to collect the effluent in a measured volume of an oxidizing solution, and to back-titrate the excess of the oxidant. In another variation, the effluent is passed into an iron(III) solution and the iron(II) produced is titrated with a standard solution of an oxidant.

Zinc is such a strong reductant ($Z^0_{Za^{2+}/Zn} = -0.76$ V) that hydrogen gas is formed when an acidic solution is passed through the column. Not only is zinc wasted by this undesired reaction, but the gas bubbles block some of the available surface area. The zinc granules are therefore usually shaken with a mercury(II) chloride solution before the column is filled. Because of the high kinetic overpotential of hydrogen on mercury (Section 28.7), the formation of hydrogen on the amalgamated surface of the zinc is decreased or even prevented.

Some reductions of analytical importance that can be accomplished with a Jones reductor include

$$\text{Fe(III)} \rightarrow \text{Fe(II)}, \quad \text{Cr(III)} \rightarrow \text{Cr(II)}, \quad \text{Cr(VI)} \rightarrow \text{Cr(II)},$$

$$\text{Ti(IV)} \rightarrow \text{Ti(III)}, \quad \text{Ce(IV)} \rightarrow \text{Ce(III)}, \quad \text{V(V)} \rightarrow \text{V(II)},$$

$$\text{Mo(VI)} \rightarrow \text{Mo(III)}, \quad \text{and} \quad \text{Cu(II)} \rightarrow \text{Cu}$$

A reductor column filled with silver granules is also frequently used. The sample solution must contain hydrochloric acid. Then the oxidation of silver involves the half-reaction*

$$\underline{\text{Ag}} + \text{Cl}^- \rightarrow \underline{\text{AgCl}} + e \qquad (\text{E}^0 = +0.22 \text{ V}) \tag{25-1}$$

---

*Throughout this chapter a half-reaction is written in the direction in which it proceeds. The E^0 value given parenthetically, however, corresponds to the particular half-reaction written as a reduction equation.

The silver reductor, sometimes called a Walden reductor, is milder than the zinc reductor and offers the advantage that no hydrogen gas is evolved. In addition, the insoluble oxidation product, silver chloride, is retained in the reductor and the effluent remains uncontaminated. The packing of the Walden reductor can be regenerated by treating with zinc metal and sulfuric acid. (We suggest that you write the reaction equation.)

Some reductions of analytical value that can be achieved in a silver reductor, somewhat dependent on the hydrochloric acid concentration established, are

$$\text{Fe(III)} \rightarrow \text{Fe(II)}, \quad \text{Cr(VI)} \rightarrow \text{Cr(III)}, \quad \text{Ce(IV)} \rightarrow \text{Ce(III)},$$

$$\text{V(V)} \rightarrow \text{V(IV)}, \quad \text{Mo(VI)} \rightarrow \text{Mo(V), Mo(III)},$$

$$\text{and} \quad \text{Cu(II)} \rightarrow \text{Cu(I)}$$

Comparing the zinc and silver reductions reveals that in the silver reductor, chromium(VI) is reduced to chromium(III) rather than to chromium(II), titanium is not reduced at all, vanadate is reduced only to vanadium(IV) [the further reduction to vanadium(III) is extremely slow], and copper(II) is reduced to copper(I), which at a sufficient hydrochloric acid concentration remains in solution as soluble chloro complexes. Molybdenum(VI) is reduced only to molybdenum(V) in 2*F* hydrochloric acid; however, reduction to molybdenum(III) is possible in 4*F* acid.

Schemes for the analysis of mixtures can be devised using the difference in the reducing power of metal reductors. For example, one aliquot of a sample solution containing iron and titanium is passed through a zinc reductor and a second aliquot through a silver reductor. The effluent from each column is titrated with an oxidizing titrant; the results of the titrations correspond to the total of iron and titanium for the first effluent, and to the iron alone for the second.

A liquid amalgam offers a further convenient form for a reducing metal since it can be readily separated from an aqueous solution once the reduction has been effected. The reducing power of an amalgam depends on the particular metal dissolved in the mercury.

Tin(II) Salts. Tin(II) salts are frequently used to reduce iron(III) in hydrochloric acid solution. The disappearance of the yellow solution color indicates that the reduction is complete. The excess of tin(II) is readily oxidized with mercury(II) chloride, which does not affect iron(II):

$$2HgCl_2 + Sn^{2+} \rightarrow \underline{Hg_2Cl_2} + Sn^{4+} + 2Cl^- \qquad (25\text{-}2)$$

The calomel formed is insoluble and under appropriate conditions does not react in the subsequent titration of iron(II). The presence of chloride ion is essential; in its absence, mercury(I) is reduced further to elemental mercury. The metal appears in a finely divided, black form, is reoxidized in the subsequent titration, and thereby interferes.

Sulfurous Acid. Sulfurous acid acts as a mild reductant in acidic solution,

$$H_2SO_3 + H_2O \rightarrow SO_4^{2-} + 4H^+ + 2e \qquad (E^0 = +0.17\text{ V}) \qquad (25\text{-}3)$$

Any excess of this reductant can be readily removed by boiling the solution, thereby volatilizing sulfur dioxide.

Other Reductants. Other reductants used before redox titrations include hydrochloric acid [especially for oxidants such as manganese(IV) oxide and lead (IV) oxide] and hypophosphorous acid, H_3PO_2, and its salts.

25.3 Preliminary Oxidation

Peroxydisulfate Ion. The ion $S_2O_8^{2-}$, now named systematically as peroxydisulfate or peroxodisulfate but commonly termed persulfate ion, used as either the potassium or the ammonium salt, is a strong oxidizing agent in acidic solution.

$$S_2O_8^{2-} + 2e \rightarrow 2SO_4^{2-} \qquad (E^0 = +2.0\ \text{V}) \tag{25-4}$$

Any excess of this oxidant can be decomposed by boiling the solution; the reactions that go on are

$$S_2O_8^{2-} + 2H_2O \rightarrow 2SO_4^{2-} + 2H^+ + H_2O_2 \tag{25-5}$$

$$2H_2O_2 \rightarrow 2H_2O + O_2 \uparrow \tag{25-6}$$

Important uses of this oxidant, often with a silver salt added as a catalyst, include the conversion of manganese to permanganate, chromium to dichromate, and vanadium to vanadate ion.

Perchloric Acid. Hot, concentrated perchloric acid (50 to 70%) is a strong oxidant:

$$2ClO_4^- + 16H^+ + 14e \rightarrow Cl_2 + 8H_2O \qquad (E^0 \sim +2\ \text{V}) \tag{25-7}$$

The acid is often used to oxidize chromium and vanadium in ores and alloys to dichromate and vanadate. It is unnecessary to remove the excess of perchloric acid, since cooling and diluting the solution effectively checks the oxidizing power of the perchlorate ion. Care must be taken, however, to exclude organic substances from the system because they may react violently, even explosively, with the hot, concentrated acid.

Sodium Bismuthate and Lead(IV) Oxide. Sodium bismuthate, $NaBiO_3$, is a fine, water-insoluble, brown powder used principally to oxidize manganese(II) to permanganate in nitric acid solution, even at room temperature.

$$\underline{NaBiO_3} + 6H^+ + 2e \rightarrow Na^+ + Bi^{3+} + 3H_2O \qquad (E^0 \sim +1.6\text{V}) \tag{25-8}$$

Chloride ion must be excluded since it would be oxidized to chlorine gas. The excess of sodium bismuthate is removed by filtration through an inert filter material such as sintered glass or porcelain. Lead(IV) oxide, PbO_2, is used similarly.

$$\underline{PbO_2} + 4H^+ + 2e \rightarrow Pb^{2+} + 2H_2O \qquad (E^0 \sim +1.5\ V) \qquad (25\text{-}9)$$

Hydrogen Peroxide. Hydrogen peroxide is a strong oxidant, even in alkaline solution.

$$H_2O_2 + 2H^+ + 2e \rightarrow 2H_2O \qquad (E^0 = +1.77\ V) \qquad (25\text{-}10)$$

This compound can also function as a reductant according to the half-reaction

$$H_2O_2 \rightarrow O_2 + 2H^+ + 2e \qquad (E^0 = +0.69\ V) \qquad (25\text{-}11)$$

Hydrogen peroxide can undergo disproportionation according to the reaction

$$2H_2O_2 \rightarrow 2H_2O + O_2 \qquad (25\text{-}12)$$

which is a combination of reactions (25-10) and (25-11). This reaction is used to decompose any excess of the reagent. The solution is simply boiled, often with a trace of either a nickel or an iodide salt added or with a piece of platinum foil placed temporarily in the solution to catalyze the reaction. Hydrogen peroxide is frequently used to assure complete oxidation of iron after its ores and alloys have been dissolved.

Nitric Acid. Nitric acid is chiefly used to attack materials, such as alloys, that are inert to hydrochloric acid, and also to ensure that metals such as iron and copper are present in the solution in their higher oxidation state.

25.4 *Permanganate as an Oxidizing Titrant*

Permanganate ion is one of the most widely used oxidizing titrants because of its high oxidizing power and because it allows a self-indicating titration. The end point can be indicated visually by the pink color imparted to the solution by the first fraction of a drop of titrant solution in excess. A redox indicator, however, can be used to advantage when very dilute solutions are involved.

In strongly acidic medium, permanganate undergoes a five-electron reduction to manganese(II):

$$MnO_4^- + 8H^+ + 5e \rightarrow Mn^{2+} + 4H_2O \qquad (E^0 = +1.51\ V) \qquad (25\text{-}13)$$

In weakly acidic, neutral, or alkaline solutions, the reduction of permanganate involves its three-electron reduction to manganese(IV), which precipitates as hydrous manganese(IV) oxide, usually written simply as MnO_2. The half-reaction can be formulated as

$$MnO_4^- + 4H^+ + 3e \rightarrow \underline{MnO_2} + 2H_2O \qquad (E^0 = +1.69\ V) \qquad (25\text{-}14)$$

or in alkaline medium as

$$MnO_4^- + 2H_2O + 3e \rightarrow \underline{MnO_2} + 4OH^- \qquad (E^0 = +0.57 \text{ V}) \qquad (25\text{-}15)$$

Permanganate has several disadvantages that has prompted a shift to other oxidizing titrants. Even though potassium permanganate is available in high purity, it cannot act as a primary standard, since freshly prepared solutions are unstable. In such solutions, permanganate decomposes by the reaction

$$4MnO_4^- + 2H_2O \rightarrow \underline{4MnO_2} + 4OH^- + 3O_2 \qquad (25\text{-}16)$$

This reaction, which goes slowly in neutral solution and more rapidly in acidic solution, is "autocatalytic," since the manganese(IV) oxide formed hastens the decomposition. Even when reagent-grade potassium permanganate is used, a trace of this oxide is present, and furthermore, organic dust particles in the water or the vessel reduce some of the permanganate. It is essential to remove this manganese(IV) oxide to make the solution stable. The initial solution is either boiled or allowed to stand at room temperature for about a week; it is then filtered. After this treatment, a neutral potassium permanganate solution (0.02F or greater), stored in the dark and protected from the entry of dust, maintains its strength for a protracted period. Storage in the dark is necessary, since decomposition is enhanced by strong light. In any event, occasional restandardization of the solution is necessary. If titrations are performed only infrequently, it is usually more expedient to prepare a potassium permanganate solution, to standardize it, and then to use it at once.

Standardizing Potassium Permanganate Solutions. Several suitable standards for permanganate solutions are available; some of the prominent ones are briefly considered.

Oxalate. Permanganate solutions can be standardized by titrating an acidic solution containing a known amount of oxalate ion. Suitable sources of oxalate include oxalic acid dihydrate, $H_2C_2O_4 \cdot 2H_2O$, sodium oxalate, $Na_2C_2O_4$, and potassium tetroxalate, $KHC_2O_4 \cdot H_2C_2O_4 \cdot 2H_2O$. The solution titrated must be acidic and is usually made at least 0.5F in sulfuric acid. At insufficient acidity, the reaction does not lead to manganese(II) ion but stops with the formation of manganese(IV) oxide. The titration should be performed promptly without allowing the acidified oxalate solution to stand long.

At the start of the titration, the pink color of the added permanganate persists for some time, indicating the slowness of the reaction. As further increments of titrant are added, the decolorization is more rapid, since the manganese(II) formed catalyzes the reaction. Heating the solution also hastens the reaction; but at too high a temperature and at too great an acidity, oxalic acid decomposes according to the reaction

$$H_2C_2O_4 \rightarrow CO_2\uparrow + CO\uparrow + H_2O \qquad (25\text{-}17)$$

Thus, precautions must be taken in the permanganate titration of oxalate.

Probably the best method of effecting the titration is to add about 90% of the required permanganate at room temperature, allowing the mixture to stand until decolorization is complete, and then to heat to 55 to 60°C, and finish the titration. It is important to run a blank determination to correct for any oxidizable substances present in the water and the reagent used and for the amount of permanganate solution required to impart a pink color under the titration conditions. (Establishing a blank is also necessary with other standards and indeed in most permanganate titrations.) The importance of a blank increases when titrant solutions of low concentrations are involved.

Arsenic(III) Oxide. Arsenic(III) oxide is an excellent primary standard. It is nonhygroscopic, contains no water of crystallization, and is readily obtainable in a purity of 99.95% or higher. The relevant half-reaction in its use in the standardization of oxidants is

$$H_3AsO_3 + H_2O \rightarrow H_3AsO_4 + 2H^+ + 2e \qquad (E^0 = +0.56\ V) \qquad (25\text{-}18)$$

In the direct titration of arsenic(III) with permanganate in acidic solution, usually a trace of an iodine compound (for example, KI or KIO_3) is added to catalyze the reaction and thereby to speed the titration.

Iron. Iron, in the form of wire of 99.90% or higher purity, is available for standardization. It must be dissolved in sulfuric acid with exclusion of air to avoid oxidation to iron(III), otherwise a preliminary reduction before the titration becomes necessary.

Other Standards for Permanganate. Other standards for permanganate occasionally used include iron(II) ammonium sulfate hexahydrate (Mohr's salt), $FeSO_4 \cdot (NH_4)_2SO_4 \cdot 6H_2O$, and potassium hexacyanoferrate(II) trihydrate, $K_4Fe(CN)_6 \cdot 3H_2O$. The Mohr's salt used should be free of iron(III). The hexacyanoferrate(II) salt, which is oxidized to hexacyanoferrate(III) in a one-electron step, has an especially favorable equivalent weight.

25.5 *Determinations with Permanganate*

Since permanganate is prominent as an oxidizing titrant, its application in inorganic analysis will be treated in greater detail than those titrants to be considered subsequently. Many of the considerations, however, apply to the other oxidants.

Determination of Iron (Zimmermann-Reinhardt Method). The most important titrimetric application of permanganate is in the determination of iron in a large variety of materials. After the sample is dissolved, all the iron is reduced to iron(II), which is then titrated. No special difficulties arise in the direct permanganate titration of iron(II) in acidic solution if chloride is absent. However, hydrochloric acid is frequently used in dissolving samples and its presence is required in some reduction techniques. Chloride ion causes difficulty because permanganate ion can oxidize it to elemental chlorine and hypochlorous acid. These reactions as such are slow but are catalyzed by iron.

The interference of chloride can be obviated by diluting the solution (thereby decreasing the acidity) and adding manganese(II). These measures decrease the oxidizing power of permanganate; this can be concluded by looking at the Nernst equation for the permanganate half-reaction equation (Section 24.5). In addition, the dilution decreases the chloride concentration. The titration is performed in the cold to keep reaction with chloride at a low rate. Further, the titrant solution is added slowly and with good stirring to avoid a high local concentration of the titrant.

During the titration, iron(III) forms; in the presence of chloride, this yields intensely yellow chloro complexes. This color makes it difficult to detect the visual end point. Phosphoric acid is added because it forms a colorless, soluble complex with iron(III). The complexation with phosphoric acid also lowers the value of the formal potential of the iron redox couple (Section 24.10) and thus improves the titration further. However, the presence of phosphoric acid also makes iron(II) more susceptible to air oxidation. In practice, manganese(II) sulfate, phosphoric acid, and sulfuric acid are combined in a single solution, known as the Zimmermann-Reinhardt solution, which is added to the sample solution after the preliminary reduction.

Determination of Manganese (Volhard's Method). In neutral or slightly alkaline solution, manganese(II) is oxidized to manganese(IV) oxide by permanganate. The course of the reaction is more complicated than is suggested by the simple aggregate equation

$$3Mn^{2+} + 2MnO_4^- + 2H_2O \rightarrow \underline{5MnO_2} + 4H^+ \qquad (25\text{-}19)$$

The manganese(IV) precipitates as a hydrous oxide, which shows acidic properties and is capable of forming an insoluble manganese(II) salt, or according to other explanations, of adsorbing manganese(II) ion. Whatever the cause, some manganese(II) is coprecipitated and thereby escapes being titrated; low results are the consequence. Volhard found that this difficulty can be circumvented by adding a zinc salt. Zinc ion interacts preferentially with the hydrous manganese(IV) oxide and the manganese(II) ion remains in solution. In practice, the zinc ion is generated by adding a slurry of zinc oxide in water to the slightly acidic sample solution. Because of the relatively high pH that the zinc oxide establishes, iron(III), titanium, aluminum, and other metals that form hydrous oxides are precipitated. On boiling, a rapidly settling precipitate is obtained, which also carries down the manganese(IV) oxide. The visual end point is then readily discernible in the clear, colorless supernatant liquid.

Metal Determinations via Oxalate Titrations. The titration of oxalate ion has been discussed relative to the standardization of permanganate solutions (Section

25.5). This titration can be used for determining oxalate, and in inorganic analysis, it receives attention for the indirect determination of metals forming insoluble oxalate salts. This approach was a standard method for determining calcium in a variety of samples until EDTA titration methods evolved (Section 20.10). Calcium oxalate is precipitated, the precipitate is separated by filtration, washed, and dissolved in acid. The oxalate is then titrated with permanganate.

Hydrogen Peroxide and Peroxides. Hydrogen peroxide is oxidized by permanganate in acidic solution. The relevant half-reaction is

$$H_2O_2 \rightarrow O_2 + 2H^+ + 2e \qquad (E^0 = +0.69 \text{ V}) \qquad (25\text{-}20)$$

Determining hydrogen peroxide involves diluting the sample, adding sulfuric acid, and titrating with permanganate to the first permanent pink color. Also here an induction period is encountered, as in the titration of oxalate. Pharmaceutical-grade hydrogen peroxide often contains a small amount of an organic compound as a stabilizer. This compound may also be oxidized by the permanganate, and then a slightly high value for the peroxide content will result. Metal peroxides, such as sodium peroxide, can be assayed by a permanganate titration through being dissolved in dilute acid, after which the hydrogen peroxide formed is titrated.

Miscellaneous Determinations with Permanganate. Nitrite is often determined by a permanganate titration. The half-reaction involved is

$$HNO_2 + H_2O \rightarrow NO_3^- + 3H^+ + 2e \qquad (E^0 = +0.94 \text{ V}) \qquad (25\text{-}21)$$

Arsenic(III) can be titrated as described for the standardization of permanganate solutions. Arsenic(V) can be reduced, for example, with sulfurous acid and then be titrated similarly. Antimony(III) and antimony(V) are determined analogously.

Titanium(IV) can be reduced to titanium(III) in a Jones reductor and then titrated to the tetravalent state. Since titanium(III) is readily susceptible to air oxidation, it is expedient to catch the effluent from the reductor in an iron(III) sulfate solution, to add Zimmermann-Reinhardt solution if chloride is present, and to titrate the iron(II) formed with permanganate. The method also applies to chromium and vanadium after reduction to chromium(II) and vanadium(II) in a Jones reductor.

Permanganate can be used in the indirect titration of oxidants. The assay of pyrolusite is an example. This mineral contains manganese(IV) oxide as the active ingredient. A powdered sample of the mineral is treated with a measured volume of an acidified solution of oxalate or iron(II), and after its dissolution is complete, the excess of oxalate or iron(II) is titrated with permanganate.

25.6 *Iodine and Thiosulfate In Redox Titrimetry*

The iodine-iodide redox couple

$$I_2 + 2e \rightarrow 2I^- \qquad (E^0 = +0.54 \text{ V}) \qquad (25\text{-}22)$$

is intermediate between strong oxidants and strong reductants. Thus in some applications, iodine is reduced to iodide ion; and in others, the reaction is the reverse. Reducing substances can be directly titrated with iodine; oxidizing substances are caused to react with iodide ion, and the iodine formed is titrated with a reducing titrant, commonly sodium thiosulfate. The two approaches are often distinguished from each other by calling them iod*i*metry and iod*o*metry. (The vowel is *i* where iodine is the titrant.)

Although iodine and thiosulfate solutions do not show good storage stability and thus require frequent standardization, iodine methods are widely applied because of the numerous substances that can be determined and because of the sharpness of the end point, which allows a reliable titration even when the titrant is only $0.001F$.

The iodine-iodide redox potential is independent of acidity over a wide range. About pH 8, however, iodine undergoes disproportionation, to form iodate and iodide. The reaction can be represented in two steps; hypoiodite ion is formed first:

$$I_2 + 2OH^- \rightarrow H_2O + I^- + IO^- \quad (25\text{-}23)$$

$$3IO^- \rightarrow 2I^- + IO_3^- \quad (25\text{-}24)$$

In acidic solution, iodide ion is oxidized to iodine by the oxygen of the air:

$$O_2 + 4H^+ + 4I^- \rightarrow 2I_2 + 2H_2O \quad (25\text{-}25)$$

Sunlight and various ions, notably copper(II) and nitrite, catalyze this reaction.

Iodine is somewhat volatile, and its solubility in water is low ($0.0013F$ in I_2 at 20°C). A more concentrated aqueous solution of improved stability can be obtained by adding an excess of iodide ion; this combines with iodine to form the soluble triiodide ion, I_3^-. For this ion, the half-reaction can be written

$$I_3^- + 2e \rightleftharpoons 3I^- \quad (E^f = +0.545 \text{ V in } 0.5F\ H_2SO_4) \quad (25\text{-}26)$$

The formal potential is displaced from the standard potential of the iodine-iodide redox couple by only 9 mV [compare equation (25-22)]. For simplicity, and to make the stoichiometry more obvious, reactions for iodine methods are usually written showing iodine instead of the triiodide ion as the product or reactant.

In neutral or weakly acidic solution, thiosulfate ion is oxidized by iodine to tetrathionate ion, which is the desired reaction in the titration. The half-reaction

$$2S_2O_3^{2-} \rightarrow S_4O_6^{2-} + 2e \quad (E^0 = +0.09 \text{ V}) \quad (25\text{-}27)$$

involves one electron per molecule of thiosulfate ion; consequently, in this reaction the equivalent weight of this ion equals its formula weight.

If the pH is too high, thiosulfate ion is converted to sulfate. The half-reaction involved is

$$S_2O_3^{2-} + 10OH^- \rightarrow 2SO_4^{2-} + 5H_2O + 8e \quad (25\text{-}28)$$

In this reaction, the equivalent weight of thiosulfate ion equals one-eighth its formula weight. However, usually the extent of the combination of reactions (25-27) and (25-28) is unpredictable, making operation at high pH values impractical.

Thiosulfate ion is not stable in acidic solution since thiosulfuric acid decomposes:

$$H_2S_2O_3 \rightarrow H_2SO_3 + S \qquad (25\text{-}29)$$

Sulfurous acid also reacts with iodine. The relevant half-reaction is

$$H_2SO_3 + H_2O \rightarrow SO_4^{2-} + 4H^+ + 2e \qquad (E^0 = +0.17\ \text{V}) \qquad (25\text{-}30)$$

This half-reaction involves two electrons, while the oxidation of thiosulfate to tetrathionate, after (25-27), involves only one electron per molecule of substance oxidized.

Whereas thiosulfate decomposes at moderate acidities, no difficulties arise during a titration, since the reaction is slow. Thus, if the thiosulfate solution is added to the sample solution slowly and with good stirring, there is no decomposition, and the reaction with iodine is stoichiometric even in strongly acidic solution. The decomposition leads to an interesting phenomenon, however. If portions of a newly prepared sodium thiosulfate solution are titrated with an iodine solution over a few days, the following will be found. The reducing power of the thiosulfate solution at first increases somewhat; then it decreases and finally reaches a stable value that is below the initial one. The increase is associated with the decomposition of thiosulfate to sulfite ion, which per molecule donates twice as many electrons as thiosulfate does. After some time, the sulfite is oxidized, by oxygen of the air, to unreactive sulfate, and the formality decreases. Consequently, it is necessary to age a sodium thiosulfate solution before its final standardization. After several days of aging, a 0.1*F* solution will maintain its strength for a protracted period. Solutions that are more dilute are less stable and are best prepared as needed from a stock solution; they should, however, be standardized before use.

Indication of the End Point in Iodine Methods. One drop of a 0.05*F* iodine solution imparts a just-visible yellow color to about 100 mL of water; hence, the appearance or disappearance of this yellow color can afford self-indication. The color is far less intense than that of permanganate, however, and even only slightly colored substances present in the sample solution interfere. Consequently, self-indication is seldom used in practice. Iodine is far more soluble in organic solvents such as chloroform and carbon tetrachloride than in water and shows an intense violet color in them. Since these nonpolar solvents are almost immiscible with water, even small amounts of iodine can be extracted from a large volume of a dilute aqueous solution by shaking with 1 or 2 mL of the solvent. Although the appearance or disappearance of the iodine color in the organic layer affords a sensitive end point, the technique is tedious since extensive shaking is necessary near the end point to assure equilibration.

Starch is the usual indicator in iodine methods. “Soluble” starch and iodine form a complex entity of intensely blue color. The color develops only in the presence of iodide and in cold solution. It fades when the solution is heated but returns on cooling. (With old starch solutions, a red-purple color is obtained rather than a blue.) The starch-iodine color reaction is quite sensitive and a distinct blue is discernible in solutions about $10^{-5}F$ in iodine. In titrating solutions containing much iodine, one must not add the starch until the end point is near, that is, until the iodine concentration has decreased to a low value; otherwise, the starch-iodine entity reacts very slowly and the end point is readily exceeded. The usual technique is to perform the titration until the yellow color of the iodine starts to fade, then to add the starch and continue the titration until the blue disappears.

When the titration of iodine is performed in strongly acidic media or in the presence of substances that catalyze the oxidation of iodide by oxygen of the air, after reaction (25-25), the colorless solution obtained when the end point is reached becomes blue again after some time. The exclusion or at least delay of this "after-blueing" requires special precautions in certain determinations.

Standardization of Iodine and Thiosulfate Solutions. Elemental iodine can be highly purified by sublimation and may act as a primary standard. A standard iodine solution can be prepared from such material by weighing and dissolving the necessary amount in a solution of potassium iodide. However, the volatility of iodine makes this approach difficult. Therefore it is preferable to prepare an iodine solution of approximately the desired concentration and to standardize it against arsenic(III) oxide, which is an excellent primary standard. The oxide is dissolved, as described for its use in the standardization of permanganate; the solution is buffered with hydrogen carbonate and titrated with the iodine solution.

A thiosulfate solution can be standardized with an iodine solution, of reliably known formality, but the preferred approach is to standardize it independently. Elemental iodine can be used for this purpose, but its volatility causes difficulties. Better results are achieved by forming a known amount of iodine in solution. Commonly, potassium iodate, which is readily obtainable in high purity and is stable in aqueous solution, is caused to react in acidic solution with an excess of potassium iodide. The reaction, leading to the formation of iodine, proceeds both rapidly and stoichiometrically:

$$IO_3^- + 5I^- + 6H^+ \rightarrow 3I_2 + 3H_2O \qquad (25\text{-}31)$$

A standard solution of potassium iodate can be prepared and used when needed to obtain an iodine solution of exactly known iodine content.

Standard solutions of strong oxidizing agents, such as potassium permanganate or dichromate, can also be used to oxidize iodide to iodine. The iodine formed is then titrated with thiosulfate.

25.7 Applications of Iodine Methods

Iodine methods can be divided into those in which iodine is the titrant (io*di*metric methods) and those in which thiosulfate is used (io*do*metric methods).

Arsenic(III) is readily oxidized to arsenic(V) by iodine. The titration with iodine is satisfactory in the pH range 4 to 9 and a medium buffered with hydrogen carbonate is often used. The reaction equation can be written

$$AsO_3^{3-} + 2H_2O + I_2 \rightarrow AsO_4^{3-} + 2H^+ + 2I^- \qquad (25\text{-}32)$$

Antimony(III) can be determined analogously.

Hydrogen sulfide reacts readily with iodine according to the reaction

$$H_2S + I_2 \rightarrow \underline{S} + 2H^+ + 2I^- \qquad (25\text{-}33)$$

and thus can be titrated directly. Because of the volatility of hydrogen sulfide, however, preferably excess iodine is added to the sample and the unreacted iodine is back-titrated with thiosulfate. Hydrogen sulfide can be determined in gaseous or liquid samples by precipitating cadmium sulfide from an alkaline medium, dissolving the isolated precipitate in acid, and causing the mixture to react with iodine.

Sulfite (and sulfur dioxide) can be titrated with iodine in acidic medium. To avoid erroneous results caused by air oxidation of the sulfite, it is also of advantage here to effect first the reaction with iodine and then to back-titrate its excess with thiosulfate.

Tin(II), in alloy analysis, can be titrated directly with iodine. The susceptibility of tin(II) to air oxidation makes exclusion of air a necessity.

Water in organic liquids, hydrated salts, and diverse samples is often determined by the Karl Fischer titration. The total reaction, simplified, can be represented by

$$C_5H_5N{\cdot}I + C_5H_5N{\cdot}SO_2 + C_5H_5N + CH_3OH + H_2O \rightarrow$$
$$2C_5H_5NHI + C_5H_5NHOSO_2OCH_3 \qquad (25\text{-}34)$$

where C_5H_5N is pyridine and CH_3OH, methanol. The titrant solution, which consists of iodine, sulfur dioxide, and pyridine in anhydrous methanol, is intensely brownish yellow. All the products of the reaction are colorless. Thus a self-indicating titration can be performed to the appearance of a permanent yellow. The titrant solution has limited stability and should be standardized frequently against a known amount of either water dissolved in anhydrous methanol or a methanol-soluble hydrated salt such as potassium sodium tartrate tetrahydrate. The titer of the Karl Fischer titrant solution is expressed in milligrams of water per milliliter.

Many oxidants (bromine and chlorine, for instance) might be directly titrated with thiosulfate, but usually side reactions do not allow a strictly stoichiometric relation. Hence, the oxidant is allowed to react with potassium iodide to form iodine, which is then titrated with thiosulfate. This approach is common in determining the "active chlorine" content of bleaching powders and bleaching solutions. Arsenic(V) reacts with iodide in highly acidic solution by the reverse of equation (25-32), which for high acidities can be formulated as

$$H_3AsO_4 + 2H^+ + 2I^- \rightleftharpoons H_3AsO_3 + I_2 + H_2O \qquad (25\text{-}35)$$

The iodine formed can be titrated with thiosulfate.

Copper(II), in alloys and ores, is determined by causing it to react with an excess of iodide and titrating the liberated iodine with thiosulfate:

$$2Cu^{2+} + 4I^- \rightarrow \underline{2CuI} + I_2 \qquad (25\text{-}36)$$

The titration is performed in the presence of the insoluble copper(I) iodide. Interference by iron, often present in samples, is obviated by complexation of iron(III) with fluoride ion.

Hydrogen peroxide oxidizes iodide ion to iodine and therefore can be determined iodometrically. In contrast to the permanganate method for peroxides (Section 25.5), organic substances used as stabilizers for hydrogen peroxide do not interfere in the determination.

The determination of iodate has been described for the standardization of thiosulfate solutions (Section 25.6). Bromate and chlorate in a strongly acidic medium react with iodide analogously and thus can also be determined.

Nitrite in acidic solution reacts with iodide according to

$$2HNO_2 + 2H^+ + 2I^- \rightarrow 2NO + 2H_2O + I_2 \qquad (25\text{-}37)$$

Titration of the iodine formed allows the determination of nitrite. The nitrogen oxide formed reacts readily with oxygen to form nitrogen dioxide, NO_2, which in turn leads to further liberation of iodine; consequently, air must be excluded.

An interesting method for determining extremely small amounts of iodide (for example, in blood serum) proceeds as follows. The iodide is oxidized to iodate by bromine in a weakly acidic medium:

$$I^- + 3H_2O + 3Br_2 \rightarrow IO_3^- + 6H^+ + 6Br^- \qquad (25\text{-}38)$$

The excess bromine is then removed (by boiling, for instance), potassium iodide is added to the solution to react with the iodate, and the iodine formed is then titrated with thiosulfate. Thus the original amount of iodide in the sample is "multiplied" by 6, and the sensitivity and precision of the determination are greatly increased.

The reaction between iodate and iodide to form iodine (25-31) proceeds stoichiometrically also with respect to hydrogen ion; consequently, an acid can be determined iodometrically or a thiosulfate solution standardized by using an exactly standardized solution of a strong acid. The calculation has as its basis that two hydrogen ions correspond to the liberation of one molecule of iodine (I_2).

25.8 *Dichromate Methods*

Dichromate ion in acidic solution is a slightly weaker oxidant than permanganate:

$$Cr_2O_7^{2-} + 14H^+ + 6e \rightarrow 2Cr^{3+} + 7H_2O \qquad (E^0 = +1.33\ V) \qquad (25\text{-}39)$$

Dichromate is used in place of permanganate in many titrations, notably that of iron(II), and in various methods in which the iron(III)-iron(II) system is used as an intermediate redox system because the presence of chloride causes fewer difficulties. Potassium dichromate, $K_2Cr_2O_7$, is readily available in a purity greater than 99.95% and can be used directly to prepare a standard solution that is indefinitely stable. An indicator is needed in dichromate titrations because the color of dichromate is not intense enough to allow self-indication. The most common redox indicator is *N*-phenylsulfanilic acid or its barium or sodium salt. (This acid is alternatively named *p*-diphenylaminesulfonic acid.)

25.9 Cerium(IV) Methods

Cerium(IV) became a popular oxidizing titrant after suitable redox indicators were discovered.

$$Ce^{4+} + e \rightarrow Ce^{3+} \qquad (E^f = +1.44 \text{ V in } 1F\ H_2SO_4) \qquad (25\text{–}40)$$

The yellow color of cerium(IV) compounds is insufficient for a self-indicating titration. The indicator most often used is the iron(II) chelate of 1,10-phenanthroline (often termed Ferroin), obtained by mixing solutions of iron(II) sulfate and of the organic compound. The intense red color of the indicator changes to a pale blue complex when it is oxidized to the corresponding iron(III) chelate. At the low indicator concentrations used in a titration, the change is from pink to almost colorless.

Cerium(IV) does not exist in acidic solutions as the simple aquo ion but forms complexes with the oxyanions present. Cerium(IV) standard solutions are often prepared in 1*F* to 8*F* sulfuric or nitric acid, and such complexes as

$$[Ce(SO_4)_n]^{-2n+4} \qquad \text{and} \qquad [Ce(NO_3)_n]^{-n+4}$$

are then present. In reaction equations, however, simply Ce^{IV} or Ce^{4+} is often written to make it easier to recognize the stoichiometric relations involved (see Section 24.3).

Cerium(IV) has salient advantages as an oxidizing titrant. Its solution in sulfuric or nitric acid is stable even on boiling and it does not oxidize chloride to chlorine so long as the halide concentration is below about 1*M* (contrast this with permanganate). The oxidizing power of cerium(IV) solutions is high, and cerium can be substituted for permanganate in most determinations. Some reactions with cerium(IV), however, are slow; for example, the direct titration of arsenic(III) is possible only if a minute amount of osmium(VIII) oxide, OsO_4, is added as a catalyst.

25.10 Bromate and Bromine Methods

Bromate ion is a strong oxidizing agent in acidic solution.

$$2BrO_3^- + 12H^+ + 10e \rightarrow Br_2 + 6H_2O \qquad (E^0 = +1.5 \text{ V}) \qquad (25\text{–}41)$$

In a bromatometric titration, so long as some of the reductant being titrated is still present, the bromine is further reduced to bromide ion.

$$Br_2 + 2e \rightarrow 2Br^- \qquad (E^0 = +1.09 \text{ V}) \tag{25-42}$$

Beyond the equivalence point, the first drop of bromate solution in excess yields free bromine, according to

$$BrO_3^- + 5Br^- + 6H^+ \rightarrow 3Br_2 + 3H_2O \tag{25-43}$$

The end point can be detected by the yellow color of the free bromine, but the use of a suitable indicator is preferable. Potassium bromate is readily obtainable in very high purity and a standard solution can be directly prepared. The solution is indefinitely stable.

The bromate titration of antimony(III) in hydrochloric acid solution is routine in analyzing alloys and ores. The sample is dissolved, the antimony brought to the tervalent state by a suitable treatment (for example, sulfurous acid reduction), and the titration is started. Antimony(III) is progressively oxidized to antimony(V) and bromate is reduced to bromide ion. At the end point, the first excess of bromate oxidizes some of the bromide formed to bromine. The end point is usually signaled by the irreversible decolorization of a dye by the bromine. Often methyl orange is used; it is not an acid-base indicator here but an *irreversible* redox indicator. Since the indicator reaction is slow, the end point must be approached slowly. Arsenic(III) can be determined likewise.

In many cases, bromine is a more suitable oxidant than bromate. But even then, a bromate solution may be used as titrant instead of a bromine solution, which is quite unstable. Potassium bromide is added to the solution to be titrated or is incorporated into the bromate standard solution. When the titrant solution is added to the acidified sample solution, bromine is formed in situ, according to equation (25-43).

25.11 Methods Based on Strong Reductants

Strong reducing agents such as titanium(III), chromium(II), and tin(II) receive attention for titrating iron(III), vanadate, molybdate, uranyl, and other ions. Since the reducing agents react readily with oxygen, air must be excluded. Chromium(II) is such a strong reductant that it reduces even hydrogen ion of water. Fortunately, this reaction is so slow that titrations with chromium(II) are still practical, but frequent restandardization of the titrant solution is necessary. Because of the special storage and buret-filling system required, these titrants are usually used only when reductometric titrations are performed routinely. It is usually more convenient to reduce the substance to be determined and then titrate it with an oxidizing titrant that is unaffected by air.

25.12 *Questions*

25-1 Explain briefly why strong oxidants are more common as titrants than as strong reductants.

25-2 What is the chief advantage in using a metal as a prereductant?

25-3 What is a disadvantage in using nitric acid as a preoxidant?

25-4 Explain briefly why a standard permanganate solution of reliably known strength cannot be prepared by simple dissolution of a known amount of reagent-grade salt in water and dilution to a known volume.

25-5 Name some standards for the standardization of a permanganate solution and discuss their relative advantages and disadvantages.

25-6 State the pH limitations for the use in redox titrimetry of (a) an iodine solution and (b) a thiosulfate solution. Explain what side reaction or other facts are causing the pH restrictions.

25-7 What are the advantages of a cerium(IV) solution as an oxidizing titrant?

25-8 Elaborate on the possibility of a consecutive redox titration of tin(II) and iron(II).

25-9 State the number of electrons involved in the two most common redox couples for permanganate and the conditions under which they are realized.

25-10 Which redox couples can be readily used in both oxidimetric and reductimetric titrations?

25-11 State some redox titrants, other than permanganate, that have more than one useful redox couple that differ in the number of electrons involved.

25-12 What possibilities can you think of for excluding the influence of oxygen of the air on solutions of strong reductants?

25-13 Suggest how you might determine both iron and copper in a solution by redox titrations; and similarly, both titanium and iron.

25-14 When an alloy containing tin is dissolved in hydrochloric acid with the rigorous exclusion of air, to what oxidation state is the tin converted?

25-15 What is an extractive end point and what are its principles of operation, advantages, and disadvantages? Give an example of such an end point.

25-16 Which redox titrants permit self-indicating visual titrations?

25-17 State which of the following ions, present in equal amounts with iron, might interfere in the permanganate titration of iron by the Zimmermann-Reinhardt method: Hg^{2+}, Ag^{+}, Zn^{2+}, Al^{3+}, Cu^{2+}, Cr^{3+}, Ca^{2+}, Mn^{2+}, Sn^{4+}, Cl^{-}, SO_4^{2-}, PO_4^{3-}, I^{-}, NO_3^{-}, and CrO_4^{2-}. Elaborate on the interferences presented and the possibilities for their exclusion.

25-18 State some substances used as primary standards for redox titrants and write the equations involved in the standardization procedures.

25-19 An acidified solution containing iron(III), copper(II), titanium(IV), mercury(II), manganese(II), zinc(II), calcium(II), sodium(I), and silver(I) is passed through a Jones reductor. Indicate which ions will be reduced and the oxidation state to which they will be reduced.

25-20 Zinc metal reduces permanganate ion readily. With this knowledge, suggest a titrimetric procedure for determining metallic zinc in zinc oxide used as a pigment. Write the reaction equation. Would you recommend that the titration be performed in a strongly acidic medium?

25-21 Highlight the use of the following terms in the description of various redox titrations: *titrant, titrant solution, back-titrant, standard solution, standardized solution, theoretical titration curve, titer, prereduction, primary standard, preoxidation, self-indication, redox indicator, irreversible redox indicator,* and *in-situ formation of bromine as an oxidant.*

26

Calculating Results in Redox Titrations

26.1 Basis of Calculations

For calculating the result of a redox titration, it is possible to write the balanced complete redox equation for the titration, establish the combining ratio, and then proceed as recommended for precipitation titrations (Chapter 19). A more convenient approach, however, is to base the calculation on the number of electrons transferred in the redox process. Reasoning as we did earlier (Chapters 15 and 19) leads to the formula

$$\text{ml}_t \times F_t \times f_t \times \frac{\text{FW}_s}{f_s} = \text{mg}_s \qquad (26\text{-}1)$$

Here f_t is the equivalence number of the titrant, and its value equals the number of electrons lost or gained by *one molecule* of the titrant. Care should be exercised in assigning a value to this number; the reason is that the titrant molecule in some cases may require a different number of electrons from what is needed by the reacting titrant species according to the titration reaction, with only ions written. When you know more about titration processes, you can go one step further with the designation of the term f_s. Previously this term referred to the species actually participating in the titration reaction; if this species and the sought-for substance were not identical, a conversion factor was an additional requirement. Here a more direct approach will be taken; by f_s will be designated the number of electrons corresponding to (that is, lost or gained by) one molecule of the sought-for substance. This more direct approach will become obvious, from Examples 26-3 and 26-4.

The equivalence number can be assigned to the titrant as well as the sought-for substances by heeding the above considerations and also either the change of the oxidation states of the reacting species or the relevant balanced half-reaction equations. The modification of equation (26-1) to calculate percentage contents is analogous to cases discussed in other titrations:

$$\frac{mL_t \times F_t \times f_t \times FW_s \times 100}{W \times f_s} = \%s \qquad (26\text{-}2)$$

As in other titrations, the chemical equivalency existing at the equivalence point takes the form

$$meq_t = mL_t \times F_t \times f_t = mL_s \times F_s \times f_s = meq_s \qquad (26\text{-}3)$$

The relation $F_t \times f_t = N_t$ holds for redox titrimetry as it does for acid-base titrimetry.

The normality of an acid (or a base) is the number of moles of hydrogen ion (or hydroxide ion) provided per liter of solution under the reaction conditions of interest. Here the normality of an oxidant (or a reductant) is the number of moles of electrons actually gained (or lost) per liter of solution under the reaction conditions of interest. Normality will not be stressed in this textbook, because, unlike the formality, the normality of a given solution depends on the reaction involved (see Example 26-5).

26.2 *Illustrative Examples*

Example 26–1 What amount of iron (*55.85*) in milligrams is present in a sample if a redox titration in acidic medium needs 42.34 mL of 0.02500*F* $KMnO_4$?

For permanganate, the electron change is five, since Mn(VII) → Mn(II); and for iron, one, Fe(II) → Fe(III). Therefore, $f_t = 5$ and $f_s = 1$. Substituting the numerical data in equation (26–1) yields

$$42.34 \times 0.02500 \times 5 \times \frac{55.85}{1} = mg_s$$

$$mg_s = 295.6 \text{ mg of Fe in sample}$$

Example 26–2 What volume in mL of 0.04000*F* $KMnO_4$ is required to titrate 25.00 mL of a 0.1000*F* solution of oxalic acid in acidic medium?

Here $f_t = 5$ as in Example 26–1; $f_s = 2$ since the oxidation of oxalic acid is a two-electron process ($C_2O_4^{2-} \rightarrow 2CO_2 + 2e$).

Substituting the numerical data in equation (26–2) yields

$$mL_t \times 0.04000 \times 5 = 25.00 \times 0.1000 \times 2$$

$$mL_t = 25.00 \text{ mL of } KMnO_4 \text{ solution}$$

Example 26–3 What is the percentage of Fe_2O_3 (*159.69*) present in an ore if a 1.000-g sample is dissolved and iron, after complete conversion to iron(II), is titrated with 0.02000*F* $K_2Cr_2O_7$, of which a volume of 32.56 mL is needed?

The reduction of dichromate to chromium(III) involves six electrons ($Cr_2O_7^{2-}$ + $14H^+$ + $6e \rightarrow 2Cr^{3+}$ + $7H_2O$); hence f_t = 6. The oxidation of iron(II) involves one electron ($Fe^{2+} \rightarrow Fe^{3+} + e$); however, there are two atoms of iron per molecule of the oxide; hence, f_s = 2. Substituting the numerical data in equation (26–7) yields

$$\frac{32.56 \times 0.02000 \times 6 \times 159.69 \times 100}{1000 \times 2} = 31.20\%\ Fe_2O_3 \text{ in sample}$$

The normality of the dichromate solution used is 0.02000 × 6 = 0.1200*N*.

Example 26–4 For standardizing a $KMnO_4$ solution, a 0.2134-g sample of As_2O_3 (*197.84*) is titrated in acidic medium. What is the formality of the permanganate solution if a volume of 39.12 mL is required?

As established in Example 26–1, we have f_t = 5. Arsenic(III) is oxidized to arsenic(V) and one molecule of As_2O_3 contains two arsenic atoms; hence, f_s = 2 × 2 = 4. The relevant formula, (26–1), when the numerical data are substituted, takes the form

$$39.12 \times F_t \times 5 \times \frac{197.84}{4} = 213.4$$

$$F_t = 0.02206F\ KMnO_4$$

Example 26–5 The $KMnO_4$ solution of Example 26–4 is used to titrate manganese(II) in nearly neutral medium. What amount of manganese (*54.94*) in milligrams is present in a sample that requires 27.70 mL of the $KMnO_4$ solution?

Under these conditions, permanganate is reduced in a three-electron reaction to manganese(IV) oxide ($MnO_4^- + 4H^+ + 3e \rightarrow MnO_2 + 2H_2O$); hence, f_t = 3. Manganese(II) is oxidized to manganese(IV); hence f_s = 2. The calculation, after (26–1), takes the form

$$27.20 \times 0.02206 \times 3 \times \frac{54.94}{2} = mg_s$$

$$mg_s = 49.45 \text{ mg of Mn in sample}$$

The normality of a potassium permanganate solution for a redox titration in acidic medium is *F* × 5, but in nearly neutral medium *F* × 3. The formality is the same regardless of the medium.

In some redox titrations, notably iodometric determinations, the sought-for substance does not react directly with the titrant but is first allowed to react with some other substance, and the product of this reaction is subjected to the titration.

Example 26–6 A 0.1480-g sample of KIO_3 (*214.00*) is treated with an excess of KI and acid. The iodine liberated is titrated with a $Na_2S_2O_3$ solution, of which a volume of 30.55 mL is required. What is the formality of this $Na_2S_2O_3$ solution?

The reactions are

$$IO_3^- + 5I^- + 6H^+ \rightarrow 3I_2 + 3H_2O \quad \text{(liberation of } I_2\text{)}$$

$$I_2 + 2e \rightarrow 2I^- \quad \text{(titration half-reaction for } I_2\text{)}$$

$$2S_2O_3^{2-} \rightarrow S_4O_6^{2-} + 2e \quad \text{(titration half-reaction for } S_2O_3^{2-}\text{)}$$

Reducing one molecule of I_2 to iodide is a two-electron process, but three molecules of I_2 result from the reaction of one molecule of iodate. Consequently, $f_s = 2 \times 3 = 6$. Alternative reasoning gives the same result—one molecule of iodate yields six atoms of iodine (five coming from the KI), and each atom is present as iodide ion at the end of the titration. Hence, as far as the iodate is concerned, the reduction is from iodine(V) to iodide ion, a change of six electrons. Thiosulfate loses two electrons for two thiosulfate ions; that is, a single thiosulfate ion is involved in a one-electron oxidation, and $f_t = 1$. The calculation, based on equation (26–1), takes the form

$$30.55 \times F_t \times 1 \times \frac{214.00}{6} = 148.0$$

and

$$F_t = 0.1358F\ Na_2S_2O_3$$

Example 26–7 Potassium iodide can be determined in the following way. The solution of the salt is acidified and treated with iodate. The first reaction is that given in Example 26–6, and the iodine liberated is distilled, collected under appropriate conditions, and titrated with a thiosulfate solution. What is the equivalent weight of potassium iodide (*166.01*) in this procedure?

Six atoms of iodine are obtained from five molecules of potassium iodide. Each iodine atom involves one electron in the thiosulfate titration. Therefore, six electrons are equivalent to five molecules of potassium iodide, and $f_s = 6/5$. The equivalent weight is given by

$$EW_{KI} = \frac{FW_{KI}}{6/5} = \frac{166.01 \times 5}{6} = 138.34$$

Some further examples show the different equivalence numbers for the same substance used in different titrations.

Example 26–8 Potassium tetroxalate, $KHC_2O_4 \cdot H_2C_2O_4 \cdot 2H_2O$, is a dihydrate double salt of potassium hydrogen oxalate and oxalic acid. A solution of this salt is prepared. When a volume of 20.00 mL of this solution is titrated in acidic medium with $0.02000F$ $KMnO_4$, a volume of 25.00 mL of the permanganate solution is required. What volume of $0.08000F$ NaOH is required to titrate 30.00 mL of the tetroxalate solution in an acid-base titration, using phenolphthalein as the indicator?

In the redox titration of the tetroxalate, four electrons are involved, since each oxalate reacts in a two-electron oxidation (Example 26–2) and there are two oxalate ions present in one molecule of the tetroxalate salt; hence, $f_s = 2 \times 2 = 4$. For permanganate, in acidic solution $f_t = 5$.

Substituting the numerical data in equation (26–2) yields

$$25.00 \times 0.02000 \times 5 = 20.00 \times F_s \times 4$$

$$F_s = 0.03125F \text{ tetroxalate}$$

When tetroxalate participates in the acid-base titration, all three available hydrogen ions react; hence, $f_s = 3$. For sodium hydroxide, $f_t = 1$.

$$\text{mL}_t \times 0.08000 \times 1 = 30.00 \times 0.03125 \times 3$$

$$\text{mL}_t = 35.16 \text{ mL of NaOH solution}$$

Example 26–9 A volume of 20.00 mL of a K_2CrO_4 solution is treated with an excess of $BaCl_2$ and the $BaCrO_4$ (*253.33*) separated by filtration, washed, and dried. The dried precipitate is found to weigh 0.1755 g. If this chromate solution is used to titrate iron(II) in acidic medium, 1 mL is equivalent to what amount of iron (*55.85*) in milligrams?

Since one mole of barium chromate is equivalent to one mole of chromate ion, the following may be written, where F_t is the formality of the chromate solution.

$$20.00 \times F_t \times 253.33 = 175.5$$

$$F_t = 0.03464F\ K_2CrO_4$$

In the titration of iron(II), the chromate is converted to dichromate, which acts as the oxidizing titrant (see Example 26–3). However, one mole of chromate ion is equivalent to one-half mole of dichromate ion ($2CrO_4^{2-} + 2H^+ \rightarrow Cr_2O_7^{2-} + H_2O$); consequently, $f_t = 3$ when related to the formality of the chromate solution. Thus

$$1.00 \times 0.03464 \times 3 \times 55.85 = 5.80 \text{ mg Fe/mL } K_2CrO_4 \text{ solution}$$

Example 26–10 The silver content of 24.00 mL of 0.1000*F* $AgNO_3$ is completely precipitated as Ag_3AsO_4. The isolated precipitate is dissolved in strong acid and the arsenate determined iodometrically, with 20.00 mL of a $Na_2S_2O_3$ solution being required. What is the formality of this thiosulfate solution?

One mole of silver yields one-third mole of silver arsenate; hence, the number of millimoles of the latter formed equals $24.00 \times 0.1000 \times 1/3 = 0.800$. In the redox titration, arsenic(V) is reduced to arsenic(III), providing by the liberated iodine two electrons per molecule; consequently from equation (26–3), 2×0.800 meq is present. For thiosulfate, $f_t = 1$.

$$\text{mL}_t \times F_t \times f_t = \text{meq}_t$$

Substituting the numerical data yields

$$20.00 \times F_t \times 1 = 2 \times 0.800$$

$$F_t = 0.0800F\ Na_2SO_3$$

Instead of expressing concentrations in formalities, one can use titer values (see Sections 3.3 and 19.2). The answer to Example 26-9 is in the form of a titer. An additional example shows how to interconvert titer values.

Example 26-11 The "iron titer" of a given $KMnO_4$ solution is 1.00 mg/mL. What is its "manganese titer"?

From Examples 26-1 and 26-5, it is evident that in the two titrations, the electrons involved are 5 and 3, respectively, for permanganate. The values of f_s for iron and manganese are 1 and 2. Two equations can be written:

$$1 \times F_t \times 5 \times \frac{AW_{Fe}}{1} = 1.00$$

$$1 \times F_t \times 3 \times \frac{AW_{Mn}}{2} = x$$

where x denotes the milligrams of manganese equivalent to 1 mL of the $KMnO_4$ solution. Dividing the first equation by the second yields

$$\frac{5}{3} \times \frac{FW_{Fe}}{FW_{Mn}} \times 2 = \frac{1.00}{x}$$

Solving for x and inserting the atomic weights for manganese (*54.94*) and iron (*55.85*) give

$$x = 1.00 \times \frac{3}{10} \times \frac{54.94}{55.85}$$

$$= 0.295 \text{ mg Mn/mL of } KMnO_4 \text{ solution}$$

26.3 Problems

26-1 A 0.2034-g sample of sodium oxalate (*134.00*) is titrated in an acidic medium with a $KMnO_4$ solution, of which a volume of 45.32 mL is needed. What is the formality of this titrant? *Ans.:* 0.01340F

26-2 What percentage of Mn (*54.94*) is present in a steel if a volume of 12.34 mL of 0.0120F $KMnO_4$ is required to titrate the manganese(II) in a solution obtained by treating appropriately a 1.800-g sample of the steel? *Ans.:* 0.67_8% Mn

26-3 A cerium(IV) sulfate solution is standardized by titrating a 0.2005-g sample of As_2O_3 (*197.84*); a volume of 42.20 mL is needed. A volume of 20.00

mL of this cerium(IV) solution is treated with KI and the I_2 liberated requires 25.68 mL of a $Na_2S_2O_3$ solution. What volume of the $Na_2S_2O_3$ solution will be needed to titrate the I_2 liberated by treating 0.1000 g of KIO_3 (*214.00*) with excess KI and excess acid? *Ans.*: 37.48 mL

26-4 A 0.2200-g sample of copper metal (*63.54*) of 99.35% purity is dissolved and the copper is determined iodometrically; a volume of 35.06 mL of a $Na_2S_2O_3$ solution is required. When a 100.0-mL volume of a H_2O_2 (*34.01*) solution is acidified, excess KI is added, and the I_2 liberated is titrated with this $Na_2S_2O_3$ solution, then a volume of 17.31 mL is required. Calculate the concentration of the H_2O_2 solution in milligrams of H_2O_2 per liter. *Ans.*: 288.8 mg H_2O_2/L

26-5 Pyrolusite, a mineral containing MnO_2 (*86.94*) as the major constituent, is assayed as follows. A solution is prepared by dissolving 3.2340 g of primary standard sodium oxalate (*134.00*) in water and diluting to a volume of 500.0 mL with water. A volume of 50.00 mL of this solution is acidified and a 0.2500-g sample of the pyrolusite is added to it. After the sample has reacted completely ($MnO_2 \rightarrow Mn^{2+}$), with oxalate being oxidized concurrently to carbon dioxide, the excess of oxalate is back-titrated with a $KMnO_4$ solution, of which a volume of 2.25 mL is required. In a separate titration in acidic medium, a volume of 16.80 mL of this $KMnO_4$ solution is found to be equivalent to 20.00 mL of the oxalate solution. What is the percentage of MnO_2 in the pyrolusite? *Ans.*: 79.43% MnO_2

26-6 What amount of $K_2Cr_2O_7$ (*294.19*) in grams must be dissolved and diluted to a volume of 500.0 mL with water to obtain a solution that has a titer of 1.50 mg of Fe_2O_3 (*159.69*) per milliliter? *Ans.*: 0.4606 g

26-7 An amount of 0.2143 g of iron wire containing 99.52% w/w Fe (*55.85*) is dissolved. All the iron is reduced to iron(II) and titrated with a permanganate solution, of which a volume of 38.22 mL is required in acidic medium. Calculate the formality of the $KMnO_4$ solution.

26-8 An amount of 0.1205 g of iron wire containing 95.0% Fe (*55.85*) is dissolved, the solution passed through a reductor, and the Fe(II) titrated with a permanganate solution, of which a volume of 21.70 mL is needed. Calculate the formality of the permanganate solution.

26-9 An amount of 2.160 g of a sample is dissolved and the calcium precipitated with an excess of oxalate. The calcium oxalate is separated from the solution, washed, and dissolved in acid. A volume of 22.0 mL of 0.0100*F* $KMnO_4$ is needed for titrating the oxalate in acidic medium. Calculate the % w/w of CaO (*56.08*) in the sample.

26-10 A material is known to be an iron oxide—FeO, Fe_2O_3, or Fe_3O_4—in pure form. A 0.1500-g sample is fused with $KHSO_4$, the melt dissolved, and the iron reduced to iron(II) and titrated with 0.00900*F* $KMnO_4$; 43.15 mL is required. The material consists of what oxide of iron?

26-11 A solution contains arsenous acid and arsenic acid. The arsenous acid is titrated with 0.01500*F* $KBrO_3$; 24.60 mL is required. After the titration, the total amount of arsenic acid present is precipitated as $MgNH_4AsO_4$, and the isolated and washed precipitate is ignited to $Mg_2As_2O_7$ (*310.46*). The weight of the magnesium pyroarsenate is 1.2531 g. How many millimoles of orthoarsenic acid, H_3AsO_4, was initially present in the solution?

26-12 When a strong acid is added to a neutral solution containing KIO_3 and KI, the following reaction takes place: $IO_3^- + 5I^- + 6H^+ \rightarrow 3I_2 + 3H_2O$. This reaction proceeds stoichiometrically with respect to H^+. Consequently a standard acid solution can be used to standardize a thiosulfate solution. Calculate the formality of each of the thiosulfate solutions from the data given.

	Acid solution	Thiosulfate solution, mL
(a)	40.00 mL of 0.1000*F* HCl	35.21
(b)	18.26 mL of 0.1023*F* HCl	21.46
(c)	23.15 mL of 0.0829*F* H_2SO_4	26.44
(d)	26.00 mL of 0.0750*F* H_2SO_4	46.86
(e)	25.00 mL of 0.1025*F* $HClO_4$	31.02
(f)	24.22 mL of 0.0993*F* $HClO_4$	18.95

26-13 An amount of 0.2350 g of a copper ore is dissolved in acid, the pH adjusted appropriately, and excess KI added. The liberated iodine requires 5.26 mL of 0.1500*F* $Na_2S_2O_3$ to reach the starch end point. Calculate the % w/w of copper (*63.54*) in the ore.

26-14 An amount of 0.01589 g of pure Cu (*63.54*) is dissolved and excess KI added. The iodine liberated requires 25.00 mL of a thiosulfate solution to reach the starch end point. Calculate the formality of the thiosulfate solution.

26-15 The arsenic present in a 0.2000-g sample of a material is volatilized as AsH_3, which is trapped in 40.00 mL of a 0.0953*F* I_2 solution. Titrating the excess iodine requires 21.50 mL of a 0.1062*F* thiosulfate solution. Calculate the % w/w As_2O_3 in the material.

26-16 Potassium tetroxalate, $KHC_2O_4 \cdot H_2C_2O_4 \cdot 2H_2O$ (*254.2*), can be titrated by an acid-base method as well as a redox method. In the following problems, one portion of a tetroxalate solution is titrated with a standard NaOH solution and an identical portion titrated with a $KMnO_4$ solution in acidic medium. For each problem, calculate the formality of the $KMnO_4$ solution.

	Tetroxalate soln, mL	NaOH soln, mL	Formality, NaOH soln	$KMnO_4$ soln, mL
(a)	25.00	30.00	0.1000	6.00
(b)	20.00	34.25	0.1100	7.25
(c)	25.00	35.00	0.0975	10.00
(d)	10.00	15.00	0.0900	4.25
(e)	30.00	25.00	0.1200	6.80

26-17 What volume in milliliters, of a solution exactly $\frac{1}{60}F$ in $K_2Cr_2O_7$ is needed for each of the back-titrations indicated? In each case, a volume of 100.00 mL of 0.100*F* $FeSO_4$ in acid solution is added to exactly 0.1 g of the substance indicated.

Substance titrated	
(a) MnO_2	*(86.94)*
(b) $KMnO_4$	*(158.0)*
(c) $K_2Cr_2O_7$	*(294.2)*
(d) K_2CrO_4	*(194.2)*

26-18 Salicylic acid, $C_7H_6O_3$ *(138.1)*, can be oxidized by permanganate (reduced to MnO_2) in alkaline solution to water and carbonate ion. How many milligrams of salicylic acid will be indicated by exactly 1 mL of a $KMnO_4$ solution that contains 7.900 g of $KMnO_4$ *(158.0)* in 500.0 mL?

26-19 Barium is determined in the following indirect way. Barium is precipitated as $BaCrO_4$ by adding a measured amount (50.00 mL) of a chromate solution of known concentration (0.0200*F*). The precipitate is separated by filtration, and the filtrate and washings are collected in a volumetric flask (250 mL) and diluted with water to mark. An aliquot (100 mL) of this solution is acidified and chromate titrated with a standard iron(II) solution (0.100*F*), of which a measured volume is required (6.00 mL). How much barium *(137.3)* was present in the original sample? *Ans.:* 68.6 mg

26-20 An iron ore sample was found to have a moisture content of 8.0% w/w. A 0.500-g sample of ore was so treated that a solution with all of the iron *(55.8)* in the +2 oxidation state was obtained. This solution was titrated with 0.0200*F* $KMnO_4$, of which 27.60 mL was needed. What is the iron content of the ore expressed as Fe_3O_4 *(231.5)* and calculated on a dry basis? *Ans.:* 46.3%

26-21 The ozone *(48.0)* content of air is determined by bubbling the air (5.0 L/min for 25.0 min) through 70 mL of a phosphate buffer to which has been added 1 g of KI and 5.00 mL of 0.0010*F* $Na_2S_2O_3$. For each mole of ozone, O_3, one mole of iodine, I_2, forms and reacts with the thiosulfate. When the airflow is stopped, the solution is acidified and titrated with $2.0 \times 10^{-4}F$ KIO_3. Calculate the ozone content of the air in ppm w/w if the titration required 3.20 mL of KIO_3 solution. Air has a density of 1.06 g/L.

Ans.: 0.21 ppm O_3

26-22 For injection, ascorbic acid (vitamin C) is available as a sterile aqueous bicarbonate solution. The product is assayed by a self-indicating titration with 2,3-dichlorophenolindophenol, which on reduction by the vitamin changes from red to colorless. For the titrant solution, 50 mg of the "indophenol" is dissolved in water to 200 mL. For standardization, 50 mg of "reference quality" ascorbic acid is dissolved in water to 50 mL, and a 2.00-mL aliquot is placed in a flask containing metaphosphoric–acetic acid buffer (the phosphate complexes and thereby inactivates metals catalyzing air oxidation of the vitamin). The solution is then titrated promptly with the indophenol solution until a pink color persists for 5 s. A blank is run, omitting the ascorbic acid. For the assay, the injection solution is titrated essentially in the same way.

In an actual case, 51.3 mg of the reference ascorbic acid was dissolved to 50 mL, and the 2.00-mL aliquot needed 16.22 mL of indophenol solution; a blank needed 0.68 mL. An ampoule was then opened; 5.00 mL was diluted with water to 100 mL, and a 5.00-mL aliquot mixed with buffer and diluted with water to 100 mL. Titrating a 4.00 mL portion of this final solution required 17.80 mL of titrant solution; an appropriate blank, containing buffer, needed 0.74 mL.

(a) What is the ascorbic acid titer of the indophenol solution in milligrams per milliliter? (b) What is the assay of the product in g/5 mL? (c) The ampoule was labeled to provide 1 g of the vitamin per 5 mL. The pharmacopeia requires that the content must be not less than 95% and not more than 115% of the label value. Was this requirement met?

Ans.: (a) 0.132 mg/mL; (b) 1.12 g/5 mL; (c) yes, close to upper limit

26-23 Titanium(IV) oxide exists in nature as rutile, anatase, and brookite; only the first-named form serves as a white pigment (look up the refractive indices in a handbook and elaborate on why rutile is preferred). As a measure to reduce the cost of a paint, rutile is frequently cut with other white pigments—and often with barium sulfate. To establish the composition of the mixed pigment, the following approach is used. A weighed sample is dissolved in hot concentrated sulfuric acid (soluble barium hydrogen sulfate, $Ba(HSO_4)_2$, and titanium(IV) sulfate form). The cooled solution is added slowly to water or water is added to it dropwise with care and stirring. Barium sulfate thereby is precipitated and can be determined gravimetrically. The filtrate is passed through a reductor column and the effluent containing violet titanium(III) is run into an iron(III) chloride solution. The equivalent amount of iron(II) formed is titrated with permanganate.

In an actual case, for a 0.5000-g sample of the mixed pigment, 21.94 mL of 0.04000*F* $KMnO_4$ was needed. (a) Calculate the percentage TIO_2 (*79.9*). Alternatively, the column effluent is protected by a blanket of carbon dioxide, and titanium(III) is titrated with an iron(II) solution. Methylene blue is the indicator and is reduced by titanium(III) to a colorless (leuco) form; at the end point, the indicator is oxidized to an intense blue species. (b) What volume of 0.1000*F* iron(III) solution would be required for the sample? *Ans.:* (a) 70.1% TIO_2; (b) 43.88 mL

26-24 Water from a mineral spring is to be analyzed for its total sulfide content. A 500-mL sample is placed in a flask and acidified with 10 mL of concentrated sulfuric acid. Carbon dioxide is then bubbled through the solution for 1 hr. The carbon dioxide leaving the flask is bubbled through two flasks, in series, containing 0.05*F* zinc acetate. A volume of 10.00 mL of 0.0125*F* iodine is added to the combined zinc-acetate solutions, and 5.0 mL of concentrated hydrochloric acid then added. The excess of iodine is back-titrated with 0.025*F* sodium thiosulfate; a volume of 6.25 mL is needed to reach the starch end point. Calculate the milligrams per liter total sulfide expressed as sulfur (*32.06*) in the sample.

26-25 The dissolved oxygen (DO) in water can be determined as follows (modified Winkler method). The sample is collected in a bottle, which is filled to overflowing before being capped, so that no air is entrapped. The sample, in its bottle, is treated with manganese(II) sulfate, potassium hydroxide and potassium iodide. The contents of the stoppered bottle are mixed; then sulfuric acid is added. The initial precipitate of manganese(II) hydroxide, $Mn(OH)_2$, combines with the dissolved oxygen to form brown, insoluble $MnO(OH)_2$.

On acidification, the $MnO(OH)_2$ is converted to $MnOSO_4$, which oxidizes iodide to iodine. The iodine, which is equivalent to the dissolved oxygen, is titrated with thiosulfate.

A 500-mL quantity of river water is treated by the procedure described and needs 32.15 mL of 0.0375F sodium thiosulfate to reach the starch end point. Calculate the DO content of the sample expressed as milligrams of oxygen (*16.00*) per liter. Also calculate the milliliters of oxygen gas, at 0°C and atmospheric pressure, that are present per liter of the water sample (1 L of oxygen gas under those conditions weighs 1.429 g).

26-26 The chemical oxygen demand (COD) is the quantity of oxygen that would be needed to oxidize the organic matter in a water sample by potassium dichromate in 50% sulfuric acid with silver(I) present as a catalyst and mercury(II) to complex chloride and thereby prevent its oxidation. The excess of dichromate is titrated with iron(II) to an end point indicated by formation of the iron(II)-1,10-phenanthroline complex. A 20.0-mL sample of waste water is placed in a reflux flask with 0.4 g of $HgSO_4$ and 10.0 mL of 0.0400F $K_2Cr_2O_7$. A volume of 30.0 mL of concentrated H_2SO_4 containing Ag_2SO_4 is added and the mixture heated under reflux for 2 hr. After the solution is cooled, it needs 3.68 mL of 0.250F $Fe(NH_4)_2(SO_4)_2$ for titration. Calculate the COD of the sample as milligrams of oxygen (*16.00*) per liter.

27

Potentiometric Titrations

27.1 Principles of Potentiometric Titrations

The limitations of using a potentiometric measurement for the direct determination of a species has been described in Section 23.8. It is difficult to reproduce conditions, but relatively easy to keep conditions constant once they have been established. In place of a single, "absolute" measurement of potential, changes in potential can be followed reliably while other parameters are held virtually constant. This approach is used in a potentiometric titration. The solution to be titrated is made the electrolyte of a half-cell, the potential of which is a function of the concentration of one or more of the species participating in the titration. In the simplest arrangement, a wire or foil of an appropriate metal is immersed in that solution to act as an indicator electrode. The half-cell thus formed is connected by salt bridge with a reference electrode. The cell established in this way is connected with a potentiometer circuit (Section 22.10) and during the subsequent titration, the cell voltage is measured as a function of titrant additions. Determining the end point graphically calls for plotting the values of the half-cell potential (E_{cell} - $E_{reference}$) versus the volume of titrant solution added. However, because the potential of the reference electrode remains constant, it is common practice to plot simply the total cell voltage instead of the half-cell potential, thus merely shifting the titration curve vertically without changing its shape or its position relative to the abscissa. A titration so conducted is described as proceeding to a potentiometric end point, or simply as a potentiometric titration. Many types of titrations can be effected potentiometrically.

27.2 Potentiometric Redox Titrations

A redox titration can be conducted potentiometrically (this has already been implied in the considerations of Chapter 24). In this case, the indicator electrode is frequently platinum wire or foil. The experimental titration curves will often closely approximate the theoretical curves calculated, as described in Chapter 24. The point of maximum slope in the curve is taken as the end point.

27.3 *Potentiometric Acid-Base Titrations*

The titration curves for acid-base titrations in aqueous solution have already been calculated (Chapter 11). These curves are plots of pH versus the volume of acid or base added. Such curves are realized experimentally in acid-base titrations to a potentiometric end point. An electrode sensitive to hydrogen ion is used as the indicator electrode. The principles underlying the electrometric determination of pH have been described in Chapter 23, and from these considerations it is obvious that a pH meter with a glass electrode can be used to follow the pH changes during the titration. The measured pH values can be plotted versus the volume of titrant solution added and the point of maximum slope taken as the end point. Alternatively, if the pH of the equivalence point is accurately known, the titration is often performed without plotting and terminated when this pH value is reached. Potentiometric acid-base titrations can be used when visual indication becomes difficult due to colorants present in the sample solution. A potentiometric titration is especially valuable for determining acids and bases that are too weak to afford a satisfactory visual titration, and it can be performed in a nonaqueous solvent system that has a sufficiently high dielectric constant (see Section 16.3).

27.4 *Other Potentiometric Titrations*

The presence of an appropriate complexing agent may markedly change the potential of a redox couple involving a metal ion (Section 24.10); consequently, the end point of some compleximetric titrations (Chapter 20) can be detected potentiometrically. Similarly, the formation of a precipitate by a metal ion that is involved in a redox equilibrium will shift the redox potential. Hence, potentiometric precipitation titrations are also feasible. In the titration of a halide with silver nitrate, a silver wire can be the indicator electrode.

Potentiometric titrations often allow the stepwise titration of two or more substances with a single titrant. For example, a mixture of bromide ion and chloride ion can be titrated with silver nitrate. In the visual titration of this mixture by the Mohr method (Section 18.4), the result corresponds to the total of the two halides. In contrast, in the potentiometric titration, two breaks will be obtained, corresponding to bromide and chloride. The accuracy is not especially high because of the closeness of the values of the respective solubility products. More accurate results are obtained in the stepwise potentiometric titration of iodide and chloride. Under favorable concentration conditions, for example, in the analysis of mineral waters, it is even possible to analyze a mixture of iodide, bromide, and chloride by titration with silver ion. Three breaks in the potentiometric titration curve are obtained.

Paralleling the use of a pH-sensitive electrode in potentiometric acid-base titrations, ion-selective electrodes of the kinds described in Section 23.7 can be used as the indicator electrode in various titrations.

27.5 *Diverse Techniques*

Instead of using the inflection point as the end point in a potentiometric titration, more precise results can be gained from the plot of the first or second difference curves. The first difference curve is readily obtained by adding the titrant solution in small, equal increments and plotting the difference between two consecutive potential (or voltage) readings versus the volume of titrant solution added. The maximum in this curve is taken as the end point. The second difference curve is obtained by plotting the difference between two consecutive differences versus the volume of titrant solution added. Here the end point corresponds to passage of the curve through the zero line. Representative curves plotted in these three ways are shown in Fig. 27-1.

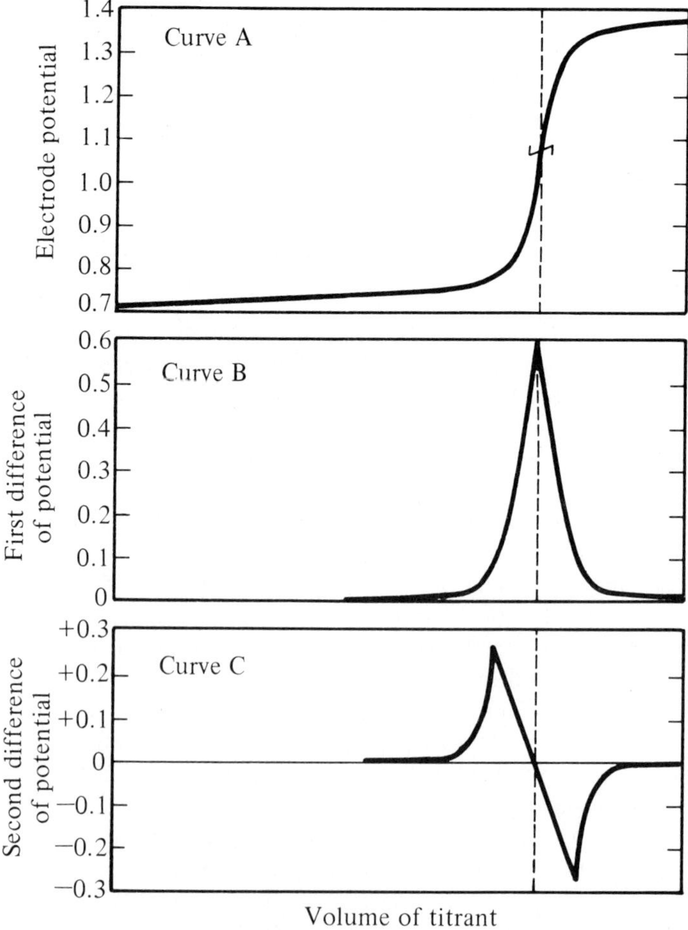

Fig. 27-1. Curve A: experimental potentiometric titration curve. Curve B: first difference curve of A. Curve C: second difference curve of A.

The curves for the first and second differences, as just defined, are often incorrectly called first and second derivative curves. Obtaining the derivative (in the sense of calculus) requires dividing the relevant difference by the volume increment causing it.

The values for the first difference can be established directly by a technique that is outdated in application but has value in clarifying principles. The technique is called a derivative potentiometric titration. Two identical electrodes are placed in the titration solution. One, however, is shielded by a glass tube with a narrow opening leading to the bulk of the solution (Fig. 27-2). Initially, both electrodes have the same potential and the cell voltage is zero. An increment of the titrant is then added and the solution is homogenized. The potential of the unprotected electrode changes because the composition of the solution has changed. However, the potential of the shielded electrode remains unchanged because it is still in contact with a solution of the composition present before the titrant was added. The voltage thus resulting is measured and recorded. Then the solution in the glass tube is flushed out by squeezing the rubber bulb, and the tube is refilled from the bulk of the solution. Both electrodes are now again immersed in identical solutions; they are therefore at the same potential, and the cell voltage is zero. This process is repeated for each increment (equal) of titrant solution. A graph of the cell voltage values, that is, of the differences in potential versus the volume of titrant solution, corresponds to the first difference of the usual titration curve. For equal increments, the differences in potential increase as the equivalence point is approached, reach a maximum for the increment in which this point falls, and become smaller again when

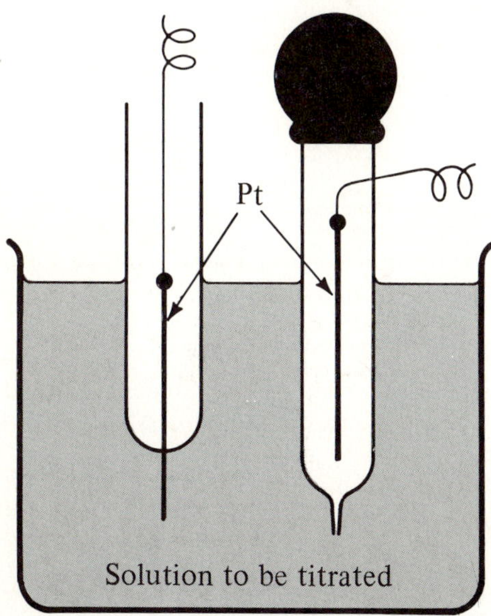

Fig. 27-2. Schematic arrangement for a derivative potentiometric titration.

an excess of titrant is added. At the equivalence point, adding the fixed amount of titrant causes the greatest fractional change in concentration and thereby in potential. Consequently, the maximum in the curve is taken as the end point.

When a potentiometric titration is routine, there is no need to plot the titration curve if the cell voltage corresponding to the end point is well established. The titrant solution needs to be added only until this voltage is reached. This modification of the potentiometric titration is readily capable of automation. Instruments are available that follow the potential and regulate the addition of titrant from a buret, the delivery of which is controlled electrically and stopped when the preselected cell voltage is reached. In other, more sophisticated "autotitrations," a double differentiation is effected by an electronic circuit; when the output signal of the titrator is zero, a relay is triggered by which the flow of titrant is terminated. With this kind of instrument, the potential of the equivalence point does need not to be established previously.

27.6 Questions

27-1 What advantages can be realized by making an acid-base titration to a potentiometric end point instead of a visual one?

27-2 Does a potentiometric titration have any advantage when more than one component of the solution reacts with the titrant? Explain.

27-3 Elaborate on the various techniques for detecting the end point in a potentiometric titration and discuss their relative advantages.

27-4 A calomel electrode is a common reference electrode in potentiometric titrations. However, any electrode that shows a constant potential during the titration can be substituted. Is it necessary to know the exact potential of such an electrode? Explain.

27-5 Adding titrant solution during a potentiometric titration in equal increments, at least near the end point, has the special advantage that the position of the end point can be calculated without plotting the data. Explain how this might be done.

27-6 In a potentiometric acid-base titration using a glass electrode and pH meter is it necessary to calibrate the latter? Explain.

27-7 What is an inert electrode?

27-8 Elaborate on the possibility of using an ion-selective electrode as the indicator electrode in potentiometric titrations of various kinds.

27.7 Problem

27-1 The following data were obtained in a potentiometric titration of an acid with a base. Plot the titration curve and both the first and second difference curves. From the last curve, read the volume of titrant corresponding to the end point and calculate the formality of the acid.

Volume of acid titrated: 50.0 mL

Concentration of NaOH: 0.0808F

NaOH, mL	Potential, scale divisions
24.00	346
0.25	349
0.50	351
0.75	355
25.00	365
0.25	379
0.50	404
0.75	451
26.00	469
0.25	481
0.50	491
0.75	496
27.00	499

Ans.: 25.62 mL; 0.0416F

28 Theory of Electrolytic Processes

The principles underlying electrolytic processes form the basis of numerous analytical methods. Electrolysis occurs when an external source of voltage is applied to an electrochemical cell to force a nonspontaneous electrochemical reaction. The electrode connected to the positive terminal of the voltage source is known as the anode, and the electrode connected to the negative terminal is the cathode. Oxidation proceeds at the anode, reduction at the cathode. (The possible ambiguities existing in the definition of the terms *anode* and *cathode* are noted in Section 22.4.) The solution in the cell is known as the electrolyte solution, or in brief, the electrolyte.

28.1 Decomposition Voltage

Consider the experimental arrangement illustrated in Fig. 28-1. The beaker contains an aqueous solution 0.10*F* in cadmium iodide, which is the electrolyte. Two identical platinum electrodes are dipped into the solution. The left and right electrodes are connected through the external circuit to the negative and positive terminals of the voltage source and hence will be the cathode and the anode, respectively. The external circuit consists of the battery B, the off-on switch S, the variable resistance, R_v, the ammeter A, and the voltmeter V. The resistance R_v allows adjustment of the voltage that is applied across the electrolysis cell.

Before the switch is closed, the voltmeter shows zero volt because the electrodes are identical and dip into the same homogeneous solution. Then the switch is closed, and a small voltage (say a few tenths of a volt) is applied to the cell, by appropriate adjustment of the variable resistor. This voltage is indicated by the voltmeter. The ammeter, however, shows that essentially no current flows, and an inspection of the electrodes reveals that no electrode reactions are taking place.

If the voltage applied to the cell is increased progressively, at first the current rises only feebly. Then at a certain voltage the current increases suddenly and electrode reactions become evident; cadmium plates at the cathode, and elemental

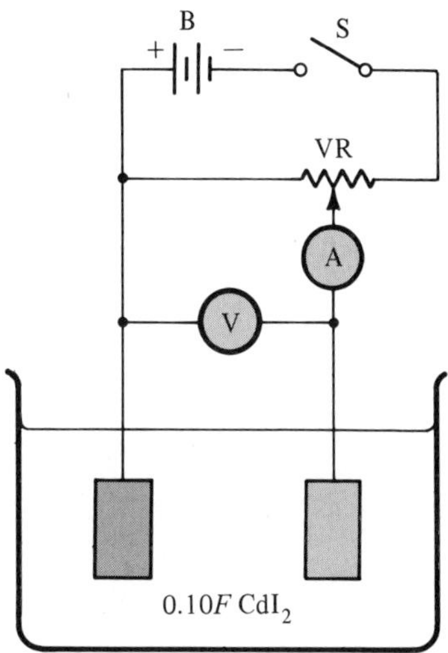

Fig. 28-1. Schematic arrangement for electrolysis.

iodine discolors the anode. The slight solubility of iodine is neglected and all of it is assumed to deposit.

When the current starts to flow and electrode reactions become evident, the voltage applied to the cell is enough to cause appreciable decomposition of the electrolyte; it is then said to be the decomposition voltage. The decomposition voltage, like any voltage across an electrochemical cell, consists of two portions—the anodic and cathodic decomposition potentials. Since there is some ambiguity in deciding when a current "just starts to flow," it is customary to define the decomposition voltage E_d as the zero-current intercept obtained by back-extrapolating the current versus applied voltage curve, as shown in Fig. 28-2.

28.2 Back Electromotive Force

If the electrolysis current is allowed to flow for some time and then the switch is opened, the voltmeter is found to indicate a voltage across the electrodes having the same polarities as during the electrolysis. This voltage arises within the cell, is in opposition to the applied voltage, and is known as the back electromotive force, or in short, the back emf. This back emf consists of two portions, the reversible back emf and the concentration of overvoltage.

The reversible back emf arises because a voltaic cell formed when the electrolytic depositions took place. One platinum electrode is now coated with cadmium and the other with iodine; together with their surrounding solutions they form a voltaic cell consisting of a cadmium(II)-cadmium half-cell and an iodine-

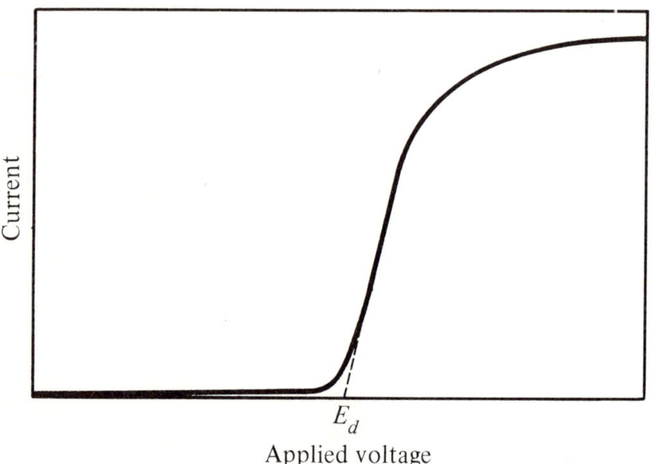

Fig. 28–2. Plot of current versus applied voltage (E_d = decomposition voltage).

iodide half-cell. Realizing that a solution 0.10*F* in CdI_2 is 0.10*M* in Cd^{2+} and 0.20*M* in I^-, we can represent the newly formed cell as

$$Cd \mid Cd^{2+}\ (0.10M) \parallel I^-\ (0.20M) \mid I_2(s)$$

This schematic presentation of the cell can readily be related to the actual electrolysis cell shown in Fig. 28–1 by considering the cell to consist formally of three parts: the left third, which envelops and includes the cathode, and for which $Cd^{2+} + 2e \rightleftharpoons Cd$ and $E^0 = -0.40$ V; the right third, which envelops and includes the anode, and for which $I_2 + 2e \rightleftharpoons 2I^-$ and $E^0 = +0.54$ V; and the middle third, which acts as an electrolytic connection.

The voltage of the newly formed voltaic cell can be calculated by the procedure described in Section 22.8. In that section the half-cells were referred to as "right" and "left" and the half-cell potentials designated E_r and E_l. However, for electrolysis cells, it is more meaningful to distinguish the anodic and cathodic half-cell potentials, E_a and E_c. The calculation takes the form

$$E_r = E_a = +0.54 + \frac{0.059}{2} \log \frac{1}{(0.20)^2} = +0.58 \text{ V} \tag{28-1}$$

$$E_l = E_c = -0.40 + \frac{0.059}{2} \log 0.10 = -0.43 \text{ V} \tag{28-2}$$

and

$$\begin{aligned} E_{\text{cell}} &= E_r - E_l = E_{\text{rev}} = E_a - E_c \\ &= +0.58 - (-0.43) = (+)1.01 \text{ V} \end{aligned} \tag{28-3}$$

It is common practice to write $E_{\text{rev}} = E_a - E_c$ to get a positive sign for the reversible emf. The positive sign of the cell voltage then indicates that the right half-cell, that is, the iodine half-cell, is the positive pole. Since this electrode is connected to the positive terminal of the voltage source and thus opposes the external voltage, and since the cell reaction $Cd^{2+} + 2I^- \rightleftharpoons Cd + I_2$ is reversible, the term *reversible back electromotive force* is appropriate.

It can be appreciated that for an electrolysis the voltage applied to the cell must at least equal the reversible emf of the voltaic cell that forms once electrolysis is initiated. For the cadmium iodide system under consideration, the decomposition voltage is essentially identical to the reversible back emf calculated from the Nernst equation. As you can deduce from looking at equations (28–1) through (28–3), the reversible back emf and consequently the decomposition voltage increases as the concentrations of the relevant ions decrease. Since the concentrations of both the cadmium(II) and the iodide ion decrease as the electrolysis proceeds, the reversible back emf increases progressively.

28.3 *IR Drop and Applied Voltage*

The decomposition voltage E_d, as defined above, is in essence the applied voltage needed to initiate electrolysis. To advance the electrolysis, that is, to plate cadmium and iodide at a reasonable rate, more voltage is needed for overcoming the electrical resistance of the cell. With an electrolysis current I, the voltage required in excess of the decomposition voltage is given according to Ohm's law: $E = IR$. If the cell resistance R is known, this voltage, called the *IR* drop, can be calculated. The applied voltage E_{app} needed for performing an electrolysis with the passage of a certain current can be given as $E_{app} = E_d + IR$. It will be shown presently that this formula does not yet fully describe the situation.

28.4 *Concentration Overvoltage and Overpotential*

From our study so far, it seems that once the decomposition voltage is reached, the current should increase in a linear fashion when more voltage is applied. Experiments show that this is true for only a moderate voltage rise. At higher voltages, the current increase becomes progressively less than what would correspond to a linear relation, and finally a point is reached beyond which an increase in applied voltage does not lead to any appreciable change in the current unless the decomposition of another system, possibly that of the solvent, is reached. The situation is illustrated in Fig. 28–2.

To explain this leveling of the current-applied voltage curve to a plateau, we must look more closely at the processes taking place at the electrodes. Consider first the situation at the cathode. Here cadmium metal is deposited, and the solution near the cathode becomes depleted of cadmium ions. In the calculation of the half-cell potential in equation (28–2), the bulk concentration of $0.10M$ Cd^{2+} was used. However, the Nernst equation applies only to the situation at the electrode surface, and here the cadmium ion concentration is lower due to the depletion. Consequently, the value based on the bulk concentration must be suspected. Several processes operate to replace the cadmium ions that leave the solution as the electrolysis proceeds. Cadmium ions are transported to the electrode region by diffusion from the more concentrated solution. Other transport mechanisms include electrostatic attraction between the positively charged cadmium ions and the negatively charged electrode, convection, and possibly agitation of the solution. If a current of, say, 0.1 A (which equals 0.1 C/s) is to be maintained, then 0.1/96,500, or about 10^{-6} equivalent, of cadmium ion must be transported to the electrode each second. So long as the applied voltage, and therefore the current, is moderate, cadmium ions are replenished fast enough by these transport processes. At higher voltages, however, the current cannot follow the voltage rise linearly, since the ion transport lags. Finally a point is reached at which the current is governed by the rate of ion transport to the electrode, and a further increase in applied voltage no longer affects the current significantly.

The situation at the anode is analogous.

To examine the voltage-current relation more closely, we must consider the "effective" back emf using the concentrations actually prevailing at the electrodes. Assume, for example, that these concentrations are $1 \times 10^{-3}M$ for cadmium ion and $2 \times 10^{-3}M$ for iodide ion. A voltage of 1.19 V is obtained (the calculation is left for you to do). Consequently, the "effective" back emf is 0.18 V greater than the reversible back emf. The cause for this additional voltage can therefore be laid to concentration effects, and the voltage is appropriately termed the *concentration overvoltage.* It is the composite of two portions, the anodic and cathodic concentration overpotentials. Some workers prefer the term *concentration polarization.*

It is now possible to present a more complete formula for calculating the voltage to be applied:

$$E_{\text{appl}} = E_d + IR + E_{\text{concn overv.}} \tag{28-4}$$

where, for cadmium iodide, E_d equals the reversible back emf. The concentration overvoltage is *not* a component of the decomposition voltage, since it arises only while current is passing. Unfortunately, the concentrations of the solutions immediately adjacent to the electrodes are not known, and consequently the concentration overpotentials cannot be calculated by the Nernst equation. This restriction holds not only for the present example, but generally. In calculations using equation (28-4), all that can be done is to neglect concentration overvoltage. The result is the value for the applied voltage required that will be approached as measures are taken to reduce concentration overvoltage. Several measures are possible to this end. Raising the temperature increases the diffusion rate and thereby the rate of particle transport to the electrode. Stirring the solution has the same effect. A decrease can be further achieved by decreasing the current density. The current density is defined as the cell current divided by the surface area of the relevant electrode and is usually expressed in amperes per square centimeter (or decimeter). The current density at a given current strength can be decreased by an increase in surface area, for example, by substituting a larger electrode. This electrode will be in contact with more solution, from which ions can be drawn.

28.5 Kinetic Overvoltage and Kinetic Overpotential

If the electrolysis experiment, described with cadmium iodide as the electrolyte, is performed with certain other types of electrolytes, the result is different. Consider the same experimental arrangement (Fig. 28-1) but with 0.10*F* sulfuric acid as the electrolyte. As we shall establish, this acid has no effect on the overall cell reaction, which is the decomposition of water according to $H_2O \rightarrow H_2 + \frac{1}{2}O_2$.

Before the external voltage is applied, no cell voltage exists since again two identical electrodes dip into the same homogeneous solution. When a certain voltage is applied, current starts to flow and the anode and cathode become coated with small bubbles of oxygen and hydrogen, respectively. Thereby again a voltaic cell is formed, here consisting of an oxygen half-cell and a hydrogen ion-hydrogen half-cell. The electrode reactions and relevant Nernst expression can be written

$$\tfrac{1}{2}O_2 + 2H^+ + 2e \rightleftharpoons H_2O$$

$$E_a = +1.23 + \frac{0.059}{2} \log[H^+]^2 \sqrt{p_{O_2} \times 9.8692 \times 10^{-6}} \tag{28-5}$$

$$2H^+ + 2e \rightleftharpoons H_2$$

$$E_c = 0.00 + \frac{0.059}{2} \log \frac{[H^+]^2}{p_{H_2} \times 9.8692 \times 10^{-6}} \tag{28-6}$$

where the concentrations of the gaseous components are expressed in terms of their partial pressures in pascals. The reversible back emf can now be calculated. The hydrogen ion concentration is $2.0 \times 10^{-1}M$, and for simplicity, it is assumed that $p_{O_2} = p_{H_2} = 1.01325 \times 10^5$ Pa (1 atm), which makes the pressure terms in both equations (28-5) and (28-6) unity. The calculation then takes the form

$$E_a = +1.23 + \frac{0.059}{2} \log(2 \times 10^{-1})^2 = +1.19 \text{ V} \tag{28-7}$$

$$E_c = 0.00 + \frac{0.059}{2} \log(2 \times 10^{-1})^2 = -0.04 \text{ V} \tag{28-8}$$

and

$$E_{rev} = E_a - E_c = +1.19 - (-0.04) = 1.23 \text{ V} \tag{28-9}$$

The term for the hydrogen ion concentration vanishes in the subtraction. Consequently, so long as the partial pressures of the two gases are both at the standard value, the reversible back emf will be 1.23 V, regardless of the pH of the solution. Thus the value of 1.23 V holds also for pure water. However, such a medium contains only an extremely small number of ions and consequently shows a very high electrical resistance. Sulfuric acid is added to lower the resistance and thus to make it easier for an electric current to flow. Any other electrolyte that does not undergo electrolytic decomposition (sodium sulfate or sodium hydroxide, for example) would do equally well.

If, in analogy to the cadmium iodide case, the decomposition voltage were equal to the reversible back emf, an applied voltage of 1.23 V should suffice to just initiate the decomposition of the water. Experimentally, however, the decomposition voltage is found to be about 1.7 V. Thus an overvoltage of 0.5 V (1.7 - 1.23) exists.

The presence of concentration overvoltage might be suspected. However, this assumption is not logical, since the decomposition potential was established at essentially zero current condition; and with no current flowing, no concentration changes around the electrode take place. In addition, the overvoltage of 0.5 V persists even on vigorous stirring of the electrolyte. Therefore, the overvoltage must be caused by other phenomena than concentration effects.

This additional type of overvoltage is called kinetic overvoltage, or activation overvoltage. Portions of this overvoltage arise at the anode and cathode and are termed anodic and cathodic kinetic overpotentials. The mechanisms involved in the establishment of such overpotentials are complex and not fully understood.

Especially large overpotentials are consistently noticed when diatomic gases are produced in the electrode reaction. Some other systems besides those involving diatomic gases also show kinetic overpotential, generally to a much lesser degree. For example, overpotential arises in the electrodeposition of cobalt, chromium, and nickel. Molybdenum shows such a high kinetic overpotential that it cannot be deposited from aqueous media under ordinary conditions. Another interesting example is the higher potential required for bismuth or silver to be deposited on a copper cathode, compared with a platinum one. Kinetic overpotentials are usually absent when the electrode reaction involves merely a change in oxidation state, without any deposition.

The size of kinetic overpotentials depends on various factors. The electrode material and the nature of the electrode surface have a great influence. Commonly, soft metals (like lead and tin, and especially mercury) cause higher kinetic overpotentials than hard metals. The overpotential increases with increasing current density and decreases with increasing temperature.

The values for the kinetic overpotential of hydrogen experienced with various electrode materials are listed as a function of current density in Table I in the Appendix. The influence of the nature of the electrode surface manifests itself clearly in the difference between the overpotentials on smooth and on platinized platinum. The second kind is platinum metal on which finely divided platinum, known as "platinum black," has been deposited. This difference results largely from the greater effective surface area of the platinized electrode and thereby the smaller current density it affords. You are now in a position to appreciate using platinized platinum in a hydrogen electrode. The values given in Table I of the Appendix and analogous data are valid under specified conditions and may change with deviations from those conditions.

28.6 Decomposition Voltage and Kinetic Overvoltage

For cases in which no kinetic overpotentials are experienced (as with cadmium iodide) the decomposition voltage equals essentially the reversible back emf and can be calculated from standard electrode potentials and concentration data by the Nernst equation. This was shown in Section 28.2. When kinetic overpotentials are encountered, these must be considered in addition to the reversible back emf.

An estimate of the decomposition voltage is of interest when an electrolysis is to be performed. Decomposition potentials are especially important when one has to decide which of two or more possible electrode reactions will occur under given conditions. The formulas necessary may be derived as follows.

If the kinetic overvoltage is denoted by the Greek letter eta, η, the relation of the decomposition voltage to the reversible back emf and kinetic overvoltage can be formulated as

$$E_d = E_{\text{rev}} + \eta \qquad (28\text{–}10)$$

Both the decomposition voltage and the kinetic overvoltage consist of anodic and cathodic portions. Thus we can write

$$E_d = E_{d,a} - E_{d,c} \qquad (28\text{–}11)$$

and

$$\eta = \eta_a - \eta_c \qquad (28\text{–}12)$$

When you use a combination of these expressions in calculations, you must pay careful attention to signs, a point the beginner often misses. The anodic kinetic overpotential is always positive. It increases the magnitude of both the anodic decomposition potential $E_{d,a}$ and the decomposition voltage. The cathodic overpotential $E_{d,c}$ is always negative. It acts to make the cathodic decomposition potential more negative, but it also increases the value of the total decomposition voltage (subtraction of a negative number is involved). Thus for the anode

$$E_{d,a} = E_{\text{rev},a} + \eta_a \quad \text{(where } \eta_a \text{ is a positive number)} \qquad (28\text{–}13)$$

and for the cathode

$$E_{d,c} = E_{\text{rev},c} + \eta_c \quad \text{(where } \eta_c \text{ is a negative number)} \qquad (28\text{–}14)$$

When the appropriate expressions are combined, one gets

$$E_d = \overbrace{E_a - E_c}^{E_{\text{rev}}} + \overbrace{\eta_a - \eta_c}^{\eta} \qquad (28\text{–}15)$$

Equation (28–15) allows evaluation of the decomposition voltage. Equation (28–13) or (28–14) can be used for deciding which of several electrode reactions will take place. Of two or more possible anodic reactions, the one with the least positive decomposition potential will occur. Analogously, at the cathode, the reaction with the least negative decomposition potential will occur.

Example 28–1 Consider a solution $1.0 \times 10^{-1} M$ in Co^{2+} and $1.0 \times 10^{-1} M$ in H^+. If this solution is electrolyzed, which will form—hydrogen gas or cobalt metal—

if the cathode is (a) bright platinum? (b) mercury? The cathodic kinetic overpotential of hydrogen gas at $p_{H_2} = 1.01325 \times 10^5$ Pa (1 atm) is –0.09 V on bright platinum and –1.04 V on mercury.

(a) At the platinum cathode, for hydrogen:

$$E_{d,c} = E_c + \eta_c$$

$$= \overbrace{0.00 + \frac{0.059}{2} \log(1.0 \times 10^{-1})^2}^{E_c} \overbrace{- 0.09}^{\eta_c} = -0.15 \text{ V}$$

for cobalt:

$$E_{d,c} = E_c = -0.28 + \frac{0.059}{2} \log(1.0 \times 10^{-1}) = -0.31 \text{ V}$$

Since the formation of hydrogen has the more positive (that is, the less negative) decomposition potential, hydrogen gas will form.

(b) At the mercury cathode, for hydrogen:

$$E_{d,c} = E_c + \eta_c$$

$$= 0.00 + \frac{0.059}{2} \log(1.0 \times 10^{-1})^2 - 1.04 = -1.10 \text{ V}$$

for cobalt:

$$E_{d,c} = -0.31 \text{ V, as in part (a)}$$

The exceptionally high kinetic overpotential of hydrogen on mercury is important for analytical chemistry. The mercury cathode allows the separation of many metal ions from aqueous solution that cannot be plated on a platinum electrode. Further, the wide applicability of the dropping mercury electrode in polarography (Chapter 29) stems in part from the high overpotential of hydrogen on mercury.

The kinetic overvoltage can never be used to supply electrical energy; it acts only as a hindrance to the electrolysis current and does not contribute to the back emf. If an electronic voltmeter (or a potentiometer) is connected across the electrodes of an electrolysis cell during an electrolysis, the applied voltage will be indicated on the meter. If the external voltage source is then disconnected from the electrodes, the indicated voltage will immediately decrease by the sum of the kinetic overvoltage and *IR* drop. The remaining voltage will be a composite of the reversible back emf and the concentration overvoltage. However, the concentration overvoltage will quickly vanish as the concentration gradients around the electrodes are eliminated by diffusion, and only the reversible back emf will be indicated on the meter.

28.7 *Estimating Applied Voltage Needed for Electrolysis*

To estimate the applied voltage necessary for an electrolysis, the following parameters should chiefly be considered: concentrations of all species involved in the electrode reactions, *IR* drop across the cell, temperature, concentration overvoltage, kinetic overvoltage, pH, and presence of complexing agents. The influences of these parameters, except for concentration overvoltage, can be evaluated with an adequate degree of reliability by calculation or can be taken from tabulated data. Estimating the concentration overvoltage, however, is difficult because of the unknown concentrations of the species at the electrode surfaces. For this reason, concentration overvoltage is usually omitted in estimating the applied voltage; the hope is that vigorous stirring during the electrolysis and possibly warming of the solution will decrease it to an almost insignificant value. The value for the applied voltage thus obtained, although not exact, is still a valuable guide in establishing the actual electrolysis conditions. From considerations and formulas given earlier, the equation used in the calculation is

$$E_{\text{appl}} = E_d + IR + (E_{\text{concn. overv.}})$$

$$= E_a - E_c + \eta_a - \eta_c + IR + (E_{\text{concn. overv.}}) \qquad (28\text{–}16)$$

Example 28–2 What applied voltage is needed to electrolyze $5.0 \times 10^{-3} F$ H_2SO_4 between bright platinum electrodes at a current of 2.0 A? The cell resistance is 0.20 Ω and the electrolyte is well stirred. Assume that η_c for H_2 is –0.44 V, η_a for O_2 is +0.40 V, and $p_{H_2} = p_{O_2} = 1.01325 \times 10^5$ Pa (1 atm).

Substituting the data in equation (28–16) yields

$$E_{\text{appl}} = 1.23 + \frac{0.059}{2} \log(1.0 \times 10^{-2})^2 - 0.00$$

$$- \frac{0.059}{2} \log(1.0 \times 10^{-2})^2 + 0.40 - (-0.44) + 2.0 \times 0.20$$

$$= +2.47 \cong 2.5 \text{ V}$$

Example 28–3 What applied voltage is needed to plate copper from a solution $1.0 \times 10^{-1} M$ in Cu^{2+} and $1.0 \times 10^{-1} M$ in H^+ between bright platinum electrodes at a current of 2.0 A? The cell resistance is 0.30 Ω. Assume that η_a for O_2 is +0.40 V and $p_{O_2} = 1.01325 \times 10^5$ Pa (1 atm).

Substituting in equation (28–16) yields

$$E_{\text{appl}} = +1.23 + \frac{0.059}{2} \log(1.0 \times 10^{-1})^2 - 0.34$$

$$- \frac{0.059}{2} \log(1.0 \times 10^{-1}) + 0.40 - 0 + 2.0 \times 0.30$$

$$= +1.86 \cong 1.9 \text{ V}$$

28.8 Formal Potentials

The concentration term for a metal ion in the Nernst equation relates to the free metal ion, that is, to the aquo complex. In many practical cases, however, the ion is not present as such and then erroneous values for potentials are calculated from the unmodified Nernst equation. For example, the metal ion may be partially hydrolyzed; that is, it forms a hydroxo complex or may form complexes with a complexing agent present in the solution. Similar reasoning holds for nonmetallic species. The changes in potential caused by complex formation can be substantial, as the following example illustrates.

Example 28–4 A solution $1.0 \times 10^{-1} M$ in Cu^{2+} and buffered to pH 5 is electrolyzed. Calculate the cathodic decomposition potential of copper with and without the solution being made $1.0F$ in EDTA. The conditional stability constant of the Cu–EDTA complex at pH 5 is 2.0×10^{12}.

Since copper shows no kinetic overpotential, the cathodic decomposition potential equals the electrode potential and in the absence of EDTA is given by

$$E_{d,c} = E_c = +0.34 + \frac{0.059}{2} \log(1.0 \times 10^{-1}) = +0.31 \text{ V}$$

With EDTA present, the calculation takes the following form. The conditional stability constant of the copper(II)-EDTA complex after equation (20–11) is given by

$$K_{st}^{H} = \frac{[CuY^{2-}]}{[Cu^{2+}][Y]^*} = 2.0 \times 10^{12}$$

where $[Y]^*$ stands for the molar concentration of the EDTA uncombined with the copper and in whatever stage of protonation it persists. Substituting the numerical data yields

$$\frac{0.10 - [Cu^{2+}]}{[Cu^{2+}](1.0 - 0.1)} = 2.0 \times 10^{12}$$

Neglecting the term $[Cu^{2+}]$ in the numerator yields

$$[Cu^{2+}] = \frac{0.10}{0.9 \times 2.0 \times 10^{12}} = 5.6 \times 10^{-14} M$$

Inserting into the Nernst equation gives the decomposition potential

$$E_{d,c} = +0.34 + \frac{0.059}{2} \log(5.6 \times 10^{-14}) = -0.05 \text{ V}$$

Thus the addition of the complexing agent decreases the cathodic decomposition potential by +0.31 - (-0.05) = 0.36 V.

Example 28-4 represents a simple case. All data necessary (that is, stability constant of the complex, pH of the solution, and concentration of the complexing agent) were known, and the processes taking place are simple and known unambiguously. Things are seldom so favorable in practice, however. Even if all data required are available, the calculations can often be involved. Frequently, highly concentrated solutions are encountered for which consideration of activity coefficients would be needed if any degree of reliability were expected from the calculated data. Often, however, the complex species and equilibria involved are unknown, or exact data about them are unavailable. Then, escape from the difficulties is possible by using the formal potential E^f of the redox couple. Instead of the molar concentrations of the various species involved, the formal concentrations are used. Consider, for example, the tin(IV)-tin(II) half-reaction in a hydrochloric acid medium. The tin(IV) is definitely not present as the species Sn^{4+}. It is known that tin(IV) tends strongly to hydrolyze and also forms chloro complexes of the composition $SnCl_n^{4-n}$, where n may be any and all numbers between 1 and 6, depending on the chloride ion concentration. To make matters even worse, polynuclear species can also exist in such a solution. Tin(II) is capable of similar behavior, although to a lesser extent. To bypass the difficulties arising from this complicated situation and to allow possibilities for calculations useful to practical applications, the various forms in which tin(IV) and tin(II) may be present are neglected and the formal concentrations of the metal in the two oxidation states are used. The Nernst equation is written

$$E = E^f + \frac{0.059}{2} \log \frac{C_{\text{Sn(IV)}}}{C_{\text{Sn(II)}}} \tag{28-17}$$

The formal potential strongly depends on the solution conditions, including the acidity or basicity of the solution, the concentration of the complexing agent, and the temperature. The value of a formal potential must therefore be given along with such relevant information; in any calculations, sensible results only if the value is applied in a case corresponding to identical or at least closely similar conditions. For example, the formal potential for the tin(IV)-tin(II) couple can be given as E^f = +0.14 V (25°C, 1F HCl). Note that the standard potential of this couple is -0.140 V. Note also the fine difference between tin(IV) and Sn^{4+}, a point you may have either missed so far or not appreciated fully. Calculations involving formal potentials are not different from calculations with standard potentials, and a single example will suffice to make the point.

Example 28-5 What will be the potential of a half-cell obtained by dissolving 7.9 g of tin(IV) chloride pentahydrate (*350.6*) in 150 mL of a 0.010F solution of tin(II) in 1F hydrochloric acid? $E^f_{\text{Sn(IV)/Sn(II)}}$ = +0.14 V (25°C, 1F HCl).

In a consolidated calculation,

$$E = +0.14 + \frac{0.059}{2} \log \frac{(7.9 \times 10^3)/350.6}{150 \times 0.010}$$

$$= +0.14 + \frac{0.059}{2} \log 15 = 0.19 \text{ V}$$

The advantages of the concept of formal potential are obvious; unfortunately many data must be established and tabulated because many values for each system are needed.

28.9 Questions

28-1 Does the decomposition voltage depend on concentration overpotential?

28-2 In your own words distinguish concentration overpotential and kinetic overpotential.

28-3 For an electrolysis cell, define the terms *anode, cathode, reversible back emf, IR drop,* and *applied voltage.*

28-4 When does the back emf arise if two identical platinum electrodes are placed into the electrolyte of an electrolysis cell?

28-5 What is a polarized electrode? Are the electrodes polarized when electrolysis is in progress?

28-6 Does any electrolysis occur below the decomposition potential?

28-7 In what types of systems should the presence of high kinetic overvoltage be suspected?

28.10 Problems

28-1 Two electrolysis cells, *A* and *B*, are filled with a solution $1.0 \times 10^{-3}F$ in $AgNO_3$ and $1.0 \times 10^{-1}F$ in HNO_3. Cell *A* has a silver cathode and a platinum anode; cell *B* has a silver anode and a platinum cathode. The following data are given:

$$\tfrac{1}{2}O_2 + 2H^+ + 2e \rightleftharpoons H_2O \qquad E^0 = +1.23 \text{ V}$$

$$2H^+ + 2e \rightleftharpoons H_2 \qquad E^0 = 0.00 \text{ V}$$

$$Ag^+ + e \rightleftharpoons Ag \qquad E^0 = +0.800 \text{ V}$$

The anodic oxygen overpotential is +0.40 V; the cathodic hydrogen overpotential is −0.09 V on platinum and −0.46 V on silver. Assume $p_{O_2} = p_{H_2} = 1.01325 \times 10^5$ Pa (1 atm). (a) Write the equation for the reactions at the anode and the cathode of the two cells, using a single arrow to indicate the direction. (b) Calculate the decomposition voltage for each cell. (c) If the electrolysis in the two cells is performed at an applied voltage of 1.00 V and the resistance of each cell is 2.0 Ω, what is the cell current in each cell? Neglect concentration overpotentials.

Ans.: (a) Cell *A*: $H_2O \rightarrow 2H^+ + \frac{1}{2}O_2 + 2e$ at anode; $Ag^+ + e \rightarrow Ag$ at cathode; cell *B*: $Ag \rightarrow Ag^+ + e$ at anode; $Ag^+ + e \rightarrow Ag$ at cathode. (b) For cell *A*, $E_d = +0.95$ V; for cell *B*, $E_d = 0.00$ V. (c) For cell *A*, $I = 0.025$ A; for cell *B*, $I = 0.50$ A.

28-2 Four electrolysis cells are filled with a solution 1.0*F* in $CuSO_4$ and 0.050*F* in H_2SO_4. Cell *A* has two platinum electrodes; cell *B*, two copper ones; cell *C*, a copper anode and a platinum cathode; and cell *D*, a platinum anode and a copper cathode. For each cell, write the equation for the electrochemical reaction at each electrode, using a single arrow to indicate the direction, and calculate the decomposition voltage. Assume that no kinetic overvoltage exists and that $p_{O_2} = p_{H_2} = 1.01325 \times 10^5$ Pa (1 atm). $E^0_{Cu^{2+}/Cu} = +0.34$ V and $E^0_{O_2/H_2O} = +1.23$ V.

Ans.: E_d values: +0.89, 0.00, 0.00 and +0.89 V

28-3 Calculate the voltage necessary to electrolyze 0.50*F* H_2SO_4 using a copper cathode and a platinum anode if the electrolysis cell has a resistance of 1.0 Ω and you want to pass a current of 0.10 A through the cell. The anodic overpotential of oxygen on platinum is +0.40 V and the cathodic overpotential of hydrogen on copper is −0.82 V. Assume $p_{H_3} = p_{O_3} = 1.01325 \times 10^5$ Pa (1 atm). The value E^0 for oxygen in acidic solution is +1.23 V.

Ans.: $E_{appl} = +2.55$ V

28-4 Calculate the decomposition potential for the formation of hydrogen from a solution of pH 10.0 when the cathodic kinetic overpotential is −0.40 V.

28-5 Two electrolysis cells are filled with a solution $1.5 \times 10^{-2}F$ in $AgNO_3$ and $1.0 \times 10^{-2}F$ in HNO_3. Cell *A* has a silver cathode and a platinum anode and cell *B* has a silver anode and a platinum cathode. The overpotential for H_2 is −0.09 V on platinum and −0.46 V on silver. (a) Calculate the decomposition voltage for each cell. (b) Write the electrode reactions in the direction in which they occur. $E^0_{Ag^+/Ag} = +0.799$ V, $E^0_{O_2/H_2O} = 1.229$ V.

28-6 Two electrolysis cells are filled with a solution 1.00*F* in $CuSO_4$ and 0.05*F* in H_2SO_4. Cell *A* has a copper anode and a platinum cathode; cell *B* has a platinum anode and a copper cathode. (a) Calculate the decomposition voltage for each cell if the O_2 overpotential is +0.40 V on either metal and the H_2 overpotential is −0.09 V on platinum and −0.60 V on copper. (b) Write the electrode reactions in the direction in which they go. $E^0_{Cu^{2+}/Cu} = 0.337$ V, $E^0_{O_2/H_2O} = +1.229$ V.

28-7 Two electrolysis cells are filled with a solution $1.0 \times 10^{-3}F$ in $CuSO_4$ and $1.0 \times 10^{-1}F$ in H_2SO_4. Cell *A* has a copper anode and a platinum cathode; cell *B* has a platinum anode and a copper cathode. The overpotential for hydrogen is −0.09 V on platinum and −0.60 V on copper. The overpotential of oxygen is +0.40 V on either metal. (a) Calculate the decomposition voltage for each cell. (b) Write the electrode reactions in the direction in which they go. $E^0_{Cu^{2+}/Cu} = +0.337$ V, $E^0_{O_2/H_2O} = +1.229$ V.

29

Polarography and Amperometric Titrations

29.1 Introduction

The relation between the voltage applied to an electrolysis cell and the current flowing through the cell was considered in Chapter 28 and presented graphically in Fig. 28-2. As you can see from that figure, the current increases sharply once the decomposition potential of the reacting substance is reached; later, because concentration overpotential becomes established, the voltage-current curve levels to a plateau. The current at the plateau is called the limiting current and is chiefly controlled by the rate at which the relevant reacting species reach the electrodes. A further increase in the applied voltage does not increase the current significantly until the decomposition potential of some other substance is reached, eventually the value for water.

Assume that the electrolysis cell contains a cathode that is small compared with the anode. In such a case, the concentration overpotential at the anode is negligible and the current is controlled by the cathodic concentration overpotential, which in turn is a function of the rate at which reducible species are transported to the cathode. If the small electrode were made the anode, the current would be controlled by the transport of oxidizable species.

If in such a cell the solution is not stirred mechanically, the mechanisms that transport ions toward the electrode are several, including diffusion, electromigration (see Section 29.4), and convection.* If migration and convection are made negligibly

*As an electrode reaction occurs, the concentration of the solution in the region of the electrode changes and consequently the density of the solution around the electrode. A density gradient becomes established and results in convection, which tends to stir the solution. The term *convection,* here and elsewhere, is restricted to mean "natural" convection. Forced convection is termed stirring, that is, agitation of the solution by a glass rod, magnetic stirring bar, shaking, and the like.

small, diffusion is the only effective mechanism by which reducible species are transported to the cathode. Under this condition, since the rate of diffusion is proportional to the concentration of the diffusing species, it should be possible to establish a relation between limiting current and concentration, and to use this relation as the basis for an analytical determination.

29.2 Polarography

Early attempts to use the principles just outlined for analytical purposes failed. With the solid electrodes used, distorted curves were obtained, which were not reproducible, since the shape strongly depended on the history, that is, the pretreatment, of the electrode. In 1922, Heyrovský succeeded in getting reproducible results by using a small electrode consisting of mercury drops formed at the immersed tip of a fine capillary. He named the technique polarography because it is based on "polarization" phenomena.

If to a cell consisting of a dropping mercury electrode and an unpolarizable half-cell, a steadily increasing voltage is applied and the current corresponding to each momentary voltage value is measured and plotted versus the applied voltage, a graph is obtained that is called a polarogram. From a polarogram, it is possible to make qualitative deductions about the species reacting and from the relation between the current and concentration, quantitative results can be obtained. Polarography is especially valuable for determining small amounts of reducible or oxidizable species; the process operates best at concentrations between 10^{-2} and $10^{-5}M$.

29.3 Polarograph

The voltage-current measurements that produce a polarogram are made with an apparatus called a polarograph, which may either be operated manually or be a recording instrument that directly plots the desired curve. Instruments vary greatly in complexity, but the principles on which their operation is based are identical. The arrangements for a manual polarograph are shown in Fig. 29-1. The polarographic cell PC is a vessel containing the sample solution and is fitted with a stopper containing holes to provide for the insertion of the following items: the inlet N, for passing nitrogen gas through the solution to remove any dissolved oxygen, which commonly interferes with the measurements (see below); the tube E, as an exit for the nitrogen; the capillary tube CT and the salt bridge SB, connecting the polarographic cell to the saturated calomel electrode (SCE), which is the unpolarizable large electrode usually used. The capillary tube CT is connected with the mercury-containing reservoir MR by the flexible tubing FT, which is rubber or plastic. A wire dipping into the mercury makes electrical connection between the mercury electrode and the rest of the circuit. The working battery WB is connected by a variable resistance VR to the slidewire resistance SW. Thus the voltage applied

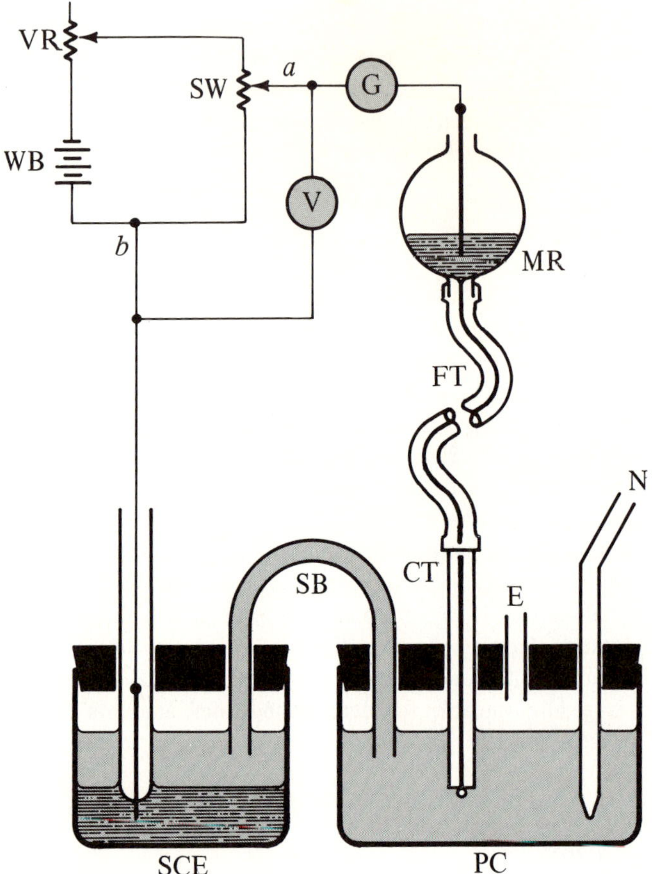

Fig. 29-1 Circuit for manually operated polarograph (WB, working battery; VR, variable resistance; V, voltmeter; SW, uniform slide-wire resistance; G, galvanometer; MR, mercury reservoir; FT, flexible tubing; CT, capillary tubing; PC, polarographic cell; SCE, saturated calomel electrode; SB, salt bridge; N, nitrogen inlet; E, nitrogen outlet.

across the slide wire can be regulated by the variable resistor and is measured by the voltmeter V. If the voltage across the slide wire is known and the wire is calibrated, the voltage taken across the contact points *a* and *b* and applied to the cell assembly is known. The current flowing through the cell assembly is measured by the galvanometer G, which may be calibrated in microamperes.

A polarogram, then can be established as follows. The sample solution is placed into the polarographic vessel, which is then attached to the stopper. Nitrogen is passed through the solution until oxygen is swept out. Then the nitrogen flow is stopped and a certain voltage is applied to the cell assembly by setting the tap on the slide wire at the appropriate position. The voltage is plotted versus the corresponding current reading taken on the galvanometer. Then the voltage is adjusted to a different value, and the current is again read and plotted. This process is continued until there are enough points within the desired voltage range to draw the curve.

29.4 Polarogram and Polarographic Wave

The graph resulting from plotting the voltage-current measurements is called a polarogram. Since commonly the SCE is used as the unpolarizable electrode, established practice in polarography is to express all potential values in volts versus this electrode instead of the standard hydrogen electrode. Fig. 29-2 is a typical polarogram. You see that increasing negative potential versus SCE is plotted to the right on the abscissa. The reason for this convention is that the number of reducible species increases as the cathode potential is made more negative.

The rise in the curve is called a polarographic wave. The upper plateau is the limiting current I_{lim}; it corresponds to the situation for which the current is governed by the amount of reducible species arriving at the cathode. (Compare the discussion directed to Fig. 28-2.) There is also a small, but not always negligible current, I_{res}. This current in an actual polarographic "run" is either established by extrapolating the portion before the wave rise or by "running" a polarogram with the species of interest absent, that is, establishing a polarographic "blank." The origin of the residual current is complicated and we shall say only that it results partly from the charging of the interface (acting as a capacitor) of the growing mercury drop and partly from polarographically active impurities in the solution.

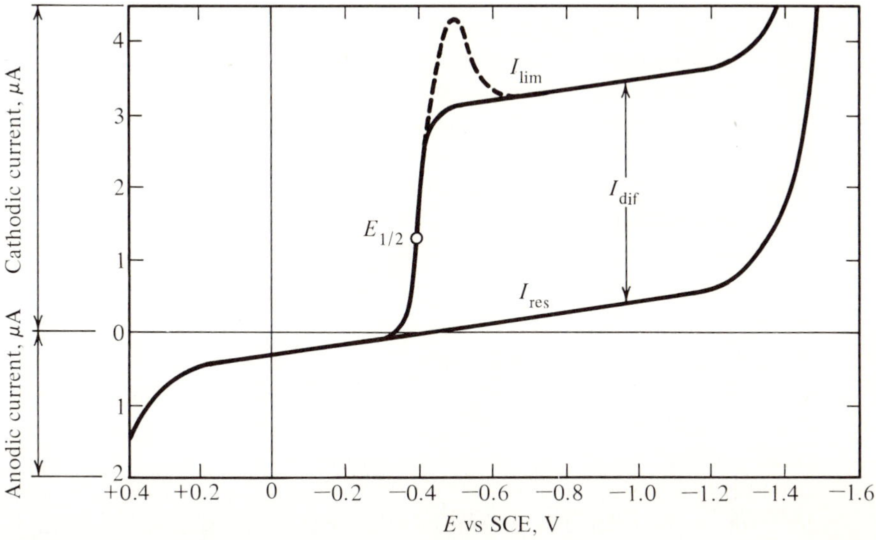

Fig. 29-2 Polarographic wave showing typical relation among residual current, diffusion current, and limiting current. (Polarographic maximum shown as dashed line.)

The difference between the limiting current and the residual current is under certain conditions almost solely governed by the diffusion of the reacting species either toward or away from the electrode. The difference is then called the diffusion current I_{dif}. These conditions will be understood from the following considerations.

When the reducible species is present alone in the solution, the limiting current also contains a further component, the so-called migration current. This current comes from the migration of reducible species to the electrode. Migration is another method of movement besides diffusion. It helps maintain electroneutrality near the electrode as electrode reaction causes charges to be added to the solution or removed from it. Since it is the diffusion current that is to be related to the concentration of the species to be determined, it is necessary to exclude migration or at least make it negligibly small. This can be achieved in the following way. Recall that a current through an electrolytic conductor flows by the motion of charged particles, that is, ions. All ionic species present contribute to this transport according to their number, mobility, and charge. If the only ions present are those undergoing the reactions at the electrodes, the total current is carried by these ions, and they will move to the electrode far more rapidly by this migration than by diffusion alone. In contrast to this, if in the solution a large concentration of ions is established that by themselves are polarographically inactive, then the transport of electricity is accomplished predominantly by the movement of these ions and only a negligibly small migration of the species to be determined takes place. Consequently, polarographic measurements are generally effected in a solution containing a so-called supporting electrolyte, that is, a polarographically inactive salt, often potassium chloride. The concentration of the supporting electrolyte is at least in the order of 0.1 to $1F$, which is large compared with the small concentration of the substances to be determined, usually of the order of 10^{-3} to $10^{-4}F$. The supporting electrolyte should not undergo an electrode reaction, at least not at a potential that is more positive than that at which the species to be determined reacts.

The supporting electrolyte has the additional function of keeping the resistance of the cell solution low (usually 100 to 500 Ω) and thereby minimizing the IR drop across the cell. With solutions too high in electrical resistance, drawn-out and distorted polarographic curves are obtained.

The potential-current relation of a polarographic wave is given by

$$E = E_{1/2} + \frac{0.059}{n} \log \frac{I_{dif} - I}{I} \tag{29-1}$$

where I_{dif} is the diffusion current and I the current at any point of the polarographic wave. The half-wave potential $E_{1/2}$ represents the inflection point of the curve and is the potential at which $I = 1/2 I_{dif}$. It is a characteristic of the species being reduced and independent of the concentration of that species. Equation (29-1) applies only to reversible electrode reactions, and how closely a system follows the equation can be taken as a measure of the degree of reversibility. Furthermore, this equation can be applied only to simple systems, for example, ones in which all the reactants and products are soluble and none of the concentration terms in the Nernst equation is of second power or higher.

Fig. 29-2 has so far been discussed relative to the cathodic portion, that is, the portion of the curve above the potential axis. There is, however, the possibility of making the dropping mercury electrode more and more positive relative to the reference electrode. Then a point will be reached at which the mercury drop becomes the anode and a species of interest can be oxidized. The wave then proceeds in a direction opposite to that in the cathodic situation. The direction of the current also changes. For an analytical determination, this is not of consequence, but for theoretical treatments it is desirable to indicate the direction of the current. This is done by assigning the cathodic current a minus sign (in parallel to the negativity of the cathode) and the anodic current a plus sign. With this convention, Fig. 29-2 would have in place of "cathodic current" and "anodic current" the current values, in microamperes, carrying the appropriate sign.*

29.5 The Ilkovič Equation

The relation between the diffusion current and the parameters pertaining to the system under consideration has been derived by Ilkovič, and the resulting equation is named after him:

$$I_{\text{dif}} = -607nCD^{1/2}m^{2/3}t^{1/6} \qquad \text{(at 25 °C)} \qquad (29\text{-}2)$$

Here I_{dif} is the average diffusion current in microamperes during the life of a drop, n the number of electrons transferred in the electrode reaction, C the concentration of the reacting species in millimoles per liter, and D the diffusion coefficient of the reacting species in square centimeters per second. The remaining two factors are characteristics of the capillary; m is the milligrams of mercury flowing from the capillary per second, and t the drop time in seconds. The numerical factor 607 derives from some constants and conversion factors. A minus sign enters because the derivation is based on a reduction (see Section 29.4 for sign convention). From the Ilkovič equation, we can deduce that the diffusion current is proportional to the concentration of the reacting species if the parameters pertaining to the electrode are kept constant.

29.6 Multiple Polarographic Waves

When more than one reducible species is present in a solution, the polarogram consists of successive waves due to the reduction of each species in turn. However,

*This convention was adopted by IUPAC in 1976. Before then, the opposite sign convention was in use; this opposite convention still exists in the literature. None of the considerations above are affected by signs. Equation (29-1) holds, for example, because not only does the diffusion current become negative for a cathodic reaction, but also the intermediate current I, and the argument for the logarithmic term will still be positive.

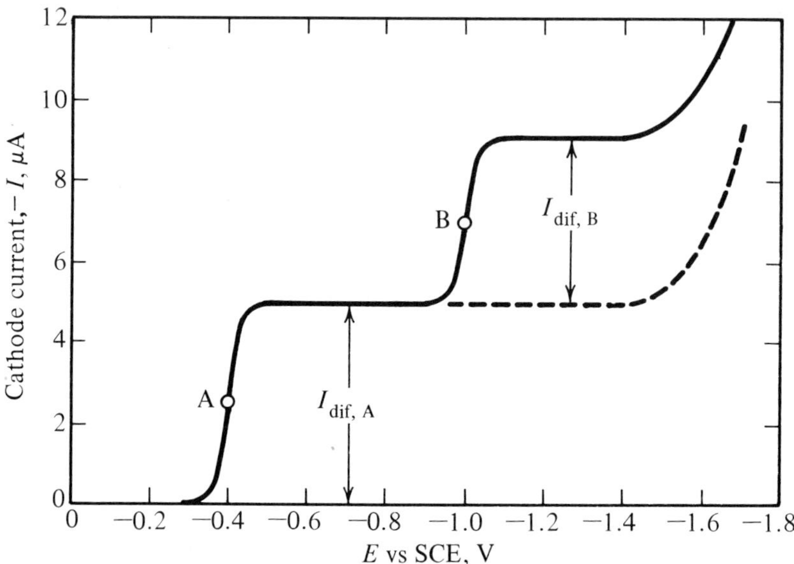

Fig. 29–3 Plot of current versus applied potential for multiple polarographic waves, idealized and simplified by assuming the residual current to be negligible.

well-separated waves are only obtained if the half-wave potentials of the various species differ by at least 0.15 V. Such a situation is illustrated in Fig. 29-3. Points A and B are the half-wave potentials of the two reducible substances. The terms $I_{\text{dif},A}$ and $I_{\text{dif},B}$ represent the corresponding diffusion currents. Applying the equation of the wave calls for measuring I from the base line of that wave, as is done for $I_{\text{dif},B}$. The dashed line indicates the shape of the polarogram that would result if the substance B were absent. The increased current at the end of the polarograph curve corresponds to the reduction of hydrogen ion and represents the beginning of the hydrogen wave.

29.7 Dropping Mercury Electrode

The dropping mercury electrode (DME) is a short length of fine-bore (~0.04-mm inside diameter) glass capillary tubing, connected by flexible tubing to a reservoir of mercury, which can be raised or lowered to adjust the mercury pressure. Each drop, when it reaches a certain size, falls from the capillary tip. The rate at which drops form and fall is determined by the capillary bore and the mercury pressure and is usually adjusted to one drop every 3 to 7 s.

The effective electrode, which is the mercury drop, is continually renewed; therefore, its surface does not become fouled by reaction products as in the case of a fixed electrode. Since all drops form identically and thus have the same history, reproducibility is achieved. The high kinetic overpotential of hydrogen on mercury allows the reduction of many metal ions that otherwise could not be reduced under the prevailing conditions. Many metal ions are reduced to the metal, which dissolves in the mercury; this removal leads to a less negative half-wave potential than what would otherwise be expected. Since the solution is stirred by the fall of the drops, each new drop comes into contact with fresh solution, and changes in the solution density are avoided. Consequently, convection is minimized as a transport mechanism.

One disadvantage of the dropping mercury electrode is the fluctuation in the current with the growth and fall of the drops. The extent of concentration overpotential, and hence of the current, is related to the electrode area. As the drop grows, this area increases and then suddenly becomes quite small as the drop falls; an oscillating current results. The usual technique is to record the average current during the life of the drop.

The region of potential over which the dropping mercury electrode can be used in an aqueous solution extends from about +0.4 to −2 V versus SCE. The positive limit is determined by the potential at which the mercury of the electrode is oxidized anodically. This value may be less positive than +0.4 V if the solution contains an agent forming either a soluble complex or a precipitate with the ions of mercury. The negative limit, as previously noted, is determined by the potential at which hydrogen ion or another component of the supporting electrolyte is reduced.

Study of the potential-current relation at a potential more positive than +0.4 V versus SCE requires that the electrode be fashioned of a noble metal, such as platinum, or another solid material; but then the difficulties mentioned in Section 29.2 must be faced. The situation can be improved by using a small platinum electrode rotating at high speed (600 revolutions per minute).

29.8 Polarographic Maxima

In many cases, an extremely large, erratic current, known as a polarographic maximum, is noticed at the potential at which a wave starts to level off (see Fig. 29-2). The maximum interferes in the measurement of the half-wave potential and the wave height but can be avoided by adding a so-called maximum suppressor to the electrolyte solution. Gelatin (about 0.005% in the electrolyte) and surface-active agents (say, dyes and detergents) are used for this purpose.

29.9 Applied Polarography

The half-wave potential is a characteristic of a reacting species, and under given conditions, is independent of the concentration of a substance. Consequently, its value can be determined for an unknown substance and the substance then identified by comparing the experimentally established value with tabulated values.

In applying polarography to quantitative analysis, one might proceed as follows. The parameters of the electrode are experimentally determined and then the diffusion current of the species is measured. These data and the value for the diffusion coefficient from tables are used together with the Ilkovič equation to calculate the concentration of the sought-for species. Generally, a result thus obtained has an intolerably large error due to the accumulation of small errors in the individual factors. In practice, the factors are combined into a single constant k, so that $I_{dif} = kC$. The value of k is determined under the actual, experimental conditions prevailing from the diffusion current of a solution of exactly known concentration. This value is then used to calculate the concentration of an unknown from its diffusion current measured under identical conditions. Alternatively, a calibration curve of diffusion current versus concentration can be established, using several solutions of known concentrations. Another technique is the so-called method of standard addition, which is best understood from an example.

Example 29–1 Exactly 100 mL of a lead(II) solution (brought to the correct pH and with an appropriate supporting electrolyte added) is introduced into a polarographic cell. Oxygen is removed by bubbling nitrogen gas. The limiting current measured at −0.60 V versus SCE is found to be 12.5 SD (scale divisions). (Calibration of the meter in amperes is unnecessary.) Then a volume of exactly 10 mL of 0.0100F $Pb(NO_3)_2$ is added. After passage of nitrogen gas and establishment of the original temperature, a current of 16.8 SD is observed at the same potential. Under identical conditions, the supporting electrolyte alone yields a current of 0.5 SD. Calculate the concentration of lead in the original solution.

Correcting for the residual current gives 12.5 − 0.5 = 12.0 SD and 16.8 −0.5 = 16.3 SD for the two diffusion currents. Let x = molar concentration of Pb^{2+} in the original solution. Then

$$12.0 = kx$$

After the standard addition

$$16.3 = k\,\frac{100x + 10 \times 0.0100}{110}$$

When k is eliminated by dividing one equation by the other, the result is

$$\frac{12.0}{16.3} = \frac{110x}{100x + 0.100}$$

$$x = 2.02 \times 10^{-3} M\ Pb^{2+}$$

Since diffusion currents change by 1 to 2% per degree celsius, it is important to keep the temperature constant during a series of related polarographic measurements.

A few practical examples taken randomly show the diverse possibilities polarography offers. In a supporting electrolyte 1F in potassium chloride, the ions of cadmium(II), lead(II), and zinc(II) have their half-wave potentials at −0.64, −0.44, and −1.00 V versus SCE, respectively. In the same medium copper(II) exhibits two waves, one at +0.04 and another at −0.22 V versus SCE. The first wave corresponds to the reduction copper(II) → copper(I) and the second to the reaction copper(I) → copper (metal). The waves of the four elements are sufficiently separated to allow their consecutive determination provided that the concentration ratios are not excessively large. Since aluminum is polarographically inactive, the four elements mentioned above can be determined polarographically when present as impurities or minor constituents in pure aluminum metal or aluminum alloys. The large separation of the half-wave potential of cadmium and zinc permits the simple determination of cadmium impurities in pure zinc metal or zinc salts. In these materials it is also possible to determine lead and copper. In an ammonia-ammonium chloride medium, cobalt can be oxidized to cobalt(III), which then gives waves at -0.28 and -1.30 V versus SCE ($Co^{III} \rightarrow Co^{II} \rightarrow Co$). Since in this medium nickel gives a wave at about -1.1 V, the first cobalt wave can be used to determine cobalt in trace amounts in nickel metal or nickel salts, thus solving in simple fashion an otherwise very difficult analytical problem. Unless special conditions are established, the alkali metals, the alkaline earths, aluminum, and a few other elements do not yield polarographic waves. Consequently, it is possible to determine small amounts of polarographically active metals in the presence of extremely high concentrations of the inactive components. Polarography besides being applicable to inorganic analysis, has become progressively important for determining reducible or oxidizable organic substances.

29.10 *Anodic Stripping Analysis*

Anodic stripping analysis (ANA) allows the determination of polarographically active species at extremely low concentrations. The usual technique is a two-step process—*depositing* one or more species at an electrode, and then *stripping* them from that electrode.

In the initial electrolysis, the working electrode is made the *cathode* and commonly involves mercury; in that case the deposited species usually form amalgams. In the stripping step, the potential of that electrode is changed so that it is made the *anode.* Since the operations take several minutes, a dropping mercury electrode is not suitable. Usually either a (stationary) hanging-mercury drop electrode (HMDE) or a thin film of mercury on a carbon surface is used.

The measurement in the stripping step is usually done by recording the current as the potential is swept slowly in the anodic direction. The entire process can then be termed *anodic stripping voltammetry* (ASV).* Some newer and advanced techniques of voltammetry are often used to improve the detection limits. The measurement in the stripping step can be accomplished by other techniques, for example, coulometry (see Chapter 31).

---

*The IUPAC nomenclature for electroanalytical chemistry recommends the term *voltammetry* for current-voltage techniques and restricts the term *polarography* to studies in which a liquid metal electrode is involved with continuous removal of deposited products; the dropping mercury electrode is the prominent example.

The increased sensitivity achieved with anodic stripping analysis is largely due to the initial enrichment of the species. Results are reproducible only if the conditions in the deposition are closely controlled, such as time of electrolysis and rate of stirring. Usually the method of standard addition is used to relate the concentration of a species of interest to the measured current (or other signal).

Anodic stripping analysis is being used more and more especially in environmental trace metal analysis. For example, lead and cadmium can be readily determined in water samples at the ng/g (ppb) level.

29.11 Amperometric Titrations

Closely related to polarography is a titration technique known as the amperometric titration, where polarographic arrangements in connection with a graphical procedure are used to detect the end point.

The method can be exemplified by the titration of cadmium ion with EDTA. Consider a solution of a cadmium salt in an ammonium nitrate-ammonia supporting electrolyte of pH 10. A polarogram of this solution is represented by the uppermost line of Fig. 29-4a. When some standard EDTA solution is added, the soluble cadmium-EDTA complex is formed. The cadmium ion complexed by the EDTA is reduced at a much more negative potential than the ion complexed by water or ammonia. Consequently, the diffusion current, which is proportional to the concentration of the "free" cadmium ion, becomes smaller. The polarogram then obtained is also shown in Fig. 29-4a. As more EDTA is added, the diffusion current decreases further. At the equivalence point, almost all the cadmium is complexed by EDTA, the diffusion current is essentially zero, and the polarogram shows only the residual current. When EDTA is added beyond the equivalence point, no appreciable change is seen in the polarogram. It is not necessary to record a polarogram after each addition of titrant. The current can be read at some point on the plateau of the cadmium wave, for example, at -0.9 V versus SCE. The data obtained are plotted versus the volume of titrant to yield the titration curve shown in Fig. 29-4b.

Generally, experimental curves will depart from strict linearity since the solution volume increases during the titration. This dilution effect can be corrected for by multiplying each reading by the factor $(V + v)/V$, where V is the initial volume of the solution and v is the volume of titrant added to the point of the observation. A plot of such corrected current versus the volume of titrant will consist of two straight lines. Alternatively, the effect of the dilution can be made negligible by using a titrant 10 or more times greater in concentration than the species being titrated. Such concentrated titrant solutions are then delivered from a microburet.

To locate the amperometric end point, one needs to find only the changes in the current. This makes simplification possible. The drop time and the temperature are not important so long as they are constant during a titration. The galvanometer does not need to be calibrated, and the readings in scale divisions can be plotted. The difficulties encountered in polarography with solid electrodes are without moment, and a rotating platinum electrode can be substituted for the dropping mercury electrode, often to advantage, since a larger diffusion current is obtained,

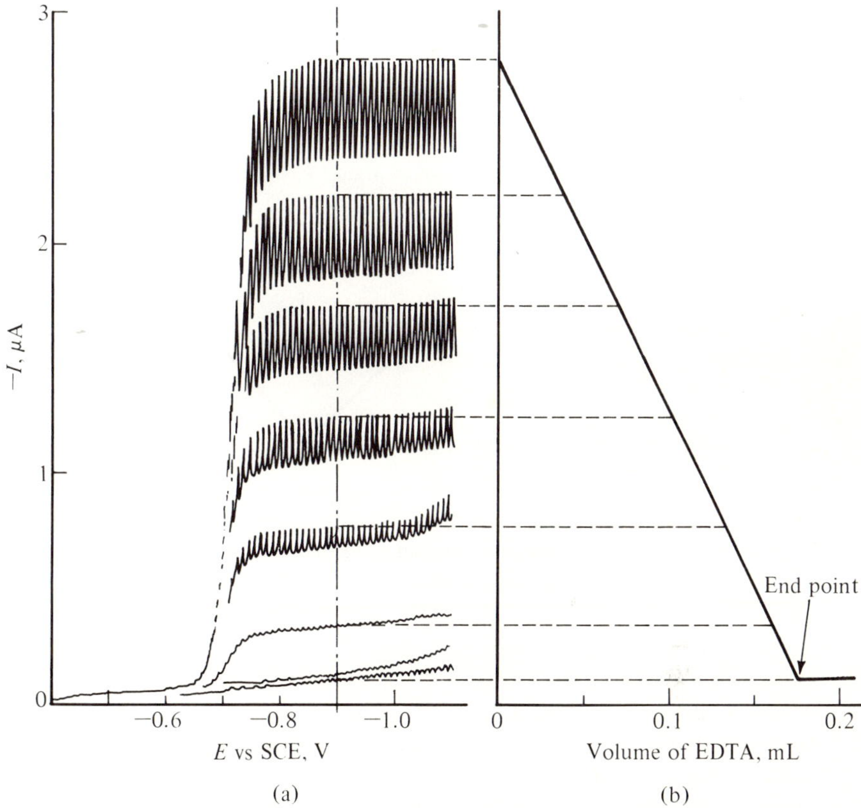

Fig. 29-4 Relation of polarograms (a) for reduction of cadmium in ammoniacal buffer of pH 10 to the amperometric titration curve; (b) for cadmium with EDTA at -0.9 V versus SCE.

and thereby greater sensitivity. The current associated with the species being titrated can be superimposed on waves of substances not responding to the titrant.

The precipitation titration of lead(II) ion with chromate ion to form lead chromate is an example of the effect of half-wave potentials and the selected potential value on the shape of the titration curve obtained. Chromate ion has a half-wave potential about +0.1 V versus SCE and lead(II) ion a potential about -0.4 V versus SCE. Two possibilities for the titration exist; a potential can be selected at which only chromate is reduced (that is, a value on the plateau of the chromate wave, say, -0.2 V versus SCE) or one at which both lead and chromate are reduced (that is, on the plateau of the lead wave, say, -0.6 V versus SCE). Corresponding titration curves at these two potentials are shown in Fig. 29-5 as curves A and B. The shape of curve A derives from the fact that up to the equivalence point no species is present that is reducible at the applied potential of -0.2 V versus SCE. Chromate, although reducible at this potential, when added reacts with lead to form lead chromate, which precipitates, and is no longer available for an electrode

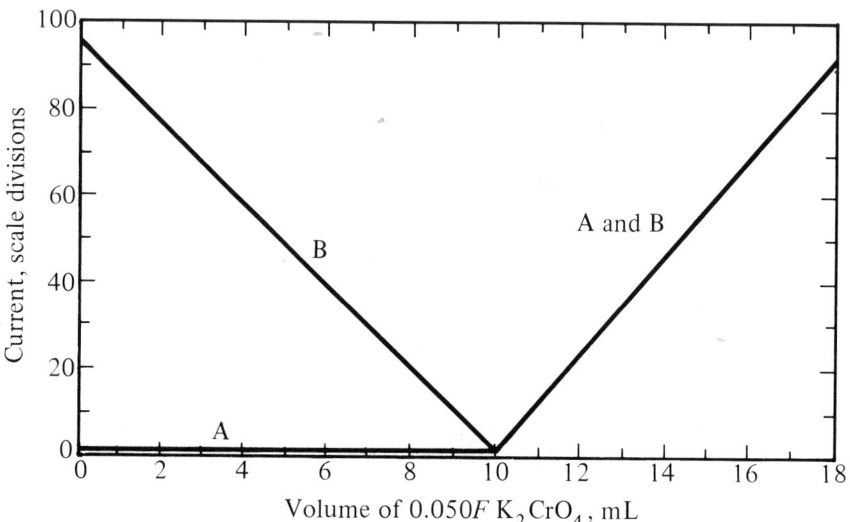

Fig. 29–5 Titration curve for the amperometric titration of 50 mL of 0.01*F* $Pb(NO_3)_2$ with 0.05*F* K_2CrO_4: curve A at −0.20 V versus SCE; curve B at −0.60 V versus SCE.

reaction. Beyond the equivalence point, free, reducible chromate is present and causes the flow of a current that increases as the concentration of chromate increases. For curve B, lead is present from the start and is reduced at the applied potential of −0.6 V versus SCE. As the concentration of lead decreases through its removal as insoluble lead chromate, the current decreases progressively. Beyond the equivalence point the situation is identical to that leading to curve A. For both cruves, the minimum value of the current corresponds to the residual current and a small current caused by the minute amounts of dissolved lead chromate.

29.12 Biamperometric Titrations (Dead-Stop End Point)

Biamperometric titrations are performed with the arrangement shown in Fig. 29–6. Two identical platinum electrodes dip into the solution to be titrated and are connected with a milliammeter and a potentiometer circuit that allows application of a constant dc voltage of the order of a few tenths of a volt. The solution should be stirred well and at a constant rate. The principles involved in such a titration are best illustrated in an example.

Consider the titration of a solution containing iodine and an excess of iodide ion with a thiosulfate solution (Fig. 29–7). At the start of the titration, iodine (present as the triiodide ion) is reduced at the cathode and iodide is oxidized at the anode. Since the reactions at the two polarizable electrodes are the reverse of each other, the value of the decomposition voltage is zero, and therefore a current flows even at the small voltage applied. The electrode reactions occur to an equal extent,

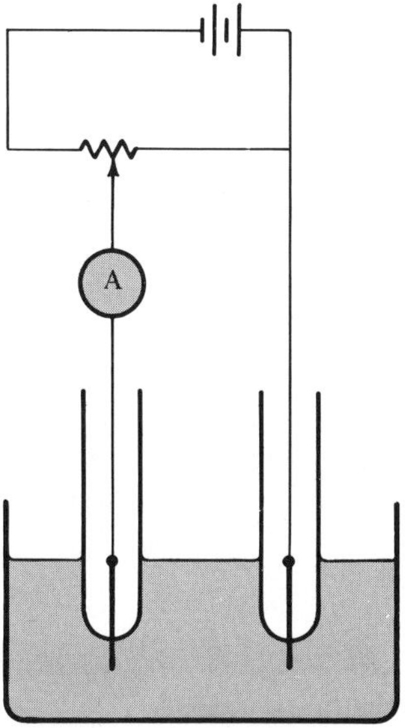

Fig. 29-6 Experimental arrangement for titration with two polarized electrodes.

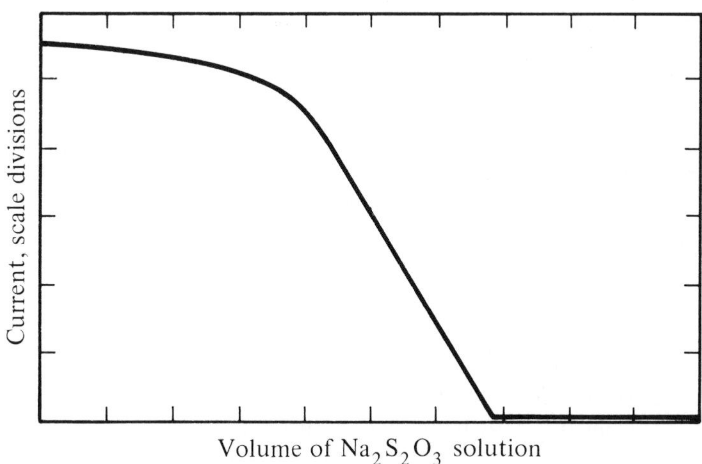

Fig. 29-7 General shape of the curve for the titration of iodine in the presence of an excess of iodide with thiosulfate using two polarizable electrodes.

and the composition of the solution remains unchanged. As thiosulfate is added, the iodine concentration decreases. At first, the change in the current is small. However, as the equivalence point is approached, the current decreases linearly with the volume of thiosulfate solution added. At the equivalence point, the concentration of iodine is essentially zero and practically no current flows, since in the absence of iodine no combination of electrode reactions having a decomposition voltage less than the small applied voltage is possible. Beyond the equivalence point, no current flows, since the irreversible thiosulfate-tetrathionate reaction has a sizable kinetic overpotential and therefore the decomposition voltage is larger than the applied voltage.

Because of the peculiar shape of the titration curve that reaches (or in other cases starts from) a "dead stop" at the end point, the method is also known as a *dead-stop titration.* The technique can be used in numerous titrations in which the following redox couples constitute at least one of the reactant-product pairs: Br_2/Br^-, I_2/I^-, Ti^{4+}/Ti^{3+}, Ce^{4+}/Ce^{3+}, Fe^{3+}/Fe^{2+}, and MnO_4^-/Mn^{2+}. Such titrations offer the advantages of simple circuitry and of a simple electrode system not even requiring a reference electrode. Since only current *changes* are of interest, the meter does not need to be calibrated. The dead-stop technique is often used for detecting the end point in the Karl Fischer titration of water (Section 25.11).

29.13 Questions

29–1 Elaborate on the difference between limiting current and diffusion current.

29–2 What are the advantages of using a mercury electrode in polarography?

29–3 In polarography what are the purposes of adding a supporting electrolyte?

29–4 What is plotted on the abscissa and the ordinate of a polarogram?

29–5 What values obtained from a polarogram can be used for the purpose of (a) quantitative analysis and (b) qualitative analysis?

29–6 What are the possible mechanisms for transporting electricity in an electrolytic cell? Which of these are undesirable in polarography?

29–7 What electrode reaction in aqueous solution limits a polarogram on (a) the negative side and (b) the positive side?

29–8 What happens to the current when the dropping mercury electrode changes from being the cathode and becomes the anode?

29–9 How much difference in the values of the half-wave potentials is generally necessary so that two polarographic waves are clearly distinguishable?

29–10 Discuss the advantages and disadvantages of polarographic determinations compared with amperometric titrations.

29–11 How can adding a complexing agent affect the value of a half-wave potential?

29–12 How does the half-wave potential depend on concentration?

29–13 Why should oxygen be removed in a polarographic determination and how is it done?

29–14 Anodic stripping analysis is sometimes called *inverse voltammetry*. Elaborate on the aptness of this term.

29–15 Sketch the titration curves when CrO_4^{2-} is titrated amperometrically with Pb^{2+} at (a) –0.1 and (b) –0.6 V versus SCE. Compare the curves with those for the reverse titration shown in Fig. 29–6.

29.14 Problems

29-1 Into a polarographic cell are placed successively exactly 50 mL of an ammoniacal buffer (pH 10), exactly 5 mL of a 0.0100*M* cadmium(II) solution, and exactly 5 mL of a solution having an unknown cadmium concentration. Following each addition, nitrogen gas is passed to remove oxygen, and the current measured at −1.0 V versus SCE is found to be 1, 18, and 25 arbitrary scale divisions, respectively. Calculate the molar concentration of cadmium in the unknown solution. *Ans.:* 0.0054*M*

29-2 A 10.0 mL portion of a solution $1.0 \times 10^{-3} M$ in a dipositive metal ion and containing supporting electrolyte is placed in a polarographic cell and deaerated. The polarographic study, in which this cation is reduced to the metal, requires 10 min, during which time the average current passed is 8.0 μA. Calculate the percentage decrease in the concentration of this ion in the cell solution (Faraday constant = 9.65×10^4 C^{mol}).

29-3 A volume of 25.0 mL of a solution containing an unknown amount of lead(II) was scanned polarographically and a diffusion current of 4.5 μA found. A volume of 5.0 mL of $5.0 \times 10^{-3} F$ $Pb(NO_3)_2$ was added to the original solution. The diffusion current was then found to be 9.7 μA. Calculate the concentration of lead in the original solution.

29-4 A volume of 50.0 mL of a $2.0 \times 10^{-3} F$ $Cd(NO_3)_2$ solution was subjected to polarography for 12 min. The average current passed was 9 μA. Calculate the weight of cadmium reduced to the metal and the percentage decrease in concentration.

29-5 A solution of potassium dichromate (*294.19*) is standardized as follows: An amount of 3.9266 g of iron (*55.85*) of 99.85% purity is dissolved in hydrochloric acid and the solution brought with water to exactly 1000 mL. A 25.00-mL portion of this iron(II) solution is titrated with the dichromate solution, of which 5.85 mL is required to reach the end point. This dichromate solution is then used in the precipitation titration of lead (*207.19*) in a well-buffered acetate medium to an amperometric end point. The titration reaction can be expressed as

$$2\,Pb^{2+} + Cr_2O_7^{2-} + H_2O \rightarrow 2\,\underline{PbCrO_4} + 2\,H^+$$

The titration is followed at a potential at which both lead(II) and dichromate ions are polarographically active. For one sample, the following data were obtained (already corrected for dilution):

Dichromate, mL	Current scale divisions
0.00	80.0
0.50	70.0
1.00	59.2
1.50	46.8
2.00	37.4
2.50	28.0
2.75	22.1

Dichromate, mL	Current scale divisions
3.00	17.8
3.25	15.6
3.50	15.0
3.75	15.3
4.00	20.1
4.25	40.0
4.50	59.8
4.75	79.9

(a) Calculate the formality of the dichromate solution. (b) For the titration of lead, plot the data and establish the titrant volume for the end point. (c) Calculate the milligrams of lead present in the sample.

Ans.: (a) 0.0500*F*; (b) 3.78 mL; (c) 78.3 mg

29-6 A sample solution contains lead(II) and cadmium ions. The diffusion current for the two ions can be measured at –0.5 and –0.07 V vs SCE, respectively. At these potentials, the supporting electrolyte alone gives currents of 0.4 and 0.6 scale divisions; a 25.00-mL portion of the sample solution diluted to exactly 50 mL with the supporting electrolyte gives currents of 5.5 and 17.0 scale divisions. After exactly 5 mL of $3.0 \times 10^{-3} F$ cadmium nitrate is added, the current at –0.7 V is 23.2 scale divisions. Calculate the concentration of cadmium in the original sample solution.

Ans.: $5.6 \times 10^{-4} F$

29-7 In an anodic stripping analysis experiment, the working electrode is a hanging-mercury drop with a volume 1.0 μL. The sample placed in the cell is 15.0 mL of $2.0 \times 10^{-7} F$ $Pb(NO_3)_2$. What is the maximum possible concentration of lead (*207*) in the mercury drop after the deposition step, expressed in micrograms per milliliter?

29-8 To determine lead in air carried by particulates, air is pulled thorough a filter at a rate of 20.0 L/min for a period of 2.0 hr. The particulate matter trapped on the filter is dissolved in 10.0 mL of dilute nitric acid, and the solution resulting is analyzed by anodic stripping voltammetry (ASV). The lead peak for this sample was 4.2 cm above the background on the recorder tracing. Then a volume of 10.0 μL of $5.0 \times 10^{-5} F$ $Pb(NO_3)_2$ was added to the cell and the deposition and stripping were repeated. This time the lead peak, corresponding to the sample and addition, was 9.7 cm above the background. Calculate the nanograms of lead (*207*) per cubic meter of air ("as is" basis).

29-9 In anodic stripping analysis, a mercury drop of 100 mg is used to deposit the cadmium from 50 mL of a $1.0 \times 10^{-7} M$ solution. Assume virtually complete deposition and calculate the "enrichment factor," that is, the ratio of the cadmium concentration in the mercury drop to cadmium in the sample solution. The density of mercury is 13.6 g/mL. *Ans.:* 7×10^3

29-10 A preliminary polarogram indicates that a sample solution contains both copper(II) and zinc. The diffusion currents of these two ions can be measured at –0.2 and –1.2 V vs SCE. At these potentials, the supporting electrolyte alone gives a current of 0.3 and 1.9 scale divisions. A 25.00-mL portion of the sample solution is diluted to exactly 50 mL with supporting electrolyte and shows 12.2 and 27.8 scale divisions at the two potentials. After addition of 5.0 mL each of $5.0 \times 10^{-3} F$ solutions of copper(II) nitrate and zinc nitrate, the current readings are 23.7 and 55.4 scale divisions. Calculate the concentration of copper and zinc in the original sample solution. *Ans.:* $7.4 \times 10^{-4} F$ copper; $6.2 \times 10^{-4} F$ zinc

30 Electrogravimetry and Electroseparation

If a product of an electrolytic reaction is plated (that is, is deposited) quantitatively on an electrode, the amount of that product can be established by weighing the electrode (dry) before and after the electrolysis. This kind of quantitative analysis is called electrogravimetric analysis, or electrogravimetry.

During the electrolysis the product, one or more, is separated from the solvent and other components of the solution. Consequently, electrolysis can be performed not only to determine a constituent of a multicomponent mixture but also with the sole intent of achieving a separation. The technique is then called electroseparation. The principles underlying electrolysis have been discussed in preceding chapters. Now we deal with their application to the two techniques mentioned.

30.1 *Requirements for Electrogravimetric Analysis*

For an electrogravimetric procedure to be feasible, the deposition of the substance of interest must be complete, and the deposit must be of known composition and must adhere firmly so that the electrode can be rinsed, dried, and reweighed without losses. In addition, the electrode must be inert; that is, it must not undergo any change in its weight in the electrolysis. All these factors have their counterpart in conventional gravimetric determinations (Chapter 6) and need no elaboration. In practice, obtaining a pure deposit is frequently complicated by the presence of other substances that may be codeposited with the substance of interest. In these cases, the conditions of the electrolysis must be adjusted and controlled so as to secure selectivity.

30.2 *Completeness of an Electrodeposition*

The Nernst equation can be used to calculate the concentration of undeposited sought-for material in the solution and hence to establish the completeness of the electrodeposition. However, this calculation requires knowledge of the electrode potential of the "working electrode," that is, of the electrode at which the process of interest occurs. The potential of this electrode can be measured by comparing it with the potential of a third, reference electrode. For this purpose, a saturated calomel electrode may be connected with the electrolysis cell by a salt bridge and the voltage of the cell consisting of the cathode and reference half-cells be measured by a potentiometer circuit. From this voltage, the potential of the cathode can be computed.

Even so, however, a concentration thus calculated is seldom more than a rough approximation; the reason is that a concentration overpotential has been established by the depletion of ions near the working electrode. This overpotential exists not because of the potential measurement, which is performed with the potentiometer under zero-current condition, but because the cathode is also part of the operating electrolysis cell.

Although the concentration overpotential can be decreased by vigorously stirring the electrolyte, elevating the temperature, and using large electrodes, the concentration calculated corresponds to the concentration around the electrode, and is always smaller than that in the bulk of the solution.

Example 30–1 In the electrogravimetric determination of copper, the reference electrode technique described above is used. The voltage of the cell consisting of a calomel electrode ($E = +0.24$ V) and the cathode of the electrolysis cell is found to be 0.05 V, with the calomel electrode as the positive pole. What is the concentration of copper(II) remaining undeposited in the acidic cell solution? If the initial concentration of copper(II) was $1.0 \times 10^{-2} M$, how complete is the deposition?

Substituting in equation (25–4) yields

$$E_c = +0.24 - 0.05 = +0.19 \text{ V}$$

From the Nernst equation for the copper(II)-copper couple,

$$+0.19 = +0.34 + \frac{0.059}{2} \log [Cu^{2+}]$$

$$\log [Cu^{2+}] = \frac{(0.19 - 0.34) \times 2}{0.059}$$

and

$$[Cu^{2+}] = 10^{-5} M$$

$$\% \text{ undeposited copper} = \frac{10^{-5} \times 10^{2}}{10^{-2}} = 0.1\%$$

Hence, the deposition is 99.9% complete.

30.3 Selectivity of Electrodepositions

For the purposes of electrogravimetry, only a single substance should be deposited at a time. Factors that determine whether a selective electrodeposition can be effected include standard electrode potentials of the species present, concentration of the species, overpotentials, efficiency of stirring, temperature, pH, and presence and concentration of complexing agents.

Consider the electrogravimetric determination of copper in a solution that is $1.0 \times 10^{-1}M$ in copper(II) ion, $1.0 \times 10^{-1}M$ in antimony(III), and $1.0M$ in hydrogen ion. Antimony(III) is present as the antimony ion in an acidic solution and has the following electrode reaction:

$$SbO^+ + 2H^+ + 3e \rightleftharpoons Sb + H_2O \qquad E^0 = +0.21 \text{ V}$$

The Nernst equation for this half-cell takes the form

$$E_c = +0.21 + \frac{0.059}{3} \log [SbO^+][H^+]^2$$

$$= +0.21 + 0.020 \log 10^{-1} = +0.19 \text{ V}$$

The potential of the copper(II)-copper half-cell is given by

$$E_c = +0.34 + \frac{0.059}{2} \log [Cu^{2+}]$$

According to the last equation (substituting $[Cu^{2+}] = 1.0 \times 10^{-1}$), copper starts to deposit on the cathode when the cathode is at +0.31 V. As the electrolysis proceeds, the concentration of copper(II) ion in the cell solution decreases and the cathode must be made more negative (that is, less positive) for the electrolysis to continue. When the concentration of copper ion falls to $1.0 \times 10^{-2}M$, the cathode has to be at least at +0.28 V; similarly, at a copper concentration of $1.0 \times 10^{-4}M$, the potential has to be at least +0.22 V. These calculations, however, neglect concentration overpotential. At current densities met in practice, the copper(II) ion concentration near the cathode may be several orders of magnitude lower than in the bulk of the solution. As a consequence, the cathode potential will be substantially lower than the potential calculated from the bulk concentration and may even be as low as +0.18 V. This is already 0.01 V lower than the potential calculated for the antimony(III)/antimony couple; hence, antimony will start to deposit with the copper. Efficient stirring will decrease the concentration overpotential greatly, but some overpotential will persist, and thus it will be impossible to deposit all the copper in a pure form.

In contrast, the difference in the values of the standard electrode potentials of copper(II) and nickel(II) is so large that copper can be deposited quantitatively before nickel begins to plate. As a consequence, copper can be determined even in the presence of considerable amounts of nickel, as the following example shows.

Example 30–2 Assume that an electrolyte solution is 0.10*F* in all three solutes: copper(II) sulfate, nickel(II) sulfate, and sulfuric acid. What will be the copper concentration in the solution when nickel is just starting to plate? $E^0_{Cu^{2+}/Cu} = +0.34$ V, $E^0_{Ni^{2+}/Ni} = -0.25$ V.

The potential at which nickel starts to plate is given by

$$E_c = -0.25 + \frac{0.059}{2} \log (1.0 \times 10^{-1})$$

This potential, by the statement of the problem, must be the same as the potential of the copper(II) half-cell, and the two Nernst equations can be equated:

$$-0.25 + \frac{0.059}{2} \log (1.0 \times 10^{-1}) = +0.34 + \frac{0.059}{2} \log [Cu^{2+}]$$

$$[Cu^{2+}] = \sim 10^{-21} M \text{ when Ni begins to plate}$$

The greater the difference in the electrode potentials of two metals, the more complete the separation that can be achieved by electrolysis. Actually, the formal potentials of the redox systems under the prevailing solution conditions are the ones that govern the possibility of an electroseparation. Thus the addition of complexing agents can greatly influence the selectivity (Section 24.10). For example, antimony(III) forms a far more stable tartrate complex in weakly acidic solution than copper(II); adding tartrate causes a greater shift to negative potentials for the antimony(III) and thus creates more favorable conditions for the selective deposition of copper in the presence of antimony.

30.4 Techniques of Electrogravimetry

Electrogravimetry embraces two principal techniques. They are electrolysis at constant (or controlled) current and electrolysis at constant (or controlled) potential. The former method, which has long been routine, involves setting the *applied voltage* to give a desired current, which is maintained during the electrolysis by occasional manual adjustments. In the latter method, which has received extensive attention only in recent decades, the *potential of the working electrode* is controlled, that is, kept constant at an appropriate value, preferably by an electronic device.

30.5 Constant-Current Electrolysis

The name *constant-current electrolysis* stems from the fact that the current is adjusted at the start of the electrolysis and then maintained roughly at that level throughout the electrolysis. The usual electrolysis arrangement is shown

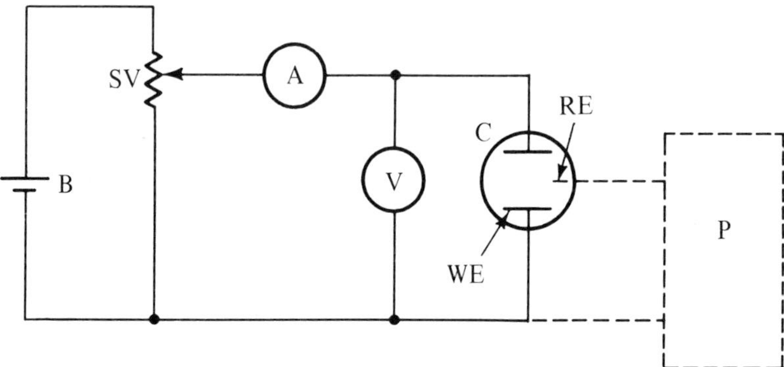

Fig. 30–1. Schematic diagram of assembly for controlled-current and controlled-potential electrolysis. A, ammeter; B, battery or other voltage source; C, electrolysis cell; WE, working electrode; SV, variable resistance to adjust applied voltage. Controlled-potential electrodeposition additionally uses the following (dashed lines): RE, reference electrode; P, potentiometer circuit (for example, that shown in Fig. 22–2).

schematically in Fig. 30-1. The mode of its function is already known from previous discussions. The constant-current technique is widely applied to the routine determination of components in many important alloys and ores. Most metals are deposited cathodically; lead in some cases is deposited anodically as lead(IV) oxide. Depending on the composition of the mixture and on which metal is to be determined, acidic, neutral, or alkaline solutions can be used, and complexing agents such as tartrate, cyanide, or EDTA can be added. Constant-current electrolysis has low selectivity; the reason for this shortcoming will become obvious from the discussion of an example, the determination of copper. At the start of the electrolysis, the current is adjusted to a certain level. As the electrolysis proceeds, the concentration of copper decreases progressively; and with less copper available, the current drops. For a while the original level can be restored by increasing the applied voltage. Sooner or later, however, such an increase will no longer be effective in bringing more copper ions to the cathode; then the level of the current can be maintained only by forcing another electrode reaction to take place. If another reducible metal is present, it will plate and the selective determination of the copper will become impossible. If no such metal is present, hydrogen ion will be reduced to hydrogen gas. This gas evolution may seem to present no problem because the hydrogen does not plate and thus cannot affect the weight of the copper deposit. However, the concurrent reduction of copper and hydrogen ions causes a flaky, poorly adherent copper deposit that may lead to losses of copper. Similar considerations hold in other cases and are valid analogously for a deposition on the anode. The situation can be improved by adequately stirring the solution, increasing the temperature, and lowering the current density using electrodes with large surface areas. But these factors are subject to limitations.

It thus becomes apparent that it is not the current that should be controlled but the potential of the working electrode. This potential, in case of a cathodic deposition, should be as negative as possible, to assure as large an electrolysis current as feasible; at the same time, it should be limited to such a value as to exclude an undesired electrode reaction. This limiting of the potential can be achieved by adding a depolarizer (Section 30.6) or by controlling the electrode potential (Section 30.7) instead of the cell current.

30.6 *Depolarizers*

A depolarizer is a substance that undergoes a reaction at a certain electrode potential and thereby limits the potential of the working electrode. The depolarizer acts as a "potential buffer" and prevents undesired electrode reactions. For example, the reduction of hydrogen ion to hydrogen gas during the electrodeposition of copper can be avoided by adding nitric acid as the depolarizer. The nitrate ion is reduced to nitrite ion ($NO_3^- + 2H^+ + 2e \rightarrow NO_2^- + H_2O$) or ammonium ion ($NO_3^- + 10H^+ + 8e \rightarrow NH_4^+ + 3H_2O$) at a potential more positive than that at which the discharge of hydrogen occurs. As long as the supply of nitrate ion lasts, the cathode potential is limited by the reduction of nitrate and little or no hydrogen gas forms. Nitrate ion here functions as a cathodic depolarizer.

Likewise, hydrazine, which is oxidized to nitrogen in acidic or neutral media, can be used as an anodic depolarizer, for example, in the deposition of lead(IV) oxide, to prevent the evolution of oxygen or the deposition of undesired metal oxides.

30.7 *Controlled-Potential Electrolysis*

Increased selectivity for an electrodeposition can be obtained by controlling the potential of the working electrode. An experimental arrangement is shown schematically in Fig. 30-1. The slide wire is adjusted so that the potential of the working electrode has the appropriate value. The manual adjustment is tedious, and in routine practice an electronic controlling device, known as a potentiostat, is used. This improvement increases markedly both the complexity of the circuitry and the cost of the apparatus.

If the electrode potential is carefully selected and controlled, electrogravimetric determinations can be achieved that are impossible with controlled-current electrolysis. A metal, for example, can be determined in the presence of a second metal when the two *formal* potentials differ by only about 0.2 to 0.3 V. Weighing the electrode (dried) between each change in potential allows two or three components of an alloy to be determined consecutively. Changing the pH, adding complexing agents, or both, between steps may further improve the selectivity.

30.8 *Electroseparation and the Mercury Cathode*

In the foregoing considerations, the final goal of the electrolysis was to single out one element at a time for the purpose of its immediate determination.

By the nature of the process, the element was thereby separated from the solution and from any other components present in the solution. Electrolysis can also be applied with the sole purpose of effecting a separation. It has the advantage, for example, over separation by precipitation or extraction that no reagents are introduced that may interfere in a later analytical procedure. After the electrolytic separation, the substances remaining in solution can be determined by any appropriate method. The deposited metals can be redissolved and then determined in a suitable way. Although platinum or other electrodes used in electrogravimetry can be used, the mercury electrode has received special attention.

A mercury cathode is of little value in electrogravimetry because of the inconvenience of weighing a liquid, and because of the volatility and high density of mercury. These factors are of no moment in electroseparations. Mercury is of special usefulness because of the high kinetic overpotential of hydrogen on this metal (Section 28.7). It offers the additional advantage that many metals are soluble in mercury through amalgam formation. In this way, many more metals can be separated at a mercury cathode. An important use of the mercury cathode is to separate large amounts of iron before the determination of aluminum or alkaline earth metals. Where the metals amalgamated with the mercury must be recovered, the electrolysis vessel can be fitted with a drain tube and stopcock. The controlled-current electrolysis technique can be used, with adequate stirring of the solution. Greater selectivity can be secured by the controlled-potential technique. In this case, polarographic measurements can be used to special advantage to ascertain the potentials most suitable for selective separations.

30.9 Questions

30–1 Explain why the mercury cathode has little value in electrogravimetric analysis but great value in electroseparations.

30–2 Why is it impossible to accomplish the electrogravimetric determination of metals such as sodium and barium?

30–3 Why is it necessary, in most cases, to use a third electrode as the reference when determining the potential of the working electrode?

30–4 Will the actual potential of the working electrode, as measured against a third electrode, be identical with that calculated by the Nernst equation, using the concentration in the bulk of the solution? Explain.

30–5 Why is the reduction of hydrogen ion to hydrogen gas to be avoided in the electrogravimetric determination of copper?

30.10 Problems

30–1 Two metal ions, one dispositive and one tripositive, are to be separated by electrolysis. It is desired to deposit the dipositive cation until its solution concentration is $1.0 \times 10^{-6} M$. If the dipositive ion has a standard potential

0.15 V more positive than that of the tripositive ion, what is the maximum concentration of the tripositive ion allowed in the solution before it starts to deposit with the dipositive metal? Assume that there is no overpotential for the deposition of either metal.

Ans.: $3 \times 10^{-2} M$

30–2 Calculate the minimum difference in standard potentials necessary so that the concentration of a unipositive metal cation can be decreased by electrolysis to $1.0 \times 10^{-6} M$ without deposition of any of a dipositive metal cation that is present at a concentration of $1.0 \times 10^{-2} M$. Assume that there is no overpotential for the deposition of either metal.

Ans.: 0.30 V

30–3 Calculate the minimum concentration of Zn^{2+} required for the deposition of zinc on a zinc electrode from a solution of pH 7.50. Assume that the kinetic overpotential of H^+ on zinc is –0.50 V. $E^0_{Zn^{2+}/Zn} = -0.763$ V.

30–4 A solution, $1.0M$ in Pb^{2+} and $1.0 \times 10^{-2} M$ in H^+, is electrolyzed with bright platinum electrodes. What is the molar concentration of Pb^{2+} in the solution when hydrogen starts to evolve? The kinetic overpotential for hydrogen is –0.09 V on platinum and –0.67 V on lead. Assume that no concentration polarization is present. $E^0_{Pb^{2+}/Pb} = -0.126$ V.

30–5 What kinetic overpotential for hydrogen evolution would be necessary to allow lead to be deposited from a solution $1.0 \times 10^{-3} M$ in Pb^{2+} and $1.0 \times 10^{-1} M$ in H^+? $E^0_{Pb^{2+}/Pb} = -0.126$ V.

30–6 A solution is $0.0010M$ in silver ion and contains gold(III) ion. At approximately how large a concentration of gold(III) ion would the two metals deposit simultaneously on electrolysis? $E^0_{Ag^+/Ag} = +0.799$ V, $E^0_{Au^{3+}/Au} = +1.50$ V.

30–7 Two metal ions, one dipositive and one tripositive, are to be separated by plating the former metal and leaving the latter in solution. What difference is necessary between the standard potentials of the two ions? Assume that the initial concentration of both metal ions is $0.01M$ and that the separation is satisfactory if only 0.01% of the dipositive ion remains in solution.

31

Coulometry and Coulometric Titrations

31.1 Current Efficiency

Coulometry is a method of analysis in which the amount of sought-for substance is calculated from the quantity of electricity (that is, coulombs = amperes × seconds) necessary to react electrochemically and completely with that substance. The calculation is based on Faraday's laws (Section 22.5) and requires knowing the efficiency of the current. The efficiency value is defined as $100N_r/N_t$, where N_r is the total number of coulombs used to promote the desired reaction and N_t is the total number of coulombs passed. Consider the electrolysis of a solution containing a copper(II) salt. If all the electrons flowing to the cathode combine with copper(II) ion to plate copper metal, the current efficiency is 100%. If, however, hydrogen ion is reduced to hydrogen gas, or if nitrate ion, acting as a depolarizer, is reduced, the efficiency of the current for the reduction of copper(II) is less than 100%. When such is the case, the exact percentage is generally variable and not reproducible; as a consequence, 100% efficiency is usually required in a coulometric method.

31.2 Measurement of Coulombs

The number of coulombs passed through an electrolysis cell can be measured in various ways. When the electrolysis is performed at constant current, the current in amperes needs only to be multiplied by the elapsed time, in seconds. When an electrode potential is maintained constant, the current decreases with time; empirical evaluation of the area under the current-versus-time curve is then possible. However, the most satisfactory technique is the use of a coulometer.

The silver coulometer, is simply an electrolysis cell that is filled with a silver nitrate solution and placed in series with the cell in which the coulometric determination is to be effected. The cathode (dry) of this coulometer is weighed before and after the electrolysis. From the weight of silver deposited, either the number of coulombs passed can be calculated, or more directly, the equivalents involved in the electrolysis. Of course, the electrolysis in the coulometer must proceed with 100% efficiency, which in this case is not difficult to attain. In the hydrogen-oxygen (or gas) coulometer, water containing, for example, potassium sulfate to reduce the electrical resistance is electrolyzed between two platinum electrodes. The total volume of hydrogen and oxygen produced is measured and related to the quantity of electricity passed.

Electromechanical and electronic devices known as current integrators are now used in place of the older-type coulometers. In principle, some of these devices operate like the watt-hour meter used to measure home consumption of electricity. The number of coulombs can be read directly from the meter or digital display.

31.3 Coulometric Determinations

Both techniques of electrolysis described in Section 30.4, constant-current and controlled-potential electrolysis, are applicable in coulometric methods. Constant-current electrolysis is seldom used, since with this technique it is very difficult to avoid undesired reactions and thus reach 100% current efficiency. The technique is the basis for coulometric titrations, however. The difficulty of maintaining full current efficiency is greatly reduced in the controlled-potential technique, which is therefore the method of choice in coulometry. The circuit described in Fig. 30-1 is used, with the addition of a coulometer in series with the electrolysis cell. Instead of noble metal electrodes, a mercury cathode is often used advantageously, since the high kinetic overpotential of hydrogen on mercury reduces the possibility of hydrogen discharge; it is thus easier to achieve 100% current efficiency for the reduction of interest. The electrolysis in controlled-potential coulometric determinations is stopped when the cell current has fallen to a very small, negligible value. Such coulometric determinations succeed, unlike electrogravimetric methods, even when there is no physical separation at the working electrode—simply a change in oxidation state. However, in such cases it is necessary to separate the anode and cathode compartments so that the reaction at the working electrode does not proceed in the reverse direction at the counter electrode.

31.4 Coulometric Titrations

In a coulometric titration, the titrant substance is generated with 100% current efficiency by a suitable electrode reaction secured by an excess of a suitable titrant precursor in the cell solution. Hence, the "addition rate" of the titrant is controlled by the current. If the current is held constant, the time can be measured, and in the plot of a titration curve is the analog of the volume of the titrant solution added in a conventional titration.

Consider as an example of a coulometric titration the determination of arsenic. The sample solution containing arsenic(III) is placed in a beaker and made about 0.1*M* in iodide ion; then some starch indicator is added. The platinum anode is inserted into this solution. The anodic half-cell thus formed is connected by a salt bridge to the cathodic half-cell (consisting, for example, of a potassium chloride solution and a platinum wire). Separating the anode and the cathode compartments is necessary to avoid the reverse of the titration reaction taking place at the anode. The solution in the beaker containing the anode is stirred, and a constant current is passed. The iodine (or triodide ion) formed at the anode by the oxidation of iodide ion reacts rapidly and stoichiometrically with the arsenic(III). When all the arsenic is converted to arsenic(V), the first trace of iodine then generated will undergo the starch-iodine reaction, thus signaling the end point visually. From the elapsed time and the value of the constant current, the amount of arsenic present in the sample solution can be readily calculated.

Most techniques for detecting an end point can be used with coulometric titrations. Instrumental methods, notably the amperometric, biamperometric, and potentiometric techniques, are usually preferred. Since a small current and a small time interval and hence a small number of coulombs can be measured more readily and precisely than a small volume of a solution, coulometric titrations are especially valuable for determining extremely small amounts of material. The method also allows the generation of titrants that are difficult to prepare or to use in ordinary titrimetry, including species readily oxidized by air (Ti^{3+} and Cu^{+}, for example).

31.5 Questions

31-1 Why is a current efficiency of 100% needed in coulometry?

31-2 What is the special value of the mercury cathode in coulometry?

31-3 What are the advantages of coulometric titrations over conventional titrimetry?

31-4 Silver ion can be generated by direct oxidation of a silver metal electrode. Elaborate on how a coulometric titration of various halide ions might be conducted.

31-5 The ethylenediaminetetraacetate ion can be freed from its mercury(II) complex at a mercury electrode ($HgY^{2-} + 2H^{+} + 2e \rightarrow Hg + H_2Y^{2-}$). Elaborate on how the EDTA titration of zinc or calcium might be effected coulometrically.

31-6 In a coulometric titration, what is the "limiting reagent" (that is, the titrant)? How is it introduced? Elaborate.

31.6 *Problems*

31-1 An As_2O_3 (*197.02*) sample is dissolved in the usual way, and a volume of exactly 1 mL of this sample solution is diluted to exactly 250 mL. A 5.00-mL volume of this dilute solution is analyzed by a coulometric titration, using iodide ion as the titrant precursor at a constant current of 4.00 mA. The end point corresponded to 300 s. Calculate the milligrams of As_2O_3 represented by each milliliter of the original sample solution. (1 F = 9.65×10^4 C.)

Ans.: 30.6 mg As_2O_3/mL

31-2 Electrically generated Br_2 is used for the coulometric titration of $HAsO_2$. The end point is detected when polarized platinum electrodes become depolarized. A current of 1.00 mA was passed for 35.6 s to react with all the $HAsO_2$ in a one-tenth aliquot of the sample. What amount, in milligrams, of $HAsO_2$ (*107.9*) was present in the total sample?

31-3 Electrically generated Br_2 is used for the coulometric titration of Sb(III). The end point is detected when polarized platinum electrodes become depolarized. A current of 1.350 mA was passed for 467.0 s before the end point was reached. What amount of antimony(III), expressed as milligrams of Sb_2O_3 (*291.5*), was present in the sample?

31-4 A gas coulometer (H_2 and O_2) is placed in series with an electrolysis cell that is filled with copper(II) sulfate solution. After electrolysis, the copper (*63.54*) deposit weighed 235.1 mg and the coulometer contained 180.0 mL of gas at 722 mm of mercury pressure and 27.0°C. Calculate the current efficiency in the electrolysis of copper. Assume that one mole of gas occupies 22.4 L at 760 mm of mercury and 0.0°C.

Ans.: 80.0% efficiency

31-5 In a coulometric titration under the conditions selected, copper(II) is reduced to metallic copper. A silver coulometer shows a deposit of 127.0 mg of silver (*107.9*). How much copper (*63.5*) is present in the sample?

Ans.: 37.4 mg of copper

32

Electrolytic Conductance and Conductometric Titrations

32.1 Introduction

Electricity is conducted through a solution, that is, through an electrolytic conductor, by the motion of charged ions under the influence of an applied electrical field. The conductance is defined as the reciprocal of the resistance of the solution and is the summation of the contributions to the conductances of all ions in the solution. Each of these contributions is usually called an ionic conductance. The ionic conductance of an ion depends on its charge and the rate at which it migrates under the influence of the electrical field. Thus, a dipositive cation contributes twice as much to the conductance as a monopositive cation if both ions migrate at the same speed. The migration rate is influenced by (1) the magnitude of the applied electrical field, (2) the charge and size of the solvated ion, (3) the temperature, (4) the viscosity and dielectric properties of the solvent, and (5) the attractive forces between the ion of interest and other ions present in the solution.

In a concentrated solution, an ion is surrounded by other ions to which it is attracted and which retard its movement toward a charged electrode. If such a solution is progressively diluted, the attractive forces diminish, vanishing at infinite dilution; thus, the ionic conductance increases with dilution, reaching a limiting value at infinite dilution. Frequently the approximation is made that the value of the ionic conductance at low finite concentrations (usually below 0.1*M*) equals the ionic conductance at infinite dilution.

32.2 Units of Conductance

Conductance is expressed in reciprocal ohms, Ω^{-1}, for which the SI unit is given the name *siemens,* with the symbol S. The conductance is proportional to the cross-sectional area a (in square meters) and inversely proportional to the length l

(in meters), of a conductor:

$$\text{Conductance} = \frac{1}{R} = \frac{\kappa a}{l} \tag{32-1}$$

The proportionality constant, denoted by the lowercase Greek letter kappa, κ, is called the conductivity, or in electrolytic conduction, the electrolytic conductivity; it has the units of $S \cdot m^{-1}$ (that is, siemens per meter).

The molar conductivity of a particular ion is designated by the Greek capital letter *lambda* with subscript *i*, Λ_i, and is defined as the electrolytic conductivity of a hypothetical solution containing one mole of only that particular ion per cubic meter of solution. The molar conductivity is expressed in $S \cdot m^2 \cdot mol^{-1}$. To allow the concentration C to be expressed in $mol \cdot L^{-1}$, a factor of 1000 must be introduced, to yield

$$\Lambda_i = \frac{\kappa}{1000C} \tag{32-2}$$

Solving equation (32-2) for κ and substituting in (32-1) gives

$$\text{Conductance} = \frac{1}{R} = \frac{1000a}{l} \times \Lambda_i C = K\Lambda_i C \tag{32-3}$$

where the proportionality constant K equals $1000a/l$. Each ion contributes independently to the conductance of the solution; hence,

$$\begin{aligned} \text{Conductance} &= \frac{1}{R} \\ &= K(\Lambda_{i,1}C_1 + \Lambda_{i,2}C_2 + \cdots) \\ &= K \sum_{n=1}^{n=n} \Lambda_{i,n}C_n \end{aligned} \tag{32-4}$$

The molar conductivity at infinite dilution, which is the limiting molar conductivity, is designated Λ_i^0. Values for some common ions are listed in Table J in the Appendix.

32.3 *Applying Ohm's Law to Solutions*

When a dc voltage is applied to two electrodes dipping into an electrolyte solution, Ohm's law is not obeyed. The causes for this have been considered earlier, and include the establishment of back emf, kinetic overvoltage, and concentration overvoltage (Chapter 28).

If the polarity of the applied voltage is changed rapidly (that is, if an ac source is used), the situation is different. For a simplified picture, assume that the polarity during the first half of a cycle is such that one of the electrodes is the cathode. Then cations are attracted to this electrode and anions are repelled; by the movement of the ions, the electric current is carried through the solution. When an electrode reaction starts to take place, however, the second half of the cycle is reached and the polarity is reversed; that is, the electrode becomes the anode. Now cations are repelled and anions are attracted. Again before there is any appreciable reaction, the polarity changes, and so forth. Consequently, the ions carry the electrical current by a reciprocating motion, that is, merely by an oscillation, and no deposition takes place. Thus no back emf or overvoltage develops, and Ohm's law is obeyed.

32.4 *Measuring Resistance*

The resistance of an electrolyte solution and therefore the conductance are usually measured with a modified Wheatstone bridge circuit, shown in Fig. 32-1. A voltage is applied across points *a* and *d*. One (or more) of the resistors R_1, R_2, R_3, and R_4 is variable and its resistance is accurately known at the various settings. If this resistance is varied, a setting can be found at which no current flows through the detector *D*, and the bridge is said to be balanced.

Since no current passes through the detector when the bridge is balanced, points *b* and *c* must be at the same potential and consequently $I_1R_1 = I_3R_3$. Similarly, it follows that $I_2R_2 = I_4R_4$. Dividing one equation by the other yields $I_1R_1/I_2R_2 = I_3R_3/I_4R_4$. Since no current passes through the detector, the current I_1 passing through R_1 must equal the current I_2 passing through R_2; that is, $I_1 = I_2$.

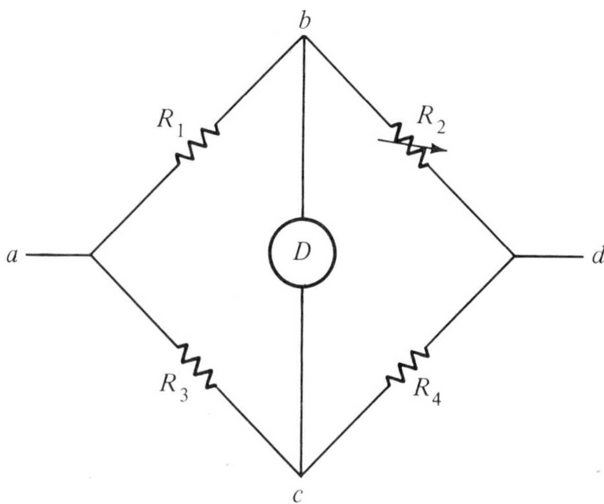

Fig. 32-1. Wheatstone bridge circuit. See the text for a detailed explanation.

Similarly, $I_3 = I_4$. Hence, for the balanced bridge, $R_1/R_2 = R_3/R_4$. If three of the resistances are known, the fourth is established and is thereby measured.

In measuring electrolytic conductance, it is customary to apply a pure sine-wave signal (10^2 to 10^3 Hz) across points *a* and *d*.

32.5 Conductance Cells

A conductance cell often consists of a pair of platinized platinum electrodes rigidly mounted in a glass vessel. For any given cell, the values of a and l are fixed, and the ratio l/a is called the cell constant K_{cell}. Although this constant might be calculated from the cell geometry, it is usually evaluated by filling the cell with a solution of known electrolytic conductivity and measuring the resistance. The value of K_{cell} is then known from the relation $K_{cell} = \kappa R$. Since conductance varies by about 2% per degree Celsius, it is necessary to control the cell temperature by a constant-temperature bath.

32.6 Conductivity Determinations

Conductivity determinations receive limited application in inorganic analysis. One of the most common uses is to check the quality of water intended for process or boiler-feed use. This measurement of conductivity is not specific for any particular impurity but is a rough indication of the total ion content of the water. Carbon in steel and in other metals and alloys is frequently determined by heating the sample in a tube furnace in a stream of oxygen. The emanating gases are passed through a solution of an alkali metal hydroxide or an alkaline earth hydroxide. The decrease in the conductance of this solution can be correlated with the amount of carbon dioxide absorbed and this with the carbon content of the sample.

32.7 Conductometric Titrations

Any reaction in which the reactants differ markedly in conductance from the products can, in principle, be the basis of a conductometric titration. In such a titration, the conductance of the solution is measured after the addition of each increment of titrant solution, and the resulting values are plotted versus volume of titrant. The end point corresponds to a break in this curve and is located by the extrapolation of two linear segments. The conductance of the solution corresponds to the sum of the contributions of all ions present. Ions that do not enter into the titration reaction give a background conductance on which the changes in conductance due to the titration reaction are superimposed. This background does not invalidate a conductometric titration unless it is so large that the small changes due to the titration are indiscernible. High concentrations of inert electrolytes are therefore undesirable in conductometric titrations.

Since the end point is obtained from *changes* in conductance, the absolute value of the cell constant does not have to be known; moreover, the conductance can be expressed in arbitrary units. However, dilution effects must be either minimized or corrected for (Section 29.10); the temperature should be kept constant throughout the titration.

32.8 Conductometric Acid-Base Titrations

For the titration of hydrochloric acid with sodium hydroxide, the reaction equation can, for present purposes, be expressed as

$$\underset{0.0350}{(H^+} + \underset{0.0076}{Cl^-)} + \underset{0.0050}{(Na^+} + \underset{0.0198}{OH^-)} \rightarrow H_2O + (Na^+ + Cl^-)$$

The number below an ion is its limiting molar conductivity Λ_i^0, in $S \cdot m^2 \cdot mol^{-1}$.

The conductance associated with the hydrogen ion starts at a value proportional to its limiting molar conductivity (0.0350) and decreases linearly to zero at the equivalence point. The conductance of the chloride ion is constant throughout the titration at a value proportional to its limiting molar conductivity (0.0076).

Hydroxide ion is not present in significant amounts until the equivalence point is passed, and then the conductance associated with this ion increases linearly to a value proportional to its limiting molar conductivity (0.0198) when 100% excess titrant is added. The contribution of the sodium ion to the conductance starts at zero, and is proportional to its limiting molar conductivity (0.0050) at the equivalence point, and to twice that value (0.0100) at the 100% excess point.

Since only the shape of the titration curve is of importance here, these limiting molar conductivities, and their sum, can be plotted as conductances, in siemens [that is, K in equation (32-4) is taken as unity]. These relations are shown graphically as dashed lines in Fig. 32-2. The point-by-point summation of these lines yields the theoretical titration curve (solid line). Only three points need to be calculated to establish this curve: the starting, equivalence, and 100% excess points. This applies whenever the reacting substances are strong electrolytes. With weak electrolytes, more points are needed and the calculations are complicated.

In the conductometric titration of a weak acid, such as acetic acid, with sodium hydroxide, the conductance at the start is small, because only a small portion of the acid is dissociated. As sodium hydroxide is added, the conductance decreases at first; the reason is that strongly conducting hydrogen ions are removed and acetate ions are formed that repress the dissociation of the acid, thereby preventing the formation of more hydrogen ion. As the titration proceeds, the conductance increases nearly linearly to the equivalence point. This increase is caused by the introduction of conducting sodium ions by the titrant solution and the formation of acetate ions. Beyond that point, the conductance increases more rapidly when excess titrant is added and thus also highly conducting hydroxide ions. The titration curve takes the form of curve A in Fig. 32-3. Since the slope of the titration curve changes only slighly at the equivalence point, it is difficult to locate the end point precisely. Analogous considerations apply to the titration of a weak base with a strong acid.

When a weak acid, such as acetic acid, is titrated conductometrically with a

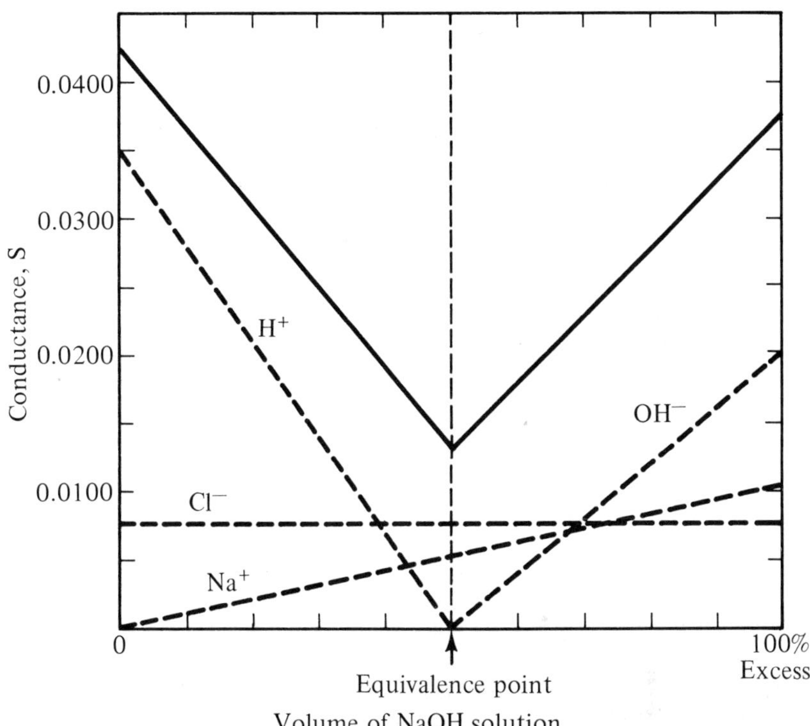

Fig. 32-2. Conductometric titration of hydrochloric acid with sodium hydroxide. The contribution of each ion to the conductance is shown as a dashed line.

weak base, such as ammonia, the titration curve (Fig. 32-3, curve B) to the equivalence point is like the curve for the titration of a weak acid with a strong base. Beyond that point, however, the conductance increases only slightly, since the ammonia is only partially dissociated. Further, its protolysis ($NH_3 + H_2O \rightleftharpoons NH_4^+ + OH^-$) is repressed by the ammonium ions present. Since the break in the weak acid-weak base curve is sharper than the break in the strong acid-strong base curve, the end point can be located far more precisely. This is quite different from the situation in a pH titration (potentiometric or visual), for which a weak acid-weak base titration is infeasible.

32.9 *Conductometric Precipitation Titrations*

In a titration of sodium chloride with silver nitrate, the reaction equation, for present purposes, can be expressed

$$\underset{0.0050}{(Na^+} + \underset{0.0076}{Cl^-)} + \underset{0.0062}{(Ag^+} + \underset{0.0071}{NO_3^-)} \rightarrow \underline{AgCl} + (Na^+ + NO_3^-)$$

The number below an ion is its Λ_i^0 value in $S \cdot m^2 \cdot mol^{-1}$.

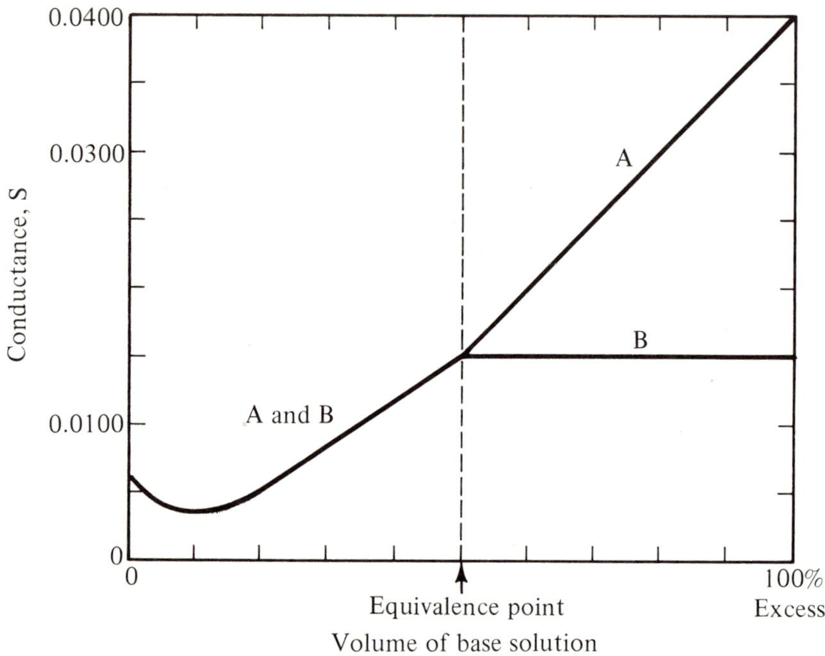

Fig. 32-3. Curve A: conductometric titration of acetic acid with sodium hydroxide. Curve B: conductometric titration of acetic acid with ammonia.

Since only strong electrolytes are involved, three points are enough to establish the shape of the theoretical titration curve. At the starting point, only sodium chloride is present, and the conductance is proportional to 0.0050 + 0.0076 = 0.0126. At the equivalence point, only sodium nitrate is present (and negligible amounts of dissolved silver chloride), and the conductance is proportional to 0.0050 + 0.0071 = 0.0121. At the 100% excess point, sodium nitrate and silver nitrate are present, and the conductance is proportional to 0.0050 + 0.0071 + 0.0062 + 0.0071 = 0.0254. The titration curve takes the form shown in Fig. 32-4.

If silver chloride were appreciably more soluble, then the conductance near the equivalence point would be increased measurably, since the concentration of silver and chloride ions could not be neglected. Then, the two straight-line portions would be connected by a curved section (dashed line in Fig. 32-4); the end point could still be located precisely by extrapolation of the straight-line segments.

32.10 Remarks

Our examples so far involve only monovalent ions and reaction mole ratios of 1:1 and are thus uncomplicated. Consider, for example, the conductometric titration of sulfuric acid with sodium hydroxide. The shape of the titration curve is that for a strong acid-strong base reaction and is like the one in Fig. 32-2. In the com-

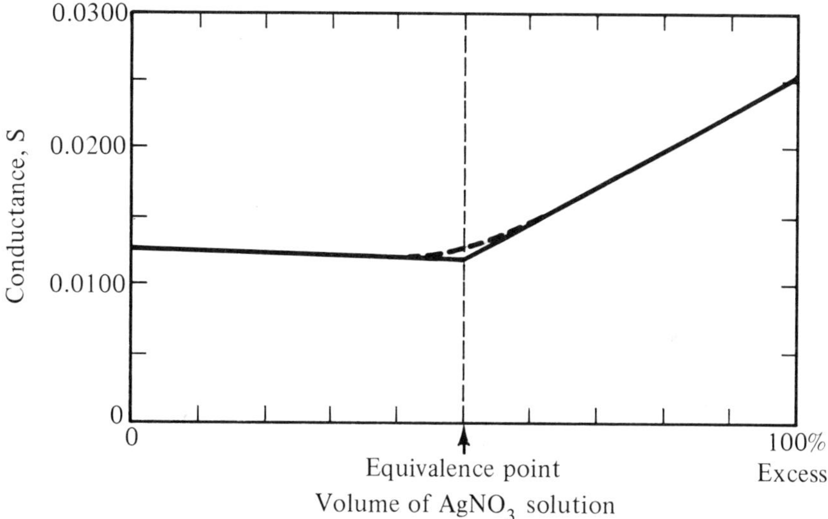

Fig. 32-4. Conductometric titration of sodium chloride with silver nitrate.

putation of that curve, however, it must be recognized that for hydrogen ion the initial concentration is twice that of the sulfate ion, and that at the equivalence point, the sodium ion has twice the molarity of the sulfate ion. (You may as an exercise use the data of Table J of the Appendix to plot the titration curve for this case.)

Conductometric titrations can be applied to many chemical systems in which an ionic reaction does not go to completion. The dissociation of the principal product causes curvature in the titration plot in the region around the equivalence point. However, the dissociation is repressed before and beyond this point by a common-ion effect, thus yielding straight-line segments; these are extrapolated to establish the end point. In addition, conductometric titrations can be performed with dilute solutions, often $0.001F$ or lower. These advantages are shared with other "linear" titrations, including amperometric titrations (Section 29.14) and some photometric titrations (Chapter 37).

32.11 Questions

32-1 Elaborate on the difference between conductance and conductivity and on the implication of the *-ivity* suffix.

32-2 Explain why the molar conductivity of a solution doesn't increase proportionally when the concentration of the solute is increased.

32-3 Does Ohm's law apply to a solution of an electrolyte? Explain.

32-4 What is the merit of determining a cell constant indirectly instead of calculating it?

32-5 Explain the need for maintaining a constant temperature in conductometric titrations.

32-6 What kind of acid-base titration can be better accomplished conductometrically than potentiometrically? Elaborate.

32-7 The silver in a silver acetate solution is to be titrated conductometrically with chloride ion. Which would be the preferred titrant—a standard solution of sodium chloride or a standard solution of hydrochloric acid? Elaborate.

32.12 Problems

32-1 A mixture of exactly 50 mL of 0.0200*F* HCl and 50 mL of 0.0200*F* acetic acid is titrated conductometrically with 1.000*F* NaOH. The following data were secured during the titration, with conductance expressed in arbitrary units:

NaOH, mL	0.0	0.40	0.80	1.20	1.60	1.80	2.20	2.60	2.80
Conductance	37.0	28.0	16.0	12.0	14.5	15.5	20.0	27.5	31.5

Plot the experimental points of the titration curve and extrapolate the line segments to locate the two end points. (a) State the values of the end points and (b) the equivalence points. (c) In your own words, try to explain the shape of this titration curve.

Ans.: (a) 1.0 mL, 2.05 mL; (b) 1.0 mL, 2.0 mL

32-2 A 1.0255-g sample of a water-soluble organic acid is weighed out, dissolved, and diluted with water to a volume of exactly 250 mL. A 25.00-mL aliquot is transferred to a conductivity cell and without further dilution is titrated with 0.1000*F* NaOH. The following data are obtained, where R is the measured resistance in ohms and V is the volume of NaOH solution in milliliters.

R	V	R	V	R	V	R	V
29.4	0.00	32.5	1.80	23.8	3.60	16.2	5.40
32.6	0.20	31.7	2.00	23.0	3.80	15.1	5.60
34.1	0.40	31.1	2.20	22.3	4.00	14.5	5.80
38.0	0.60	30.1	2.40	21.7	4.20	13.8	6.00
36.9	0.80	29.1	2.60	21.0	4.40	13.1	6.20
35.8	1.00	27.7	2.80	20.4	4.60	12.6	6.40
34.9	1.20	26.7	3.00	19.6	4.80	12.0	6.60
34.1	1.40	25.7	3.20	18.5	5.00	11.5	6.80
33.3	1.60	24.7	3.40	17.3	5.20	11.1	7.00

Convert the resistance values to conductance, correct for dilution, and plot the titration curve. From the curve, conclude whether the acid is monoprotic or diprotic. Calculate the formula weight of the acid.

Ans.: diprotic acid; FW = 427

32-3 The conductivity of 0.100*F* HCl is 3.94 $S \cdot m^{-1}$. What is the molar conductivity of the solution? *Ans.:* 0.0394 S

32–4 What is the molar conductivity in $S \cdot m^2 \cdot mol^{-1}$, at infinite dilution and 25 °C, of a solution of:

(a) HCl (c) Na_2SO_4 (e) $NaKSO_4$
(b) NaCl (d) Ag_2SO_4 (f) H_2SO_4

32–5 Using the limiting molar conductivities of the relevant ions, sketch the conductometric titration curves for the following titrations.

(a) HNO_3 with NaOH (d) $AgNO_3$ with KCl
(b) H_2SO_4 with $BaCl_2$ (e) $LaCl_3$ with NaOH
(c) Ag_2SO_4 with $BaCl_2$ (f) KCl with $NaB(C_6H_5)_4$

33

Review Questions on Electroanalytical Methods and Redox Titrations

33-1 Compare the relative advantages and disadvantages of electrometric titration methods involving linear plots of data (conductometric and amperometric titrations) and methods that involve logarithmic plots (potentiometric redox and acid-base titrations).

33-2 Which electroanalytical methods are based on concentration overpotential? In which should such overpotential, as a nuisance, be minimized as much as possible? In which is such overpotential absent?

33-3 Which methods do not require complete reaction or complete consumption of the sample solution taken for the determination?

33-4 Which methods have the greatest sensitivity, that is, can respond to very small concentrations of the substance to be determined?

33-5 Which electroanalytical methods need the least amount of equipment? Which a considerable amount?

33-6 Which methods operate at constant current? Which at constant potential?

33-7 For which electroanalytical methods is it necessary to know the absolute value of (a) the current? (b) the voltage? and (c) the resistance? For which methods is it necessary to know only (d) differences in parameters a, b, c, or (e) a function proportional to one of these parameters?

33-8 Which methods require that the temperature be known precisely? Which require only that the temperature be constant during the period of measurements? Which do not require temperature control so long as changes are not extreme?

33-9 Which electroanalytical methods require close control of pH? Under what circumstances is this close control necessary?

33-10 How does the presence of a complexing agent affect (a) a redox titration?

(b) polarographic measurements? (c) a determination by controlled-potential electrogravimetry?

33–11 Compare the formal potential of a redox indicator with the pK_a value of an acid-base indicator, and elaborate on the parallels in using these indicators, respectively, in redox and acid-base titrations.

33–12 What is an inert electrode? Name some methods in which it finds application.

33–13 Which electroanalytical methods require the presence of a high concentration of inert salts and for what purpose? In which method should a high concentration of inert electrolytes be avoided?

33–14 Which electrochemical methods can be used for analytical purposes by the direct measurement of voltage, current, and the like, and can be modified to titration methods based on the measurement of differences in one of the parameters?

33–15 In which analytical methods can theoretical predictions about feasibility, performance, conditions, and so on, be based on applying the Nernst equation?

33–16 Which electroanalytical techniques and measurements can be used to obtain data or perform operations for an *immediate* purpose that is not detection or determination of a substance?

33–17 In which methods is mercury important, and why does it offer special advantage?

33–18 In which methods and techniques of measurement is connecting half-cells by a salt bridge a common practice?

33–19 In which electroanalytical methods is measuring time involved?

33–20 What is a junction potential? How is it avoided, or at least minimized? In which methods does its occurrence impose a limitation on the accuracy and the precision of the measurement?

33–21 What ways can you suggest for testing whether an electrode reaction is reversible or irreversible?

33–22 In a certain supporting electrolyte, two metals have their polarographic waves essentially coinciding. Elaborate on the possibility and mechanism of resolving the coincidence by adding a complexing agent.

33–23 In which redox titrations and electrometric determinations must oxygen be excluded? Explain why and discuss how the exclusion can be achieved in practice.

33–24 When is a small electrode area desirable in an electroanalytical process? When is a large electrode area desirable? Explain your answers.

33–25 How would you make a silver coulometer work on a titrimetric basis?

33–26 How can you decrease concentration overpotentials?

33–27 Some substances can be used as primary standards for both a redox and an acid-base titration. Would such a substance have the same equivalent weight in both titrations? Name a few substances that would fall in this category.

33–28 Why is sulfuric acid or sodium hydroxide added to the electrolyte in a gas coulometer, although only water is decomposed?

33–29 What solutions should not be electrolyzed with a gold or a platinum foil as the anode? Explain why. What electrode material would be suitable in such cases?

33–30 How do you measure a voltage without drawing a current. Which methods of analysis call for this technique?

33–31 Name some autocatalytic redox reactions.

33–32 Describe some redox titrations and thereby exemplify the use of the terms, *redox titration, titrant, titrant solution, standard solution, standardized solution, selfindication, redox indicator, back-titrant, extraction end point, iodimetry,* and *iodometry.*

34

Electromagnetic Radiation and Matter

34.1 General Remarks

You remember from earlier courses that phenomena involving "light" can be used for analytical purposes. The color of a solution or of a flame and the scattering of light associated with the appearance of turbidity can be used in detecting and identifying substances. These phenomena have counterparts in quantitative analysis.

For the purposes of analytical chemistry, electromagnetic radiation does not need to be restricted to light, which in the narrow sense refers to radiation in the visible region. Analytical chemists commonly allow a looser usage and include the near ultraviolet and infrared regions in speaking of light. In this chapter and subsequent ones, the term *light* is favored whenever the context is primarily directed to radiation in these three regions; the term *radiation* will be used when other spectral regions are involved and when the considerations are relatively general. Matter and radiation interact by various mechanisms, some of which will be examined when we study the relevant methods of analysis. A full understanding of how matter and radiation interact requires knowing more physics and physical chemistry than you probably do now. However, only superficial knowledge is required for appreciating how an analytical instrument operates and how it can be applied. Historically, several analytical methods involving radiation were developed before their theoretical basis was elucidated; indeed, the methods were important subsequently in elaborating underlying theory.

Radiation can be treated as a transverse wave, that is, a wave in which the plane of vibration is perpendicular to the direction of propagation. For many purposes, however, it is more advantageous to treat radiation as a "stream" of photons, that is, discrete particles of energy that have no rest mass.

34.2 Definitions

Analytical methods based on the interaction of radiation with matter can be called optical methods. Many of the terms used with such methods have different nuances and degrees of restriction; and for one and the same phenomenon or property, more than one term may be assigned. Different authors prefer different terms and may apply them with slightly different connotations. It is therefore appropriate to define some of the general terms at this point. You are probably familiar with them, but thinking about them will refresh your memory and apprise you of the sense in which they are used here.

Wavelength. Wavelength, denoted by the lowercase Greek letter *lambda*, λ, is the distance along the direction of propagation between two points that are in phase on adjacent waves. Various units can be used, depending on the wavelength region of interest. For the ultraviolet and visible regions, the nanometer, abbreviated nm, is convenient (1 nm = 10^{-9} m). For this unit, the term *millimicron*, mμ, was formerly used and may still be encountered. The angstrom (1 Å = 0.1 nm = 1 $\times$ 10^{-10} m) will also be seen, but it does not conform to the SI program (see Table K in the Appendix). For longer wavelengths, the micrometer, abbreviated μm, is an appropriate unit (1 μm = 1 $\times$ 10^3 nm = 1 $\times$ 10^{-6} m); the former term *micron*, μ, is not to be encouraged.

Wavenumber. The shorter the wavelength, the greater the energy of the corresponding photons (see Section 34.3). For some purposes, it is convenient to have a quantity proportional to energy, namely, the reciprocal of wavelength; this is termed the wavenumber and designated by the lowercase Greek letter *nu* with a superior tilde $\tilde{\nu}$. Under preferred SI practices, the wavenumber is to have the unit reciprocal meter, m^{-1}; however, for analytical chemistry, the convenient unit is the reciprocal centimeter, cm^{-1} (hence, 1 ν = 1/cm).

Frequency. Frequency, denoted by the lowercase Greek letter *nu*, ν, is the number of wave cycles per second. In SI units, the unit of frequency is the hertz, which equals 1 cycle per second (1 Hz = 1/s). Frequency in one sense is more fundamental than wavelength. The frequency of radiation (monochromatic) remains constant regardless of the medium in which it is propagated or transmitted. In contrast, the wavelength varies inversely with the velocity of radiation in the medium; that is, $\nu = c_{rad}/\lambda$, where c_{rad} is the speed of radiation (light) in the medium.

Monochromacity. *Monochromatic radiation* (or *light*) is a term used to denote radiation of a single frequency, or less precisely, of a single wavelength. Compound, or heterochromatic, radiation implies the presence of radiation of many frequencies (wavelengths).

Spectrum. The total spectrum of radiant, electromagnetic energy extends from cosmic, gamma, and X rays (<0.1 to ~10 nm) through the ultraviolet (10 to ~380 nm), the visible (380 to 780 nm), the infared (~0.78 to ~300 μm), and the microwave regions (~0.1 to 1000 cm), to the radio wave region (>1 m). In analytical chemistry, the term *spectrum* is usually combined with a descriptive word as follows. An *emission spectrum* is the array of all wavelengths obtained from compound radiation emitted by a radiation source; the array comprises the separated, individual wavelengths (see Chapter 40). An *absorption spectrum* is the array of lines, or more frequently, bands of those wavelengths that are removed from compound radiation when it passes through an absorbing medium. The array is usually displayed as a curve in a plot of absorbance versus wavelength (see Fig. 35-8, for example).

Radiant Power. Radiant power (also known as radiant flux), denoted by P, is the rate at which radiant energy is transported by a beam of radiation; in simple terms, it is the intensity of the radiation.

34.3 *Remarks, and Classification of Optical Methods*

In thinking of radiation as dualistic (wave versus particles), we have to realize that the shorter the wavelength and the greater the frequency, the greater the energy of the corresponding photons. The relation between the energy E of a photon and frequency is given by

$$E = h\nu \tag{34-1}$$

where h is Planck's constant. In SI units, with E in joules and ν in hertz, the value h is 6.63×10^{-34} J•Hz^{-1}. From previous courses, you may be more familiar with the units erg, s^{-1}, and 6.63×10^{-27} erg•s, respectively, for the three terms.

The equivalent relation between energy of a photon and wavelength and wavenumber are

$$E = \frac{h}{\lambda} = hc_{\text{rad}}\tilde{\nu} \tag{34-2}$$

The beginner sometimes incorrectly assumes that the energy of the photons parallels the intensity of the radiation beam. For a certain effect (for example, elevation of electrons from a lower energy level to a higher one), absorption of photons of a certain energy or above is required (see Section 35.1). In other terms, radiation at a certain wavelength or below is needed. The effect cannot be achieved by replacing high-energy photons (short-wave radiation) by a more intense beam of low-energy photons (long-wave radiation). The distinction can be clarified by an analogy, admittedly imperfect. To penetrate a thick oak board, a high-energy bullet from a rifle (high-energy photon) is needed. Even a very large number of low-energy pellets from an air pistol (high-intensity beam) will not accomplish the result sought.

Various interactions of radiation and matter will be considered in detail in the following chapters. A brief exploration of some of the possibilities will help in classifying many of the important optical methods of analysis. Classification can be undertaken in different ways, depending on the point of view and the objective.

One scheme focuses on the phenomena observed when radiation interacts with matter. If the radiation impinges on matter and the photons pass through without interaction, the situation is trivial and does not allow an analytical method to be developed. In contrast, the radiant beam may be refracted (change its direction), reflected ("bounced back"), dispersed (have the component wavelengths separated with compound radiation), or absorbed. With plane-polarized radiation (that is, with the wave vibration occurring in a single plane), rotation of the plane of polarization may take place under certain circumstances (see Section 42.3).

The emergence of radiation from matter is often termed *luminescence.* If emission of radiation results from the absorption of radiant energy, the term *fluorescence* is used when emission follows absorption immediately. When emission is delayed, the term *phosphorescence* is applied. Radiation of very short wavelength can be produced when particles, such as electrons, impinge on matter; the radiation is then in the X-ray range. High-energy gamma radiation can produce the same result.

Luminescence can result from other processes than the impingement of radiation on matter. *Incandescence* is the emission of radiation by the thermal excitation of matter (for example, in the wire of a light bulb or in a flame). *Triboluminescence* can result from the interaction of mechanical energy and solids and can occur in the breaking and grinding of a substance (Greek *tribein,* "to rub"). When a sugar cube or candy is broken, for example, a short-lived green flash is emitted; this is so low in intensity that it can be seen only in total darkness with dark-adapted eyes. *Chemiluminescence* results when part of the energy of a chemical reaction leaves the system as radiation. If the chemical reaction takes place in a living organism, the special term *bioluminescence* can be used; fireflies have this trait.

Often more than one of the phenomena mentioned above may be concurrent. Reflection and absorption may occur simultaneously, for example. If "white" compound light impinges on a surface, part may be absorbed and part reflected; the reflected light has an altered composition, and appears colored to an observer.

Other possibilities for classifying optical methods of analysis exist. A distinction can be made about whether molecules or atoms are involved (thus the terms *atomic spectroscopy* and *molecular spectroscopy*). Another possibility is to group phenomena or methods according to a common property or approach (for example, *flame* emission and *flame* absorption techniques). The wavelength range used can be made a distinguishing factor (for example, visible, ultraviolet, and infrared absorption photometry).

Although it is possible to develop a single, unified classification scheme for optical methods of analysis, such a scheme has little practical value when the primary purpose is to introduce the mere principles of the methods. It is more advantageous to use a mixed approach involving common features in instrumentation, practical importance and frequency of application, historical significance, as well underlying phenomena. We use this eclectic approach.

34.4 *Questions*

34–1 Define in your own words the terms *luminescence, frequency, monochromatic radiation, compound light,* and *optical methods of analysis.*

34–2 In what senses can the term *light* be used in connection with optical methods?

34–3 The wavenumber is sometimes falsely called "frequency." Elaborate on this and establish the relation between $\tilde{\nu}$ and ν.

34–4 What do you understand by the "dualistic" approach to the behavior of light?

34–5 Elaborate on the difference between "an intense radiation" and "high-energy waves."

34–6 What happens to the speed of light on passing from one medium to another (having different refractive indices)?

34–7 What is a spectrum? What is the plural of the term?

34–8 Name some optical methods you are already familiar with from previous courses. Elaborate on the wavelength range involved and the underlying principles as best you can.

34–9 The text of a literature review reads, "For the consideration of the far infrared, we have included the 300–310 cm^{-1} region." What quantity is implied by these numerical limits? What are the corresponding wavelength limits in micrometers (1 $\mu m = 10^{-6}$ cm)?

35

Photometric Methods of Analysis

35.1 *Absorption of Radiation*

Absorption of radiation by matter implies the absorption of photons. From the standpoint of both theory and practice, it is convenient to differentiate between atomic processes of absorption and molecular ones.

The orbital electrons of an atom can persist only in discrete energy levels, of which the lowest is called the *normal* or *ground state* E_0; higher ones are called *excited states,* $E_1, E_2, \ldots, E_n$ (see Fig. 35-1a). For an electron to be elevated to a higher state, the atom must absorb energy. Absorption occurs only if the energy is "tailored" (that is, is quantized) to the energy difference E between two levels of the atom, say, $a + 1$ and a. This energy difference is given by

$$E = E_{a+1} - E_a = h\nu \tag{35-1}$$

The energy E is also required of the absorbed photon and is related to the frequency of the radiation ν as indicated [see equation (34-1)]. In atomic absorption only electronic levels are involved, in contrast to molecular absorption (see below). Consequently, an atomic spectrum consists of discrete lines (see Chapter 39).

An atom remains in the excited state usually for only a short period ($\sim 10^{-8}$ s). Then it relaxes; that is, the absorbed radiant energy is released, usually as thermal energy. A photon, however, may be emitted; then the term *atomic fluorescence* is applied (Section 39.8). In solids, the excited state may have a long lifetime and the phenomenon *phosphorescence* is involved. Widely different amounts of energy may be needed to achieve excitation, depending on the energy levels involved. The range of radiation involved is from X rays, for the electrons in the innermost shell of multielectron atoms (Chapter 40), to visible light, for weakly held valency electrons (Chapter 39).

For absorption of radiation by molecular processes, the situation is similar but more complex; in addition to electronic transitions, excitation of vibrational and rotational levels is often possible. The quantized energy involved for the last two levels is smaller, but an interplay between the transition types exists, as is delineated by Fig. 35-1b. Each of the three vertical arrows shown indicates an

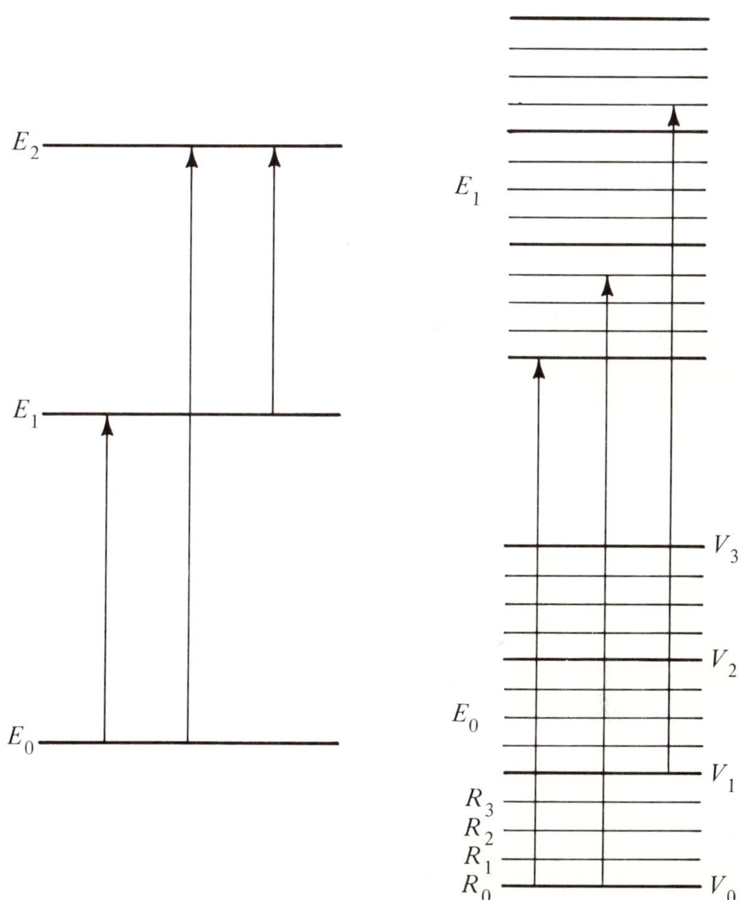

Fig. 35-1. (a) Atomic absorption from ground state E_0 to first and second excited electronic states E_1 and E_2, and from first to second excited states. (b) Molecular absorption from ground state E_0 to first excited electronic state E_1, with interplay of vibrational V and rotational R levels indicated.

electronic transition from the ground state E_0 to the first excitated state E_1, but the vibrational and rotational levels are different. According to quantum mechanics, some transitions are allowed, others are forbidden, and some are more probable than others. Because of the many possibilities and probabilities, molecular absorption spectra consist not of lines but of bands that often overlap. This situation applies to the absorption of radiation corresponding to the ultraviolet and visible regions. When photons of low energy are absorbed, no electronic transitions are possible, but vibrational and rotational levels may still be excited; this is the case for infrared absorption (Section 35.19).

35.2 Definitions

Some definitions beyond those given in Section 34.2 are needed in studying the absorption of radiant energy.

Color. Color is the psychophysical sensation experienced when light of one wavelength or more in the visible region is viewed by the eye. The term *light,* for some purposes, is used restrictively to denote radiant energy in the visible region only. Since the eye cannot distinguish whether visible light is monochromatic or compound, the term *monochromatic light,* as here used, is not necessarily identical with "one" color or "pure" color.

Radiant Power. Radiant power (also known as radiant flux), denoted by P, is the rate at which radiant energy is transported by a beam of light. In simple terms, the radiant power is the intensity of the light.

Transmittance. Transmittance, denoted by T, is the ratio of the radiant power P transmitted by a sample to the radiant power P_0 incident on the sample. Hence, $T = P/P_0$. (Terms formerly used in place of *transmittance* include *transmittancy* and *transmission.*) The *percent transmittance* is given by $T \times 100$, and may be denoted by $\% T$.

Absorbance. Absorbance, denoted by A, is defined as the negative logarithm (to the base 10) of the transmittance; that is, $A = -\log T = \log(1/T) = \log(100/\% T)$. (Terms formerly used in place of *absorbance* include *optical density, extinction,* and *absorbancy.*)

Absorptivity. Absorptivity, denoted by a, is a measure of the ability of a material to absorb light. When the main interest is in solutions, absorptivity can be more closely defined as the ratio of the absorbance to the product of the actual concentration of the absorbing species and the length of the light path in the absorbing medium. Consequently, a is the specific absorbance, that is, the absorbance per unit concentration and unit thickness. By convention, for the ultraviolet and visible regions concentration c is expressed in grams per liter; and length of light path b is expressed in centimeters. Hence, $a = A/bc$ and a has the units $L \cdot g^{-1} \cdot cm^{-1}$. (Absorptivity was formerly called *absorbancy index, extinction coefficient,* and *specific extinction;* a new term proposed is *specific absorption coefficient.*)

Molar Absorptivity. Molar absorptivity, denoted by the lowercase Greek letter *epsilon,* ϵ, is the absorptivity with concentration C of the absorbing species expressed in moles per liter and length of light path expressed in centimeters. Hence, the molar absorptivity has the units of $L \cdot mol^{-1} \cdot cm^{-1}$. (Molar absorptivity was formerly called *molar absorbancy index* and *molecular extinction coefficient;* a new term proposed is *molar absorption coefficient.*)

Absorptivity and molar absorptivity depend on the wavelength of the radiation and the nature of the absorbing species.

35.3 *Laws of the Absorption of Light*

Since the student and practicing analyst usually encounters the laws governing the absorption of radiation in terms of the ultraviolet and visible regions, the term *light* is used in our current study of the subject. The laws considered nevertheless have wider application, covering all electromagnetic radiation. For some spectral regions, different terms and units may be met.

Suppose the sample under study is a solution of a light-absorbing solute in a nonabsorbing solvent placed in a rectangular glass cell. Suppose further that a beam of parallel, monochromatic light strikes the cell at a right angle to one of its sides. The general manifestation is that only a portion of the radiant power of the incident light beam will be transmitted, that is, will leave the cell on the opposite side. A portion of the radiant power is lost by reflections at the air-glass and glass-solution interfaces. A second portion is lost by scattering caused by dust particles and inhomogeneities in the solution and the glass. And a third portion is consumed in an absorption process by the solute.

For analytical purposes, as will become more evident in the subsequent development of the topic, interest is centered on the decrease in the radiant power caused by absorption. Consequently, it is necessary to allow for the light lost by scattering and reflection. How this is done will be considered later. For the presentation of the laws of the absorption of light, which follows, scattering and reflection are assumed not to exist.

Lambert's Law. Lambert's law, also known as the Bouguer-Lambert law, describes the relation between the radiant power of the incident light P_0 and the power of the transmitted light P as a function of the length of the light path at a constant concentration. Suppose light of 80 arbitrary energy units is incident on a solution of given concentration and thickness. Assume that for the particular solution (reflection and scattering are negligible) the amount of light absorbed is such that light of only 40 units emerges. Then the transmittance of the sample is 40/80 = 0.50, or 50%. The situation is shown in Fig. 35-2. If the transmitted beam of 40 units falls upon an identical "layer" of solution, again 50% of the radiant energy will be absorbed, and the transmitted beam will have 40 × 0.5 = 20 units. After passing through a third identical layer the energy will be 10 units, and so forth.

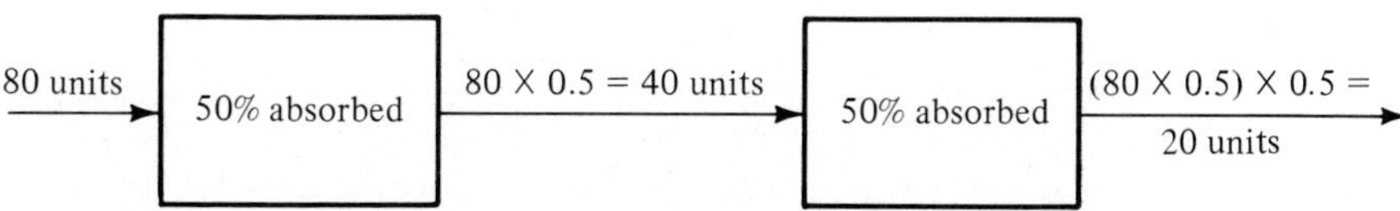

Fig. 35-2. Arrangement of absorbing layers to demonstrate derivation of Lambert's law.

Thus the law may be stated: The transmitted radiant power decreases in a geometric progression as the length of light path increases in an arithmetic progression. The mathematical expression of this law takes the form

$$\log \frac{P}{P_0} = -kb \tag{35-2}$$

where k is a proportionality constant and b the length of the light path. As described above, the ratio of the radiant power of the transmitted light to the power of the incident light, P/P_0, is the transmittance T. Transmittance is related to the absorbance by the equation $-\log T = A$. Thus, the law can be written in several forms, all fully equivalent:

$$\log T = -kb, \qquad \log \frac{1}{T} = kb,$$

$$\log \frac{100}{\%T} = kb, \qquad A = kb \tag{35-3}$$

Beer's Law. Beer's law describes the relation of the radiant power of the incident and transmitted light as a function of the concentration of the absorbing species for a light path of constant length.

Reference is again made to Fig. 35-2. Each "layer" contains a solution of the same concentration and consequently contains the same number of absorbing species (which are assumed not to react with each other). If the absorbing particles of the second layer were transferred to the first layer, the concentration there would be doubled. Light now transmitted by the first layer would have an intensity equal to the light that has passed through two layers of the solution of original concentration. Consequently, the law may be stated as: The radiant power of a beam decreases in a geometric progression as the concentration increases in an arithmetic progression. The mathematical formulation of the law takes the form

$$\log \frac{P}{P_0} = -k'c \tag{35-4}$$

where k' is a proportionality constant and c the concentration. As with Lambert's law, the ratio of the radiant powers can be expressed as transmittance or absorbance, and hence the law can be written in several forms, which are fully equivalent:

$$\log T = -k'c, \qquad \log \frac{1}{T} = k'c,$$

$$\log \frac{100}{\%T} = k'c, \qquad A = k'c \tag{35-5}$$

Lambert-Beer's Law. Lambert-Beer's law, as the name implies, is the combination of the two laws and considers the relation between the radiant power of the incident light and the power of the transmitted light as a function both of the length of the light path and of the concentration of the absorbing species. Its mathematical formulation takes the form

$$\log \frac{P}{P_0} = -k''bc \tag{35-6}$$

where k'' is a proportionality constant, b the length of the light path, and c the concentration.*

---

*Calculus allows derivation of the formulation of Lambert-Beer's law. When light passes through a solution, the change in radiant power dP is greater as the radiant power P and the number N of absorbing particles in the beam are greater. The mathematical relation is

$$dP = -kPdN$$

where k is a proportionality constant, and the minus sign implies that P decreases as N increases. Rearrangement and integration yield

$$\int \frac{dP}{P} = -k \int dN$$

$$\ln P = -kN + \text{constant}$$

When no absorbing particles are present, the radiant power is unchanged, that is, for $N = 0$ $P = P_0$. Consequently, the integration constant can be evaluated:

$$\ln \frac{P}{P_0} = -kN$$

The number of absorbing particles is directly related to their concentration c in the solution and to the length of the light path b through it. Hence,

$$N = k'bc$$

where k' is a proportionality constant. Substituting this equation in the previous one yields

$$\ln \frac{P}{P_0} = -kk'bc$$

Converting to base 10 logarithms by multiplying by 2.3 and combining the three factors into a new constant k″ yield

$$\log \frac{P}{P_0} = -k''bc$$

which is identical with equation (35-6).

Note that the unit of the radiant power, whatever it may be, cancels in the fraction; hence, the left side of the equation is a dimensionless number. Thus it follows that the value and the units of the proportionality constant depends on the units in which b and c are expressed. The light path b is usually expressed in centimeters. The concentration is usually expressed either in grams per liter, denoted by c, or in moles per liter, denoted by C. The proportionality constant k'' accordingly becomes the absorptivity a or the molar absorptivity ϵ (see the definitions of these terms given in Section 35.2). Thus Lambert-Beer's law can be written in numerous forms, which are fully equivalent:

$$\log \frac{P}{P_0} = -abc, \qquad \log T = -abc, \qquad \log \frac{1}{T} = abc,$$

$$\log \frac{100}{\% T} = abc, \qquad A = abc \tag{35-7a}$$

$$\log \frac{P}{P_0} = -\epsilon bC, \qquad \log T = -\epsilon bC, \qquad \log \frac{1}{T} = \epsilon bC,$$

$$\log \frac{100}{\% T} = \epsilon bC, \qquad A = \epsilon bC \tag{35-7b}$$

The physical content of the law is the same irrespective of whether the concentration is expressed in grams per liter or moles per liter. One form of expression is convertible to the other by the formula weight of the absorbing species.

The proportionality constant of Lambert-Beer's law, whether given as a or ϵ, is a characteristic of the absorbing species, and the numerical value generally depends on the wavelength at which it is measured. Lambert-Beer's law is strictly obeyed only with monochromatic light. The implications of this restriction will be considered later.

35.4 Illustrative Examples

A few illustrative examples will advance your appreciation of Lambert-Beer's law and acquaint you with calculations based on it.

Example 35–1 The transmittance of a sample is 30.0%. What is its absorbance?

$$A = \log \frac{100}{\% T} = \log \frac{100}{30.0} = 2 - \log 30 = 2 - 1.477 = 0.523$$

Example 35–2 A sample shows an absorbance of 0.70. What is the transmittance and percent transmittance?

$$-\log T = A = 0.70$$

$$T = 10^{-0.70} = 10^{0.30-1} = 10^{0.30} \times 10^{-1} = 2.0 \times 10^{-1} = 0.20$$

$$\% T = 0.20 \times 100 = 20\%$$

Example 35-3 A solution containing 2.5 mg per 100 mL of a light-absorbing solute of formula weight 200 shows a transmittance of 20% when measured in a cell with a light path of 1.0 cm.

(a) Calculate the absorptivity of the solute.

$$a = \frac{-\log T}{bc} = \frac{-\log (20/100)}{1.0 \times 2.5 \times 10^{-3} \times 10}$$

(10^{-3}: for conversion of mg to g; 10: for conversion of 100 mL to 1L)

$$= \frac{-\log 0.20}{2.5 \times 10^{-2}} = \frac{0.70}{2.5 \times 10^{-2}} = \frac{70}{2.5} = 28 \text{ L} \cdot \text{g}^{-1} \cdot \text{cm}^{-1}$$

(b) Calculate the molar absorptivity of the solute.

$$C = \frac{2.5}{200 \times 100} M$$

$$\epsilon = \frac{-\log T}{bC}$$

$$= \frac{-\log (20/100)}{1.0 \times [2.5/(200 \times 100)]}$$

$$= \frac{0.70 \times 2.00 \times 10^4}{2.5}$$

$$= 5.6 \times 10^3 \text{ L} \cdot \text{mol}^{-1} \cdot \text{cm}^{-1}$$

The result can also be obtained immediately by multiplying the value found for the absorptivity by the formula weight:

$$28 \times 200 = 5.6 \times 10^3 \text{ L} \cdot \text{mol}^{-1} \cdot \text{cm}^{-1}$$

(c) What percent transmittance will be observed when the solution is measured in a 2.0-cm cell?

For the 1.0-cm cell, by Lambert-Beer's law

$$-\log 0.20 = a \times 1.0 \times c$$

and for the 2.0-cm cell,

$$-\log T = a \times 2.0 \times c$$

Dividing one equation by the other yields

$$\frac{-\log 0.20}{-\log T} = \frac{1.0}{2.0}$$

$\log T = 2.0 \times \log 0.20 = 2.0 \times (0.30 - 1) = 2.0 \times (-0.70) = -1.40$. Hence,

$$T = 10^{-1.40} = 10^{0.60-2} = 4.0 \times 10^{-2} = 0.040$$

and $\quad \% T = 4.0\%$

Doubling the length of the light path does *not* halve the transmittance, since there is a logarithmic relation existing between C and T. It is the absorbance that is directly proportional to the light path.

Since the path length is here simply an integer, as in common practice, the calculation can be simplified.

As previously established,

$$\log T = 2.0 \times \log 0.20$$

Hence,

$$\log T = \log(0.20)^{2.0}$$

and

$$T = 0.20^{2.0} = 0.040$$

or

$$\% T = 4.0\%$$

(d) What is the absorbance when 0.50 mg of the solute is present in 50.0 mL of solution and the measurements are made in a 3.0-cm cell?

The absorptivity has been calculated in part (a). The concentration here is

$$\frac{0.50 \times 10^{-3}}{50.0} \times 10^{3} = 0.010\text{g/L}$$

Hence,

$$A = abc = 28 \times 3.0 \times 0.010 = 0.84$$

35.5 Colorimetric Determinations

One of the simplest forms of colorimetry operates as follows.* The sample solution and a series of standard solutions containing the sought-for species in different, known concentrations (and equal concentrations of any required chromo-

---

*A remark on terminology is appropriate. In physics, *colorimetry* refers to the measurement or specification of colors. In the usage of chemical analysis, a colorimeter, unless otherwise specified, is a device for comparing the color of a sample solution with the color of a standard for establishing a concentration. Devices such as the Duboscq colorimeter (Section 35.6) are more appropriately termed *color comparators.*

genic agent) are individually placed in identical clear tubes. The standards are arranged in the order of their concentration, and the sample solution is compared visually with the standards for a color match. The concentration is then evaluated from the concentration of the most closely matching standard or standards. This approach is simple and rapid but of limited accuracy. It is widely used, however, when the concentration of a trace species needs to be known only with an accuracy of say 50 to 75%.

An important example of a comparative color determination is estimating the iron content of diverse chemicals by means of the red color that iron(III) forms in acidic solution with thiocyanate ion. The "heavy metals (as lead)" test, often used in the quality control of chemicals and pharmaceuticals, affords another important example. The sample solution is brought to pH 3 to 4 and is then treated with hydrogen sulfide. The sulfides of relevant metals (Ag, Hg, Pb, Bi, Cu, Cd, As, Sb, Sn) form but remain colloidally dispersed. The faint color of the treated sample solution is compared with standards containing known amounts of lead ion that have been similarly treated. Before the test is made, liquids are evaporated and organic samples ignited. By this proximate test, a rapid, aggregate impression can be reached for this group of metals at the 0.5 to 50 $\mu g/g$ (ppm) level.

For *limit analysis,* the determination may involve merely a single standard. For example, if the color developed with a sample is *less* than or equal to the color of a standard corresponding to the *maximum* content allowed by a product specification, then the sample meets that specification. Another example is afforded by the *o*-tolidine method for free chlorine in water (Section 36.2). As a limit test, the yellow color developed by the sample preparation should be equal to or *greater* than the color of a standard corresponding to the *minimum* free chlorine concentration that adequately inhibits microbial growth.

A variation of the color comparison technique involves replacing the technique of the standard solution by synthetic standards, including properly tinted glass or plastic or even a chart printed in true colors. For example, in the *o*-tolidine test for chlorine, artificial standards are often used because the oxidation product of this reagent is unstable; such standards are either yellow-colored plastic or glass or sealed tubes filled with a mixture of copper(II) sulfate and potassium dichromate solutions. Visual comparison, especially with artificial standards, is still popular, in spite of the availability of instrumental photometric methods. Visual comparative testing has application in the field testing and rapid laboratory "semiquantitative" screening and assessment of such diverse samples as waters, wastes, soils, chemical process samples, brines, and urine.

35.6 Hehner Cylinders and the Duboscq Colorimeter

Hehner cylinders, although they are no longer used, are instructive in their operation. They are identical cylinders having an optically flat bottom and on the side, near the bottom, a drain tube fitted with a stopcock. The sample solution is placed in one cylinder and a standard solution in the other. The two cylinders are placed on white paper or over a uniformly illuminated frosted plate and are viewed from above. From the cylinder presenting the darker color, liquid is drained progressively until a color match is attained. The heights of the liquid columns in the two cylinders are then read, using gradations etched on their walls. With color match, the absorbance of the solutions must be identical because the radiant power of the incident light is the same. From Lambert-Beer's law,

$$A = ab_x c_x = ab_s c_s \tag{35-8}$$

where the subscripts x and s refer to the sample and standard solutions. The absorptivity a has no subscript, since the absorbing species is the same in both solutions. Cancellation of a and rearrangement yields

$$c_x = \frac{b_s}{b_x} \times c_s \tag{35-9}$$

Before the broad availability of photoelectric detectors, visual colorimeters were used. The Duboscq colorimeter is a more elaborate arrangement than the Hehner cylinders for individually varying the length of the light path in the solutions so that a visual color match can be secured. In brief, the light beams pass through vertical cells containing the sample and standard solutions and enter vertically mounted glass plungers having flat ends. These beams are brought by means of prisms and lenses to an eyepiece, where they are viewed as semicircles in juxtaposition. Each cell is mounted on its own platform carried by a rack and pinion movement. When the platforms are moved up or down, the dividing line between the two half-fields virtually disappears if a color match is secured. The lengths of the light path through the sample and reference (or standard) are then read from calibrated scales attached to the racks, and the concentration of the unknown is obtained by using Equation (35-9).

35.7 Limitations of Visual Colorimetry

The human eye is quite capable of detecting minute differences in the intensities of the colors of two solutions if these solutions exhibit colors of the same shade (that is, hue) and if they are viewed concurrently. If, however, a second color is present in one of the solutions the situation becomes different. The reason is that the eye is unable to "analyze" colors, that is, for example, to differentiate whether a certain green is a single color or whether it is a mixture of yellow and blue. (Note the difference between eye and ear; the ear can, for example, isolate the sound of a violin from the total acoustic impression provided by a full orchestra.) The practical consequence of this fact is best understood by an example. Manganese is readily determined colorimetrically if it is oxidized to permanganate and the color developed is compared with that of a standard. This method is sensitive because of the high absorptivity of permanganate, and it can be applied to the analysis of steel. Suppose a steel also contains chromium, which under the conditions of the determination is oxidized to dichromate. The standard solution shows only the purple color of permanganate. In contrast, the sample solution presents a mixture of the colors of permanganate and of dichromate (red orange). This color mixture to the eye is a new color. Consequently, in a Duboscq comparator, no color match (either of color shade or of color intensity) can be attained, whatever the adjustment of the path lengths. One remedy for overcoming the difficulty is to add to the standard solution an amount of dichromate equal to that present in the sample solution. However, such an addition is impractical because it requires the operator to know the concentration of dichromate in the sample solution. Another and far better approach is to place, in both light beams of the comparator, a filter that "cuts out" the color of dichromate. This filter has a color of its own, and what is viewed now is a new color that is a mixture of the colors of the permanganate and the filter. Since the intensity of the filter color is the same in both beams, variation of the path lengths and thereby of the intensity of the permanganate color can now lead to a color match the eye can detect.

The use of such a filter confers an additional advantage. Lambert-Beer's law is strictly valid only for monochromatic light. The lower the degree of monochromacity, the lower the concentration to which the law is followed with reasonable closeness. The color filter cuts out from the compound light a narrower wavelength band and thus provides light of monochromacity.

35.8 Photometric Determinations

Photometric determinations are based on measuring the ratio of the radiant power of the light entering a sample to the power of the emergent light.* In chemical analysis, the terms *photometry* and *photometric determination,* unless otherwise modified, are understood to imply the measurement of molecular absorption in solutions. Although visual photometry is possible, a photoelectric detector replaces the eye in current analytical practice. Photometric methods using *white* light are feasible, but light of a restricted wavelength region is used almost exclusively. When restricted spectral regions are secured by filters, the instrument may be termed a *filter photometer,* and measurements using it, *filter photometry.* When light of a narrow wavelength range is obtained by using a portion of a prism or grating spectrum, the instrument is known as a *spectrophotometer,* and measurements using it, *spectrophotometry.* When a distinction between filter photometry

---

*A photometer in the strict sense of physics is a device that measures the radiant power of light, regardless of any previous interactions of that light. In chemical analysis, unless otherwise qualified, the term *photometer* refers to an apparatus that allows measuring the absorbance or transmittance of a system. Consequently, for the usual term *photometric determination,* the more exact expression "absorption photometric determination" should be understood.

and spectrophotometry is unnecessary, the general terms *photometer, photometry,* and *photometric determination* are appropriate.

The constructional and operational details of a photometer are the concern of the practice of chemical analysis and outside our scope. However, the essential components of a photometer and their function must be described since you need to know about them to understand the basic aspects of photometric methods.

The principal components of a typical spectrophotometer for the ultraviolet and visible regions are shown schematically in Fig. 35-3. The light emerging from the source is collimated to a parallel beam, which passes through a monochromater (here, a prism and slits) and enters the cell, where absorption may take place. The beam emerging from the cell strikes a photoelectric detector, which generates an electrical signal proportional to the radiant power of the incident light beam. The detector output, amplified if necessary, is fed to a meter, the sensitivity of which can be varied by a shunt resistor.

The performance of a photometric determination with such an instrument can be described in terms of a sample solution consisting of a single absorbing species in a nonabsorbing solvent. The cell is first filled with the solvent and placed in the light beam, and an appropriate wavelength is set. The needle of the meter is brought to the 100% T position by adjusting the resistor. Then the cell is emptied, filled with the sample solution, and positioned in the light beam exactly as before. Or more conveniently, the sample is placed in the beam in a second, identical cell. The meter now indicates a value of less than 100% T, since a portion of the radiant power is absorbed. The value read corresponds to the percent transmittance of the sample solution. If the cell length and the absorptivity are known, the concentration of the absorbing species can be calculated by applying Lambert-Beer's law.

Several consequences of this method of operation should be appreciated. As previously mentioned (Section 35.3), other processes besides absorption—scattering

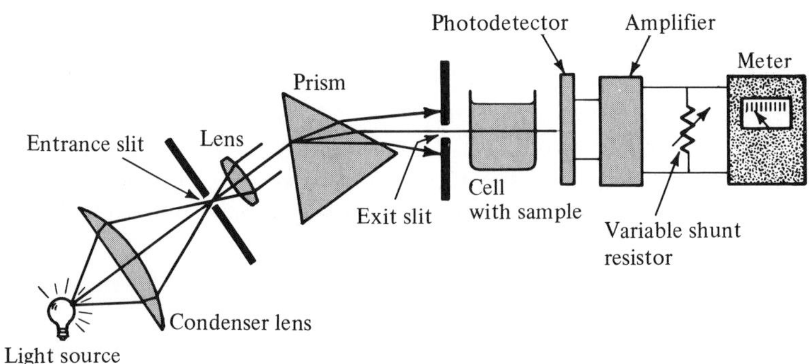

Fig. 35-3. Schematic arrangement of the principal components of a spectrophotometer. (For simplification, the narrow beam of parallel light present after the second lens is shown as a single ray. The prism can be rotated so that light of any desired wavelength can pass through the exit slit, which is usually adjustable to allow variation.)

and reflection—can reduce the radiant power of the light beam. Scattering, due to inhomogeneities or dust particles in the solution, can be minimized by appropriate treatment (thorough mixing and filtration). Reflection is taken care of as follows. The meter reading is proportional to the radiant power of the beam striking the detector. With the solvent in the cell, the radiant power of the beam is reduced by reflection, and the initial adjustment of the meter to 100% T in effect compensates for reflection losses. This compensation is fully successful if the indices of refraction of the solution and the solvent are identical or nearly so, and they usually are in photometric practice.

From an understanding of how reflection losses are compensated for, you can appreciate that any absorbance due to the solvent itself is also compensated for concurrently. Consequently the restriction to a nonabsorbing solvent, previously assumed, is unnecessary. This additional compensation is fully achieved only in quite dilute solutions, since the concentration of the solvent in the sample solution must not depart markedly from the concentration of the pure solvent. (The possibility of compensating for impurities in reagents and solvents used for the treatment of the sample will be discussed in Section 36.7.)

In principle, comparing the radiant power of the beam transmitted by the solvent and the sample solution under identical conditions corresponds to measuring the absorbance of the solute as if it alone were distributed homogeneously throughout the space actually occupied by the sample solution.

A further important conclusion is that if two or more absorbing solutes are present in the solution, the absorbance value (but not the transmittance value) obtained will be the sum of the individual absorbances due to the various absorbing species.

35.9 Transmittance and Absorbance Curves

Selecting a wavelength appropriate for a photometric determination calls for establishing the wavelength region in which absorption is significant. You do this by examining the curve obtained by plotting either transmittance or absorbance versus wavelength.

A transmittance curve can be obtained as follows: A cell filled with solvent is placed in the light beam and the instrument at a wavelength of, say, 400 nm is adjusted to read 100% T. Then the solvent is replaced by the solution and the transmittance read. This process is repeated at, say, 410, 420, 430, . . . nm. The interval between the wavelength settings does not need to be constant; indeed, the values selected will depend on the shape of the curve and the degree to which finer details of the curve must be secured. The values of % T found are plotted versus the wavelength values and a smooth curve is drawn through the points. Either substituting in the formula $A = -\log T$ or reading the absorbance scale provided additionally on most instruments leads to the corresponding absorbance values; and these can be plotted versus the wavelength values. A typical transmittance curve and its equivalent absorbance curve are shown in Fig. 35-4. You can see that an absorbance maximum corresponds to a transmittance minimum, and vice versa.

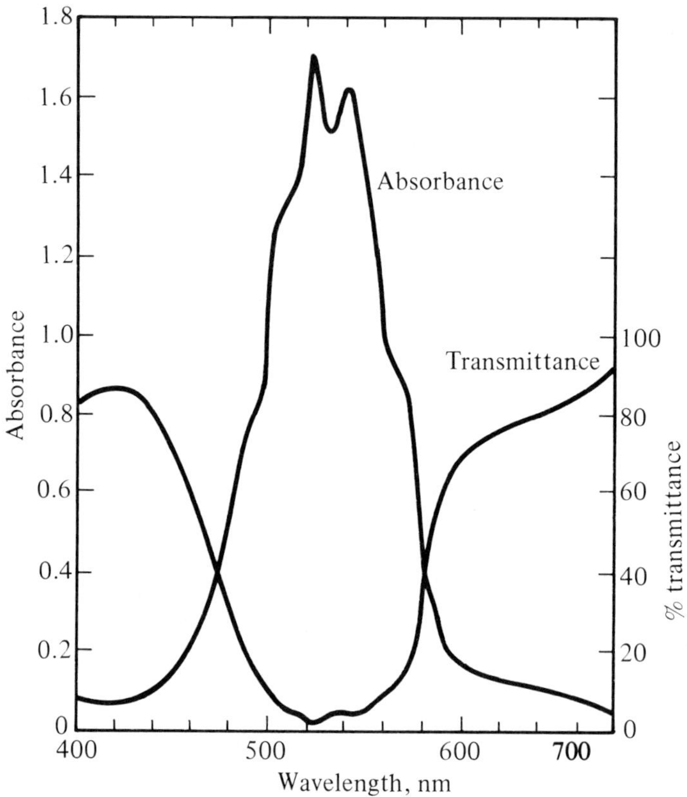

Fig. 35-4. Absorbance and transmittance curves for $KMnO_4$ in water (40 mg/L, 1-cm cell) for the region 400 to 680 nm.

Note the following relation, which may be a source of confusion. The transmittance scale is commonly limited by the values zero and unity and the percent transmittance scale by zero and one hundred. In contrast, the absorbance, by definition, extends from zero to infinity, since $-\log 1 = 0$ and $-\log 0 = \infty$.

35.10 Calibration Curves

It may seem feasible to obtain the result of a photometric determination as follows. The absorbance of the absorbing solute is secured by experiment; the length of the light path in the cell used is measured and the absorptivity of the solute is taken from a table of listed values. From these data, the concentration of the solute is then calculated by applying Lambert-Beer's law. Unfortunately, such a procedure seldom affords an acceptable result. Although the absorptivity at a given wavelength is a characteristic of a substance, the absorptivity values determined will be different with different photometers, principally because the degree of monochromacity will vary from instrument to instrument. Even with a single instrument and at an unchanged wavelength setting, different absorptivity values may be obtained for the same substance at different slit widths; the degree of monochromacity decreases as the slit width increases. It is possible to determine the absorptivity of a substance for a given instrument at a given wavelength setting and slit width and then to use that value for calculating results with the absorbance measurement secured under identical operating conditions. But there still remains the problem of determining the length of the light path. For rectangular cells, the distance between the inner surfaces of two opposite cell faces may be taken as the path length provided that the beam consists of parallel light and traverses perpendicularly to the two faces. These conditions, however, are not always met. Assigning a definite path length becomes even more difficult when cylindrical cells are used.

Consequently, it is more convenient to use a calibration curve, which is obtained as follows. A series of standard solutions containing the species of interest in different but known concentrations is prepared and the transmittance or a function related to it is plotted versus the concentration. Then the transmittance of the sample solution is determined under identical instrument settings and the concentration of the sought-for substance is obtained from the calibration curve. Various methods of plotting a calibration curve are feasible; several are considered below.

35.11 Plot of Transmittance versus Concentration

Since transmittance and concentration are related by a logarithmic function, a straight line is not obtained when these two variables are plotted on ordinary (linear) graph paper (see Fig. 35-5). This fact does not hinder the practical use of the plot as a calibration curve, but it is difficult to decide from such a curve whether Lambert-Beer's law is obeyed and to judge whether the conditions for a photometric measurement could be improved. Consequently, such a plot is seldom used in analytical practice.

35.12 Plot of Absorbance versus Concentration

If Lambert-Beer's law is obeyed, a straight line is obtained on plotting either transmittance versus concentration on semilogarithmic graph paper, or what amounts to the same thing, absorbance versus concentration on linear graph paper. The second method is usually preferred, because interpolation is simpler for an equal-increment scale. For these two plots, the slopes of the lines correspond to $-ab$ or $+ab$, respectively, in the expression of the law. Indeed the slope of the straight-line portion can be read and used to calculate the concentration directly.

The absorbance-versus-concentration curve corresponding to the curve in Fig. 35-5 is shown in Fig. 35-6. Beer's law is obeyed to a concentration of about 0.67 mg/100 mL. Beyond that point, deviation from a straight line becomes clearly evident. The nonlinear portion can still be used as a calibration curve, but the curva-

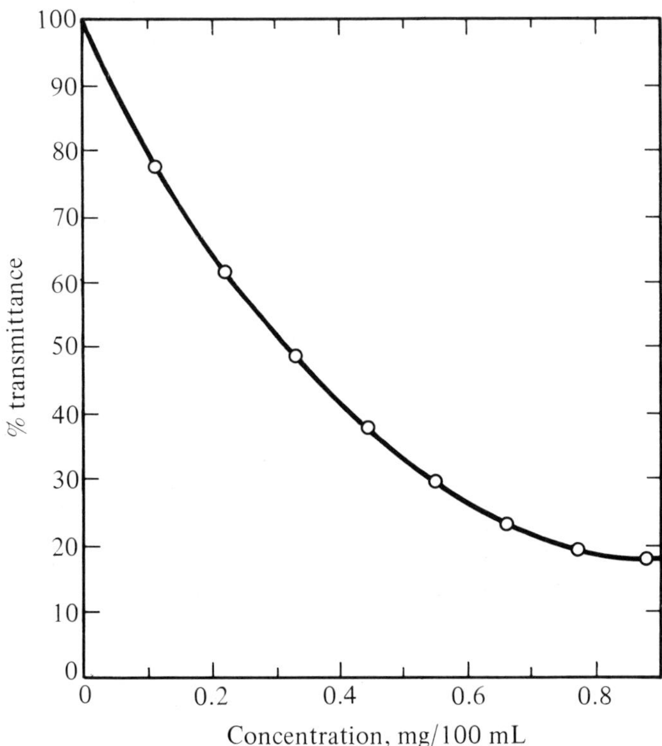

Fig. 35-5. Typical calibration curve, percentage transmittance versus concentration. (See also Fig. 35-6.)

ture is a warning that the working limits for the system are close and that caution should be exercised in interpreting the results (see below).

35.13 Deviations from Lambert-Beer's Law

An absorbance-versus-concentration curve is usually a straight line at a sufficiently low concentration range. At some higher concentration, however, the curve starts to deviate from a straight line and may bend toward either the ordinate or the abscissa. Various factors may operate singly or together to produce a deviation. Frequently, however, the deviation is not because of a true failure of the law but because the conditions prevailing do not conform to the premises of the law.

Lambert-Beer's law is strictly valid only for monochromatic light. Strict monochromacity is not attainable with a practical photometer. When a filter is the monochromating device, the limitations are usually pronounced, and for a given filter little can be done to improve the situation. Consequently, with a filter photometer, deviations are frequent, even at a relatively low concentration. In a spectrophotometer, the prism or grating produces a continuous spectrum from which a

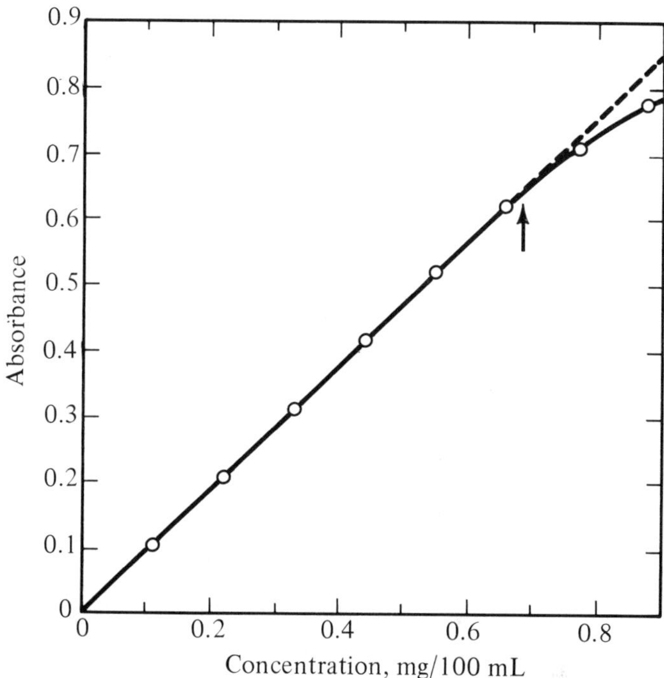

Fig. 35-6. Typical calibration curve, absorbance versus concentration. Arrow indicates point at which deviation from Lambert-Beer's law becomes pronounced. (Same data as Fig. 35-5.)

small wavelength region is selected by a slit. The degree of monochromacity attained is the greater the narrower the slit. As the slit width is progressively decreased, however, the radiant power of the beam passed decreases, and eventually a point is reached at which the instrument can no longer be adjusted to 100% *T* with the solvent in the beam. Consequently, monochromacity is also limited here. Insufficient monochromacity is the most frequent reason for deviations from Beer's law and always causes the absorbance-versus-concentration curve to bend toward the concentration axis. Fortunately, the requirement of strict monochromacity can often be relaxed in practical photometry (see Section 35.16).

In the normal absorption process, radiant energy is consumed by the absorbing species, which is brought to an excited state. As this species returns to the ground state, the absorbed energy is dissipated as heat. Sometimes, however, the absorbed energy, or a part of it, is emitted as light. This phenomenon is known as fluorescence and is undesirable in photometric determinations because large deviations from straightline calibration curves result. In theory, fluorescence could be compensated for by having the same degree of fluorescence in the reference and sample solutions. It is difficult to achieve this condition, however, because fluorescence is extremely sensitive to impurities and solution conditions.

Sometimes it is held that deviations from the law are the result of chemical effects, especially when dissociation-association equilibria are involved, since the position of the equilibrium will depend on the concentrations of the participating species. Often, however, the deviations are not true ones but result from inappropriate plotting of the calibration curve. Here is an example.

Suppose the formal concentration C_{In} of an acid-base indicator of the weak-acid type is to be determined photometrically. The dissociation equilibrium of the indicator and the relevant material balance may be written

$$\mathrm{H}In \rightleftharpoons \mathrm{H}^+ + In^-$$

$$C_{In} = [\mathrm{H}In] + [In^-]$$

Suppose further that the anion In^- is the only species that absorbs at the wavelength selected. If the photometric measurements are made on strongly alkaline solutions, the calibration curve is a straight line when absorbance is plotted versus C_{In}, because in this medium dissociation is essentially complete, practically no HIn is present, and consequently $C_{In} \cong [In^-]$. Next, suppose that the photometric measurements are made under less alkaline conditions, where the indicator is, say, only about 50% dissociated. Under these conditions, the exact degree of dissociation (at a certain pH maintained by a buffer) will depend on the formal concentration of the indicator. The higher this concentration, the lower the degree of dissociation. Nevertheless, if the absorbance were plotted against the molar concentration of the absorbing species, $[In^-]$, a straight line would be the outcome. However, the conventional plot is one of absorbance versus C_{In}, since it is the last-mentioned quantity that is of analytical interest. Because an equality no longer exists between C_{In} and $[In^-]$, a straight line cannot be expected. The two quantities are related through the expression for the acidity constant, and in fact, this constant can be computed from such an apparent deviation from Lambert-Beer's law. In a photometric determination, therefore, it is necessary to differentiate clearly between absorbing species and sought-for substance, and when the two are not identical, the relation between their concentrations under the prevailing conditions becomes of utmost importance.

Finally, deviations (usually erratic ones) from a straight-line plot may result from temperature changes during the measurements. Variations in temperature cause expansion or contraction of the solution and thus changes in concentration. The novice sometimes fails to appreciate that the solution under measurement is exposed to a very intense light beam and the radiant power absorbed by the solution is converted to heat. Thus, significant warming can occur when the solution is allowed to remain in the cell compartment of the instrument for a protracted period. Such warming can result in evaporation losses, especially if nonaqueous, volasolvents are used. Additionally, the warming frequently causes gas bubbles to form from the release of absorbed gases (commonly air). If the bubbles cling to the cells walls and are in the light path, the analytical results can be highly erratic. Temperature effects become especially serious when the absorbing species participates in a temperature-sensitive equilibrium.

35-14 Blank and Reference Solutions

Although photometric determinations can be made with highly concentrated solutions of substances of low absorptivities, the principal analytical application is to dilute solutions. Often the substance to be determined is transformed by a suitable chemical reaction to a species that absorbs highly. The reagents and solvents used in this process are seldom pure enough. Impurities present will have a greater influence on the result in the smaller concentration of the substance to be determined. For example, in the photometric determination of small amounts of silica in water, silica impurities may be present in the reagents and are introduced from the glass vessels used in the operation. It is necessary to compensate for these impurities. Such compensation is secured by using a "blank" instead of the solvent in the initial setting of the meter to 100% *T*. A blank is commonly prepared by carrying out all the treatments to which the sample is subjected (except that the sample is not added). If the sample is a liquid, an equal volume of the solvent is usually added in the preparation of the blank.

A blank of the type described does not compensate for any species present in the original sample that absorbs at the wavelength used for the photometric measurements of the substance of interest. In this situation and preferentially when higher accuracy is desired, comparison is made with a solution of a standard. Such a solution is usually prepared from material closely resembling the sample material in composition but having an exactly known content of the component to be determined. For example, in the photometric determination of manganese in steel, a steel of known manganese content is used as a standard. Solutions of the sample and standard are prepared identically. A cell filled with water is placed in the light beam when the meter is set to 100% *T*. Then the transmittances (or absorbances) of the sample and standard solutions are measured. From the two values obtained, the concentration of the manganese in the sample solution is calculated and thus the manganese content of the steel sample. The closer the transmittances of the sample and standard solutions, the better the result of the determination.

Another way of obviating the interference by absorbing impurities in the sample is the technique of standard addition. A known amount of the substance to be determined is added to one aliquot of the sample solution. This aliquot and one of the original solution are then treated in an identical way. The transmittance of the two final solutions is measured with the solvent or a blank used to set 100% *T*, and the values are used to calculate the unknown concentration. The calculations involved in operating with a standard or with the standard addition technique are illustrated in Section 36.3.

35-15 Photometric Error

As with any other experimental measurement, the final transmittance or absorbance value in a photometric determination is limited in accuracy and precision. The reliability of the result depends on the quality of the instrument, the conditions of the determination, the reproducibility of the instrumental setting, and the like, and particularly, the care of the operator. It is especially interesting to consider how the precision of the photometric measurement, that is, the so-called photometric error, affects the result of the determination. The relation between this photometric error and the precision of the final concentration value is complex. The mathematical relation can be derived by differentiating the expression of Lambert-Beer's law, but this derivation is beyond our scope. Such a formula, however, was used to calculate the data leading to the curve shown in Fig. 35-7. This plot allows reading the relative deviation in the concentration caused by an absolute photometric error of 1% *T*. The curve has an interesting shape. It can be seen that the relative deviation in concentration is minimum at a transmittance of about 37%. Here a photometric error of 1% *T*, that is, of 1 division on a 100-division transmittance scale, causes a relative deviation of 2.7% in the concentration. The same photometric error at 80% *T* produces a relative deviation of 5.6% in concentration. For a good instrument, the actual photometric error experienced is often only 0.2% *T*, which at the optimum condition of 37% *T* yields a concentration value precise within 0.5%.

You can appreciate that when possible, the conditions should be adjusted so that when the sample solution is measured, the transmittance falls between about

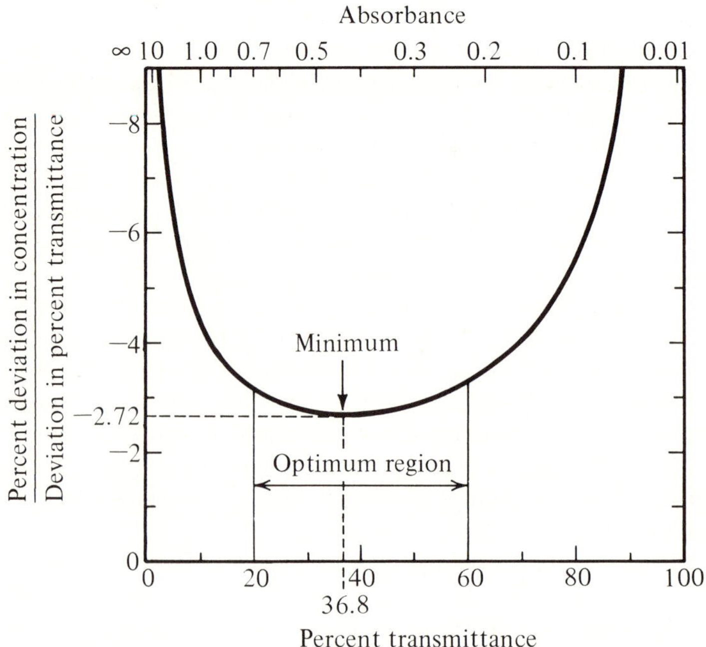

Fig. 35-7. Plot of the relative deviation in concentration caused by a 1% photometric error versus percentage transmittance (lower abscissa scale) and absorbance (upper abscissa scale). Calculated from $100\ \Delta c/c\ \Delta T = 0.434/(T \log T)$, for $\Delta T = 1\%$.

10 and 80%, or better, between 20 and 60%; that is, the absorbance lies between 0.7 and 0.2. How these conditions can be secured is considered below.

The above considerations apply fully only to simple instruments. Each type of instrument has its particular point or range of optimum performance, which either is given in the manufacturer's instructions or can be established experimentally. An attempt should be made to operate as closely as possible to the optimum condition or within the range prescribed.

35.16 Optimum Conditions for Photometric Methods

Before the transmittance or absorbance is read, photometric methods may involve several preliminary steps. Consequently, the selection of the optimum conditions must be based on the evaluation of many facts. Two groups of steps can be differentiated: (1) those involving chemical operations and (2) those related to instrumental considerations.

Only rarely is the substance to be determined a highly absorbing species per se. Chemical reactions of various kinds are used to produce such a species. Some sample approaches are considered in Chapter 36. Often the substance to be determined is a minor constituent, or even a trace constituent, so that separation and enrichment steps are necessary, involving such techniques as precipitation, extraction, distillation, and chromatography. (The principles of such methods are discussed in other chapters of this textbook.) Here, we take up the second group of steps, instrumental considerations.

One of the basic considerations is choosing an appropriate wavelength. To delineate some of the relevant aspects, we refer to the absorbance curve shown in Fig. 35-8. This curve is hypothetical, drawn to emphasize various features relevant to the discussion. The photometric determination will have the highest sensitivity if the wavelength used corresponds to the largest absorbance peak, that is, to 450 nm for the assumed curve. The choice must often be tempered, however, by practical considerations. The situation involving the presence of some other substance that absorbs at that wavelength needs no elaboration. The main difficulty experienced is the inability to reproduce in actual determinations the exact wavelength setting used in the earlier establishment of the calibration curve. A slight change in the wavelength setting brings about a considerable change in the absorptivity and consequently in the slope of the calibration curve. This difficulty is avoided if a reference solution is used and the wavelength setting is left untouched during all the relevant photometric measurements, or if the calibration and determination are performed consecutively without changing the setting.

An additional point is important in relation to peaks. For each wavelength in the definite band passed by the slit, the absorptivity of the substance is different and the light has a low degree of monochromacity. Consider in Fig. 35-8 the flat peak around 525 nm or especially the long shoulder stretching between 550 and 600 nm. Along the shoulder, the absorptivity at each wavelength is essentially the same. Thus the substance, so to speak, does not differentiate between light of 560 and, say, 590 nm. For the substance, either of the wavelengths are absorbed to the

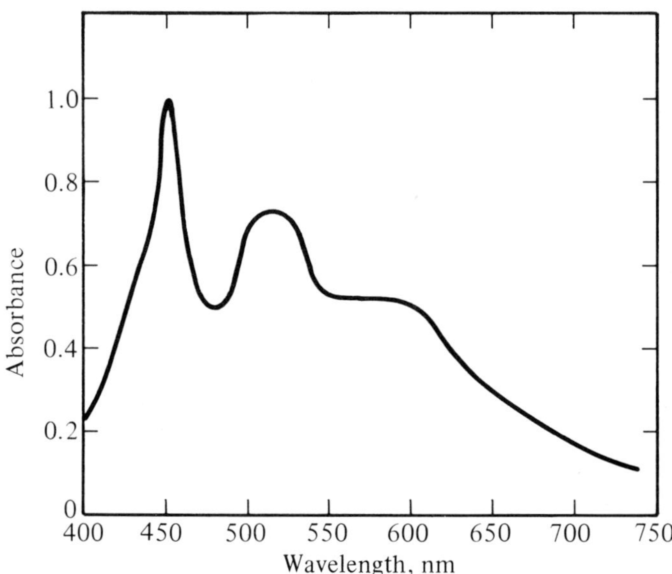

Fig. 35-8. Hypothetical absorbance curve.

same extent, and consequently light at any wavelength range along the shoulder is monochromatic for the substance, although not monochromatic in terms of wavelength. Thus selecting a wavelength band covering a well-rounded peak or a flat shoulder will yield quasimonochromatic light; then the strict requirements set forth in the derivation of the Lambert-Beer law can be relaxed. For the hypothetical substance considered in Fig. 35-8, selecting a wavelength along the shoulder would resolve the problem of monochromacity, but sensitivity would be reduced by a factor of about 2.

As concluded in Section 35.15, the photometric error will have a relatively small effect on the precision of the concentration value if the transmittance of the sample solution falls between about 20 and 60% *T*. It is important to consider this fact in establishing a photometric procedure, so that the most favorable portion of the calibration curve will correspond to the optimal range of the error curve.

When the transmittance of the sample solution falls outside the optimal range, the situation can be remedied as follows. If the transmittance is too *small,* either the sample solution can be diluted to provide a lower concentration of the absorbing species or the photometric measurement can be made either in a cell affording a shorter light path or at another wavelength at which the absorptivity is smaller. The last remedy is less convenient because it requires that a new calibration curve be established at the new wavelength. If the transmittance is too *great,* the photometric measurement can be made with a cell that affords a longer light path. If such a cell is not available or its use does not bring about the desired result, change may be possible to another wavelength at which the absorptivity is larger. The only other approaches, which are less convenient, are to prepare the final solution anew either using a large amount of sample or diluting to a smaller volume. Evaporating solvent from the final solution helps in some cases.

35.17 *Ultraviolet Spectrophotometry*

The treatment of photometry so far, with emphasis on the molecular aspects, (Sections 35.8 through 35.15) has been largely directed to the visible region of the spectrum and can be extended, with mere minor changes, to the ultraviolet region. Absorption in the ultraviolet is essentially identical to what has been described for molecular species in Section 35.1, but with photons of higher energy involved. As the light source in a spectrophotometer for the ultraviolet region, a mercury vapor discharge lamp or a hydrogen or a deuterium discharge lamp is used. Since ordinary glasses do not allow ultraviolet radiation to pass, lenses, prisms, cells, mirrors, and other relevant optical components are fashioned from quartz or high-silica materials. (For measurements in the far ultraviolet, below about 200 nm, oxygen of the air absorbs, and consequently, operation in a vacuum is necessary.)

Commercial spectrophotometers often operate in both the ultraviolet and visible regions and are termed ultraviolet-visible spectrophotometers. In addition to single-beam manual instruments, already described, double-beam manual and automatic recording units are used (see Section 35.18).

The laws, procedural details, use of a blank, and evaluation of absorbance curvès and data for the ultraviolet region are identical to what pertains to the visible region. Impurities and contaminants are, however, not visible to the eye and special precautions must be taken. An "invisible" fingerprint left on a cell, for example, may contain organic materials that absorb strongly in the ultraviolet; thus an apparently clean cell can introduce a serious error.

The number of materials that in research studies or practice are either identified or determined by ultraviolet spectrophotometry is legion and encompasses all classes of compounds. Ultraviolet spectrophotometry is widely applied in the quality control of products in the pharmaceutical and chemical industries and also finds use in clinical and enzymatic analysis. In various forms of chromatography, absorbance measurements in the ultraviolet region are used for the detection and quantitation of the species separated.

35.18 *Automatic Recording Spectrophotometry*

In a photometric determination, as considered in the preceding sections, measurements are made at a single wavelength, which is secured by a monochromator, or in simple instruments, by a filter. For developing a photometric method, it is important to establish the optimum wavelength and also to investigate the possibility of interferences. For these purposes, it is advantageous to study the absorbance behavior over a broad wavelength range, that is, by an absorbance curve of the type of Fig. 35-4. Such a curve can be obtained by making absorbance measurements at various wavelengths, plotting the results, and drawing the best-fitting curve. If only a relatively few points are considered, the danger is great of missing an absorption band or a "shoulder" on such a band. Consequently, the more measurements the better. This situation is especially important when the absorbance curve, is used not for evaluating the conditions for a determination, often the case, but for identifying a compound or elucidating its structure. For these purposes, the ideal situation is to have an infinite number of points, which in practice amounts to a continuous tracing of the absorbance curve by an automatic recording spectrophotometer.

Such instruments vary greatly in complexity and cost, depending on the wavelength region covered, number of scanning speeds provided, quality of the monochromator, accuracy and precision attainable, and the like. Common automatic recording spectrophotometers cover the ultraviolet and visible regions from 200 or 210 nm to 700 or even 800 nm. (Spectrophotometers for the infrared region are almost exclusively of the recording kind, for reasons that will be made clear in Section 35.19).

Whatever the range of the recording instrument, the underlying design principles are similar. The monochromator is a scanning one, that is, the dispersion device (prism or grating) is moved through an arc by a motor, which is also coupled to drive a drum on which the recording chart paper is held. Consequently, distance along the side of the chart in the direction of the drum rotation (the longer side) is related proportionally to wavelength.

Automatic recording spectrophotometers are usually double-beam instruments. The light from a *single* source is split into two beams by a half-silvered mirror, an arrangement of prisms, or other devices. One beam passes through the reference cell, the other through the sample cell. For a given source, even without fluctuation in its output, the intensity of the light emitted will vary with the wavelength, and so will the optical efficiency of the cell and other components, and also the sensitivity of the detector. Consequently, a compensation device is provided to get a straight "baseline" when the cells are filled with the same liquid and the wavelength is scanned. This regulation is secured, to name only two possibilities, by changing the slit width in unison with the rotation of the monochromator prism or grating, or by moving a comb or flag into the light beam or out of it as appropriate.

Most important in the recording spectrophotometer is a rotating disk with holes so located that the sample and reference beams fall intermittently onto the detector. The intensities detected for the two beams are processed electronically in such a way that their ratio is derived as an electrical output (that is, with appropriate settings and calibration). The signal is fed to the recorder pen. Often additional circuitry is provided that can be activated by moving a switch so that the signal fed to the recorder corresponds to absorbance instead of transmittance.

The requirements for source, optical components, and detectors are the same as previously mentioned. To cover the entire ultraviolet-visible region, a switch in source (and sometimes in other components) is needed. This usually happens at about 400 nm. In some instruments, this is accomplished manually by moving a lever; in others, automatically.

35.19 Infrared Spectrophotometry

For the infrared region, the energies of the absorbed photons are not enough to excite electronic transitions in molecules; only vibrational and rotational levels are excited. However, because of the numerous vibrational modes possible for a polyatomic molecule, infrared absorption spectra are more complex than spectra for the ultraviolet-visible region.

The many modes for vibration can be appreciated by studying two identical atoms (for example, hydrogen) covalently bonded nonlinearly to a third atom (carbon), which is assumed to be a stable reference point. For this situation, the principal modes for vibration are delineated in Fig. 35-9 and involve bending and stretching. As indicated, two atoms, each of which is covalently bonded to a third, can swing back and forth in a plane perpendicular to the paper either in the same direction (wagging) or in opposite directions (twisting). Additionally, in the plane of the paper the two atoms can swing in phase (rocking) or out of phase (scissoring). In stretching modes, the two atoms move along the valence bond direction either together (symmetrical stretching) or in opposition (asymmetrical stretching). Each of these vibrational modes involves the absorption of radiation at different wavelengths. The absorption is further complicated because the two atoms involved in a given vibration may be different and the molecule may be complex. Also, "harmonic" and "beat" frequencies for vibration can be generated. As a consequence, the spectrum recorded consists of bands, and often many of them; this can be seen in Fig. 35-10, which is a reproduction of the infrared absorption spectrum of the simple hydrocarbon hexane, $CH_3CH_2CH_2CH_2CH_2CH_3$.

Usually the wavelength region (or maximum) of each absorption band corresponds to a small group of atoms within even a complex molecule, and such

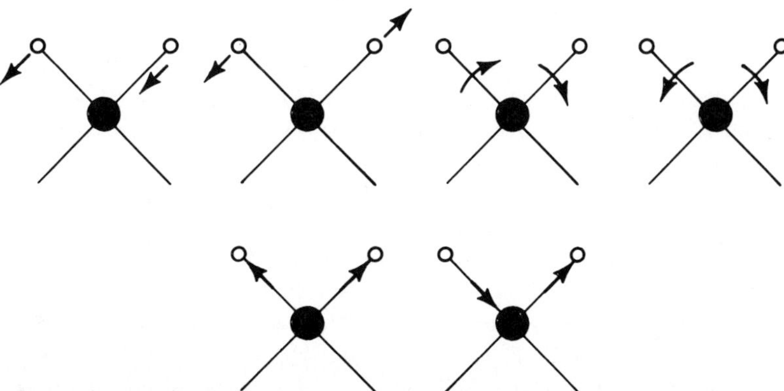

Fig. 35-9. Principal modes of vibration of hydrogen about carbon in an organic compound: bending (top line, left to right, wagging rocking, twisting, and scissoring) and stretching (bottom line, left to right, symmetrical and asymmetrical; see text).

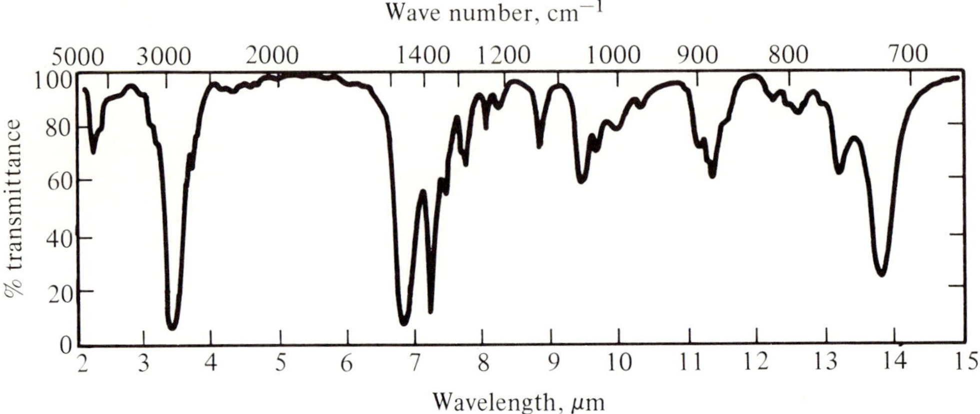

Fig. 35-10. Infrared absorption spectrum of hexane, $CH_3CH_2CH_2CH_2CH_2CH_3$, a typical alkane.

atomic groupings tend to have the same wavelength regardless of the molecular structure in which they are present. These so-called *group* or *characteristic frequencies* can be used to "fingerprint" a compound and help in its detection and identification. For example, for alkanes, methylene groups, $-CH_2-$, show strong absorption due to stretching at 3.42 μm and bending at 6.83, 7.8, 8.8, and 14 μm; methyl groups, $-CH_3$, show stretching at 3.38 μm and bending at 6.9 and 7.3 μm. You should be able to locate these group frequencies for the spectrum of hexane in Fig. 35-10. As you might suppose, the infrared absorption-spectrum for 1-hexanol, $CH_2CH_2CH_2CH_2CH_2CH_2OH$, shows absorption at the group frequencies for methylene and methyl groups, as in hexane, and additionally, at the group frequencies for the primary alkanol group, $-CH_2OH$ (that is, stretching at 2.9 μm and bending at 7.2 and 9.8 μm).

Infrared spectrophotometers are almost exclusively of the double-beam recording type (described in Section 35.18). Glass and quartz are not transparent above 3.5 μm. Consequently cells and prisms are usually fashioned of metal salts, such as sodium chloride (transparent to 15 μm), potassium bromide, or thallium(I) bromide. Prisms of these salts are attacked by moisture and must be kept dry. (Some instruments have a heater to achieve this; it remains on when the unit is in a standby mode.) These and some other shortcomings are avoided by use of a reflecting grating ruled on a metal such as aluminum.

The radiation source is a bar of thermally stable material electrically heated to 1000 to 1800°C. The Nernst glower is a mixture of the oxides of zirconium, cerium, and thorium, and the Globar source is a sintered silicon carbide.

A thermocouple, thermistor, or bolometer can be used as the detector. The last-named detector involves a wire resistance that is exposed to the infrared beam and forms one arm of a Wheatstone bridge. Any change in the temperature of the wire causes a change in its resistance, thereby causing the bridge to be unbalanced. The electrical signal created is amplified and fed to the recorder.

Liquid samples (water-free) can be examined by infrared techniques when they are placed in a sodium chloride or a potassium bromide cell. Solid and liquid samples can also be examined as solutions. Water cannot be used as the solvent because it absorbs strongly in the infrared region; it also dissolves the cell. A well-dried organic solvent is selected that is transparent in the infrared region of immediate interest. Solids can be finely ground and then mixed in known proportion with dried, powdered potassium bromide; the mixture is then pressed into a disk, commonly 1 cm in diameter and up to 2 mm in thickness. This disk is placed in the "sample" beam. Another disk containing only potassium bromide is placed as the blank into the "reference" beam.

For qualitative purposes, the finely ground sample can also be slurried with paraffin oil or a halocarbon grease. The resulting *mull* is spread between sodium chloride or potassium bromide plates and the "sandwich" is positioned in the sample beam.

Because gases are of relatively simple structure and thereby have simple infrared absorption spectra, infrared methods are of special interest in analyzing gases. Because gases have a low density and thus few absorbing molecules in the light path, a cell of conventional length does not provide adequate absorption for measurement. Consequently, cells having path lengths of 10 cm or greater are used; they are fitted with windows made of the appropriate metal salt. Compounds such as air pollutants present in the gas phase in trace concentrations can be determined by using multipass cells with mirrors; these allow path lengths of up to 40 m.

Inorganic compounds with covalent bonds (for example, sulfate and phosphate salts) can be examined by infrared methods. The main applications for infrared, however, include identifying organic compounds and elucidating structures, notably through assessing group frequencies (see above).

Quantitative applications for infrared methods of analysis lag partly because they have poorer reliability compared with visible and ultraviolet photometry (short path length, uncertain thickness of disks, narrow bands with deviation from Lambert-Beer's law encountered, and so on).

35.20 Questions

35–1 What is the relation between the absorbance of a solution and the concentration of the absorbing species?

35–2 Would doubling the concentration of a solution cause the transmittance to decrease by a factor of 2, other conditions being unchanged? Explain your answer.

35–3 By what processes is light generally lost when a beam passes through a cell containing an absorbing solution? Which of these processes is of primary concern for analytical applications?

35–4 What is monochromatic light? Is monochromatic light identical with light viewed by the human eye as a single or a pure color?

35–5 On what does the numerical value of absorptivity depend?

35-6 Does the concentration of the absorbing species influence the absorptivity?

35-7 What is the relation between absorbance and transmittance?

35-8 Within what wavelength region does visible light fall?

35-9 Would the introduction of some scattering dust particles into a solution change the absorbance of the solution? Would it change the molar absorptivity of the absorbing solute? Explain your answers.

35-10 With reference to Fig. 35-2, assume that the two layers contain solutions of different substances and show different absorbances. Show that the total absorbance will be the sum of the absorbances of the two layers. Will the total transmittance be the sum of the two transmittances?

35-11 How would the absorbance value for a solution change if the concentration were doubled and the path length halved?

35-12 Define *absorbance* and *transmittance.*

35-13 In your own words, state what is meant by the terms *spectrum, monochromacity,* and *compound light.*

35-14 What is stated by Lambert's law? Beer's law?

35-15 Explain why the degree of monochromacity of the light is important in photometric determinations.

35-16 Describe in your own words the difference between *colorimetry* and *photometry* as the terms are used in chemical analysis.

35-17 What is the approximate range of the visible portion of the electromagnetic spectrum in nanometers? in meters? in centimeters?

35-18 What is a "blank" in a photometric determination? What is its function?

35-19 Explain what difficulties may arise if the wavelength selected for a spectrophotometric determination corresponds to a sharp peak in the absorbance curve?

35-20 What are the advantages of a plot of absorbance versus concentration compared with a plot of percentage transmittance versus concentration as a calibration curve for a photometric determination?

35-21 When the concentration of an absorbing species is doubled and other conditions are kept constant, what is the effect on the transmittance? on the absorbance?

35-22 List the causes of some of the so-called deviations from Lambert-Beer's law.

35-23 What are the advantages of using a standard sample as a reference solution in a photometric determination?

35-24 Why are results unsatisfactory when a practical photometric determination is based on a tabulated absorptivity value?

35-25 Is light of "single color" or "pure color" identical with monochromatic light? Explain.

35-26 Absorbance data can be presented as a plot of the logarithm of the absorbance versus wavelength. In such a plot all curves for a single substance will show the same shape, regardless of the concentration of the substance, and will be displaced from each other only with regard to the ordinate. By a mere parallel shift, two such curves may be superimposed. Derive this finding as a mathematical consequence of Lambert-Beer's law. (*Hint:* Express log A in terms of the expression for Lambert-Beer's law.)

35-27 Compare the production of radiant energy by the sources used in ultraviolet, visible, and infrared photometric methods.

35-28 Compare methods of detecting radiation in ultraviolet, visible, and infrared spectrophotometry.

35-29 What is achieved by using a so-called multiple-reflection cell in photometry?

35-30 Using information provided in Sections 35.8 and 35.18, sketch the arrangements for a double-beam spectrophotometer and elaborate on its advantages and disadvantages compared with a single-beam instrument.

35.21 *Problems*

35-1 Calculating the absorbance value of each of the following samples from the measured % T value.

	% T	
(a)	20.0	(*Ans.:* A = 0.699)
(b)	50.0	
(c)	35.0	
(d)	65.0	
(e)	43.0	
(f)	61.0	

35-2 Calculate the % T of each of the following samples from the measured absorbance value.

	A	
(a)	0.235	(*Ans.:* 58.2% T)
(b)	1.72	
(c)	0.689	
(d)	0.055	
(e)	0.830	
(f)	0.425	

35-3 The absorbance of a solution, containing 4.0 mg of a light-absorbing solute (*150.0*) in a volume of 250.0 mL is measured in a 5.00-cm cell and found to be 0.400. Calculate the molar absorptivity of the solute.

Ans.: 7.5×10^2 $L \cdot mol^{-1} \cdot cm^{-1}$

35-4 Calculate the molar absorptivity for a particular solute from the following data: formula weight, 200.0; % *T*, 50.0; *c*, 1.3 mg/L; cell path length, 2.00 cm.

35-5 A solution containing 10.0 mg of a light-absorbing solute (*125.0*) in 250.0 mL shows 25.0% *T* in a cell having a 5.00-cm light path. Calculate the absorbance of the solution and the absorptivity and molar absorptivity of the solute.

35-6 A solution contains 4.0 mg of a light-absorbing compound (*50.0*) in 300.0 mL. The solution shows 35.0% *T* when measured in a 5.00-cm cell. Calculate the absorbance of the solution and the absorptivity and molar absorptivity of the solute.

35-7 A solution containing 5.0 mg of a light-absorbing substance in 250.0 mL shows 50.0% *T* when measured in a cell with a 3.00-cm light path. What will be the % *T* of a solution containing 1.0 mg of the substance in exactly 1 L when measured in a 10.0–cm cell?

35-8 The molar absorptivity of a substance is 2.3×10^4 $L \cdot mol^{-1} \cdot cm^{-1}$. A solution of the substance is placed in a 1.00-cm cell and the transmittance is found to be 0.200. Calculate the molar concentration of the solution.

Ans.: $3.0 \times 10^{-5} M$

35-9 A solution shows 64.0% *T* when measured in a 2.00-cm cell. Calculate the % *T* when the same solution is measured in a 1.00-cm cell.

35-10 A solution containing 10.0 ppm of $KMnO_4$ (*158.0*) shows an absorbance of 0.80 in a 2.00-cm cell at 520 nm. Calculate the molar absorptivity of $KMnO_4$.

35-11 The "heavy metals (as lead)" proximate test (see Section 35.5) was included in assessing a lot of pharmaceutical-grade sodium bicarbonate. To 5.0 g of a representative sample of the lot were added 10 mL of water and then 9.5 mL of 10% hydrochloric acid. The mixture was heated to boiling and kept at that temperature for 1 min. A drop of phenolphthalein solution was added and then aqueous ammonia dropwise until the solution became faint pink in color. The solution was transferred to a tube with a ring mark at 25 mL, was treated with 2 mL of 1*F* acetic acid, and diluted to mark with water. To three other identical tubes the necessary volume of lead nitrate standard solution was added to introduce, respectively, 0.02, 0.04, and 0.06 mg of lead; then phenolphthalein, aqueous ammonia, and acetic acid were added as above, and the solution diluted with water to 25 mL. To all *four* tubes was then added, with mixing, 10 mL of water freshly saturated with hydrogen sulfide gas. The tubes were allowed to stand 10 min with occasional swirling. Each tube was then viewed downward over a white surface. The "gray" developed by the sample was about midway between the color of the 0.02 and 0.04 mg lead standards. (a) What is the heavy metals (as lead) value for the lot expressed in micrograms per gram? Would this lot meet a specification for this test of 5 ppm maximum?

Ans.: (a) 15 μg/g; (b) yes

36

Applied Colorimetric and Photometric Determinations

Relatively few substances are so highly colored, or more generally, have such a high absorptivity that their direct colorimetric or photometric determination is possible. Of the inorganic cations, copper(II) and nickel(II), for example, form colored aquo ions, but the color is so feeble that relatively high concentrations are needed for a reasonably successful determination. The principal domain of photometry, however, is the determination of small amounts and low concentrations. Therefore it is usually necessary to transfer the cation or any other species to be determined into a strongly absorbing entity. Sometimes oxidation (or reduction) will yield a strongly absorbing species. For example, manganese is oxidized to permanganate ion and chromium to chromate ion. The most satisfactory approach for obtaining strongly absorbing species, however, is to use a chromogenic agent, that is, a color-forming agent.

36.1 Chromogenic Agents

For inorganic analysis, chromogenic agents are usually complexing agents that form a strongly absorbing complex with the species of interest. Only a few inorganic substances are suitable as chromogenic agents for quantitative purposes. Common examples include ammonia, which forms an intensely blue complex with copper(II), and thiocyanate ion, which forms an intensely red complex with iron(III) and a red orange one with molybdenum(V). Most chromogenic agents are organic substances. Several requirements must be fulfilled if a chromogenic agent is to be suitable for a given photometric purpose. The agent itself should not absorb light in the wavelength region in which the complex absorbs. If the agent does absorb slightly, compensation is possible by a proper reagent blank; where absorption by the agent is significant, special measures are needed (see the consideration of dithizone in Section 36.2).

The absorptivity of the colored complex formed should be high and remain essentially constant when changes in conditions take place within a reasonable range. Such conditions include temperature, pH, and the presence of inert salts.

The color should develop rapidly and once reached should be stable at least long enough to permit the photometric (or colorimetric) measurement. The complex formed should have a high value for its stability constant. As was explained in Section 35.12, a straight-line relation, according to Lambert-Beer's law, is possible only if the absorbance is plotted versus the concentration of the absorbing species; in practice the concentration of the sought-for substance is usually plotted on the abscissa. As long as the conversion of the species to be determined to the colored complex is essentially complete, the plot will still be a straight line. If this complex dissociates, however, deviation from a straight line will result. Fortunately, the equilibrium can often be shifted favorably by adding an adequate excess of the chromogenic agent. Such an excessive addition can be used only if the reagent itself does not absorb appreciably at the wavelength used in the determination.

High selectivity is a further desirable feature so that there will be few interferences and hence a limited need for preliminary separations or resort to masking. From a practical standpoint, it is also preferable for the solution of the chromogenic agent to be stable in storage so that frequent preparation is avoided.

Many complexes used for photometric determinations are readily extractable from an aqueous solution into an organic solvent nonmiscible with water. This behavior offers an advantage in permitting the concentration of small amounts of the colored complex and often confers added selectivity on the determination.

36.2 Examples of Photometric Determinations

For inorganic substances, very many photometric (and colorimetric) procedures are known, and additional ones are being introduced. Only a few representative examples can be given here to illustrate the principles involved.

Manganese is often determined by oxidation to permanganate ion with potassium periodate or ammonium persulfate as the oxidant. The absorbance of the permanganate ion can be measured at about 530 nm (the molar absorptivity is 2.3×10^3 $L \cdot mol^{-1} \cdot cm^{-1}$). The method is used routinely for determining manganese in steel and other alloys. Iron, if present, is decolorized by adding phosphoric acid, which yields a colorless iron(III) complex (Section 25.5).

Titanium, in small amounts in alloys, rocks, minerals, and the like, is frequently determined in a strongly acidic medium by adding hydrogen peroxide and measuring the orange peroxotitanate ion at 410 nm (molar absorptivity, 5×10^2 $L \cdot mol^{-1} \cdot cm^{-1}$).

So-called isopolyacids result when two or more molecules of a simple oxygen acid condense to form an aggregate; their structure involves a central ion linked to surrounding acid anhydride molecules through oxygen. If two or more different acids are involved, the resulting condensed acid is known as a heteropoly acid; the molybdophosphoric acids, $P_2O_5 \cdot xMoO_3 \cdot yH_2O$ ($x = 12$ or 24 commonly) are well-known examples. Heteropolyacids and their anions are important in chemical analysis, especially in photometric and colorimetric methods and in solvent extraction. The yellow polymolybdophosphate complex (also known as phosphomolybdate) is frequently used for determining phosphate in rocks and fertilizers. Photometric measurements are made at 380, 400, or 420 nm. Some heteropolyacids are reduced readily to blue-colored species; these are known as heteropoly blues. With molybdophosphoric acid, the product is known as phosphomolybdenum blue. The development of such a deeply colored species affords greater sensitivity and allows the determination of extremely small amounts of phosphorus. The reduction is effected with tin(II) or ascorbic acid, for example.

Silicon and arsenic form silicophosphoric and arsenophosphoric acids; these are also reducible to heteropoly blues. These reactions are often used for determining small amounts of these elements. For example, silica in water is often determined photometrically in this way.

An important procedure for the photometric or colorimetric determination of small amounts of ammonia, is known as Nessler's method. When it is conducted as a visual comparison method, the ammonia-containing sample solution and a series of ammonia standards are placed in special flat-bottomed test tubes known as Nessler tubes. A solution of potassium tetraiodomercurate(II) is added. The total reaction can be expressed as

$$\underset{\text{Nessler's reagent}}{2HgI_4^{2-}} + 2NH_3 \rightarrow \underset{\text{intense orange}}{\underline{NH_2Hg_2I_3}} + NH_4^+ + 5I^- \qquad (36\text{-}1)$$

(The colored product can be formulated in other ways.) A suspending agent is added, usually gum arabic, to retard flocculation and settling of the colloidal precipitate. The solutions are all diluted to the same volume and the visual comparison effected promptly. Alternatively, standards and a sample solution, prepared similarly, may be measured photometrically. For the determination of small amounts of nitrogen, a Kjeldahl digestion can be performed (see Section 14.9) and after sodium hydroxide is added, the ammonia is distilled (or diffused) and "nesslerized."

Nickel can be determined as the red complex formed with dimethylglyoxime, $CH_3C(:NOH)\,C(:NOH)\,CH_3$ (see also Section 7.2). The complex can be extracted into chloroform and determined at 375 or 327 nm (molar absorptivity 3.5×10^3 and 5.0×10^3 $L{\cdot}mol^{-1}{\cdot}cm^{-1}$, respectively). This separation improves the selectivity, since for example, copper can be kept in the aqueous phase by masking it as the copper(I)-thiosulfate complex. The sensitivity is excellent since less than 2 μg of nickel present per milliliter of the aqueous solution can be determined accurately. Alternatively, the nickel-dimethylglyoxime complex can be oxidized (with bromine, for example) in an alkaline aqueous medium to an intensely red entity, which allows direct photometric measurements at 465 nm (molar absorptivity, $\sim 1.5 \times 10^4$ $L{\cdot}mol^{-1}{\cdot}cm^{-1}$).

Of the reagents forming colored complexes with metal ions, dithizone (that is, diphenylthiocarbazone) is frequently used. Its formula can be written as follows, to emphasize the possibility of the formation of five-membered rings on chelation with metal ions:

```
                NH   N
               / \  / \\
C6H5—NH    C    N—C6H5
                ||
                S
```

The structure of the neutral (uncharged) complexes formed depends on the metal, its oxidation state, and the pH of the solution. Complexes having 1:1 and 1:2 metal-to-reagent molar ratios form, and for many metals, the ligand atoms are nitrogen and sulfur. The reagent itself as well as the complexes are only slightly soluble in water but readily soluble in carbon tetrachloride and chloroform. In these solvents, the "free" dithizone is green, but most of the complexes are red or orange.

Dithizone is one of the most sensitive reagents for metals and is often used in trace analysis, notably in separating and determining lead, mercury, bismuth, and copper. Unfortunately, the reagent absorbs markedly at the wavelength maximum of the metal complexes and vice versa, although to a different degree. Consequently, it is necessary either to remove the excess reagent or to add it in an exact, reproducibly measured amount. In the second case, either the increase in the absorbance can be measured at a wavelength at which the metal complex absorbs strongly or the decrease in absorbance at a wavelength at which dithizone itself absorbs strongly. If dithizone is represented by HDz, the extraction of a metal-dithizone complex from an aqueous solution can be represented by

$$\underset{\text{in water}}{M^{n+}} + \underset{\text{in solvent}}{n\text{HDz}} \rightleftharpoons \underset{\text{in solvent}}{\text{MDz}^{m}} + \underset{\text{in water}}{nH^{+}} \quad (36\text{–}1)$$

From this equation, you can see that the extraction efficiency of a metal depends on the pH and the stability of the metal-dithizone complex. The stronger the complex, the lower the pH at which it can be effectively extracted. This behavior confers some degree of selectivity on dithizone for practical analysis. (Solvent extraction is considered further in Section 44.4.)

Diethyldithiocarbamate ion, usually added as the sodium salt, forms water-insoluble, solvent-extractable complexes with numerous metal ions (notably those that are sulfophilic and form insoluble sulfides). This reagent has the structure

```
CH3CH2          S
      \        //
       N—C
      /        \
CH3CH2          S-
```

The complexes formed have four-membered chelate rings involving the two sulfur atoms, and for many metal ions, M^{n+}, the composition corresponds to $M(Dt)_n$, where Dt^- represents the diethyldithiocarbamate ion. Selectivity in using this reagent is achieved by adjusting the pH and using masking agents (citrate, tartrate, cyanide, EDTA, and the like). Copper can be determined photometrically in many materials, including blood serum, as the yellow-to-brown copper(II) complex formed with the diethyldithiocarbamate anion. The selectivity of the extraction is often augmented by adding citrate and EDTA, which mask many metals. Photometric measurements are made at 436 nm (molar absorptivity in carbon tetrachloride, 1.3×10^4 $L \cdot mol^{-1} \cdot cm^{-1}$).

Various oxidants are determined by allowing them to react with essentially colorless substances to form colored products. The absorbance measured is proportional to the amount of the oxidant. Redox indicators often find use for this purpose. Probably the best known of these procedures is the colorimetric or photometric determination of "active" chlorine in water in the parts per million range using *o*-tolidine, $(4{-}NH_2{-}3{-}CH_3C_6H_3{-})_2$. The oxidation product of this compound in acidic solution is yellow, and a photometric determination can be made at 438 nm. Very small concentrations of chlorine can thereby be determined, depending on the instrument, down to 0.01 μg/g (ppm).

Absorption photometry is also possible for the gaseous state. This topic is briefly mentioned in Section 35.19 and Chapter 39.

36.3 *Kinetic Analysis by Photometry*

Many quantitative analytical methods require that some reaction go practically to completion. In contrast, in kinetic analysis, the change in the concentration of a reactant or product is measured as the reaction proceeds. The calculation is based on an underlying reaction-rate relation. Many analytical approaches are used in kinetic analysis, and photometry is widely applied. The rate of a catalyzed reaction depends on the concentration of the catalyst; consequently, its concentration can be determined by kinetic measurements. The principles of kinetic analysis can be appreciated from a few examples of catalytic methods.

Determining trace iodine is of practical importance and is used, for example, in assessing thyroid activity. For the so-called protein-bound iodine (PBI) determination, a serum sample is first digested with acid to destroy organic matter, under conditions that prevent volatilization of iodine. The *slow* reaction between cerium(IV) and arsenic(III),

$$2\ Ce^{IV} + As^{III} \rightarrow 2\ Ce^{III} + As^{V} \qquad (36\text{–}2)$$

can be catalyzed by iodine/iodide. The catalysis is represented by the following two *fast* reactions:

$$Ce^{IV} + I^- \rightarrow Ce^{III} + I^0 \qquad (36\text{–}3)$$

$$As^{III} + 2I^0 \rightarrow As^{V} + 2I^- \qquad (36\text{–}4)$$

The iodine formed by the first reaction oxidizes arsenic(III), to re-form iodide, which reacts with more cerium(IV), and so on. Since iodine is present only at trace levels, the only noticable colored species present are those formed by cerium(IV). The progress of the reaction can therefore be followed by the decrease in color intensity. For the determination, the sample preparation is added to a freshly prepared mixture of cerium(IV) and arsenic(III) in dilute sulfuric acid. Either the decrease in the absorbance at 420 nm is measured at a known elapsed time, or the time needed for complete decolorization is recorded. The calculation of the iodine content of the sample is based on carrying known amounts of iodide through the procedure, often as a standard addition to the sample.

Trace copper in semiconductor materials can be determined photometrically by means of the slow reaction between iron(III) and thiosulfate ion catalyzed by copper. The rate of bleaching of the violet complex or complexes formed by iron(III) and salicylic acid (*o*-HOC_6H_4COOH; HSal) by thiosulfate is a linear function of the trace copper concentration. The catalyzed reaction can be written

$$2Fe(Sal)_x^{3-x} + 2S_2O_3^{2-} \rightarrow 2Fe^{2+} + S_4O_6^{2-} + 2x\text{Sal}^- \qquad (36\text{-}5)$$

Kinetic methods of analysis are very important for the biosciences. Enzymes are biological catalysts; consequently, enzymatic methods are largely based on kinetic measurements. For example, phosphatases catalyze the hydrolysis of phosphate esters, and depending on the optimum pH for their activity, are termed *acid* (pH 5-6) or *alkaline* (pH 8-10). The determination of phosphatase activity in blood serum is an important diagnostic tool. For a photometric approach, the phosphate ester of a phenol having a favorable absorbance is used as the reactant (substrate). For example, when *p*-nitrophenyl phosphate, disodium salt, *p*-$NO_2C_6H_4OPO(ONa)_2$, is used, the rate of formation of *p*-nitrophenol, *p*-$NO_2C_6H_4OH$, is measured at 405 nm and is a linear function of phosphatase activity.

36.4 Calculations in Photometric Determinations

Some examples of calculations involving Lambert-Beer's law have already been given in Section 35.4. Some additional examples will further elucidate the principles involved in applied photometric determinations.

Example 36-1 A substance S is determined in the following way. A 0.7520-g amount of the sample material containing S is dissolved, a chromogenic agent is added, and the solution is diluted to a volume of exactly 250 mL. As a reference, a standard known to contain 0.345% of S is used; a 0.3500-g amount of the standard is dissolved, the chromogenic agent is added, and the solution diluted to a volume of exactly 50 mL. A 1-cm cell is used for all photometric measurements. The instrument is adjusted to 100% transmittance with the pure solvent in the cell; then the transmittance is read for the sample solution and for the standard solution; the values found are 52.5% and 45.0%. Calculate the percentage of S in the sample material.

For the reference solution, the following equation applies:

$$-\log 0.450 = ab \times \frac{0.3500 \text{ g} \times 0.345\%}{50.0 \text{ mL}}$$

And for the sample solution where x is the percentage of S in the sample material:

$$-\log 0.535 = ab \times \frac{0.7520 \text{ g} \times x}{250 \text{ mL}}$$

By division,

$$\frac{-\log 0.450}{-\log 0.525} = \frac{0.3500 \times 0.345}{50.0} \times \frac{250}{0.7520 \times x}$$

and

$$\frac{0.653 - 1}{0.720 - 1} = \frac{0.802_5}{x}$$

Solving gives

$$x = 0.802_5 \times \frac{0.280}{0.347} = 0.648$$

Hence, the percentage of S in the sample material is 0.648%.

Example 36-2 A 0.500-g sample of a substance, after appropriate treatment, yielded a solution of exactly 100 mL. The solution showed a transmittance of 5.0% in a 1-cm cell. What volume of this solution in milliliters must be further diluted to exactly 250 mL to obtain a solution having a transmittance of 40%, which is close to the optimum value (see Section 36.3), when measured in the 1-cm cell?

Let C and C' represent the original and final concentrations. Then

$$-\log 0.050 = abC$$
$$-\log 0.40 = abC'$$

The relation between the two concentrations is obviously

$$C' \times 250 = C \times x$$

where x is the milliliters of the initial solution to be diluted to exactly 250 mL. Hence,

$$C' = \frac{C \times x}{250}$$

Inserting this expression for C' in the expression of Lambert-Beer's law, dividing one expression by the other, and substituting the values for the logarithms yield

$$\frac{0.70 - 2}{0.60 - 1} = \frac{a \times 1 \times 250}{a \times 1 \times x}$$

$$x = \frac{250 \times 0.40}{1.3} = 77 \text{ mL}$$

Example 36–3 What transmittance will be read in a 5.00-cm cell for a 0.00200% solution of a substance of formula weight 250, possessing at the wavelength used a molar absorptivity of 5.00×10^3 L•mol^{-1}•cm^{-1}?

$$\log T = -\epsilon bc = -5.00 \times 10^3 \times 5.00 \times \frac{0.00200 \times 1000}{100 \times 250} = -2.00$$

Hence

$$T = 0.010$$

Example 36–4 The concentration of a substance S in a solution is determined photometrically by the technique of a standard addition. The transmittance of the solution is found to be 45.1%. Then to the solution a known amount of S is added corresponding to an increase of 0.0040 g of S per liter of the solution. The transmittance is now found to be 29.5% under identical conditions. What is the concentration of S in the initial solution assuming that Lambert-Beer's law is obeyed?

Let x be the concentration of S in grams per liter in the initial solution. Then

$$-\log 0.451 = ab \times x$$

The concentration of S after the "standard addition" is $x + 0.0040$. Hence,

$$-\log 0.295 = ab \times (x + 0.0040)$$

Dividing, and substituting the value of the logarithms, yield

$$\frac{0.654 - 1}{0.470 - 1} = \frac{x}{x + 0.0040}$$

$$\frac{0.346}{0.530} = \frac{x}{x + 0.0040}$$

On rearrangement,

$$0.184x = 0.0040 \times 0.346$$

And solving gives

$$x = 0.0075_2$$

Hence, the concentration of S in the sample solution is 0.0075 g/L.

36.5 Questions

36-1 What is a chromogenic agent and what properties should it have for advantageous use in photometry?

36-2 Explain why inorganic complexing agents are seldom used as chromogenic agents.

36-3 When a photometric determination is performed at a wavelength corresponding to a sharp peak in the absorbance curve, what special precautions must be taken?

36-4 What are the advantages of combining a solvent extraction with a photometric determination?

36-5 Outline in your own words some general principles and possibilities for transferring a substance to be determined into a strongly absorbing species.

36-6 Elaborate on the possibilities of using a system that does not fully comply with Lambert-Beer's law for a spectrophotometric determination.

36-7 How can oxidation and reduction reactions be used in photometric analysis? Elaborate on the principles and give examples.

36-8 What might be the consequences of having a turbidity present in a solution subjected to a photometric measurement?

36-9 A pair of cells is badly matched and cell A has a light path 1% longer than cell B. Elaborate on the effect on the results if cell A is used for the blank in the standardization but B is used for the blank in the actual determination. Will the result be high or low?

36-10 How would you cope with a permanent absorbing background present in a solution to be subjected to a photometric determination?

36-11 Silver ion is found to interfere in the determination of ammonia by Nessler's method. Explain.

36-12 Care should be exercised in cleaning, storing, and using photometric cells. Explain.

36-13 Elaborate on the possibility of determining the formal solubility of a sparingly soluble electrolyte MA when an accurate photometric method is known for determining the cation M.

36.6 Problems

36-1 What is the absorbance when the transmittance is 69.9%? *Ans.:* 0.156

36-2 What is the percentage of transmittance corresponding to an absorbance of 0.025? *Ans.:* 94.4%

36-3 Calculate the molar absorptivity of a substance of formula weight 273 showing in a 0.0100% solution a transmittance of 24.6% in a 2-cm cell. *Ans.:* 831 $L \cdot mol^{-1} \cdot cm^{-1}$

36-4 To 50 mL of an aqueous $CuSO_4$ solution, an excess of aqueous ammonia is added to form the copper(II)-ammine complex, and the solution is diluted to exactly 100 mL with water. A volume of 50 mL of a standard $CuSO_4$ solution known to contain 2.0 mg of copper per milliliter is analogously treated and diluted. Portions of the two solutions are placed in the cells of a Duboscq comparator. When a color match is secured, the lengths of the light paths in the sample and standard solution are found to be 43 and 50 mm. Calculate the copper concentration in the original sample solution. *Ans.:* 2.3 mg/mL

36-5 A standard contains 0.620% of a substance S. When a 0.5000-g amount of the standard is dissolved, the color developed, and the solution after dilution to exactly 100 mL is measured photometrically in a 1-cm cell, a transmittance of 19.3% is observed. A similar material having an unknown content of S is to be analyzed. A 0.6402-g amount of it is analogously treated and dissolved to a volume of 50 mL, and under the same conditions a transmittance of 55.3% is observed. What is the percentage of S in the unknown? *Ans.:* 0.087% S

36-6 What is the molarity of a solution that shows in a 2-cm cell the same transmittance as a $3.0 \times 10^{-4} M$ solution of the same substance in a 5-cm cell? *Ans.:* $7.5 \times 10^{-4} M$

36-7 Solution *A* shows a transmittance of 20.0% in a 1-cm cell. Solution *B* of the same substance shows a transmittance of 60.0% in a 1-cm cell. What volume in milliliters of solution *B* must be added to 100 mL of solution *A* for the final solution to show a transmittance of 40.0% in a 1-cm cell? *Ans.:* 171 mL

36-8 The aqueous solution of a substance shows a transmittance of 28.7% in a 1-cm cell. A volume of 40.0 mL of this solution is diluted with 100 mL of water. What transmittance will the dilute solution show when measured in a 5-cm cell? *Ans.:* 16.8% *T*

36-9 In the determination of inorganic ions, it is difficult to reach a molar absorptivity greater than 2×10^4 $L \cdot mol^{-1} \cdot cm^{-1}$ when operating in the visible region. The largest cells that most instruments will accommodate without modification have a 5-cm light path. Such instruments, with meticulous attention to possible sources of error, have a minimum detectable absorbance value of about 0.005. For an inorganic ion, calculate the molar concentration of a solution that will give a significant response under these favorable conditions. *Ans.:* $5 \times 10^{-8} M$

36–10 A solution containing 4.65 mg of a light-absorbing substance (*486.0*) in a volume of 260.0 mL shows 34.8% transmittance when measured in a 1.00-cm cell at 432 nm. When a solution, containing 1.02 mg of the same substance in 100.0 mL, is measured in a 2.00-cm cell at 521 nm, 44.0% transmittance is observed. How much larger is the molar absorptivity at 432 nm than at 521 nm? Express the result as a percentage of the molar absorptivity at the latter wavelength.

36–11 In an acid-base titration, 30.0 mL of a $1.0 \times 10^{-2} F$ NaOH solution is found to be equivalent to 40.0 mL of a solution of potassium tetroxalate, $KHC_2O_4 \cdot H_2C_2O_4 \cdot 2H_2O$. In the titration of 20.0 mL of this tetroxalate solution under acidic conditions, a volume of 5.0 mL of a $KMnO_4$ solution is required. Calculate the percent transmittance of a solution containing 0.50 mL of this $KMnO_4$ solution per liter when measured in a 3.00-cm cell. The molar absorptivity of $KMnO_4$ is 6.7×10^3 $L \cdot mol^{-1} \cdot cm^{-1}$ at the wavelength used in the measurements.

36–12 Of a standard steel sample containing 1.67% w/w chromium, an amount of 0.5000 g is dissolved in acid, the chromium is suitably oxidized to dichromate, and the solution is diluted to a volume of 250.0 mL. A volume of 10.0 mL of this solution is diluted with water and acid to 100.0 mL; the resulting solution shows 40.7% transmittance in a 1.00-cm cell. When 0.7500 g of an unknown steel is dissolved, oxidized, and diluted to 200.0 mL, the resulting solution shows 61.3% transmittance under identical instrument settings. What is the chromium percentage in the unknown?

36–13 A soluble complex MY dissociates according to the reaction $MY^+ \rightleftharpoons M^+ + Y$, where M^+ is a metal ion and Y is a complexing agent. At 500 nm, the species Y and MY^+ do not absorb light, but a $1.0 \times 10^{-3} M$ solution of M^+ in a 3.00-cm cell shows 50.0% transmittance. When 2.0×10^{-4} mol of MY is dissolved in water and diluted to exactly 100 mL, the solution shows 76.0% transmittance at 500 nm. Calculate the stability constant of the complex.

36–14 The molar absorptivities of compound A are 341 $L \cdot mol^{-1} \cdot cm^{-1}$ at 510 nm and 692 $L \cdot mol^{-1} \cdot cm^{-1}$ at 625 nm. The molar absorptivities of compound B are 722 $L \cdot mol^{-1} \cdot cm^{-1}$ at 510 nm and 488 $L \cdot mol^{-1} \cdot cm^{-1}$ at 625 nm. A solution containing only A and B as light-absorbing species shows 32.7% transmittance at 510 nm and 16.2% transmittance at 625 nm when measured in a 5.00-cm cell. Calculate the molar concentrations of A and B in the solution.

36–15 Iron(III) gives a red complex with 5-sulfosalicylic acid; this is a sensitive reaction used for the photometric determination of iron. A 0.100-g sample of material to be analyzed is dissolved in 20.00 mL of water and a volume of 5.00 mL of 1% 5-sulfosalicylic acid solution is added. When this mixture is measured in a 1.00-cm cell against water, the % *T* reading is 54.3. When 10.00 mL of water and 5.00 mL of the reagent solution are mixed and measured in a 5.00-cm cell, the reading is 90.1% *T*. An amount of 0.200 g of a standard with a content of 0.15% Fe is dissolved in 30.00 mL of water, and 5.00 mL of reagent is added. When the transmittance is measured in a 2.00-cm cell, it reads 60.2% *T*. Calculate the percentage Fe in the sample.

36–16 A solution containing a known substance at unknown concentration shows a transmittance of 40.0%. To 10.00 mL of that solution is added 4.00 mL of a $2.00 \times 10^{-3}F$ standard solution of that substance. The transmittance of the resulting solution under identical conditions is 50.0%. Calculate the concentration of the substance in the original solution.

Ans.: $1.36 \times 10^{-2}F$

36–17 The SO_2 (*64.1*) content of air can be determined by passing it through about 10 mL of tetrachloromercurate(II) solution at a rate of 1.0 L/min for exactly 30 min. The resulting solution is transferred to a 25-mL volumetric flask, solutions of pararosanilin and formaldehyde are added, and the flask is filled with water to mark. The intensity of the color developed is proportional to the SO_2 content. For a single air sample, the resulting transmittance was found to be 70.0%. A standard was prepared by mixing 1.0 mL of a solution containing 20 μg of SO_2/mL with the reagents in a 25-mL flask and diluting with water to mark. The transmittance of the standard under identical conditions was 58.0%. What is the SO_2 content of the air in ppm w/w? The density of air is 1.06 g/L.

Ans.: 0.41 ppm w/w

36–18 You are to perfect a photometric procedure for determining nickel (*58.7*) present to the extent of 1% in a material. The weighed sample is suitably dissolved, the chromogenic agent added in excess, and the solution diluted to 100 mL. The resulting solution is measured in a 10-mm cell against a reagent blank at a wavelength at which the colored nickel complex has a molar absorptivity of about 500 $L \cdot mol^{-1} \cdot cm^{-1}$. What approximate sample weight should you specify so that the absorbance measurements are within the preferred range around 40% *T*? *Ans.:* About 0.5 g

36–19 Dissolved silica is determined photometrically in waters by adding molybdate to a well-mixed, filtered sample after its acidification. A green yellow silicomolybdate complex is formed. To a 50.0-mL sample of filtered river water was added in rapid succession 1.0 mL of 1:1 hydrochloric acid, 2.0 mL of ammonium molybdate solution, and 1.5 mL of oxalic acid solution. After 2 min, the transmittance at 410 nm versus a reagent blank was 42.0% while the transmittance of a standard containing 0.80 mg/L of SiO_2 (*60.09*) was 52.6%. Calculate the milligrams per liter of dissolved silica in the river water.

36–20 Orthophosphate in waters is determined by converting it to the stibinophosphomolybdate complex and making photometric measurements at 880 nm, using a reagent blank as the reference. A 50.0-mL sample of lake water was made just colorless to phenolphthalein and then a volume of 8.0 mL of a combined antimonylmolybdate reagent solution was added. After 10 min, the transmittance was found to be 62.5% while the transmittance of an orthophosphate standard solution containing 0.500 mg/L of phosphorus (*30.97*), which had been treated exactly like the sample, was 46.7%. Calculate the orthophosphate content of the lake water expressed as milligrams per liter of phosphorus.

36–21 The ammonium content of a lot of sodium oxalate was determined by Nessler's method. To a solution of 1.0 g of the salt in 50 mL of ammonia-free water was added 2 mL of a solution of potassium tetraiodomercurate(II) (Nessler's reagent). The color developed matched that given by a standard prepared by mixing a solution of 0.010 mg of ammonium ion in 50 mL of water with 2 mL of Nessler's reagent. What is the ammonium ion content of the salt in percentage w/w and in micrograms per gram?

36–22 To 1.8 mL (2.0 g) of 30% hydrogen peroxide was added 0.1 mL of concentrated sulfuric acid, and the mixture was evaporated to dryness. The residue was dissolved in 45 mL of water and 3 mL of 10% sodium hydroxide solution and 2 mL of Nessler's reagent were added. The color developed was compared with a series of ammonium ion standards, each containing in 50 mL the amounts of the reagents used with the sample. The color match was intermediate between a standard of 10 and 5 μg of ammonium. (a) What is the ammonium content of the hydrogen peroxide in micrograms per gram (ppm)? (b) Would this sample meet the ammonium specification for reagent grade 30% hydrogen peroxide of 0.0005% w/w?

Ans.: (a) 4 μg/g

37 Photometric Titrations

A titration to a photometric end point, or simply a photometric titration, is performed by measuring the absorbance of the solution after each addition of titrant and then plotting the absorbance values versus the milliliters of titrant. The volume of titrant solution needed to reach the end point is evaluated from the resulting titration curve. Compared with the conventional titration technique, a photometric titration needs more effort and additional instrumentation. However, these disadvantages are often more than counterbalanced because of the higher accuracy and precision possible and because in many cases a titration can be performed photometrically under conditions and with systems for which a visual titration is not possible. Some of these conditions will be made evident by discussion and examples; thereby also the field of application and the advantages of photometric titrations will become clear.

Two kinds of photometric titrations can be differentiated: (1) titrations using an added indicator and (2) self-indicating titrations. With these two kinds, not only are titration curves of different shapes obtained, but there are also some other points that warrant their separate treatment.

37.1 *Titrations with Indicator Added*

Consider the titration of hydrochloric acid with sodium hydroxide using methyl orange as the indicator. The typical titration curve obtained when the absorbance is followed at a wavelength at which only the yellow (base) form of the indicator absorbs is shown in Fig. 37-1 as the solid-line curve. The three clearly distinguishable segments of the curve are joined by curvatures. Such curvatures are not detrimental, since the straight portions can be extrapolated, as shown, to the respective intersects *B* and *C*. At the start of the titration, the indicator is in essence completely in its nonabsorbing acid form, and the absorbance is zero. As the titration proceeds, hydrochloric acid is neutralized and the indicator is unaffected until point *B* is approached. This point corresponds to about pH 3, at which value the indicator begins to change to its absorbing base form; consequently the absorbance

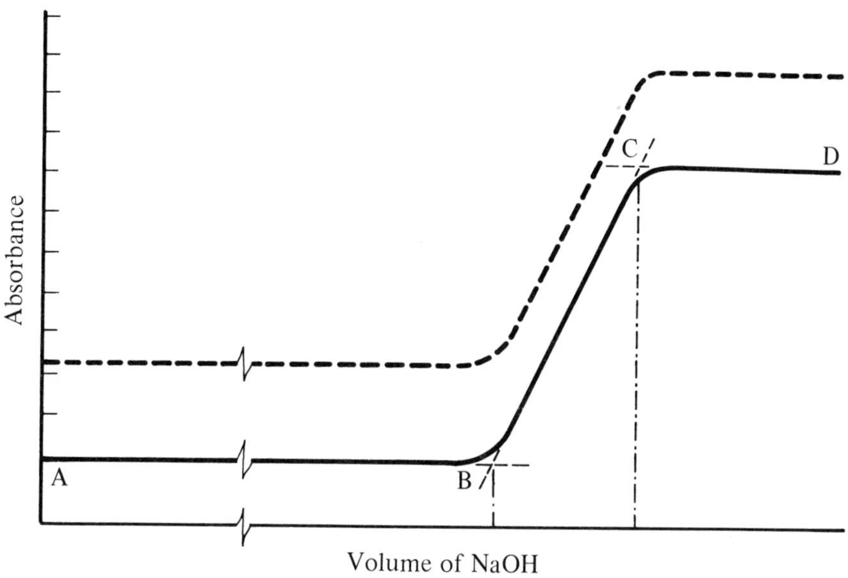

Fig. 37-1. Idealized curves for the photometric titration of HCl with NaOH using methyl orange as the indicator at a wavelength at which the yellow (base) form absorbs. Solid line curve: other absorbing species absent. Dotted line curve: other absorbing but nonreacting species present.

increases. This change continues throughout the inclined segment until point *C* is reached, which corresponds to about pH 5. Here the conversion of the indicator to the base form is complete, and further addition of titrant causes no change in absorbance.

The question arises about the point to be taken as the end point. Is it *B* or *C* or some point in between? Point *B* corresponds to about pH 3, a value too small for the titration of even a strong acid in sizable concentration. Point *C* corresponds to about pH 5, which is close enough to pH 7 to suggest that it should be taken as the end point. However, there is a further consideration.

The segment *BC* is largely related to the "titration" of the indicator from its acid form to its base form. It is necessary, in establishing the straight-line portion of this segment, to add a relatively large amount of the indicator. Consequently, in contrast to a visual titration, in which a minute amount of indicator is needed, titration of the indicator may introduce an error that cannot be neglected. It thus becomes clear that point *C* is also not the end point, since it corresponds to the titration of both the hydrochloric acid and the indicator. The difficulty can be circumvented by adding the indicator as its sodium salt, that is, in a pretitrated form. Under these circumstances, point *C* can be taken as the end point.

In practice, a sample solution may contain an absorbing substance that does not react with the titrant. In a visual titration, the presence of this substance will often obscure the indicator color change. In contrast, a photometric titration will usually be possible if the substance does not absorb strongly at the wavelength at which the titration is monitored. A constant "background" absorbance would be added to each point in the titration curve, bringing about the dotted line curve of Fig. 37-1. Note that the shift is parallel to the abscissa and does not affect the position of point *C* with respect to that axis. Consequently, the photometric titration still yields a correct result.

Obviously, the inverse titration of sodium hydroxide with hydrochloric acid can also be conducted photometrically to the methyl orange end point. It is good practice for you to draw the typical titration curve and to reason about the form in which methyl orange should be used and the position of the end point.

37.2 Self-Indicating Titrations

There are many titrations in which the system during the titration changes color although no indicator has been added. Only a few of these systems, however, show a sufficiently abrupt and intense change in color to be suitable for visual indication. Notable examples are titrations with permanganate.

Consider the titration of iron(II) in an acidic, phosphoric acid-containing medium with permanganate. In a visual titration, permanganate is added until the first drop in excess imparts a pink color to the solution. This small excess causes no appreciable error if a large amount of iron is being determined. In contrast, in the titration of a small amount of iron, the excess permanganate necessary to indicate the end of the titration may amount to a small percentage error. Of course, an indicator blank may be applied, but it becomes difficult or impossible to make such a correction when the sample solution is colored; this is often the case with natural waters or solutions from dissolution of alloys. Here a photometric end point is especially expedient. Assume that a wavelength is selected at which permanganate absorbs strongly (535 nm) and at which no other species absorbs significantly. The titration curve resulting is shown as curve A in Fig. 37-2. Up to the equivalence point, the absorbance remains zero. Beyond that point, the absorbance increases linearly with the amount of permanganate added. The best line through the few points secured after the end point is drawn and extrapolated to intersect with the zero line; the intersection is taken as the end point. Curve B illustrates the situation when a colored but nonreacting impurity is present. This "background" absorbance is to be added to the titration curve. The shape of the curve is not changed thereby nor is its position changed relative to the abscissa. All that happens is a shift of the curve parallel to the ordinate.

Only a few titration systems show color changes as abrupt as those involving permanganate. In most cases, the color change is too gradual for visual self-indication. But many such systems lend themselves readily to photometric titrations. Consider as an example the titration of copper(II) in ammoniacal solution with EDTA using a yellow filter. The titration curve for this case is shown in Fig. 37-3. Initially the absorbance reading is high, because the copper(II)-ammine com-

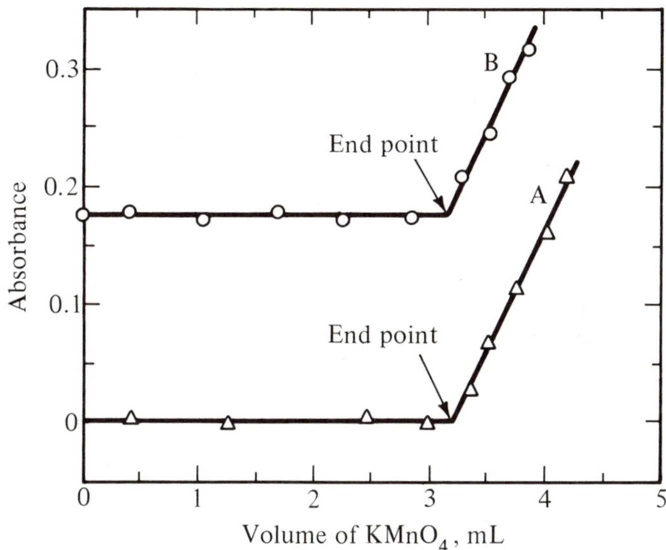

Fig. 37-2. Titration curves for the photometric titration of an acidic, dilute Fe(II) solution with $KMnO_4$, at a wavelength at which only MnO_4^- absorbs (ca 530 nm). Curve A, pure solution; curve B, in the presence of foreign material absorbing at the selected wavelength but not reacting with $KMnO_4$.

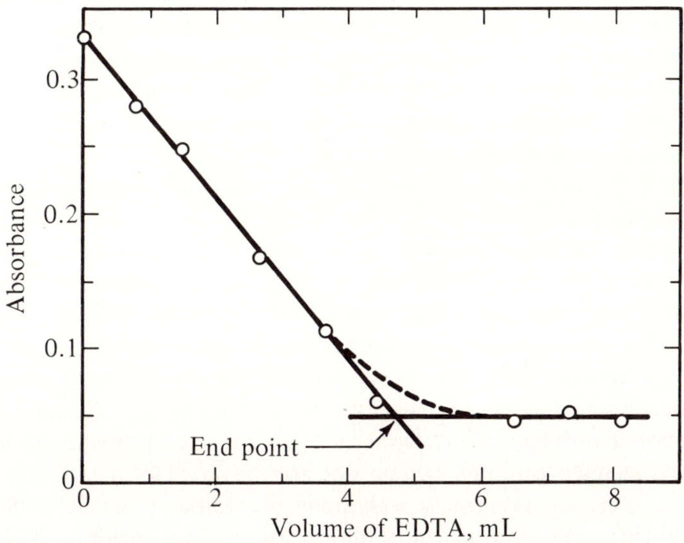

Fig. 37-3. Titration curves for the photometric titration of an ammoniacal copper (II) solution with EDTA using a yellow filter.

plexes absorb strongly at the wavelength region transmitted by the filter. On progressive addition of EDTA, the copper is transferred to the more weakly absorbing copper(II)-EDTA complex and the titration curve descends linearly. Beyond the equivalence point, the addition of nonabsorbing EDTA causes no change in absorbance, and a horizontal line results. Extending the two straight lines yields an intersect that is taken as the end point.

37.3 Some General Considerations

The two segments in the curve for a self-indicating titration will be straight only if no distortion results from the dilution of the solution during the titration. Such distortion is obviated if the titrant solution has a concentration far greater (50 to 100 times) than that of the species being titrated. The volume of titrant then required is so small that the resulting dilution has no significant effect. A buret that can deliver a small volume precisely (that is, a microburet) is necessary. Another possibility is to correct the absorbance for dilution by multiplying each reading by the factor $(V + v)/V$, where V is the initial volume of the solution to be titrated and v is the volume of titrant solution added, both expressed in the same unit, usually milliliters.

Dilution effects are usually of no moment if an indicator is used because the region of interest on the titration curve is close to the end point and restricted to a very small range of volume.

Titration curves, especially when self-indication is used, may show a pronounced curvature in the vicinity of the end point. This is the case if the titration reaction does not go to completion, for example, when a weak complex is formed or extremely dilute solutions are involved. Such curvature, unless it extends extremely far, is no hindrance for the precise location of the end point, since this point is obtained by extrapolation of the straight-line segments corresponding to values removed from that point. In these remote regions, either the substance to be titrated or the titrant is present in excess and causes a shift of the equilibrium position toward completion.

The photometric technique is not restricted to the visible region but can equally well be applied with ultraviolet light or less frequently, near-infrared. Consequently, the technique extends the possibilities for titrations far beyond those with visual indication. When the wavelength is selected by a monochromator, the methods may be called a spectrophotometric titration, if the distinction is desired.

For the occasional performance of a photometric titration any filter photometer or spectrophotometer can be used. The titration is performed outside the instrument in a beaker as usual. But after each addition of titrant, a portion of the solution is withdrawn and placed in a cell, and the absorbance measured. The portion is reunited with the main portion and the next increment of titrant added, and so forth. More convenient is a photometer modified to accommodate the titration vessel in the cell compartment. This light-tight compartment has a hole for inserting the buret tip and provision for stirring the solution. Magnetic stirring is convenient. Special devices called phototitrators are available for photometric titrations. Indeed, automatic titrations are possible by having the change in absorbance at the end point generate an electrical signal to operate a relay that stops the flow of titrant solution.

37.4 *Questions*

37–1 In your own words define the terms *self-indication, titration curve,* and *dilution effect.*

37–2 Would you expect extraneous light (remaining at a constant level) to affect the accuracy and precision of a photometric titration? Explain your answer.

37–3 Substances A and B are nonabsorbing and react completely to form a stable absorbing species AB_2. Both a photometric determination and a photometric titration are possible. Elaborate on the theoretical and practical advantages and disadvantages of the two methods for determining either A or B.

37–4 Sketch the photometric titration curves for the following self-indicating systems; the numbers given are the approximate effective absorptivities. Assume a 1:1 reaction in all cases. Neglect dilution effects, plot three points (starting point, equivalence point, and 100% excess titrant added), and connect them by straight-line segments.

	Species titrated	Titrant	Reaction product
(a)	5000	5000	0
(b)	0	5000	5000
(c)	0	0	5000
(d)	5000	0	5000
(e)	1000	5000	0
(f)	5000	5000	5000
(g)	1000	3000	8000

37–5 Bismuth ion, EDTA, and the bismuth-EDTA complex are colorless; the copper-EDTA complex is colored. Bismuth ion can be titrated in acidic medium with EDTA if some copper(II) is added as the "indicator." Elaborate on the shape of the titration curve. Could a mixture of bismuth and copper(II) be titrated photometrically so that both metals are determined? Explain.

37–6 A self-indicating photometric titration and an amperometric titration (Chapter 29) have similar titration curves. Compare the two techniques in their application, limitations, and advantages. Name cases for which one would succeed, the other fail. What is plotted on the ordinate and abscissa to obtain the titration curves? Name interferences for one that would be of no consequence in the other.

37–7 A photometric titration curve has the shape of an inverted V that rests on the abscissa. From this information, what can you deduce about the absorptivities of titrant, species titrated, and reaction product?

37-8 Some organic acids show large differences in their ultraviolet absorbance spectra when dissociated and not dissociated and thus lend themselves to photometric titration. Are such acids very strong, do you think, or very weak? Explain your answer.

37.5 Problem

37-1 A photometric titration is based on the reaction of species M with titrant N to form MN. For three cases, I–III, the absorbance values of 0.100 M solutions of M, N, and MN are given in the accompanying table for three wavelengths. Assume that the reaction goes virtually to completion and that corrections for dilution are applied. Sketch on graph paper the titration curve for each of the cases at each wavelength. Under the assumptions, three points suffice, namely, the starting point, the equivalence point, and the point corresponding to 100% excess of titrant (N).

		Absorbance Value		
Case	*Species*	*420 nm*	*540 nm*	*660 nm*
I	M	0.500	0.400	0
	N	0	0.400	0.600
	MN	0	0	0
II	M	0.500	0.400	0
	N	0	0.400	0.600
	MN	0.200	0.200	0.200
III	M	0	0	0.600
	N	0.600	0	0
	MN	0	0.600	0

38

Turbidimetry, Nephelometry, and Fluorimetry

Three diverse methods are closely related to photometric and colorimetric determinations in either principle or instrumentation, or both. Two of these methods, turbidimetry and nephelometry, are based on the scattering of light. The third method, fluorimetry, is based on the measurements of the fluorescence emitted by some substances on appropriate excitation.

38.1 *Turbidimetry*

When a turbid solution, that is, a suspension of solid particles in a liquid, is brought into the light path of a photometer, less radiant power reaches the photodetector than if the clear (nonabsorbing) liquid were in the light path. This reduction results from the scattering of light due to reflection on the suspended particles and refraction by them. The scattered light is dispersed in all directions, and consequently the radiant power of the beam directed toward the detector is diminished. Some light may truly be absorbed by the particles. A turbidimetric measurement can be performed like a photometric measurement and the result expressed in absorbance units. A relation between the absorbance and "concentration" of the suspended material can be derived that is analogous to Lambert-Beer's law. However, this formula holds, if at all, only for a limited degree of turbidity and a very short light path, because the relation among number, size, and shape of the particles and the amount of light scattered is complicated. Consequently, calibration curves for turbidimetric determinations usually are far from linear.

It is not difficult to understand that for a given amount of material suspended in a given volume of liquid, the scattering due to reflection increases with a decrease of particle size. But this reasoning holds only to a certain particle size, below which no reflection occurs. A body can reflect light only when its size is at least one-half of the size of the wavelength of the light striking it. Consequently, blue light is scattered more than, say, yellow or red light and thus affords higher sensitivity in turbidimetry. These facts, coupled with the knowledge that a uniform particle size can seldom be obtained, suggests the involved situation that exists.

It is usually extremely difficult to adjust the conditions in a turbidimetric determination so that the particle size is reproducible. Some of the factors exerting a decisive influence include the manner, order, and rate of mixing the reactants, agitation, temperature, presence or absence of inert electrolytes, and concentration of the solutions mixed. Large particles are unwanted, in contrast to the requirements for a gravimetric determination, since they cause inhomogeneities by settling. Protective colloids such as gum arabic and gelatin are frequently added to reduce flocculation and settling. In a turbidimetric determination, low accuracy and precision are usually accepted in exchange for speed and simplicity.

As you can see in A at the left of Fig. 38-1, the arrangement of the components in a turbidimeter is the same as in a simple photometer. Indeed, any photometer can be used as a turbidimeter. Although monochromacity is desirable, it is not mandatory, and operation with a filter of moderate quality or even with white light is possible. Turbidimetric measurements can also be made by visual comparison with standards.

Possibly the most frequent application of turbidimetric methods in inorganic analysis is the determination of chloride and sulfate in water through the formation of silver chloride and barium sulfate respectively. Adequate accuracy often is secured by comparing the turbidity developed in the sample with that of standards. The procedures are satisfactory at concentrations as low as a few parts per million. The determination of ammonia by Nessler's method, considered in Section 36.3, is a hybrid between a colorimetric (or photometric) method and a turbidimetric one.

38.2 Turbidimetric Titrations

Turbidimetric titrations are performed like photometric titrations (see Chapter 37). The "absorbance" is plotted versus the volume of titrant solution added. The "absorbance" increases as the precipitate forms and the curve levels off (or declines slightly) when all the sought-for substance is precipitated. Thus the end point corresponds to an abrupt change in slope. The change, for a given amount of substance titrated, occurs at the same volume of titrant added, regardless of the size of the particles formed. The particle size influences the slope of the curve only before the leveled portion. In addition, the end point is obtained from the two lines drawn through a series of points, thus affording an averaging process that reduces the influence of random errors. Consequently, the accuracy and precision attained are often significantly better in a turbidimetric titration than in the corresponding turbidimetric determination.

38.3 Nephelometry

In a turbidimetric measurement, the ratio of the intensities of the transmitted and the incident light is measured. It is, however, possible to directly measure the

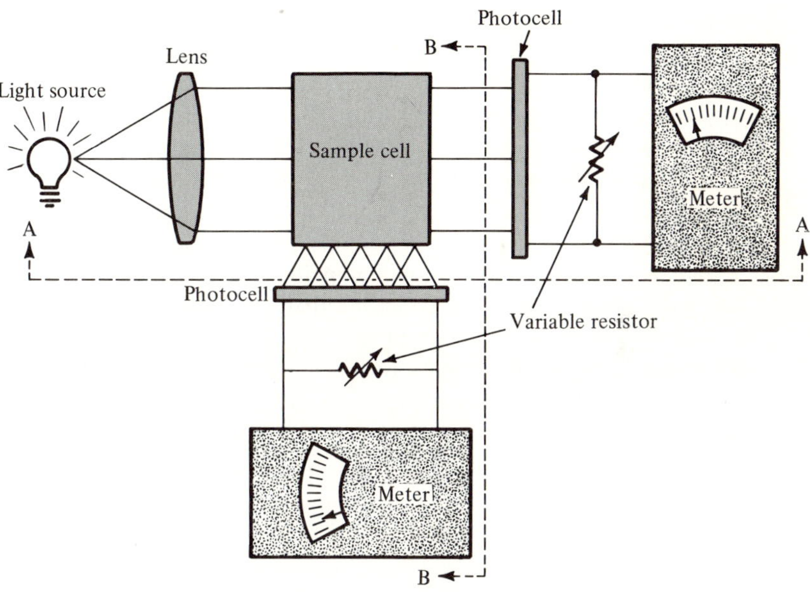

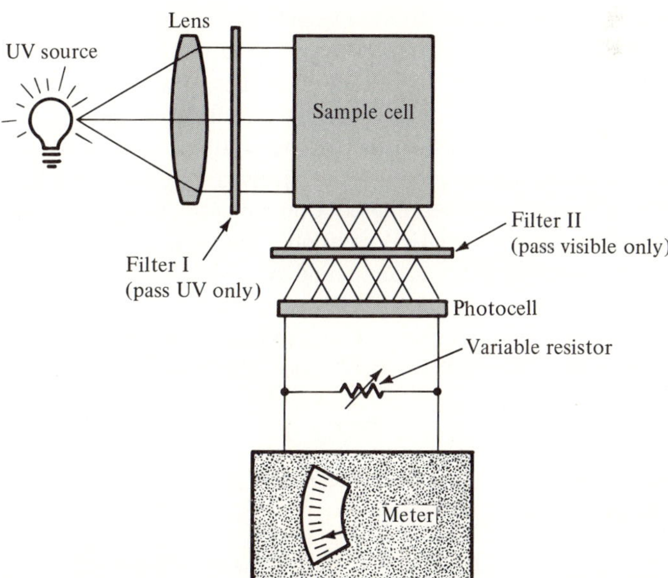

Fig. 38-1. Schematic arrangement of the components of a turbidimeter (A), a nephelometer (B), and a fluorimeter (C), viewed from above.

intensity of the scattered light, or more exactly, a portion of that light, and thereby to gain sensitivity. The technique using this principle is known as nephelometry. The instrument used is termed a nephelometer, and the schematic layout of its principal components is shown in B of Fig. 38-1. The photodetector is usually, but not necessarily, mounted at right angles to the incident beam.

Most of the requirements put forth for turbidimetry also apply for nephelometry, and the difficulties experienced are similar. The calibration curves here are also far from linear, unless very low concentrations of the sought-for substances prevail. A nephelometric determination for small quantities of a sought-for substance commonly yields better results than its turbidimetric counterpart because directly measuring a small light intensity is more accurate and precise than measuring a small change in the intensity of a strong light. However, this advantage is rarely of such moment that buying a special instrument is justified. Consequently, nephelometry is not significant in routine analysis, although the technique is an important tool for evaluating the size and shape of macromolecules and determining their molecular weight.

38.4 Fluorimetry

The mechanisms leading to fluorescence by absorption of radiation in the visible and ultraviolet regions have already been outlined (Sections 34.3 and 35.1). Now we shall emphasize the fluorescent properties of molecules (see Section 39.8 for atomic fluorescence). It suffices here to state that the fluorescence emitted is usually of longer wavelength than that of the exciting radiation. As with molecular absorption processes, the electronic levels involved in fluorescence have vibrational and rotational transitions superimposed. Consequently, fluorescence spectra consist of connected bands. For identical solution conditions a plot of fluorescence intensity versus wavelength for a compound often approximates a mirror image of the absorption spectrum.

The relation of fluorescence intensity to concentration can be derived readily. One form of Lambert-Beer's law (see Equation 35-7a) is

$$P = P_0 10^{-abc} \tag{38-1}$$

where the symbols have the meaning assigned in Section 35.3. The radiant power of the absorbed radiation P_{abs} is given by

$$P_{abs} = P_0 - P = P_0 - P_0 10^{-abc} = P_0(1 - 10^{-abc}) \tag{38-2}$$

The *fluorescence quantum efficiency* (also known as the *fluorescence quantum yield*) is defined as the ratio of the radiant power of the emitted fluorescence F to the radiant power of the incident radiation P_0 and is denoted by the capital Greek letter phi, Φ. Consequently,

$$F = k'\Phi P_{abs} = k'\Phi P_0(1 - 10^{-abc}) \tag{38-3}$$

where k' is a constant taking instrumental parameters into consideration, including the fact that only a portion of the total fluorescence is measured.

Recall from mathematics that $10^{-abc} = e^{-2.3abc}$ and that the second exponential term can be expanded as a convergent series, of which the first two terms are $1 - 2.3abc$; higher terms can be neglected if $2.3abc$ is of the order of 0.05, depending on the error acceptable (see Question 38-7).

Substituting this approximation in equation (38-3) yields

$$F \cong k'\Phi P_0(2.3abc) \tag{38-4}$$

or by a change in the constant

$$F \cong k\Phi P_0 abc \tag{38-5}$$

For a given fluorescent compound and conditions (solvent, temperature, cell dimensions, specific instrument, and so on), all the factors except concentration are constant, and equation (35-5) reduces to

$$F \cong Kc \tag{38-6}$$

where K is the "combined" proportionality constant.

The approximation requires that the term abc be small and not, as often stated, that the concentration must be small. Relatively high concentrations may yield a straight-line relation between fluorescence intensity and concentration if both the absorptivity a and the path length b are small. The approximation made is such that when a and b are constant, F/c decreases if c increases beyond the valid range; in other words, the curve for fluorescence intensity versus concentration levels at higher concentrations and becomes progressively less useful.

From equation (38-5), you can see that the radiant power (intensity) of fluorescence is proportional to the radiant power of the incident radiation; consequently, compared with photometry, this method has an additional parameter that can be controlled to adjust and to improve the sensitivity. Additionally, for a fluorescent species, the wavelength of the incident radiation must be small enough to provide the energy needed to excite the fluorescence. Note also that only a relatively few compounds that absorb ultraviolet radiation are fluorescent. From these various considerations, it is clear that fluorimetry can often provide greater selectivity than photometry.

Fluorimeters of various designs exist but all operate on the same principle. The light from the exciter lamp (often a mercury vapor lamp) passes into the sample. The fluorescence is emitted in all directions; its intensity is measured at an angle to the incident beam, usually at 90°, using a photodetector and a meter. In simple instruments, such as that shown schematically in *C* of Fig. 38-1, two filters are used. The filter between the source and the cell isolates the exciting radiation of the desired wavelength region. The filter between the cell and the photodetector excludes any exciting radiation and allows only the fluorescence to pass. More elaborate instruments, known as spectrofluorimeters, are capable of greater selectivity and sensitivity. A monochromator is used to select a particular wavelength for the fluorescence measurements; often another monochromator is used to secure a selected wavelength of monochromatic radiation for excitation.

Like recording spectrophotometers, recording spectrofluorimeters provide a tracing for spectra. A fluorescence excitation spectrum is a display, for a selected emission wavelength, of fluorescence intensity versus the wavelength of the exciting radiation. A fluorescence emission spectrum is a display, for a selected excitation wavelength, of fluorescence intensity versus the wavelength of the emitted fluorescence.

In a fluorimetric determination, the concentration of the sought-for species is found from a calibration curve, which is a plot of the intensity of the fluorescence versus the concentration of that species. The sought-for species must be fluorescent or be relatable to a species that is. The conditions for the calibration and the conduct of samples should match closely. This is an important consideration since the intensity of fluorescence may change due to changes in the fluorescence quantum efficiency that are related to temperature, pH, solvent, and presence of species that interfere in the excitation-relaxation-fluorescence processes. Such species are known as *quenchers,* and their effect in reducing fluorescence is called *quenching.* This phenomenon can be applied to determining a quencher by measuring the reduction of fluorescence intensity as a function of quencher concentration for a constant concentration of the fluorescent compound.

Fluorimetry provides not only high selectivity but also high sensitivity; that is, concentrations down to 10^{-4} to $10^{-7} M$ can be determined. This higher sensitivity of fluorimetry, compared with photometry, can be appreciated. With photometry, for low concentrations a small difference between two large signals is measured (operation near 100% T with almost the total incident beam impinging on the detector). In contrast, with fluorimetry a small signal is measured against no background or almost no background.

Practical applications for fluorimetry are extensive both in inorganic trace element analysis and in the determination of organic substances, including vitamins, steroids, epinephrine and related compounds, enzymes, and aromatic hydrocarbons. The technique is important for the biosciences, in diagnostic laboratories, and for drug analysis.

A substance to be determined that is itself not fluorescent can often be converted to a fluorescent species by causing it to react with a suitable *fluorogenic agent.* For example, amino acids can be transformed into fluorescent derivatives by reaction with *o*-phthalaldehyde, 1,2-$C_6H_4(CHO)_2$, and can thereby be determined fluorimetrically at trace levels.

Various fluorogenic agents, termed *fluorescent probes,* react with proteins, tissue, enzymes, biological membranes, cells, and the like. The fluorescence that can be excited is measured, often over a specific period, thereby allowing function and "fine" structure to be studied, even in vivo.

Fluorogenic agents are important for inorganic analysis, since inorganic species do not show intrinsic fluorescence in solution (uranium and thallium are important exceptions). Fluorogenic agents for metals are usually complexing agents (and often chelating agents). For example, beryllium in the submicrogram range can be determined fluorimetrically after chelation with the organic reagent morin (that is, 2′,3,

4′,5,7-pentahydroxyflavone). The substance 8-quinolinol forms fluorescent chelates with various metals including aluminum, gallium, zinc, and magnesium. The gallium chelate can be extracted into chloroform, in which it is highly fluorescent while the "free" 8-quinolinol shows negligible fluorescence. In this way, gallium can be determined fluorimetrically down to 3×10^{-8} g/mL of extract. Fluoride acts as a quencher for the fluorescence of the aluminum-morin chelate; this quenching effect has been applied to the fluorimetric determination of trace fluoride.

Fluorimetry is not restricted to solutions. Uranium in trace amounts can be determined by melting in a mold a known amount of the sample with a fixed amount of sodium fluoride. A flat button results when the mixture is cooled. Under ultraviolet illumination, a yellow-green fluorescence is visible even to the eye. With a suitable fluorimeter, even 10^{-9} g of uranium can be determined with an error of ±10% maximum.

Fluorescence can be applied qualitatively for detection purposes, for example for locating and identifying zones and spots in chromatography, both paper and thin-layer, and as a detector permitting quantitation in high-performance liquid chromatography. Fluorescence is used in assessing radioactivity by a technique known as liquid scintillation counting.

38.5 *Questions*

38-1 Many fluorimeters can readily be used as nephelometers. Discuss the modifications needed in switching from fluorimetry to nephelometry.

38-2 Explain why a turbidimetric titration usually yields better results than a turbidimetric determination.

38-3 Fluorescence can be detected by the eye if the sample is located and irradiated in a black box having a tube attached through which the observation is made. Elaborate (a) on the possibility of a fluorescence analog to visual colorimetry and (b) on using an acid-base indicator fluorescent only in its base form in titrating the acidity of a highly colored sample (red wine, for example).

38-4 In the turbidimetric or nephelometric determination of a suspended material that is colored, would you recommend that the incident beam should be of a wavelength that is strongly absorbed? Explain your answer.

38-5 The continuous turbidimetric monitoring of smoke or dust in an air stream is possible. Speculate on the instrumental arrangements.

38-6 The intensity of the fluorescence excited in a fluorescent substance is proportional to the intensity of the exciting light, but only to a certain intensity. Explain why. Operating within the proportionality range allows an increase in the sensitivity of a determination to be achieved. How? Compare this situation with what happens in a photometric determination. How would an increase of the intensity of the incident beam affect the sensitivity of a photometric determination?

38-7 In the derivation of equation (38–4), the approximation $e^{-x} \cong 1 - x$ was applied. If x is 0.05, the resulting error is about 0.1%. To understand the situation better, calculate the percentage error in e^{-x} for various values of x (1, 0.8, 0.6, 0.4, 0.2, 0.1, 0.05, 0.02, 0.01) and plot x versus the percentage error. Since one of the variables ranges in value over two orders of magnitude, what type of graph paper is the most convenient?

38-8 From equation (38–5), it might be concluded that any level of fluorescence intensity F can be achieved by increasing the radiant power of the exciting light P_0 sufficiently. Actually, a plot of F vs P is a curve that initially is a rising straight line and then bends and levels to a horizontal line. Elucidate this relation, taking into account what happens when more photons enter a solution then there are molecules present in the ground state.

38.6 Problem

38-1 In equation (38–4) if the term $2.3abc$ has a value of about 0.14, the error caused by using the approximation for $e^{-2.3abc}$ is about 1% (cf. Question 38–7). If a cell of 1-cm path is used and the molar absorptivity of a fluorescent compound is 1.5×10^3 L·mol^{-1}·cm^{-1}, at approximately what molar concentration will there be error from this source? *Ans.:* $6 \times 10^{-5} M$

39 Atomic Spectroscopy

39.1 *Introduction*

Having examined the interaction of radiant energy and *molecular* species, we now focus on *atomic ions* and *atoms* and their emission and absorption of radiant energy; the field is atomic spectroscopy, broadly defined. The analytical techniques can be appreciated from some considerations of atomic spectra.

39.2 *Atomic Spectra*

When radiation emitted from excited atoms is passed through a narrow slit and dispersed by a prism or a grating, a spectrum results that under certain conditions consists of discrete lines. Such a spectrum is known as a line spectrum, or from its origin, as an atomic-emission spectrum. The excited atoms can be produced by exposing the material of interest to an appropriate energy source, including flame, high-voltage electric spark, electric arc, and the plasma from subjecting an ionizable gas to a high-energy electric field. The energy provided by such sources disrupts the chemical bonding, to leave atoms or atomic ions; it then excites orbital electrons, that is, brings them to high energy levels. Within a brief time, the electrons return to lower energy levels or even to the ground state, and the energy is released as radiation. Only definite, quantized transitions of electrons are possible; consequently, radiation of specific wavelengths is emitted. The greater the energy of the photons provided by the exciting source, the higher the possible level to which electrons are excited and the shorter the wavelength of the radiation emitted. As a result, with an increase in the energy supplied by the source, more spectral lines are observed; the added ones are particularly in the shorter-wavelength (ultraviolet) region. The wavelengths of the spectral lines are a characteristic of the emitting element, and therefore atomic spectra can be used for qualitative purposes. The intensity of the emission lines can be correlated with the amount of the elements present and can be used for quantitative analysis.

Two principal forms of atomic emission can be differentiated: flame (emission) spectrometry and arc-spark emission spectroscopy. Their separate treatment

(Sections 39.3 and 39.4) is warranted because they differ markedly in the energy levels excited, instrument design, and practical procedures.

In a flame, most of the atoms are in the ground state and only a small portion are thermally excited. Consequently, the atomic vapor is ready to absorb energy. If a strong beam of compound radiation (that is, "white light" containing all wavelengths as a continuum) is passed through the atomic vapor persisting in the flame, the atoms absorb light, but only light of the wavelengths characteristic for the particular element. As a result, the beam of light leaving the flame has these characteristic wavelengths at a lower intensity than in the original beam. If this beam is then diffracted and the spectrum displayed, lines appear that are dark compared with the unaffected bright spectral background. Measurement of the intensity of the radiation at these specific wavelengths with and without the absorbing atoms present in the flame allows flame absorption photometry to be established analogously to the molecular photometry for solutions.

Molecular species have broad absorption bands. In contrast, atoms absorb at more sharply defined wavelengths. Consequently, in the arrangement mentioned above, only an extremely minute portion of the incident compound light is involved, and not enough radiant energy corresponding to these wavelengths is present in "white light" beams to be practical. It is necessary to use a beam that contains radiation of the required wavelength at high intensity. How this is accomplished in practice is explained when we take up atomic absorption spectrometry (Section 39.5). Methods not involving a flame for atomization are also applicable in atomic absorption spectrometry (see Sections 39.6 and 39.7).

The energy from a radiant beam that is absorbed by atoms in a flame elevates electrons to higher energy levels (see Section 35.1). When the electrons return to lower levels or to the ground state, the energy is emitted as radiation of discrete wavelengths, and in all directions in space. Only a minute fraction of this emitted radiation proceeds in the direction of the incident beam. The possibility exists therefore of measuring the intensity of the re-emitted radiation at an angle to the incident beam and relating it to the amount of the involved species in the flame. This technique is called atomic fluorescence spectrometry (Section 39.8).

39.3 Emission Flame Spectrometry

It has been known since antiquity that when salts of certain metals are introduced into a flame, they will impart a color to it. Flame tests have long been applied in qualitative inorganic analysis. Early attempts to use flame colors for quantitative purposes failed, however, mainly because of difficulties in introducing the sample appropriately and evenly into the flame. These difficulties were resolved about 1929 by Lundgårdh, who developed the burner-atomizer, which since then has been modified and perfected. The technique of emission flame spectrometry, which in its simple form is by tradition often called flame photometry, is attractive because liquid samples can be used directly. In addition, flame spectra are simple, that is, only comparatively few lines are produced; consequently, the use of a relatively inexpensive instrument is permitted; furthermore, the technique is rapid and shows high sensitivity for some metals.

The operation of a simple flame photometer becomes obvious from Fig. 39-1, which shows the principal parts schematically. The sample solution is sucked up by an atomizer operated by one of the flame-producing gases and is fed as a fine spray into the flame. The light emitted is collected by a parabolic mirror and a lens and passed through a monochromator (prism or grating) or simply a filter. The light of the appropriately selected wavelength strikes a photodetector and the magnitude of the electrical signal developed is read on a meter. Not shown in this figure are pressure regulators and flow meters, which are installed in the gas lines to permit reproducible adjustment of flow rates and of the ratio of fuel to oxidizer gas. In addition to simple city gas-air flames (1900 °C), flames of higher temperature are used, including gas-oxygen (2800 °C), hydrogen-oxygen (2800 °C), and acetylene-oxygen (3000 °C).

The temperatures involved in emission flame photometry are low compared with what prevails in an arc or spark (see Section 39.4). Consequently, only spectral lines corresponding to relatively low excitation energies are produced. Thus the flame spectrum of an element is simple, and lines in the visible and near ultraviolet regions are the only ones excited in an ordinary gas flame. For many elements the energy available in a flame is not enough to excite any line; this confers a favorable degree of selectivity on flame photometry.

Although a prism or a grating monochromator can be used, a filter suffices

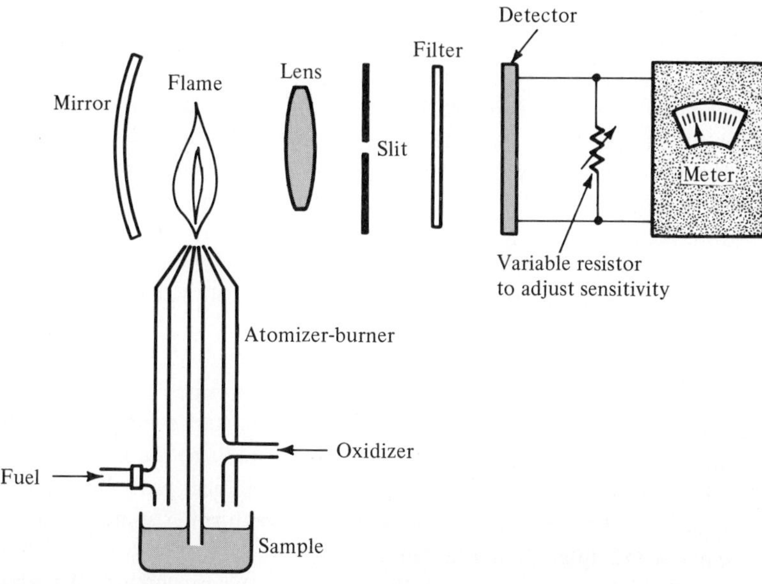

Fig. 39-1. Schematic arrangements of the components of a simple flame photometer.

for many analytical purposes, because of the simplicity of the flame spectra. Special filters are often used that have a narrow "band pass" and therefore allow the selection of a single prominent wavelength.

One advantage of flame photometry is that it permits using a small volume of sample solution. Often a single milliliter of the solution can supply the flame with material long enough to permit changing filters several times and to secure readings for several elements. The filters are frequently mounted on a disk so that they can be changed simply by its rotation.

As you can conclude from any experience you have had with the flame tests for sodium and potassium, the sensitivity of emission flame photometry is high for these and several other metals. Thus 0.01 mg/mL of sodium or 1 mg/mL of barium can be detected readily and somewhat higher concentrations can be determined with favorable accuracy. This high sensitivity is especially valuable for sodium and potassium and certain other metals for which other analytical methods are either tedious or inapplicable to minute amounts.

Calibration Curve. A calibration curve can be used to determine the concentration of the element in question. Such a curve is established with solutions containing the sought-for substance in known concentration and under identical instrument settings, including the rate of sample introduction, gas pressure, and gas ratio. A correction may be necessary for the flame background, which is the radiation emitted over a wide spectral range due to the combustion of the fuel and the hot or burning solvent or other substances present in the sample. The reading of the output meter is plotted versus concentration.

Internal Standard. An internal standard offers another approach for establishing the unknown concentration. Lithium is frequently used for this purpose because it shows a readily excitable spectral line and is rarely present in practical samples. First, the ratio of the intensities of the lithium line compared with the line of the element to be determined is established experimentally at various concentration levels for both. Then in the actual determination, a known concentration of lithium is established in the sample solution and the ratio of the line intensities is measured again. From the two ratios and the known concentration of the lithium, the concentration of the sought-for substance is obtained by proportion, or graphically by a calibration curve.

Standard Addition. Standard addition presents a further possibility. In this technique, known amounts of the sought-for substance are added to aliquots of the sample solution. These aliquots and an untreated one are fed to the flame, and for each solution the reading is taken. Either a graphical procedure can be used or the unknown concentration can be calculated, as in other standard additions (for example, see Chapter 29 and Section 35.7).

Sources of errors exist with emission flame photometry as with other analytical methods. The following discussion will aid your understanding of the causes of such errors and the possibilities for remedies. The excitation of electron transitions in a metal with subsequent emission of radiation takes place only if the metal is in the gaseous state. Since the metal to be determined is present as a dissolved salt, several processes must take place to have the metal eventually present as an atomic vapor in the flame. The sequence of events can be presented as follows:

Liquid sample → nebulization forming droplets of liquid → vaporization of solvent and formation of a residue of solute or solutes →

decomposition of residue, disruption of bonds, and formation of atoms in gas phase → excitation of atoms in gas phase → loss of excitation energy with emission of light

Unless each of these processes proceeds appropriately, the result may be faulty.

Chemical Interference. Chemical interference is one kind of disturbance possible and comes about as follows. In a flame, the energy available is only enough to vaporize metals of high volatility. As the first step, the salt decomposes. The rate for this process is low for some salts and high for others. Consequently, the anions present in the solution are important in flame photometry. The highest line intensities are usually obtained with chlorides and nitrates. Phosphates and sulfates are less suitable. When, for example, sulfate is added in increasing amounts to a solution containing calcium chloride, the line intensity for calcium decreases progressively, but eventually a point is reached at which adding more sulfate has no further effect on the line intensity. When the reduced intensity still affords enough sensitivity, sulfate ion (or the relevant interfering anion) can be added deliberately to the sample solution to such an extent that there is no further effect on the line intensity. Obviously the same addition must also be made to the standard solutions used to establish the calibration curve.

The addition of organic solvents or complexing agents, or both, to the sample solution frequently reduces or even obviates chemical interferences and indeed may sometimes increase the sensitivity of a method. When huge amounts of interfering anions are present, a separation may become necessary. Such a separation is often achieved by simply passing the sample solution through a column of an anion exchange resin in the chloride or nitrate form (see Chapter 45).

Excitation Interference. Excitation interference is a phenomenon encountered with the metals of the first group of the periodic table. An alkali metal may show higher line intensity when another alkali metal is present than if it is alone in the solution. This effect can be explained in the following way. The desirable sequence of events is to elevate an electron in the metal atom to a higher energy level and have it return at once to the ground state with simultaneous emission of light. Once the atom is returned to that state, it is ready to repeat the process. Consequently, there should be as many atoms as possible in the ground state. With alkali metals, because of their low ionization energy, a substantial portion of the atoms dissociate according to $M \rightleftharpoons M^+ + e$. The *ions* are not able to participate in the emission mechanism. If another alkali metal is added, it too ionizes. Thereby the electron concentration in the flame is increased and the equilibrium is shifted to the left. Consequently, more *atoms* of the metal to be determined are present and the line intensity is increased.

Radiation Interference. Radiation interference needs only brief mention. If another species is present that emits light of a wavelength very close to that of the metal to be determined, the monochromating device may not be able to discriminate between the lines. The interfering species must then either be removed or be added in identical quantities to the solutions used in establishing the calibration curve.

Flame Temperature. Flame temperature has a great influence on a determination. If the temperature is too low to cause dissociation of the salt, to effect vaporization, and to excite the atoms of the metal, no lines, or only very weak ones, are obtained. But too high a flame temperature may also have detrimental effects. Under this condition, electrons not only may be elevated to higher energy levels but may completely dissociate, leaving an atomic ion behind. Then immediate return to a lower energy level or to the ground state is not possible and a loss of intensity of the emitted light is the consequence.

Flame photometry represents an extremely valuable approach for the determination of readily excitable elements in small samples and at low concentrations, especially when many samples must be analyzed routinely. The determination of sodium, potassium, and calcium in such diverse samples as blood serum, urine, soil extracts, and industrial and natural waters is standard procedure in many laboratories. With flames of a higher temperature (notably acetylene-oxygen and acetylene-nitrous oxide), magnesium, iron, and certain other elements can also be determined.

39.4 Arc-Spark Emission Spectroscopy

The low energy provided by flames is enough to excite only a limited number of elements. It is necessary to provide greater excitation energy to obtain spectral lines for other elements. This additional energy can be provided by a dc or an ac arc or a high-voltage spark. In the dc arc, for example, struck between two electrodes, the electrons emitted from the negative electrode bombard the positive one, and the temperature there is raised to possibly 6000 °K. The positive electrode may either be fashioned from the material to analyzed (for example, an alloy) or impregnated with it. Spectra thus produced contain a multitude of lines, most of which lie in the ultraviolet region. The number of lines depends on the excitation conditions, the elements present, and the temperature. For transition metals, up to 1000 lines have been identified under certain conditions.

These facts evoke appreciation that the requirements for the apparatus are much greater than under the simple conditions of flame emission photometry. Lines of different elements are frequently close together and monochromators of considerable resolution are mandatory. But even then, for the identification or determination of an element, characteristic lines must be selected that are free of coincidences.

The principal parts of a typical spectrograph and their arrangement are shown schematically in Fig. 39-2. The light emitted from the arc (or spark) is focused by a lens on the entrance slit of the monochromator; the light then passes through a collimating lens (that is, one that renders the light parallel), on through the prism, and by a further lens is finally projected on a photographic plate or film. This plate or film can be moved up or down so that several spectra can be recorded consecutively but separately. In a grating spectrograph, the prism is replaced by a grating, which is usually ruled on a concave mirror.

If the material to be analyzed is electrically conducting, for example, an alloy, it may be shaped to rods, which are used as electrodes. Alternatively, a graphite electrode may be used with a small crater at its tip. The powdered sample material, often with graphite added to make it electrically conducting, is introduced into this crater or a sample solution is placed there and evaporated. The arc or spark is then fired and the spectrum photographed. With a dc arc, it is necessary to have the sample material as the anode or within its crater.

From the location of lines at certain wavelengths, it is readily possible to evaluate the spectrum qualitatively. Since at the extreme temperature conditions chemical bonds are completely disrupted, only elements can be detected, and no deductions can be made about their combination. Several possibilities exist for applying spectrography to quantitative analysis.

For the analysis of an alloy, for example, the following procedure is a possibility. First a spectrogram is obtained from the sample. Then the plate or film is moved and spectrograms made for one or more standard alloys of similar and exactly known composition, under identical instrument settings, including the gap between the electrodes, electrical current, and time of exposure of the plate to the spectrum. The plate is then developed and fixed under specified and reproducible conditions, and dried. For each element to be determined, one or several lines are selected and their intensities, that is, the degree of blackening on the plate, are compared for the sample and the standard. For this comparison, an instrument known as a densitometer can be used; it functions like an absorption photometer and measures the absorbance of the spectral line on the plate relative to a clear portion of the emulsion. From the knowledge of the ratio of the line absorbances for

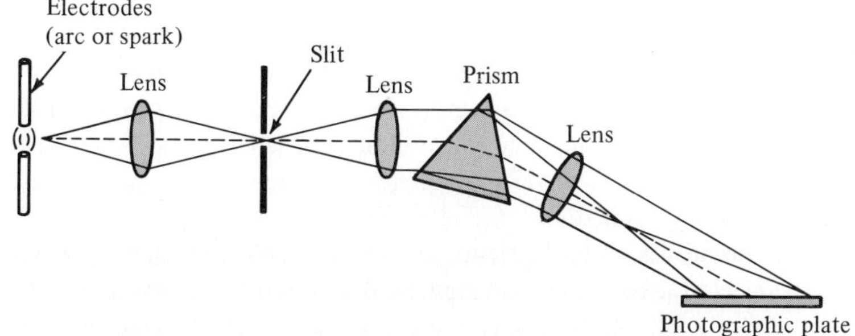

Fig. 39-2. Schematic arrangement of the components of a spectrograph.

sample and standard, together with the known content of the element to be determined in the standard, the percentage of a sought-for element in the unknown can be calculated. The method of an internal standard can be used analogously to the method mentioned for flame photometry. It has the advantage that the results are relatively independent of small variations in the exciting source and in the photographic development and fixation. The internal standard method is most readily applicable to sample solutions but can be used with powdered samples as well.

The method of "last lines," or "persistent lines," is also often used with sample solutions. This technique is based on the following. The spectral lines of an element differ in intensity since the various electron transitions vary in their probability. If a spectrum is obtained from each solution of a dilution series, first the weakest lines disappear, and finally the most intense, most persistent lines. For purposes of establishing the "last lines" for an element, a measured volume of a stock solution of exactly known concentration is brought onto the electrode, and after sparking or arcing has gone on for a definite time, the spectrum is obtained. A portion of the stock solution is then diluted by a known ratio, and a spectrum is again secured under identical conditions. This process is repeated until the last lines have disappeared. Then the concentration at which this disappearance took place is established. The spectra of a dilution series are obtained in identical fashion, so that a sample can be analyzed. From the dilution ratios and the concentration at which the last lines disappear, it is possible to calculate the concentration of the sought-for substance in the original sample solution. The technique can be used for the concurrent determination of more than one component in a mixture. The method can also be applied to powdered samples with graphite as the diluent.

Other approaches are also used for quantitative results, but the methods described will suffice to provide an understanding of the principles involved. With close attention to operating details and the use of a good spectrograph, the result of a determination can be precise and accurate to within ±5% (relative).

Instruments also are used in which the photographic recording is eliminated by photodetectors. A single detector can be used, moved so that it slowly scans the spectrum. The detector output is amplified and fed to a strip chart recorder. There the spectrum is displayed as an array of sharp peaks, the locations and heights of which can be related to wavelengths and intensities of the corresponding lines. Alternatively, a number of photodetectors are used, each is mounted exactly at the location of a particular spectral line. Each output is fed to a separate counting device, which after appropriate calibration allows the content of the relevant element to be read directly. With such emission spectrometers, many elements can be determined concurrently in a relatively short period. Such direct-reading instruments are expensive but pay for themselves when numerous samples must be routinely and rapidly analyzed for many constituents—notably in the process control of metallurgical melts.

Some advanced instruments use an ionizable gas, such as argon, in a high-voltage electrical or radiofrequency field to produce a plasma with temperatures from 6000 to 10,000°K. On aspiration of the sample solution into the plasma region many more lines are excited, including ionic ones, than with an arc or a spark, and interferences are reduced or eliminated. Such inductively coupled plasma (ICP) excitation with a direct-reading atomic emission spectrometer and computer processing allows the determination of up to 50 elements in liquid samples within a few minutes and with concentrations ranging over five or more decades of content.

39.5 *Atomic Absorption Spectrometry*

The passage of a beam of compound radiation through a flame leads to a reduction in the intensity at the wavelength for an energy-absorbing atomic species present in the flame. For practical application, the beam must present the required wavelength at high intensity. For many elements, so-called hollow cathode lamps are used. Their construction is shown schematically in Fig. 39-3. The glass envelope E, with a flat end window of quartz Q, contains argon or helium at a reduced pressure. Across the circular electrode (the anode, positive) A and the cylindrical tube (the cathode, negative) C is placed a high direct-current voltage. Atoms of the rare gas are ionized and the positively charged ions formed move toward the cathode and strike it forcefully. The cathode is fashioned of the metal of interest or lined with it. The atoms of this metal are thereby excited, and as a consequence, an intense beam B, containing the wavelengths of the spectral lines of the metal, is emitted through the window Q. The beam is collimated and directed toward and through the flame. The emitted light from the source has enough intensity for the purpose at hand and offers the advantage that it is, so to speak, tailored to the particular element to be determined, because the particular wavelengths are specific for the element and unable to be absorbed by any other element.

An atomic absorption determination of a metal proceeds as follows. An intense beam of light from a hollow cathode lamp for that element is passed through the flame, with the solvent or blank being aspirated, and falls on a photodetector. The electrical signal created by the detector is led to a meter or a digital display. A reading of 100% transmittance, or zero absorbance, is secured by a circuit adjustment. Now the sample solution containing the metal is aspirated. A portion of the light is absorbed, and as a result, a lower transmittance, or higher absorbance, reading is shown.

This process, in principle, is analogous to molecular photometry as described

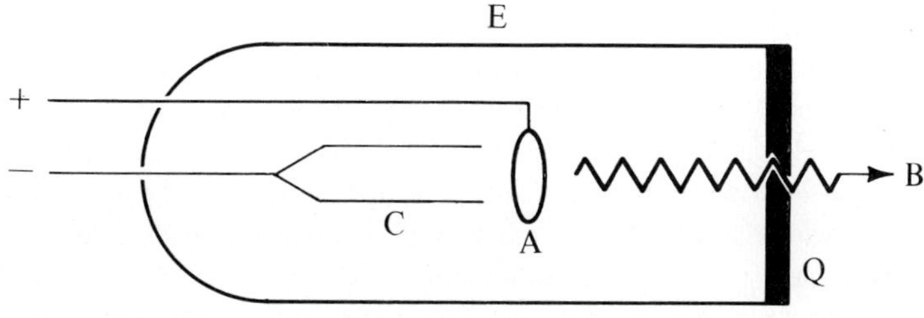

Fig. 39-3. Schematic presentation of a hollow cathode lamp.

for solutions (Chapter 35). However, the actual operation is somewhat different. The light incident on the detector consists of several portions: (1) light from the source beam (less whatever intensity was absorbed), (2) light from the heat excitation of neutral atoms of the metal, and (3) light from the flame background. The two last contributions are unwanted and can be excluded.

The light beam, on emerging from the flame, reaches a monochromator, which separates the spectral line of interest from others emitted by the hollow cathode lamp; it also eliminates the background continuum of the flame, except for the small portion within the wavelength region that is allowed to pass. The arrangement contrasts with that for molecular photometry of solutions, in which the monochromator is located *before* the sample solution, to avoid exposing it to the full intensity of the source beam.

A further measure toward eliminating unwanted light is to *modulate* the system. In early designs, this was achieved mechanically by placing a rapidly rotating disk with slots between the light source and the flame. The "chopped" beam creates an alternating signal in the detector output while the flame emission and background cause essentially a steady signal. Electronically it is simple to separate the two types of signal and to have only the component corresponding to the source beam (after rectification) fed to the meter or digital display. In more recent designs, an intermittent signal is secured from the hollow cathode lamp itself by applying either an alternating current or an intermittent direct current. The detector circuit is tuned to the frequency of the intermittent signal.

Because the flame background is readily excluded in atomic absorption spectrometry, the technique has a distinct advantage over flame emission spectrometry. The alternating current created with atomic absorption spectrometry has an added benefit in that such a signal can be amplified by less expensive means.

Since the source beam presents largely only the wavelengths of the particular element to be determined, a high degree of selectivity is secured and radiation interference is hardly of any consequence in atomic absorption spectrometry. Chemical and excitation interferences, however, arise as in flame photometry. Interferences from molecular absorption bands may also require correction.

The operation of the aspiration and the flame, and regulation of the gases and the like, are essentially identical with those steps in flame emission spectrometry. The burner usually consists of a narrow slot up to 10 cm or more in length in the center of a flat plate that forms the top of the spray chamber. The source beam is passed longitudinally through the flame. The evaluation of results by a calibration curve or standard additions proceeds as for flame photometry.

Atomic absorption spectrometry was introduced for analytical purposes by Walsh about 1955. It rapidly achieved an important place in analytical practice because of its high sensitivity, excellent selectivity, and relative simplicity. The technique is now widely used for determining various metals at minor and trace concentrations in alloys, chemicals, and environmental samples, to name only a few examples.

Extraction is a possible technique where increased sensitivity is needed and to remove the element or elements to be determined from an undesirable matrix.

The one or more species of interest are extracted, commonly as chelates, from an aqueous solution into an organic solvent and the extract is aspirated directly to the flame. The increased sensitivity is due partly to the enrichment contingent on concentration into a smaller volume. In addition, the selected organic solvent is combustible, thereby often yielding a higher flame temperature; this usually helps in the formation of metal atoms instead of refractory metal oxides.

Since the burden of selectivity can fall on the atomic absorption spectrometer, the chelating extractant can be relatively unselective, but it should provide essentially complete extraction of metals and over a broad pH range. The most widely used extractant with atomic absorption procedures is 1-pyrrolidine-carbodithioate, as the ammonium salt (APCD). This reagent has the structural formula

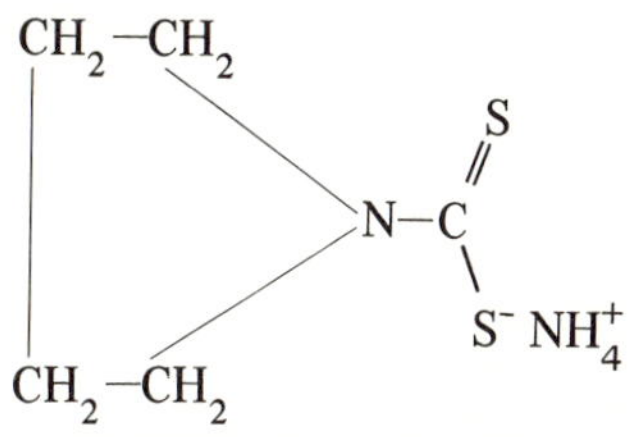

$$\begin{array}{l} CH_2-CH_2 \\ \quad | \qquad\quad \searrow \qquad\quad S \\ \quad | \qquad\qquad\quad N-C \!\!\diagup\!\!\diagup \\ \quad | \qquad\quad \nearrow \qquad\quad \backslash \\ CH_2-CH_2 \qquad\quad S^- NH_4^+ \end{array}$$

Under another name, tetramethylenedithiocarbamate, the reagent is better recognized as closely related to diethyldithiocarbamate (see Section 36.2). Extraction with this reagent, usually into 4-methyl-2-pentanone (that is, methyl isobutyl ketone) has been applied in the atomic absorption determination of over twenty elements and to such practical samples as blood, tissue and plant materials, milk, brine and waters; and cements, rocks, and soils.

39.6 *Electrothermal Atomization in Atomic Absorption Spectrometry*

The flame offers a convenient device for nebulizing and atomizing a sample solution to give an atomic vapor. Its use, however, may require a relatively large volume of sample solution, and interferences may require special attention for quantitation. Also the population of atoms in the flame is relatively low. Techniques for obtaining an atomic vapor without use of a flame are therefore of interest. High-temperature atomization often involves using an "oven" or "furnace," consisting of a graphite tube through which light can pass. The sample solution is injected into the tube. The beam from a hollow cathode lamp passes coaxially through the tube and absorption takes place as in a flame. At first, the tube is electrically heated to evaporate the solvent; and then further heating to a higher temperature volatilizes some matrix elements and chars organic compounds. Finally, the tube is heated to an even higher temperature, and in the reducing environment provided by the graphite walls, a high population of neutral atoms is produced. This extremely sensitive technique is applied, for example, to the determination of trace lead in microliter volumes of blood.

39.7 Cold-Vapor Atomization in Atomic Absorption Spectrometry

In some cases, an atomic vapor can be secured at relatively low temperatures, but examples are limited. The so-called cold-vapor atomic absorption determination of mercury is extremely sensitive and has attained importance in analyzing foods, tissue, industrial wastes, and waters. This method is relatively simple. The sample is appropriately treated to give a solution; this may mean homogenization of organic samples, acid digestion, or both. The acidic preparation is placed in a vessel that functions essentially as a gas scrubber. A solution of a reducing agent, such as tin(II) chloride or sodium borohydride. $NaBH_4$, is added to reduce mercury(II) to elemental mercury. A stream of nitrogen is passed through the liquid and carries the mercury, as a vapor, to a cylindrical absorption tube, through which a beam of light from a mercury vapor lamp is passed coaxially. The tube is placed in an atomic absorption spectrometer just as a burner and flame would be. The processes of absorption, measurement, and calibration are followed analogously. Separate instruments are also available that are optimized for this determination of mercury. By cold-vapor atomic absorption spectrometry, mercury present at the nanogram-per-gram (ppb) level can be determined by a procedure that is almost completely free from interference.

39.8 Atomic Fluorescence Spectrometry

Atomic fluorescence is another approach to atomic spectroscopy. (The principle has been explained in Section 39.2.) The arrangements are like those for flame atomic absorption spectrometry, but the source-flame axis is usually at a right angle to the flame-monochromator axis. Hollow cathode lamps and electrodeless discharge tubes are the sources.

39.9 Questions

39-1 Elaborate on the difference between line spectra and continuous spectra.

39-2 What parameters in emission flame spectrometry govern the number of spectral lines excited and the intensity of a single line?

39-3 Explain why relatively few lines are excited in a city gas-air flame.

39-4 Elaborate on why a simple filter instrument suffices for the flame photometric determination of sodium or potassium but an instrument with a monochromator is best for determining magnesium or aluminum. (*Hint:* Consider the differences in the flames required.)

39-5 Some oxyanions reduce the line intensity of metals in a flame. For example, the emission due to barium is markedly reduced by the presence of sulfate or phosphate. Elaborate on the basis of this phenomenon and suggest how it might be applied for the indirect determination of an anion.

39–6 Organic solvents influence the intensity of flame emission of metals, partly because of their effect on the flame temperature. Elaborate on what solvents and solvent classes might be expected to increase or decrease flame temperatures. For what metals would an increase in flame temperature by solvent addition be probably unwanted?

39–7 Elaborate on the difference in principles of emission and absorption flame spectrometry.

39–8 What methods can you name that could be used to determine the content of a sought-for element in a sample subjected to arc emission spectrography.

39–9 What is "flame background"? Elaborate on the different ways by which it is reduced or eliminated with the various flame analytical techniques.

39–10 Contrast emission spectrography with wet analysis methods (gravimetry, visual titrimetry, and solution absorption photometry) for advantages and disadvantages, the useful concentration range of the sought-for substance in the sample, speed, instrumental requirements, necessity of separations, ease of sample preparation, and accuracy and precision.

39–11 Explain why the graphite furnace provides a higher population of neutral atoms than a flame does.

39–12 Many elements exhibit fewer chemical interferences when an acetylene-nitrous oxide flame is used for atomic absorption spectrometry; however, at the same time, the sensitivity is reduced. Explain.

39–13 Many halide salts are volatilized at the temperatures provided by a flame and then show absorption in the region from 200 to 300 nm. What effect could these salts have on the atomic absorption determination of zinc at 213.9 nm?

39.10 Problems

39–1 In the flame photometric determination of sodium, water was aspirated and the instrument was set to zero. Then the sample solution was aspirated and the reading was 25.0 scale divisions. To 25.00 mL of the sample solution was added 5.00 mL of a standard sodium chloride solution containing 7.0 μg of sodium per mL. When the resulting solution was aspirated, the reading was 55.0 scale divisions. What is the sodium concentration in the sample solution expressed in milligrams per liter (ppm)? *Ans.:* 1.0 mg/L

39–2 For the atomic absorption determination of zinc, five 25-mL volumetric flasks were readied. To each was added 10.0 mL of the sample solution and 2 mL of a buffer. Then to the flasks was added a zinc standard solution containing 10.0 μg of zinc per milliliter in the volumes indicated in the accompanying table. Each flask was then filled with water to mark. The instrument was set to zero with a blank. On aspiration of the solutions, the readings were as shown. Calculate the zinc content of the sample solution in milligrams per liter.

Flask no.:	1	2	3	4	5
Std., mL:	0	2.00	4.00	6.00	8.00
Reading:	25.1	44.9	65.2	83.8	96.2

Ans.: 2.5 mg/mL

39–3 For the determination of copper, atomic absorption spectrometry was used with the standard addition method. Four 50-mL volumetric flasks were taken and to each were added 20.00 mL of the sample solution and 10 mL of a buffer. Then to each of flasks 2, 3, and 4 were added exactly 5, 10, and 15 mL, respectively, of $1.00 \times 10^{-3} F$ copper sulfate. All flasks were then filled with water to mark and the contents mixed. The readings with the settings initially adjusted to zero with a reagent blank were 36.0, 47.5, 59.0, and 69.0 scale divisions. Calculate the copper (*63.54*) content of the original sample solution in micrograms per milliliter (ppm). *Ans.:* 52 μg/g (ppm)

39–4 The flame photometric determination of sodium is conducted with lithium as an internal standard. With given instrument settings and filters, and zero adjustments against water, identical concentrations (in milligrams per liter) of sodium and lithium gave readings of 48.0 and 40.0 scale divisions. A volume of 20.0 mL of a sample solution is placed in a 50-mL volumetric flask. Then 10.0 mL of a solution containing 25.0 mg/L of lithium is added and the solution is diluted with water to mark and mixed. The resulting solution is aspirated and readings for sodium and lithium are given as 36.0 and 28.0 scale divisions. Calculate the concentration of sodium in the original sample solution, in milligrams per liter, assuming that measurements for both elements are made within the linear response range. *Ans.:* 13.4 mg/L

39–5 With a copper hollow cathode lamp in place, an atomic absorption instrument is adjusted to zero with water. A sample solution is aspirated and the reading is 50.0 scale divisions. Then 20.0 mL of the sample solution and 10.0 mL of a standard containing 2.00×10^{-4} g/L of copper are mixed. On aspiration of this mixture, the reading is 40.0 scale divisions. Calculate the concentration of copper, in grams per liter, in the sample solution.

Ans.: 5.0×10^{-4} g/L of Cu

40 X-Ray Methods

40.1 Introduction

X-rays are electromagnetic radiation in the wavelength range 0.01 to 10 nm. They are emitted during certain types of radioactive decay and can be produced in an X-ray tube (see Fig. 40-1a). Such a tube consists of an evacuated glass envelope containing a wire filament (cathode) and a "target" (anode). A voltage difference up to several thousand volts is established between the filament and the target. The filament is heated and the electrons then emitted are attracted by the anode and greatly accelerated in the strong electric field. The high-velocity electrons, on striking the target, are capable of knocking electrons from the inner orbitals (for example, the *K* or *L* shell) of the target atoms. Each electron thus removed is then replaced by one from a higher orbital, and the energy lost by this second electron is emitted as an X-ray.

The X-ray spectrum produced, after suitable dispersion, consists of "white" noncharacteristic radiation constituting a continuous spectrum, with superimposed sharp lines characteristic of the element used as the target and related to the atomic number by Moseley's law:

$$\frac{c}{\lambda} = a(Z - \sigma)^2 \qquad (41\text{-}1)$$

Here c is the velocity of light, λ the wavelength, a a proportionality constant, and Z the atomic number. The constant σ has a value dependent on the series of lines, that is, dependent on whether the line originates from an electron falling from the L to the K shell (K_α lines), the M to the K shell (K_β lines), the M to the L shell (L_α lines), and so on.

X-ray line spectra are simple, that is, they consist of only a few lines and are similar for all elements, but the actual wavelengths of each of these lines varies from element to element. The X-ray lines are not significantly influenced by the state of the element (gas, liquid, or solid) or whether it is present as the free element or chemically bonded in a compound.

Lenses and prisms in the usual sense cannot be used with X rays. In a wavelength-dispersive X-ray spectrometer (Fig. 40-1), in which a sample undergoes excitation by X rays, the collimator necessary to obtain a parallel beam is a long narrow gap between two metal blocks or several gaps between closely spaced metal plates or concentrically arranged tubes. The dispersion of the emitted spectrum is effected by a crystal, which acts as a diffraction grating, since the interatomic distances in the crystal are of the same order as the wavelengths of the X rays. A detector scans the arc, as shown in the figure, and the signal created is fed to a

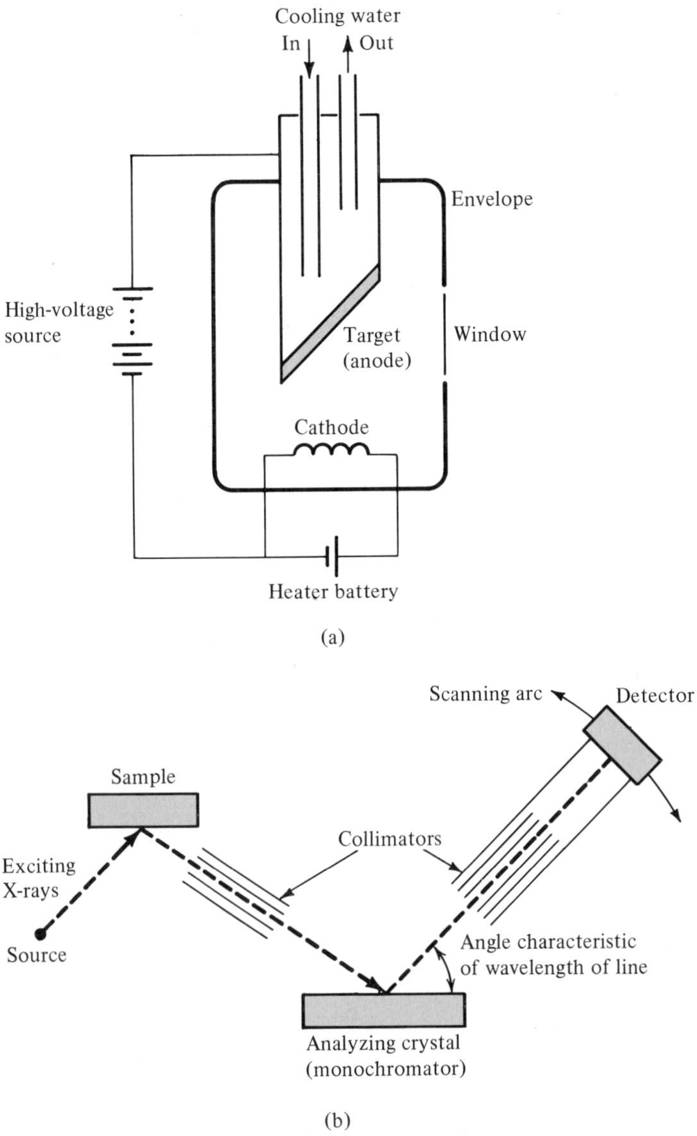

Fig. 40-1. (a) X-ray tube; (b) wavelength-dispersive X-ray spectrometer.

recorder. Radiation detectors including the Geiger counter and related devices based on the ionizing effect of X rays on a rarefied gas can be used (see Section 41-2 and an elementary textbook on chemistry or physics). Scintillation counters receive special attention. Such a counter is a photomultiplier tube, the window of which is covered by a thallium-activated sodium iodide crystal. If the crystal is struck by an X ray, fluorescence occurs in the visible region, and its intensity is measured by the photomultiplier in connection with a relevant circuit and meter.

X-ray methods of interest to analytical chemists are based on emission (especially secondary "fluorescent" emission), diffraction, and absorption.

40.2 X-Ray Emission and Fluorescence Methods

If the substance to be analyzed is made the target in an X-ray tube, a spectrum is produced containing the characteristic lines of all the elements present. The wavelength of a line (or simply the angle between the beam of the line and the incident beam) can be measured and used for qualitative purposes. Measuring the intensity of the lines gives a quantitative evaluation. This technique of X-ray emission analysis is restricted because tubes that can be dismantled are needed to introduce the sample material as the target. Since the tube is operated under vacuum, readily volatized substances cannot be analyzed. In addition, on bombardment with the electrons, the target reaches high temperatures even when cooled by the flow of water through airtight tubes; many substances cannot stand such conditions.

Another approach X-ray fluorescence, is of greater general applicability. X-rays can be produced not only by bombarding a substance with electrons but also by irradiating the substance with short-wavelength X rays. The secondary, or fluorescent, X-rays thus produced are less intense than X rays obtained by electron bombardment by a factor of about 1000. Consequently, a primary X-ray beam of high intensity and detectors of high sensitivity are mandatory. The primary components of a wavelength-dispersive X-ray spectrometer and the arrangements are shown in Fig. 40-1b. Solid or liquid samples can be used. The sample containers are fashioned from polymers such as polyethylene or aluminum since these materials show low absorptivities for X rays. Elements with an atomic number of 12 and higher can be analyzed in air; the elements with atomic number 5 through 11 emit X-rays of longer wavelengths that are absorbed in air. These lower elements require special equipment that allows operation in a vacuum or in a helium atmosphere.

Alpha and beta particles from a radioactive source can excite inner shells of atoms, leading to emission of the characteristic X rays of the atoms struck. A nondispersive, energy resolution, X-ray spectrometer uses a radioactive source with a proportional counter and pulse-height selection circuits. The detector is based on a semiconductor that produces a signal proportional to the incident photon energy.

An important domain of X-ray fluorescence methods is the rapid analysis of alloys, minerals, and other solid materials. The method is nondestructive. Depending on the quality of the instrument and care of the operator, even minor constituents can be determined with an accuracy and a precision of 1 to 2% or even better.

40.3 *X-Ray Diffraction Methods*

When a monochromatic X-ray beam strikes a crystalline substance, diffraction takes place. Depending on the wavelength of the X rays, the interatomic spacing of the layers in the crystal, and the position of the crystal relative to the beam, extinction will occur in some directions and reenforcement in others. Thus, the radiation leaving the crystal will, on striking a photographic plate, create an arrangement of spots, called a "diffraction pattern." The pattern is characteristic of the crystal system and can be used to identify a crystal. In analytical chemistry, often a powdered sample, with random distribution of the crystallites, is placed in the X-ray beam and the diffraction pattern recorded on a narrow strip of photographic film as a series of lines.

Studying the position of the lines (and their resolution) allows estimating the ratio of crystalline to amorphous material, and with the aid of known spacing values, makes possible the identification of the crystalline phase or phases present. Thus, for example, X-ray powder diffraction can allow a decision about whether a sample is a mixture of sodium chloride and potassium sulfate or of potassium chloride and sodium sulfate. Wet methods of analysis would reveal only the presence of sodium, potassium, chloride, and sulfate ions but not the pairing of the ions in the solid sample. Simple mixtures can also be analyzed semiquantitatively by measuring the relative intensities of X-ray diffraction.

In general, however, the application of X-ray diffraction to analytical chemistry is limited. The method is applied for other purposes, including the determination of stresses and imperfections in materials such as alloys or plastics, and the measurement of particle size. The method is one of the principal tools in crystallographic studies.

40.4 *X-Ray Absorption Methods*

When a beam of X rays, either polychromatic or monochromatic, is passed through a sample, the X rays interact with the matter, and energy is absorbed. Measuring the intensity of the beam with and without the sample in it allows a determination of the absorbance in a photometer arrangement, as in the ultraviolet and visible regions. But here either a Geiger or a scintillation counter is used as the detector. The higher the atomic number of an element, the greater its absorptivity for X rays. Consequently, lead of high atomic number can be determined by X-ray absorption in leaded gasoline; the major elements present—carbon and hydrogen—have low atomic numbers.

40.5 *Electron Probe Microanalysis*

Electron probe analysis had its first practical application by Castaing in 1951. In this procedure, the specimen is bombarded with a thin electron beam. The X-ray spectrum excited contains the wavelengths characteristic of the elements present in the minute area struck by the electron probe. The intensities of these wavelengths are proportional to the mass concentration of these elements. The operation of an electron probe microanalyzer can best be explained by referring to the admittedly much simplified schematic diagram, Fig. 40–2. The electron beam from the electron "gun" by passage through two (or more) magnetic fields, that is, through "magnetic lenses," is focused and demagnified onto a region, about 1 μm in diameter, of the electron-opaque specimen. The X rays generated are diffracted by a crystal and the wavelengths of interest are isolated by a slit and fall onto a detector (Geiger counter, scintillation counter, and the like). To direct the probe to the exact point of interest, observation by a mirror system (not shown in the figure) and a visual optical microscope is often used. The whole arrangement (except the optical microscope) is enclosed, and under high vacuum. The specimen can be moved relative to the electron probe, and point-by-point analysis is thus possible. In this way the distribution of a selected element can be established rapidly. The electron probe microanalyzer is elaborate and complex but allows the analysis of exceedingly minute segregations, inclusions, and so on, at the surface of materials. Only surface layers can be investigated, because the electron beam does not penetrate deeper than about 1 μm. In a region about 1 μm in diameter it is possible to detect elements

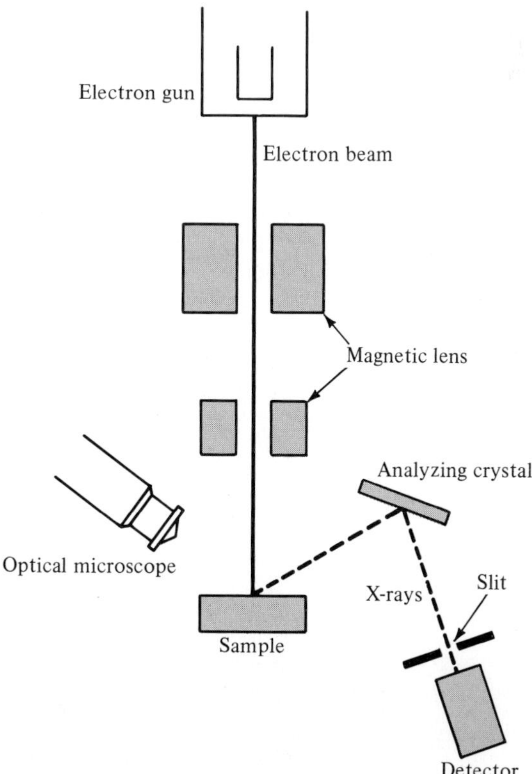

Fig. 40–2. Schematic diagram of an electron probe analyzer.

present at small percentages of concentration, and the actual amount of an element excited is often only 10^{-14} to 10^{-15} g. When appropriate standards are available, the determination of elements is possible with a relative percentage error, that is amazingly small for the effective sample size.

40.6 Questions

40–1 What relation exists between the wavelength of an X ray and the atomic number of the element emitting the ray?

40–2 In wavelength-dispersive X-ray spectrometer, what is used as a collimator and what as a monochromator?

40–3 Why is a simple X-ray spectrometer not suitable for measuring the X-ray fluorescence of elements with atomic numbers lower than 12?

40–4 What are "white" X rays?

40–5 Compare the application of X-ray methods with other nondestructive techniques in application, instrumentation, speed of analysis, and precision.

40–6 X-ray lines are designated by a capital English letter and a Greek letter subscript. To what do these two letters refer?

40–7 By what means can X rays be produced?

41

Analytical Methods Based on Radioactivity

41.1 Radioactivity and Radioactive Decay

Radioactive isotopes in their spontaneous decay eject one or more energetic particles or radiate energy, including alpha particles (high-speed, dipositive helium ions, He^{2+}), beta particles (high-speed electrons), positrons, and gamma rays (electromagnetic radiation of very short wavelength). When the decay involves electron capture, X rays are produced. Formerly, since only a few elements are naturally radioactive, analytical methods based on radioactivity were limited in their application. When more intense sources of high-energy particles were developed, notably the nuclear reactor, with the resulting availability of artificially produced radioactive isotopes of many elements, the possibilities for the analytical use of radioactivity increased enormously.

The rate of disintegration of a radioactive isotope follows the so-called exponential decay law; that is, the rate of disintegration is proportional to the number of nuclei of active isotope present. In the notation of calculus, this takes the form

$$\frac{dN}{dt} = -\lambda N \tag{41-1}$$

where N is the number of nuclei of the radioactive isotope present and t is the time; and the Greek letter lambda, λ, is a proportionality constant known as the decay constant. The value of this constant is a characteristic of the isotope. The minus sign enters the equation because the number of nuclei of the isotope decreases with time. The disintegration rate dN/dt is usually expressed in curies (1 Ci = 3.7×10^{10} disintegrations per second) or a submultiple, such as millicuries (1 mCi = 1×10^{-3}

NOTE: In this chapter we assume that you have an understanding of the fundamentals of the phenomena of radioactivity on the level as they are presented in elementary textbooks on physics and chemistry. Only a brief recapitulation of some basic facts can be given here.

Ci).* The specific activity of a sample is usually placed on a weight basis and expressed in curies or millicuries per gram of sample.

Rearranging and integrating equation (41-1) between $t = 0$ and $t = t$ yields

$$N_t = N_0 e^{-\lambda t} \tag{41-2}$$

where N_t and N_0 are the number of nuclei of the radioactive isotope present at time t and time zero; e is the base of natural logarithms.

It is also convenient to consider radioactive decay by the half-life of the isotope, that is, the time required for the number of nuclei of the isotope present in a sample to be reduced to one-half the original number. The relation between the half-life, denoted by the Greek letter tau, τ, and the decay constant can be readily derived. From the definition of the half-life, $N_t = N_0/2$; substitution in equation (41-2) yields

$$\tfrac{1}{2}N_0 = N_0 e^{-\lambda\tau}$$

$$\tfrac{1}{2} = e^{-\lambda\tau}$$

Conversion to logarithms to the base 10 yields

$$\log \tfrac{1}{2} = -\lambda\tau \log e$$

$$-0.3010 = -0.4343\ \lambda\tau$$

$$\lambda = \frac{0.693}{\tau} \tag{41-3}$$

Combining the integrated and differential forms of the decay law, equations (41-1) and (41-2), yields

$$\frac{dN}{dt} = -\lambda N_0 e^{-\lambda t} \tag{41-4}$$

The rate of disintegration at time $t = 0$ is given by

$$\left(\frac{dN}{dt}\right)_{t=0} = -\lambda N_0 \tag{41-5}$$

Consequently,

$$\frac{dN}{dt} = \left(\frac{dN}{dt}\right)_{t=0} e^{-\lambda t} = \left(\frac{dN}{dt}\right)_{t=0} e^{-0.693t/\tau} \tag{41-6}$$

---

*The curie has been accepted temporarily for use with the International System of Units (SI); however, the "activity of a radionuclide," that is, its disintegrations per second, is termed the becquerel and given the symbol Bq (1 Ci = 3.7×10^{10} Bq = 3.7×10^{10} disintegrations per second).

Conversion of this equation to the logarithmic form yields

$$\log\left(\frac{dN}{dt}\right) = \log\left(\frac{dN}{dt}\right)_{t=0} - \frac{0.301}{\tau}\,t \tag{41-7}$$

A semilogarithmic plot of disintegration rate (logarithmic axis) versus time (linear axis) should therefore yield a straight line, the slope of which is $-0.301/\tau$. Thus, the half-life of a radioactive isotope can be established by measuring the disintegration rate at various times. A sample containing more than one radioactive species will give a plot showing several essentially straight-line portions if the half-lives of the isotopes are different enough.

41.2 *Detecting and Counting Radioactivity*

Several devices are available for detecting and counting the energetic particles and radiant energy produced in radioactive decay. Which device is best applied will depend on the type of emission to be counted, its intensity, and whether or not differentiation is sought between emissions of different energies.

The Geiger-Müller counter (or simply Geiger counter) consists of a glass tube lined with a metal cylinder and a central wire that is insulated from the cylinder. The tube is filled with appropriate gases at low pressure. A voltage of at least 600 V is applied between the wire and cylinder. When a high-speed particle or an ionizing radiant energy (gamma or X rays) enters the tube, often through a permeable window, some of the gas is ionized, a discharge between tube and wire occurs, and the voltage drops momentarily. The electrical "pulse" thus produced is amplified and recorded by counting devices. In the Geiger-Müller counter, the impulse size is essentially independent of the energy of the incident particle or ray; thus the number of disintegrations is counted. Somewhat similar counters operating at higher voltages have an impulse response that is proportional to the energy of the incident ionizing particle or ray; they are called proportional counters.

A scintillation counter consists of a fluorescent material mounted on a photomultiplier tube. The material shows fluorescence in the visible region with an intensity proportional to the energy of the incident particles or ray. Since the photomultiplier responds linearly to the intensity of the fluorescence, the response of the entire scintillation counter is proportional to the energy of the incident particles or rays. The output of the counter is amplified and fed to a counting device. Solid-state semiconductor detectors, such as silicon or germanium doped with lithium, are being adopted because of their improved resolution of energy peaks.

An exposition of counting devices is beyond the scope of the chapter. When the number of counts exceeds about 50 per second, mechanical counters are replaced by so-called scaling circuits connected to electromechanical counters; the entire unit is called a scalar. If a so-called discrimination circuit (discriminator) is placed between the amplified output of a proportional or scintillation counter and a scaler, selectively counting pulses exceeding a preset size is possible.

In the selection and use of counters, the geometry of the sample must be considered since radioactive decay is random in all directions. Often it suffices to place the sample and standards in the same orientation and position relative to the counter. It is also necessary to subtract any background count caused by low-level radioactivity of the instruments and their surrounding and by cosmic rays. Adequate shielding of the counting chamber by lead reduces the background markedly.

41.3 Tracer Methods

When a radioactive tracer (that is, a radioactive isotope) of one of the elements present is incorporated in a sample, the sample is said to be "labeled," "spiked," or "tagged." Only a very small amount of the isotope has to be added to secure a disintegration rate that can be counted easily. Frequently the amount added is negligible compared with the amount of the element present in the sample. Tracer methods receive diverse application in chemistry and other fields of science and technology, notably the biosciences. For example, they allow a complex chemical separation to be followed readily and the efficiency of the process to be evaluated. They also allow the determination of physical constants, such as solubilities.

41.4 Liquid Scintillation Spectrometry

Liquid scintillation counting (LSC) is an important technique for measuring substances labeled with a beta emitter, notably tritium (T or ^{3}H) or carbon 14 (^{14}C). The sample containing the labeled species is dissolved, solubilized, or dispersed in what is known as an LSC cocktail. The resulting liquid counting system, in a screwcap glass or plastic vial, is placed in an LSC spectrometer, where the disintegrations are detected and counted. The counting is usually done at a subambient temperature (10 or 15°C) to reduce "background." For quantitative purposes, an uncomplicated approach involves adding a known amount of a standard to one or more portions of the sample-cocktail mixture, thereby providing known numbers of disintegrations per minute. The simplest LSC cocktails consist of an aromatic solvent, such as toluene, in which a highly fluorescent substance is dissolved at low concentration. In this use, the substance is known as a scintillator, or a fluor. So that aqueous samples can be incorporated, the cocktail may also contain hydrophilic substances.

In the LSC sample-cocktail mixture, the energy of the beta particle formed on each atomic disintegration is transferred by solvent molecules to fluor molecules, which are excited and then return to the ground state by the emission of fluorescence (see Section 38.4). Ideally each disintegration should produce a detectable flash of light. In practice, however, the ratio of the number of counts detected to the actual number of disintegrations may range for tritium from 0.2 to 0.8, depending on the cocktail and type of sample involved. The fluorescent events are detected by photomultiplier tubes, placed outside the scintillation vial, and counted by the associated circuits.

Each disintegration counted should be relatable to the energy of the beta particle involved. By discrimination and scaling circuits (see Section 41.2), the counting can be restricted to beta particles of a narrow energy range, and more than one radioactive isotope can thereby be counted in a single sample.

In the biosciences, liquid scintillation spectrometry is widely applied in studying the metabolism of substances and their transport within living systems. The main requirement is that the substance of interest can be synthesized with the radioactive atoms incorporated in adequate numbers. Organic samples that are unfavorable for direct LSC counting can be oxidized in solution or combusted and the evolved water (with T_2O present) or carbon dioxide (with $^{14}CO_2$), or both, can be incorporated into an LSC cocktail for counting.

41.5 *Isotope Dilution Methods*

The principle of isotope dilution can be readily appreciated from an example. Assume that a certain substance is to be determined in a mixture. Exactly m grams of a radioactively labeled form of the substance having a specific activity of A millicuries per gram is added to the sample solution taken. The substance of interest is then precipitated and a known weight of the isolated precipitate is subjected to counting. The specific activity of the precipitate is thereby established as A' millicuries per gram. Simple reflection reveals that the value of A' will decrease as the amount of the (unlabeled) substance in the sample solution is increased and as the amount of labeled substance added is decreased. The required relation, therefore, is

$$A' = \frac{m}{m + x} A \tag{41-8}$$

where x is the grams of sought-for substance in the sample solution taken. Solving for x yields

$$x = m\left(\frac{A}{A'} - 1\right) \tag{41-9}$$

The precipitation does not need to be quantitative! In a similar way other separation processes can be used, including solvent extraction and ion exchange. The only requirement is that the amount of the separated material taken for counting be established (by gravimetry, photometry, and so on).

41.6 *Activation Analysis*

In activation analysis, the elements to be determined are converted to radioactive isotopes by bombarding the sample with high-energy particles; the activity thus produced is measured. Activation by various charged particles is feasible, but the capture possibilities are few for such particles, because they are readily deflected by the electrical field of the orbital electrons or of the nucleus itself. Consequently, activation by neutrons, which are uncharged, is usually preferred. Although a neutron source can be secured by appropriate use of a cyclotron, a van de Graaff generator, or a radioactive isotope source, the preference is for a nuclear reactor because of the high neutron flux it provides. On the capture of a single neutron by a nucleus, an isotope of the same element is formed with the atomic weight increased by one unit. The sample, after sufficient exposure (minutes to weeks), is removed from the reactor and the disintegration rate is recorded over some period. From the semilogarithmic plot Equation (41-7) then constructed, it is possible to ascertain half-lives and thereby to detect what elements are present in the sample.

In principle, the concentration of an element could be calculated from the knowledge of the decay constant and decay pattern of the relevant radioactive isotope, the particular geometry and efficiency of the counter, and the pertinent data relating to the possible nuclear reactions. However, this is not the procedure adopted in practice. Uncertainties in the counter geometry and other factors make it preferable to irradiate and then count standards under identical conditions and to use the data thus secured to make quantitative conclusions about the composition of the sample.

If necessary, in the study of complex materials by activation analysis, chemical separations may be effected before the various counting operations. Often some of the nonradioactive form of an element is added as a "carrier" to allow the separations to proceed on a reasonable scale or to permit determination of the yield during the chemical processing.

Activation analysis is especially valuable in trace analysis. In favorable cases, samples in the milligram range can often be analyzed containing 1 ng (that is, 10^{-9} g) or less of a detectable element. The sensitivity of activation analysis for a given element depends on the intensity of the particle source and the sample exposure time. It also depends on the ability of the element to capture a particle and on the half-life of the radioactive isotope produced.

41.7 Questions

41-1 Define the terms *curie* and *specific activity.*

41-2 Explain why there is a higher "yield" in radioisotopes secured on irradiation with neutrons than there is with charged particles.

41-3 What two principal methods of analysis use radioactivity?

41-4 Explain the term *half-life.*

41-5 Suggest possible experimental procedures for using radioactive tracers in determining (a) a solubility product, (b) the degree of coprecipitation of a substance, and (c) the completeness of a separation by precipitation and by solvent extraction.

41-6 The following isotope dilution technique can be applied. Two *equal* aliquots of the sample solution are taken and "tagged" by adding *different* but known amounts of the radioactively labeled substance to be determined. The sought-for substance (the amount originally present plus the labeled portion added) is precipitated in each aliquot and the specific activity of each of the two precipitates is determined. From the ratio of the two values, the amount of sought-for substance originally present in an aliquot can be calculated without knowing the specific activity of the labeled substance added. Demonstrate this fact mathematically.

42

Miscellaneous Instrumental Methods

Some additional instrumental methods receive attention in quantitative analysis. We shall briefly delineate the background of these methods, suggesting their usefulness in the practice of analytical chemistry. Several of the methods are predominantly applied to specific problems, often related to organic chemistry.

42.1 *Refractometry*

When a beam of radiation passes obliquely from one medium to another of different "density," the beam changes direction, that is, the radiation is refracted. You may recall from physics that the refractive index (also known as the index of refraction), designated by n, is given by the ratio of the speed of the radiation in the two media or by the relation $n = (\sin i)/(\sin r)$, where i and r are the angles of incidence and refraction. These angles are measured from the line perpendicular to the interface between the media. The value of a refractive index depends on wavelength and temperature. Values for the refractive index for the passage of light from air (or a vacuum) into a substance are recorded in the literature. Values are often given for 20°C or 25°C and the D line of the sodium atomic spectrum (that is, the yellow doublet at 589.0/589.6 nm). The notation n_{D}^{20} implies the refractive index for 20°C and the D line. For water, n_{D}^{25} has the value 1.33250; note six significant figures are given. With a relatively simple refractometer, refractive index values can be measured with a precision of ±2 or ±3 in the fourth decimal place.

The index of refraction is valuable as an analytical tool, since it can be measured rapidly, precisely, and nondestructively. The refractive index under stated conditions (wavelength, temperature, pressure, reference medium) is a characteristic of a compound and is an intensive property (like density, boiling point, and melting point).

For liquids, the measurement of refractive index is simple and straightforward, but temperature must be controlled within a narrow range for precise results.

For solids, the procedure is tedious; and for some crystals additional steps are needed because the refractive index may differ along different crystallographic axes. Refractive indices of solids can be estimated on a microscope stage, allowing the identification of crystals separated from solutions or melts.

The refractive index is often the method of choice for analyzing binary homogeneous systems of two miscible liquids or a solid solute dissolved in a solvent. The only restriction is that the two components must differ sufficiently in their refractive indices. When the components do not interact, the relation between concentration and refractive index is direct and the calibration curve may approximate a straight line over the range of interest. When the components interact, the plot of concentration versus refractive index is usually curved and even may show a maximum or a minimum. For example, an ethanol-water mixture shows a maximum value for the refractive index at 79.3 % w/w ethanol.

Refractometry is applied routinely in a variety of fields, as some examples will show: water content (adulteration) of milk; total sugar in soft drinks, beer, and tree sap; and the content of unsaturated oils in butter, fats, vegetable oils, seed extracts, and the like. The refractive index does not respond to a specific component and in this respect is analogous to electrical conductivity. Refractive index can thereby be used in an unselective or nonspecific way, for example, to estimate the total content of aromatic compounds in petroleum hydrocarbons or the salinity of seawater or butter.

In some cases, a differential refractometer is used to measure the difference in the refractive indices of two substances or materials. One advantage is that precise temperature control is then unimportant as long as the two media are at the same temperature. Applications for such a refractometer include use as a detector in liquid chromatography (see Section 48.2) and for on-line controls in chemical plants.

To secure a "refractometric function" relatively independent of temperature, the so-called specific refraction r can be used as defined by the Lorentz and Lorenz equation:

$$r = \frac{n^2 - 1}{n^2 + 2} \cdot \frac{1}{d} \qquad (42\text{-}1)$$

Here d is the density, which is temperature-dependent. Since the refractive index is a dimensionless quantity, the specific refraction has the units of volume per unit weight. The *molar* or *molecular refraction* R is obtained by multiplying the specific refraction by the molecular weight of the compound.

42.2 *Interferometry*

We have just seen how the refractive index can be obtained by measuring the change in direction of a light beam when it passes obliquely into a second medium. By interferometry, the refractive index can be measured in a different way.

If two coherent light beams are allowed to meet and their waves are out of phase, they interfere. For monochromatic light, with suitable experimental arrangements, a "zebra-stripe" pattern of dark and light "fringes" is obtained. If one of the beams is now brought into a medium of different refractive index (that is, one in which the light has a different velocity), the fringes are displaced. This displacement can be measured with high reliability, allowing calculation of the refractive index to 7 decimal places if the temperature is held within ±0.5°C. Obviously, the extent of the fringe movement depends not only on the refractive index of the medium brought into one beam but also on the length of the light path through that medium. Interferometry is thereby distinctly different from refractometry, in which refraction at a single point is measured.

Interferometry is well suited for applications that involve only small changes in refractive indices like gas analysis. By interferometry, for example, carbon dioxide or methane can be determined in air down to a few hundredths of a percent using a cell with a 1-m light path. Application to liquid systems is also common, for example, the determination of ethanol in blood, the activity of ferments (enzymes), and the concentration of heavy water (deuterium oxide) in ordinary and enriched water.

42.3 Polarimetry

Ordinary light (radiation) consists of an infinite number of waves vibrating in an infinite number of planes about the axis of propagation. In contrast, plane-polarized light has all its vibrations in a single plane. Various anisotropic crystals and also dichroic crystals with their axis aligned on a surface (Polaroid® film) can convert ordinary light to plane-polarized light. Some substances, predominantly organic ones and notably sugars and amino acids and their derivatives, are optically active, that is, the plane of the polarized light is rotated when the light passes through a solution of such a substance. The direction and the magnitude of the rotation are characteristic of the substance. Measuring the rotation identifies the compound, can be applied as a criterion of its purity, and allows determination of its concentration in solution.

The principles of polarimetry can be elucidated in terms of the schematics for a typical polarimeter, Fig. 42-1. Emerging from the light source, a beam of ordinary, monochromatic light is first collimated and then passed through a Nicol double-diffracting prism (the polarizer). The light, which is then polarized, is passed through the cylindrical sample cell, then through a second Nicol prism (the analyzer) and into the eyepiece. The analyzer is mounted on a graduated circle and can be rotated. Also placed in the beam at the polarizer and analyzer positions are two so-called half-shade prisms, which cause the observer to see the field of vision as a circle of light divided into two halves. As the analyzer is rotated, the brightness of the halves change, and a position can be found at which the two half-circles are identically illuminated and the line dividing them virtually dis-

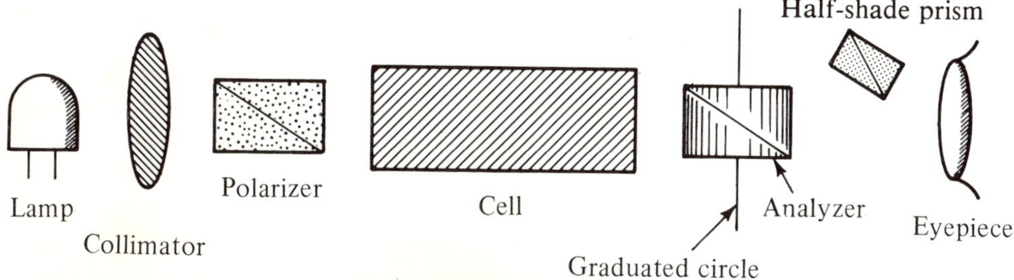

Fig. 42-1. Schematic arrangement of a polarimeter.

appears. By this equal-illumination technique, with the solvent in the cell the instrument is set to zero and then with the sample solution in the cell the angle of rotation is established. The calculation of the specific rotation, $[\alpha]$, is based on the relation

$$[\alpha]_\lambda^t = \frac{a}{dc} \tag{42-2}$$

Here d is the length of the light path in the cell measured in decimeters (1 dm = 10 cm), c is the concentration of the optically active solute in the solution in grams per milliliter, and a is the experimentally established angle of rotation in degrees. The symbol $[\alpha]$ is the specific rotation (optical) in degrees of rotation for unit length, in decimeters, and unit concentration, in grams per milliliter. (This function is also known as the specific rotatory power.) The units for specific rotation are generally omitted in tables and text. Since the optical rotation is both temperature-dependent and wavelength-dependent, the first condition is indicated by a superscript and the second by a subscript. The symbol has been used almost universally for many years, and with experience you will not interpret the brackets as a designation of molarity. The sodium D line was used almost exclusively on polarimeters until recent years, but is now often replaced by the yellow-green mercury line, at 546.1 nm, obtained from a mercury arc lamp with filters.

Once the specific rotation of a substance has been determined or taken from a table, equation (43-2) can be used to calculate the concentration in solutions from the measured angles of optical rotation.

In the refining and processing of cane sugar (sucrose), the concentration of the sugar in solutions is commonly determined polarimetrically. Usually a 2-dm cell is used, and equation (42-2) reduces to

$$c = \frac{a}{d[\alpha]_D^{20}} = \frac{a}{2 \times 66.5} = \frac{a}{133} \tag{42-3}$$

Applied to sugars, the technique is sometimes known as saccharimetry. In the study, determination, and application of amino acids and many other biochemicals, polarimetry is an important analytical tool.

42.4 Microscopy

The optical microscope is a valuable tool in the chemical laboratory, predominantly for qualitative purposes. Even minute amounts of substances and materials can be identified by a study of their form, color, appearance, crystal habit, and optical properties under ordinary and polarized light; in some cases, chemical tests performed on the microscope stage complement the examination. The purity of some solid substances can be assessed on a microscope "hot stage" by following the rate of crystallization from the melt. Noble metals in their compounds can be reduced to the metal and melted into tiny spheres; measuring the particle diameter with the known density of the metal allows estimating amounts of the metal too small for conventional weighing. Minute amounts of evolved gases can be measured by the size of suitably formed bubbles. The number of particles of each discrete component of a mixture can be counted under the microscope; from this information and the total number of particles counted, a "number" percentage of each component can be calculated. In mineralogy, thin slices of rocks, for example, are cut, ground to transparency, and polished; areas of identical optical properties are measured and related to particular components, and an estimate of the composition is established. However, all these methods serve for special purposes only and require special training and equipment. A full discussion of chemical microscopy is a subject within itself.

42.5 Mass Spectrometry

Mass spectrometry involves the acceleration of gaseous (positive) ions in an electrical field and then their separation in a magnetic field into ion beams according to the mass-charge ratio of the ions. Each ion beam can be detected and its intensity measured. The instrument involved is known as a mass spectrometer. This technique for separating isotopes and establishing their atomic masses is well known. It is also possible to measure the relative abundance of an ionic species in the established gas phase. Organic molecules, on cleavage in a beam of electrons, yield ionic fragments. The mass spectra (that is, the fragmentation patterns) secured are sufficiently different for the analysis of mixtures, even for closely related molecules. For example, mixtures of 10 or more homologous hydrocarbons can be analyzed by mass spectrometry, often with an accuracy and a precision of 1%. The use of a stable isotope of unusual mass number (for example, ^{2}H, ^{13}C, ^{15}N, ^{18}O) in conjunction with a mass spectrometer offers one technique of so-called tracer analysis. In the determination of an element by the isotope dilution approach, for example, a known amount of the trace isotope is incorporated in some way into the sample or its reaction products. Mass spectrometric measurement of the relative amounts of the isotopes of the element present allows calculation of the amount or percentage of that element in the original sample.

Spark source mass spectrometry is a technique for studying inorganic solids. It uses a vacuum spark, in which a high-voltage radiofrequency discharge (20 to 100 kV) is produced between two closely spaced electrodes of the material to be analyzed. The constituents of the sample that are vaporized and ionized are accelerated and separated. In this way, information on over 50 elements at contents of 100 ppm to a few ppb can be secured; the accuracy is ±10 to 25% when suitable standards are used. Organic materials are ashed before being examined by this technique.

42.6 *Questions*

42–1 In the introduction to refractometry, the term *density* was enclosed in quotation marks to indicate some special or unusual meaning. Elaborate on what was intended, consulting a basic physics textbook if necessary.

42–2 Is it necessary in measuring a refractive index to base the measurement on the phenomenon of refraction? Elaborate.

42–3 What is meant by the statement that a compound is optically active? What conditions are necessary for optical activity to be present in a molecule? Consult an introductory textbook on organic chemistry if necessary.

42–4 What is saccharimetry?

42–5 What are some factors that can influence the value of the refractive index of a liquid? a gas?

42–6 What forces effect the separation of ions in a mass spectrometer?

42–7 What is an interferometer? What does it measure?

42–8 What is understood by the term *fragmentation pattern*?

42–9 Elaborate on some properties and phenomena that could be used to identify a substance on a microscope stage.

42–10 What is indicated by (a) the subscript and (b) the superscript with the symbol for refractive index? The symbol for specific rotation?

42–11 What are the units for molar refraction (density, in grams per milliliter) and specific rotation (cell path, in decimeters)?

42–12 The U.S. National Bureau of Standards provides special-purity sucrose certified to a specific rotation of $+26.997^\circ$ in water at a concentration of exactly 26% w/v at exactly 25°C and 589.3 nm. Elaborate on how this standard reference material might be used to assure both the accuracy and the precision of angles of rotation measured with a given polarimeter.

42.7 *Problems*

42–1 At 20°C and the sodium D line, the refractive index of water is 1.33299 and of a sugar solution having exactly 20% sugar content is 1.33639. Assume a linear relation between the sugar concentration and the refractive index and calculate the value of the index for a solution exactly 12% in sugar.
Ans.: 1.33503

42-2 The dextrorotatory and levorotatory (*d* and *l*) forms of an optically active compound have identical specific rotations but opposite signs. One of the essential amino acids is *l*-leucine, which has a specific rotation of −14.0 $deg \cdot cm^3 \cdot dm^{-1} \cdot g^{-1}$ in water. An experimental synthesis yielded a mixture of the *d* and *l* forms with a net specific rotation of +2.3 $deg \cdot cm^3 \cdot dm^{-1} \cdot g^{-1}$ in water. Calculate the percentages of the two forms in the mixture.

Ans.: 58% *d* and 42% *l*

42-3 What is the milk sugar (lactose) content in grams per liter in an aqueous solution to be used in candy manufacture when the rotation in a 10-cm cell at room temperature and for the sodium D line is +7.780? The specific rotation for lactose under the conditions mentioned is +52.63 $deg \cdot cm^3 \cdot dm^{-1} \cdot g^{-1}$.

Ans.: 148 g/L

42-4 The refractive index of benzene is 1.5011 and of ethanol is 1.3611. A mixture of the two liquids has a refractive index of 1.4943. What is the approximate percentage of benzene present? (Note the loss of significant figures.)

Ans.: 5%

42-5 At 20°C, 2-propanone (that is, acetone) has a refractive index of 1.3588 and a density of 0.7908 g/mL. Calculate the specific refraction.

Ans.: 0.2898

42-6 The refractive index of common water, H_2O, is 1.3328 and of heavy water, D_2O, is 1.3283. Calculate the percentage of deuterium oxide in a mixture of refractive index 1.3290. (Note the loss of significant figures.)

Ans.: 84%

42-7 A 1.7530-g sample of an optically active compound is dissolved in water in a 50-mL volumetric flask and diluted with water to mark. The rotation is measured in a 2-dm cell several times to improve the precision; the values found are +19.367°, +19.374°, and +19.366°. As the alternative to setting the polarimeter to zero, the cell is filled with water; again three readings are taken, and the values found are −0.012°, −0.009°, and −0.015°. Calculate the specific rotation of the compound.

Ans.: $+276._4$ $deg \cdot cm^3 \cdot dm^{-1} \cdot g^{-1}$

43

Review Questions on Optical and Miscellaneous Instrumental Methods

43–1 Name various units in which wavelength is often expressed.

43–2 Name and briefly describe some modes of interaction of radiant energy and matter that are of interest in analytical chemistry.

43–3 What is the difference in the mechanism of absorption of radiant energy in the ultraviolet, visible, and infrared regions?

43–4 What difference might you expect in the materials used in an ultraviolet spectrophotometer and in an infrared spectrophotometer for prisms, lens, cells, and so on?

43–5 What finally happens to the radiant energy absorbed by a solution containing a nonfluorescent light-absorbing solute?

43–6 Elaborate on the difference between fluorescence and phosphorescence.

43–7 What is reflection of light? refraction? diffraction?

43–8 Elaborate on the complexity of the processes leading to light scattering.

43–9 How may light scattering be used for analytical purposes?

43–10 What is the effect of light scattering on a spectrophotometric determination?

43–11 What is polarized light and how may it be used in chemical analysis?

43–12 What is understood by the term *spectral line*?

43–13 In your own words define *blank* as the term is used in optical methods and elaborate on the difference between a reference and a blank.

43–14 Elaborate on the method of "internal standards." Name optical methods in which this technique is chiefly applied.

43–15 Elaborate on the advantages of photometric, nephelometric, and turbidimetric titrations over photometry, nephelometry, and turbidimetry, respectively.

43-16 Compare photometric methods with gravimetric methods and visual titrimetric methods in accuracy, concentration range of sought-for substance, interferences, instrumental requirements, and speed.

43-17 Explain why it is feasible to use a filter in flame emission photometry whereas a monochromator is necessary in arc-spark emission spectrography.

43-18 Elaborate on the difference in principles, but not on the differences in instruments, between atomic absorption flame spectrometry and conventional solution photometry.

43-19 Explain why a high-intensity incandescent lamp is an inadequate light source for atomic absorption spectrometry.

43-20 What optical methods can be considered essentially nondestructive when applied to a solid sample? What fields of science and technology might be interested in them?

43-21 Briefly define or explain the following terms and name optical methods or apparatus with which they are associated: *atomizer, scintillation counter, visual comparison, spark, target, hot stage, densitometer,* and *flame background.*

43-22 In a precipitation titration using a radiometric end point, the activity of the supernatant liquid over the precipitate can be followed and the counts per minute plotted versus the volume of titrant solution added. For the simplest radiometric titrations, the resulting curves consist of two intersecting straight lines. The position of the lines will depend on whether the sample solution is active, or the titrant solution, or both. Sketch your concept of the three shapes for the curves. Based on other titrations involving linear plots of data that you have encountered, suggest possible advantages for radiometric titrations.

43-23 The salt used in the "infrared KBr disk" technique must be essentially free of moisture, and it should be low in nitrate content. Explain.

43-24 Elaborate on why infrared methods are applied less frequently for quantitative determinations than ultraviolet or visible-region photometric methods.

43-25 Which would have the ruled lines closer together, a grating to produce a spectrum in the infrared region or one for the visible region? Could both gratings be ruled on glass? Elaborate.

43-26 Assume that two hydrogen atoms are bound to a carbon atom in a compound. If one of the hydrogens is replaced by a deuterium atom, would the scissoring frequency in the infrared be shifted? If it is shifted, would it be to a longer or a shorter wavelength? (If necessary, consult a physics textbook about Hooke's law as applied to harmonic oscillators.)

43-27 In a so-called colorimetric titration, a standard solution of the species to be determined is added from a buret with stirring into a tube containing solvent and the relevant chromogenic agent, until the color visually matches that in a second tube containing the sample solution with an identical amount of the chromogenic agent. The final volumes of the two solutions should be similar. From your knowledge of colorimetry (Section 35.5), color formation, and photometric titrations (Chapter 37), elaborate on this method. Can you see any advantage over comparing the color of the sample preparation with a series of standards?

43-28 Various amounts of a salt MX_2 are dissolved in a fixed volume of water to form a series of colored solutions in which the ion M^{2+} is electroreducible. Single out which of the following parameters depend on the concentration of the salt in the solution: formula weight of the salt, electrode potential of the ion M^{2+}, electrical conductance of the solution, formality, solution density, solution color, wavelength of the absorbance maximum for the solution, absorbance at that maximum, solution volume, solution weight. Elaborate and extend the considerations to other parameters and properties of interest for analytical chemistry.

43-29 For optical methods of analysis involving the interaction of radiant energy with matter, state (a) the spectral region of the radiation and any special requirements, (b) the nature of the interaction, and (c) the origin or nature of the signal that is measured. For example, for molecular photometry in solution, (a) a beam of light (ultraviolet or visible) (b) is absorbed by the sample and (c) the decrease in the intensity of the beam is measured.

44 Extraction Methods

44.1 Distribution Law

When a solute is allowed to distribute itself between a pair of immiscible solvents in contact, the equilibrium established is governed by the distribution law: At constant temperature (and constant pressure in the case of a gas), the ratio of the equilibrium concentrations of a species distributed between two immiscible solvents is constant. In organic analysis, one of the solvents is usually water and the other an organic liquid.

The distribution equilibrium for a species A may be written

$$A_w \rightleftharpoons A_o \tag{44-1}$$

where the subscripts *w* and *o* refer to the aqueous and organic phases. It is here assumed that the molecular weight of the species is identical in both phases; that is, no association of molecules occurs. The distribution law can then be expressed as

$$P = \frac{[A]_o}{[A]_w} \tag{44-2}$$

where *P* is known as the partition coefficient or the distribution constant. Its value depends on the temperature and the nature of the solute and the solvents involved. Sometimes this relation is written as the reciprocal.

When the two phases are equilibrated in the presence of an excess of a sparingly soluble solute, the expression of the partition coefficient reduces simply to the ratio of the solubilities of the compound in the two phases:

$$P = \frac{S_o}{S_w} \tag{44-3}$$

44.2 *Distribution Ratio*

In analytical practice, the distribution ratio is often more important; it is defined by

$$D = \frac{C_{A,o}}{C_{A,w}} \tag{44-4}$$

where C_A is the formal concentration of the species A. The relation between the partition coefficient and the distribution ratio in the extraction of a metal ion can be appreciated from a simple example.

Assume that a metal ion M^+ on reaction with a complexing agent HZ forms the uncharged complex MZ, which is the only species extractable into the organic phase. The two equilibria involved may then be displayed as

$$\begin{array}{c} M^+ + HZ \rightleftharpoons H^+ + (MZ)_w \\ \upharpoonleft\downharpoonright \\ (MZ)_o \end{array} \tag{44-5}$$

Inspecting this scheme reveals qualitatively that an increase in the concentration of HZ and also a decrease in the concentration of the hydrogen ion shifts the complexation equilibrium to the right. As a consequence, the position of the extraction equilibrium will also be shifted, and more of the metal will be extracted as its complex.

The quantitative treatment proceeds as follows. The expression for the partition coefficient takes the form

$$P = \frac{[MZ]_o}{[MZ]_w} \tag{44-6}$$

When the extraction is performed, the interest is not in how much of the metal *complex* is extracted, but in how much of the metal ion is removed from the aqueous phase; and thus the distribution ratio of the metal ion D must be considered:

$$D = \frac{C_{M,o}}{C_{M,w}} \tag{44-7}$$

The formal concentration of the metal in the aqueous phase is given by the following material balance:

$$C_{M,w} = [M^+]_w + [MZ]_w \tag{44-8}$$

Since only the complex is extracted into the organic phase,

$$C_{M,o} = [MZ]_o \tag{44-9}$$

The expression for the distribution ratio then takes the form

$$D = \frac{[\mathrm{MZ}]_o}{[\mathrm{M}^+]_w + [\mathrm{MZ}]_w} \tag{44-10}$$

Dividing the numerator and the denominator of this equation by $[\mathrm{MZ}]_w$ yields

$$D = \frac{\dfrac{[\mathrm{MZ}]_o}{[\mathrm{MZ}]_w}}{\dfrac{[\mathrm{M}^+]_w}{[\mathrm{MZ}]_w} + \dfrac{[\mathrm{MZ}]_w}{[\mathrm{MZ}]_w}} = \frac{P}{\dfrac{[\mathrm{M}^+]_w}{[\mathrm{MZ}]_w} + 1} \tag{44-11}$$

In the aqueous phase, the expression for the stability constant of the complex MZ, as considered in Section 20.5, is

$$K_{\mathrm{st}} = \frac{[\mathrm{MZ}]_w}{[\mathrm{M}^+]_w [\mathrm{Z}^-]_w} \tag{44-12}$$

In addition, for the aqueous phase, the expression for the dissociation constant of the weak acid HZ can be written

$$K_a = \frac{[\mathrm{H}^+]_w [\mathrm{Z}^-]_w}{[\mathrm{HZ}]_w} \tag{44-13}$$

By use of the expressions for these two constants, the term $[\mathrm{M}^+]_w/[\mathrm{MZ}]_w$ can be derived from

$$K_{\mathrm{st}} K_a = \frac{[\mathrm{MZ}]_w}{[\mathrm{M}^+]_w [\mathrm{Z}^-]_w} \times \frac{[\mathrm{H}^+]_w [\mathrm{Z}^-]_w}{[\mathrm{HZ}]_w} \tag{44-14}$$

Hence,

$$\frac{[\mathrm{M}^+]_w}{[\mathrm{MZ}]_w} = \frac{[\mathrm{H}^+]_w}{[\mathrm{HZ}]_w} \times \frac{1}{K_{\mathrm{st}} K_a} \tag{44-15}$$

Introducing this equality into the expression for D, (44-11), yields

$$D = \frac{P}{\dfrac{[\mathrm{H}^+]_w}{[\mathrm{HZ}]_w} \times \dfrac{1}{K_{\mathrm{st}} K_a} + 1} \tag{44-16}$$

At a given temperature, P, K_{st}, and K_a are constants for a given metal ion, solvent, and complexing agent. Inspecting equation (44-16) reveals that the value of the distribution ratio can be increased either by decreasing the concentration of the hydrogen ion in the aqueous phase or by increasing the concentration of the complexing agent in that phase. An increase in the value of the distribution ratio means that the efficiency of the extraction of the metal is enhanced. Thus, the previous qualitative conclusions are verified and placed on a mathematical basis.

Although the above example represents a simple case, the conclusions drawn have general validity; in more involved cases, the mathematical effort becomes considerable.

44.3 Selectivity of Extraction

The complexation can be viewed as a competition between the hydrogen ion and the metal ion for the complexing agent, and other factors being equal, the stronger the complex formed, the lower the pH at which the extraction is still efficient. This fact is important if more than one metal ion is present. To secure selectivity with a given agent and solvent, a second complexing agent is added that competes with HZ for the metal. The amount of MZ formed is then decreased; thereby the second agent acts to mask that metal. Adjusting the concentrations of the masking agent and extractant can sometimes be of added value.

44.4 Application of Extraction

Extraction can be applied in two ways: either to remove the desired species from other substances or to remove these substances and leave the desired species in one of the phases. Extracts, after dilution to known volume, can often be used for a photometric determination (Chapter 36). An extracted nonvolatile material is readily recovered by evaporation of the organic phase, and any organic complexing agent remaining can be destroyed by ashing.

Separations by solvent extraction have distinct advantages over precipitation. In the extraction, there is no analog to coprecipitation, used in the restricted sense of the term (Section 6.7). A substance that is not extracted alone under the prevailing conditions will not be "coextracted" with another substance. Hence, separation by extraction is usually "cleaner" than separation by a precipitation. If an extraction is efficient, the technique can be used to concentrate trace amounts of a substance from a large volume of an aqueous solution. Separations can even be effected where the efficiency is low. All that needs to be done is to repeat the extraction with fresh portions of the solvent until only a negligible amount of the substance remains unextracted. Such a repeated extraction is illustrated by example 44-1.

Example 44–1 The partition coefficient of iodine between carbon tetrachloride and water has the value 85. Exactly 100 mL of an aqueous 0.0010*M* iodine solution is extracted twice with 20-mL portions of the organic solvent. What percentage of the iodine initially present remains in the aqueous phase after each extraction step?

Let x be the molar concentration of iodine in the aqueous phase after an extraction. After the first extraction, the number of millimoles of iodine in the organic phase equals $100 \times 0.0010 - 100x$; hence, the molar concentration of iodine in that phase equals

$$\frac{100 \times 0.0010 - 100x}{20}$$

The expression for the partition coefficient yields

$$85 = \frac{100 \times 1.0 \times 10^{-3} - 100x}{20x}$$

$$x = 5.6 \times 10^{-5} M$$

Hence,

$$\begin{matrix}\text{\% iodine remaining} \\ \text{in the aqueous phase}\end{matrix} = \frac{5.6 \times 10^{-5} \times 100}{1.0 \times 10^{-3}} = 5.6\%$$

Similarly, for the second extraction step:

$$85 = \frac{100 \times 5.6 \times 10^{-5} - 100x}{20x}$$

$$x = 3.1 \times 10^{-6} M$$

$$\begin{matrix}\text{\% iodine remaining} \\ \text{in the aqueous phase}\end{matrix} = \frac{3.1 \times 10^{-6} \times 100}{1.0 \times 10^{-3}} = 0.31\%$$

The application of the extraction technique to inorganic compounds per se is limited because relatively few compounds have sufficiently favorable partition coefficients. Notable examples are the extraction of elemental iodine into various solvents and the extraction of metal-halide complexes, including iron(III) as $HFeCl_4$, into ether. In most cases, organic extraction agents are used.

Dithizone, in the extraction and photometric determination of metals, has been considered in Section 36.3. This compound forms colored chelates with about 20 metal ions and hence has low selectivity as far as complex formation is concerned. The chelates formed by dithizone with various metal ions differ in stability. Since dithizone is a weak acid, some selectivity in the extraction can therefore be secured by appropriate adjustment of the pH. Additional complexing agents may be added for the purpose of masking (see Section 20.9). Thus, if potassium cyanide is added, lead(II), which does not form a cyano complex, can be extracted selectively into carbon tetrachloride, in the presence of copper(II), zinc(II), cadmium(II), mercury(II), silver(I), and certain other cations that form stable, nonextractable cyano complexes.

The reagent 8quinolinol, which is also known by the trivial name *oxine,* offers another example of a complexing agent that is used as an extractant for metal ions. The leftmost structure shown herewith is the structural formula of 8-quinolinol, Hox; and at the right is the formula of the electrically neutral, solvent-extractable 1:2 chelate, $M(ox)_2$, formed with a divalent metal ion, M^{2+}.

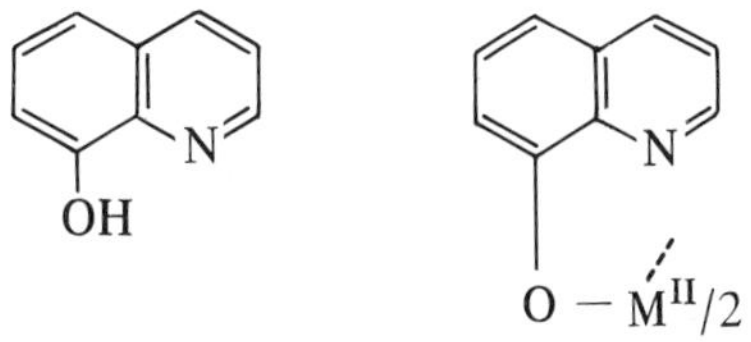

N
OH

N
$O - M^{II}/2$

In the rightmost formula, the M/2 is a concise way of implying that a second molecule of the same reagent is attached similarly to the same metal ion ("central atom"). Note that the metal ion is bound in two five-membered chelate rings, a structure conferring considerable stability on the species (cf. Section 20.4). For some metals, one or more molecules of water may be additionally coordinated, $M(ox)_2 \cdot nH_2O$.

Some further examples of extraction are considered in Chapter 36 as a preliminary step to photometric determinations and in Section 39.5 in the group separation of metals for atomic absorption spectrometry. Many complexing agents can be supplied as extractants for metal ions; all, however, operate in the same way. The selectivity achieved can often be improved by appropriately adjusting the pH of the aqueous phase, adding one or more masking agents, and selecting the solvent judiciously.

Extraction is not restricted to application with inorganic analysis. The technique is often applied for separating, identifying, and determining organic compounds, including drugs, vitamins, and hormones. Extraction into different solvents and from media of different acidity or alkalinity is an important approach to isolating and determining organic compounds; this you will learn from relevant laboratory course work in organic chemistry. From a mixture of compounds dissolved in water and made alkaline with sodium bicarbonate, for example, many "neutral" compounds can be extracted with ether. If the aqueous layer is then made acidic, many acidic substances are converted from their anionic form to their "free" acid form and can then be extracted with ether.

Multistep countercurrent extraction has proved of value for separating complex organic mixtures. The process is briefly described in Section 47.3.

44.5 *Leaching*

Extraction in the broadest sense of the term also applies to the selective removal of one or more substances from a solid phase by treatment with a suitable liquid; the process is then termed *leaching*. For example, the components of a mixture of sodium perchlorate and potassium perchlorate can be separated by extraction with 1-butanol, which dissolves only the sodium perchlorate. From a mixture of the chlorides of calcium, barium, and strontium, only calcium chloride is extracted by a mixture of ether and ethanol, which is free of water. In soil analysis, leaching is an important initial step in determining "available" ions and other nutrients.

44.6 Questions

44–1 Elaborate on the significance of masking in connection with extraction equilibria.

44–2 Discuss the extraction end point in redox titrations involving iodine (Section 25.7) in relation to the extraction technique and especially to Example 44–1.

44–3 With the data of Example 44–1, calculate the molar concentration of iodine remaining in the aqueous phase if 5.0-mL and 50.0-mL portions of the organic solvent are used in the two extraction steps. Discuss the influence of the volume of the organic phase on the amount of iodine unextracted.

44–4 Discuss the influence of pH on the extraction of chelates.

44–5 State why relatively few inorganic compounds are extractable per se into organic solvents.

44–6 In practice, a buffer is sometimes added to the aqueous phase of an extraction system. Elaborate on the possible functions of the buffer.

44–7 A metal ion M reacts with HZ and HY to form the complexes MZ and MY, which are the only species extractable into a given solvent. If $K_{st,MZ} = K_{st,MY}$ and $P_{MZ} = P_{MY}$, and HZ is a far weaker acid than HY, which complex will be extracted preferentially at a given pH?

45

Ion-Exchange Methods

45.1 *The Nature of Ion Exchangers*

Ion exchange can be defined as a reversible process in which ions are exchanged between a liquid and a solid. The solid is known as an ion exchanger, and it undergoes no substantial change. Although complex inorganic substances can function as ion exchangers, water-insoluble organic polymers are usually used in analytical practice. An ion-exchange resin is a special type of polyelectrolyte and consists of a three-dimensional polymeric hydrocarbon network to which are bonded many electrically charged groups, such as the sulfonate group, $-SO_3^-$, or a quaternary ammonium group, $-N(CH_3)_3^+$. Ion-exchange resins are commercially available, usually as small, uniform beads, in diameters ranging from 0.04 to 1.2 mm.

45.2 *Strongly Acidic Cation-Exchange Resins*

Ion-exchange resins having sulfonic groups as the exchange sites are termed *strongly acidic cation-exchange resins.* When such a resin is treated with a strong acid (often 6*F* hydrochloric acid), all the sulfonate groups are converted to the acid form (that is, the hydrogen form, or more fully, the hydrogen ion form). This resin may be represented by $(RSO_3^-)H^+$, where R is the resin network. The resin in this form behaves as an insoluble strong acid. The sulfonate group has less attraction for the hydrogen ion than for the sodium ion; consequently, if the resin in the hydrogen form is placed in a solution of sodium chloride, ions are exchanged. The equilibrium established can be represented by

$$(RSO_3^-)H^+ + Na^+ \rightleftharpoons (RSO_3^-)Na^+ + H^+ \qquad (45\text{-}1)$$

For each equivalent of sodium ion, one equivalent of hydrogen ion is freed. The position of the equilibrium depends on the acid strength of the resin and the concentration of the sodium ion in the solution in contact with the resin. For a strongly acidic resin, the exchange affinity for cations depends on the charge of the cation; tripositive cations are held more firmly than dipositive cations, and these more firmly than unipositive ones. Thus, if the resin in the sodium form is placed in a dilute solution of an aluminum salt, ions are exchanged, and for each tripositive aluminum ion exchanged three unipositive sodium ions are freed. Within a given group of simple cations of equal charge, the affinity of the resin tends usually to be greater as the atomic number of the cation is greater. Thus, for the second column of the periodic table, the order of the affinity is $Be^{2+} < Mg^{2+} < Ca^{2+} < Sr^{2+} < Ba^{2+}$.

The affinity between an ion and an ion-exchange resin may be so great that a water-insoluble salt can be dissolved. For example, if enough strongly acidic cation-exchange resin is placed into an aqueous suspension of barium sulfate, this compound is eventually dissolved. The process is represented by

$$2(RSO_3^-)H^+ + \underline{BaSO_4} \rightleftharpoons (RSO_3^-)_2\, Ba^{2+} + 2H^+ + SO_4^{2-} \qquad (45\text{-}2)$$

Note that in this example the exchange equivalence corresponds to the displacement of two moles of hydrogen ion by one mole of barium ion.

45.3 Batch Operation with an Ion-Exchange Resin

When a strongly acidic cation-exchange resin is placed in a dilute solution of sodium chloride, exchange according to equation (45-1) takes place but the reaction does not go to completion. If more complete conversion of the sodium chloride to hydrochloric acid is desired, the equilibrated resin may be separated by filtration or decantation and a further portion of fresh resin added to the solution. This so-called batch operation is tedius and hardly ever used in quantitative analysis.

45.4 Column Operation with an Ion-Exchange Resin

A column operation applied to ion-exchange processes is far more efficient than a batch operation. The resin is placed on top of a glass wool plug in a vertical tube, suitably a buret in many analytical applications. A strongly acidic cation-exchange resin can then be converted completely to the desired ionic form by causing a sufficiently concentrated solution of the desired cation to pass, usually by gravity flow, through the resin column (often called the resin bed). Thus, the hydrogen form is obtained by passing 6*F* hydrochloric acid, followed by distilled water until the effluent is neutral. At no time should the liquid level be allowed to drop below the top of the bed; otherwise, air bubbles become entrapped and reduce the active surface available for ion exchange. Then the sample solution is placed on top of the bed and allowed to percolate through the column at an effluent rate of about 3 to 5 $mL \cdot min^{-1}$ for each square centimeter of resin-bed cross section. Thus, in the conversion of sodium chloride to hydrochloric acid, the solution as it passes through the column encounters fresh, unequilibrated layers of resin, and conversion to hydrochloric acid is complete. Attention must be paid to certain details. The exchange capacity of the bed must exceed the amount of sodium ion to be exchanged. Further, the sodium chloride solution must be dilute enough; otherwise, a concentrated solution of hydrochloric acid would result and the equilibrium would not favor full exchange of sodium for hydrogen. (Recall that the resin is converted to its hydrogen form by treatment with concentrated hydrochloric acid.) Evenually after prolonged use the column becomes exhausted and is then regenerated by passage of a concentrated solution of an appropriate ion. Strongly acidic cation-exchange resins are available with various exchange capacities, commonly within the range 0.6 to 2.3 milliequivalents of cation per milliliter of wet resin.

45.5 *Strongly Basic Anion-Exchange Resins*

Strongly basic anion-exchange resins function analogous to strongly acidic cation exchangers. Thus, the conversion of sodium chloride to sodium hydroxide by such a resin in its hydroxide form can be represented by

$$(RN(CH_3)_3^+)OH^- + Cl^- \rightleftharpoons (RN(CH_3)_3^+)Cl^- + OH^- \qquad (45\text{-}3)$$

In practice, complete anion exchange is not as easily attained as cation exchange. Anion-exchange resins, especially in the hydroxide form, show lower chemical and thermal stability.

45.6 *Applications of Ion Exchange in Analytical Chemistry*

Ion-exchange methods have diverse applications in analytical chemistry. A salt, preferably one with a high equivalent weight, can be used for standardizing a base or preparing a standard acid solution. A weighed amount of the salt (dried) is dissolved in distilled water. The solution is then allowed to percolate through a cation-exchange column in the hydrogen form, and the acid formed is completely removed from the column by the passage of distilled water. The combined effluent, since it contains an amount of hydrogen ion that can be calculated from the amount of the salt taken, can be used directly to standardize a base. Alternatively, the effluent can be diluted to a known volume, to obtain a standard acid solution.

Cation exchange using a resin bed in the hydrogen form can be applied to determined by titration with a standard base; thereby, the number of equivalents of the sample solution, the number of equivalents of hydrogen ion in the effluent is determined by titration with a standard base; thereby the number of equivalents of cations in the sample solution is established. If the sample solution contains free acid, and hydrogen ion is to be distinguished from other cations, an identical aliquot of the sample solution is titrated with the base without being passed through the column. The volume of base required in this titration is subtracted from the volume needed for titrating the effluent. The difference corresponds to the equivalents of other cations besides hydrogen ion. The method cannot be applied if the sample contains cations that precipitate as hydroxides during the titration.

Sodium hydroxide solutions containing carbonate may be passed through an anion-exchange column in the hydroxide form. The carbonate is exchanged for hydroxide, and the carbonate-free sodium hydroxide solution that emerges is suitable for use as a titrant.

Ion exchange offers a convenient and efficient method of removing from a sample solution cations that interfere in the determination of an anion, or conversely of removing anions that interfere in the determination of a cation.

A very significant analytical use of an ion-exchange resin is the concentration of a trace element to a point that it can be determined by an available technique (for example, the concentration of trace copper(II) in biological samples).

Ion-exchange resins are now in general use for deionizing water in the laboratory and in plant operations. Both cations and anions are removed by use of a mixed resin bed containing cation-exchange resins in the hydrogen form and anion-exchange resins in the hydroxide forms.

Some metals that form anionic chloro complexes can be separated on anion exchange columns. Iron(III), cobalt(II), and nickel(II), for example, can be separated by passing a solution of the metals in 9*F* hydrochloric acid through the column in the chloride form and rinsing with the 9*F* acid. Nickel does not form a chloro complex and passes out of the column. The other ions are held by the resin as anionic complexes (such as $FeCl_4^-$ and $CoCl_4^{2-}$). Then 4*F* hydrochloric acid is passed, with the result that cobalt is eluted. Finally, iron(III) is eluted with 1*F* hydrochloric acid. The metal content of each fraction from the column can be determined, for example, by an EDTA titration. This separation, which has proved useful in analyzing high-temperature alloys, depends largely on the selectivity afforded by the complexation reaction, rather than on differences in the affinity of the various chloro complexes toward the resin.

45.7 Ion-Exchange Chromatography

So far, we have examined ion-exchange applications involving gross changes, such as the removal of ions from a solution or separations based on simple replacement or conversion of a species on the ion-exchange column. The goal of chromatography (as described in Chapters 46 through 48) is also to secure separations, but in a subtle manner. Chromatography involves passing a mobile phase along a stationary phase and allowing even small differences in the affinity of two or more species toward the latter phase to "multiply," so that separations are effected gradually and progressively. For ion-exchange chromatography, the mobile phase is a liquid passed continuously through the stationary phase, the packed ion-exchange column. An example will elucidate the point.

Consider that a neutral solution containing lithium, sodium, and potassium, as salts, is placed on a column of a strongly acidic cation-exchange resin in the hydrogen form. When the column is washed with water, these three elements remain on the column. According to what we have mentioned before, if 6*F* hydrochloric acid were now passed, these elements would simply be displaced and pass from the column. In contrast, if 0.7*F* hydrochloric acid were passed, the position of the equilibrium represented by equation (45-1) would not be at an extreme, and there would be neither complete retention nor displacement from the resin. Consequently, the ions are distributed between the mobile and stationary phases according to their affinity toward the resin. As the acid continues to pass, these small differences are "magnified," and the three elements emerge from the column at different times. The fractions could be caught in separate vessels and evaporated to small volume, and the contained alkali metal then determined, for example, by emission flame photometry.

Although extremely small differences in affinity toward the ion-exchange resin can allow successful separations, the situation can sometimes be improved by introducing a competing equilibrium. For example, a complexing agent can be added to the solution passed (the eluent). This agent then competes with the resin for polyvalent metal ions. This approach is often used for separating metals that show similar chemical behavior, such as the rare earth elements (lanthanides).

Ion-exchange chromatography can also be applied to organic compounds that persist in an ionic form under appropriate solution conditions. For example, ion-exchange chromatography is widely applied to the analysis of mixtures of amino acids, and hence the analysis and study of proteins and polypeptides.

In the most elementary form of ion-exchange chromatography, the fractions are collected in separate vessels and analyzed by appropriate methods, as exemplified above. The emergence of a species can be detected in various ways, including use of a conductance cell (Section 32.5), a differential refractometer (Section 42.1), or a photometer. For photometric purposes, a chromogenic agent can be added continuously to the eluate; for example, ninhydrin (that is, 1,2,3-indantrione) is used to form strongly absorbing blue products with amino acids.

Direct quantitation becomes possible if the detector response is plotted versus time (under constant flow conditions) using a chart recorder. The area under a response peak (or the height of a narrow peak) can be related to the amount or concentration of the eluted species.

45.8 *Questions*

45-1 Define the term *ion exchanger.*

45-2 What is the difference between a batch operation and a column operation? Which would be expected to be most valuable in analytical practice?

45-3 What will be the composition of the effluent when a dilute solution of each of the following is percolated through a cation-exchange column in the hydrogen form: (a) KBr, (b) $Al(NH_4)(SO_4)_2$, (c) HNO_3, and (d) NaCl + H_2SO_4?

45-4 What will be the effluent obtained when a dilute solution of each of the following is passed through an anion-exchange column in the hydroxide form: (a) NaI, (b) $BaCl_2$, (c) $Ba(NO_3)_2$ + HNO_3, and (d) HCl?

45-5 Write the equation representing the exchange reactions occurring when a solution containing $Ca(HCO_3)_2$ and NaCl is passed through a *mixed* bed with the cation and anion exchangers in the hydrogen and hydroxide forms, respectively.

45-6 What will be the effluent obtained when a dilute $BaCl_2$ solution is passed through a column of a cation-exchange resin in its hydrogen form? in its sodium form? in its iron(III) form?

45-7 Phosphate seriously interferes with the cation separations according to the hydrogen sulfide scheme. Elaborate on the possibility of using ion exchange to overcome this interference.

45-8 Discuss the difference in the concept of equivalence in ion-exchange and acid-base titrations.

45.9 Problems

45-1 An ion-exchange column has an inside diameter of 2.0 cm, and in its use in a 3-min period, a volume of 50 mL of effluent is collected. What is the flow rate of the column? *Ans.:* 5.3 mL/min for 1 cm^2 of bed cross section

45-2 In a column of 2.0-cm^2 cross section, a 10-cm bed of a cation-exchange resin in the hydrogen form is established. This resin has an exchange capacity of 2.4 meq/mL of wet resin. If the column is completely converted to the aluminum form, what is the weight of aluminum (*27*) retained? *Ans.:* 0.43 g

45-3 The sodium ion from 50 mL of a 1.5% solution of NaCl (*58*) is to be removed by a cation-exchange column in the hydrogen form having an exchange capacity of 5.0 meq/g of dry resin. If we assume that all the capacity of the resin is used, what weight of resin (dry) is required to exchange this amount of sodium ion? *Ans.:* 2.6 g

45-4 A volume of 25 mL of 0.10F $Al_2(SO_4)_3$ is passed through a cation-exchange column in the sodium form. If complete exchange of aluminum ion for sodium ion is assumed, what weight of sodium ion (*23*) is present in the effluent? *Ans.:* 0.34_5 g

45-5 A 0.3728-g sample of KCl (*74.56*) is dissolved in about 50 mL of distilled water and passed through a cation-exchange column in its hydrogen form. The column is washed with water and the combined effluent is titrated with a NaOH solution, requiring 50.00 mL. Calculate the formality of the base. *Ans.:* 0.1000F

45-6 A mixture containing only KCl (*74.56*) and $BaCl_2 \cdot 2H_2O$ (*244.3*) is dried and the content of water (*18.02*) found to be 10.00% w/w. An amount of 0.8000 g of the dried mixture is dissolved in water and passed through a cation-exchange resin in the hydrogen form. The effluent and washings are collected in a 200.0-mL volumetric flask and diluted to mark. Assume 100% exchange efficiency and calculate what volume of 0.0250F $Ba(OH)_2$ will be needed to titrate a 25.00-mL portion of this acid solution.

45-7 A material is known to contain Na_2SO_4 (*142.04*), NaCl (*58.44*), $NaNO_3$ (*84.99*), and some inert material. A 2.000-g sample of this material is dissolved in water and diluted to mark with water in a 250-mL volumetric flask. A one-tenth aliquot is taken and passed through a cation-exchange resin in the hydrogen form. The column is then rinsed with water. The combined effluent and washings require 25.77 mL of 0.1000*F* NaOH. Another one-tenth aliquot requires 12.82 mL of 0.0800*F* $AgNO_3$ in a precipitation titration. A one-fifth aliquot is caused to react with an excess of $BaCl_2$ solution, and 197.2 mg of $BaSO_4$ (*233.40*) is yielded. Calculate the % w/w composition of the sample.

45–8 A solution contains per milliliter 0.120 g of NaCl (*58.4*), 0.150 g of $MgSO_4$ (*120.4*), and 0.090 g of H_3PO_4 (*98.0*). If 10.00 mL of this solution is passed through an ion-exchange column in the hydrogen form and the column is rinsed with water, how many milliliters of 0.5000*F* NaOH will be needed to titrate the combined effluent and rinsings to the phenolphthalein end point? Enough exchange resin is taken to assure complete conversion.

Ans.: 128 mL

45–9 A solution is known to be about 50% w/w in either NaCl (*58.4*) or KCl (*74.6*). So that it can be decided which salt is present, a sample of exactly 0.5 g is diluted to exactly 500 mL with water, and a 100-mL aliquot is passed through an ion-exchange column in the hydrogen form. The combined effluent and water washings are titrated with 0.1250*F* NaOH, of which a volume of 7.08 mL is required. (a) Which of the two salts is present? (b) What is the exact percentage composition of the solution? (c) Which indicator should be used in the titration, methyl orange or phenolphthalein?

Ans.: (a) NaCl; (b) 51.7% w/w; (c) either

46 Introduction to Chromatography

46.1 Chromatography Defined

Chromatography is a term applied to a large variety of procedures that differ markedly in their conduct and the degree of instrumentation and automation. Chromatographic techniques allow the separation, identification, and determination of an array of chemical species, ranging from simple metal ions to high-molecular-weight compounds of complex structure, such as proteins.

In spite of the seeming diversity, all chromatographic processes effect the separation of substances by selectively altering the speed of their migration with a mobile phase as it moves in contact with a stationary phase. Various substances have different affinities for the two phases; those substances that are more strongly attracted to the stationary phase will migrate more slowly, thereby affording separation. Only slight differences in affinity are needed to accomplish a complete separation, since even minute effects "multiply" as the mobile phase moves relative to the stationary one. You may appreciate the analogy with fractional distillation. In that technique, the boiling material is repeatedly vaporized and condensed as it passes through the packing before fractions are taken into receivers.

Chromatography allows separations that it is either extremely difficult or impossible to accomplish by other techniques. For example, benzene (bp, 80°C) and cyclohexane (81.4°C) can be separated and individually determined by gas-liquid chromatography. The substances α- and β-carotene, $C_{40}H_{56}$, can be separated by liquid-solid chromatography even though they have almost identical molecular structures, differing only in the shift of a single double bond from one carbon to an adjacent one.

46.2 Classification of Chromatographic Methods

In chromatography, three stages of matter can be involved in the mobile-stationary phase relation: solid, liquid, and gas. Consequently, six different combinations exist for the two phases; two, the solid-solid and gas-gas, can be excluded for obvious reasons. The four remaining possibilities have all been realized in practice. Each combination can be designated by an acronym, wherein the intial letter indicates the mobile phase. For example, gas-liquid chromatography is assigned the acronym GCL. The other combinations (and acronyms) you may figure out as an exercise.

Other approaches to classifying chromatographic techniques are feasible. For example, focus can be placed on a key feature, leading to such terms as *gas, paper, thin-layer,* and *ion-exchange chromatography.* Another possibility is classifying according to the process involved in the distribution of species between the two phases; this leads to the terms *partition, adsorption,* and *ion-exchange chromatography.*

46.3 *Frontal and Displacement Modes of Analysis*

Three main modes for chromatographic operations can be distinguished, namely frontal, displacement, and elution analysis. Their elaboration will provide further insight into chromatographic processes. For simplicity, we assume that only two species are to be separated. As in most forms of chromatography, the stationary phase will be considered as a column of material within a tube through which the mobile phase is passed.

Frontal Analysis. In frontal analysis, the two species (compounds) are dissolved in an inert solvent, that is, one that does not interact with the stationary phase. The two species, however, have an affinity for that phase, although it is to a different degree. As the solution is passed through the column continuously, the liquid that initially emerges contains only the "faster-moving" species; then the mixture emerges. Consequently, by frontal analysis, only the faster-moving species can be separated in a pure form and then only with partial recovery. Although this mode of operation is of little value for analytical purposes, it does provide thermodynamic and other basic data on chromatographic processes.

Displacement Analysis. In the simplest form of displacement analysis, the sample solution containing the two species is introduced as a shallow layer on top of the column, and a solvent is passed that is active toward the stationary phase, to a *greater* extent than the two species. Consequently, the species are progressively displaced along the stationary phase. The one with the lower affinity emerges first from the column, followed by the second. For analytical purposes, there are drawbacks to this technique. The two "zones" immediately follow each other, and often overlap. Also, the column is fully "loaded" with the active material of the mobile phase and before re-use must be "regenerated," a task that is not always accomplished easily.

46.4 Elution Analysis

The most important mode for chromatography for analytical purposes is elution analysis. The underlying principles will become clear from looking at Fig. 46-1. The arrangement is like the one that M. S. Tswett, a Russian botanist, used in the first decade of this century to separate plant pigments extracted from leaves and petals. Tswett is often credited with making the first successful application of a complete chromatographic technique, although some aspects of the approach were realized earlier by other workers. Tswett coined the name *chromatography* (Greek *chroma,* "color"; *graphein,* "to write"). (It is a coincidence that *tswet* is the Russian word for color.)

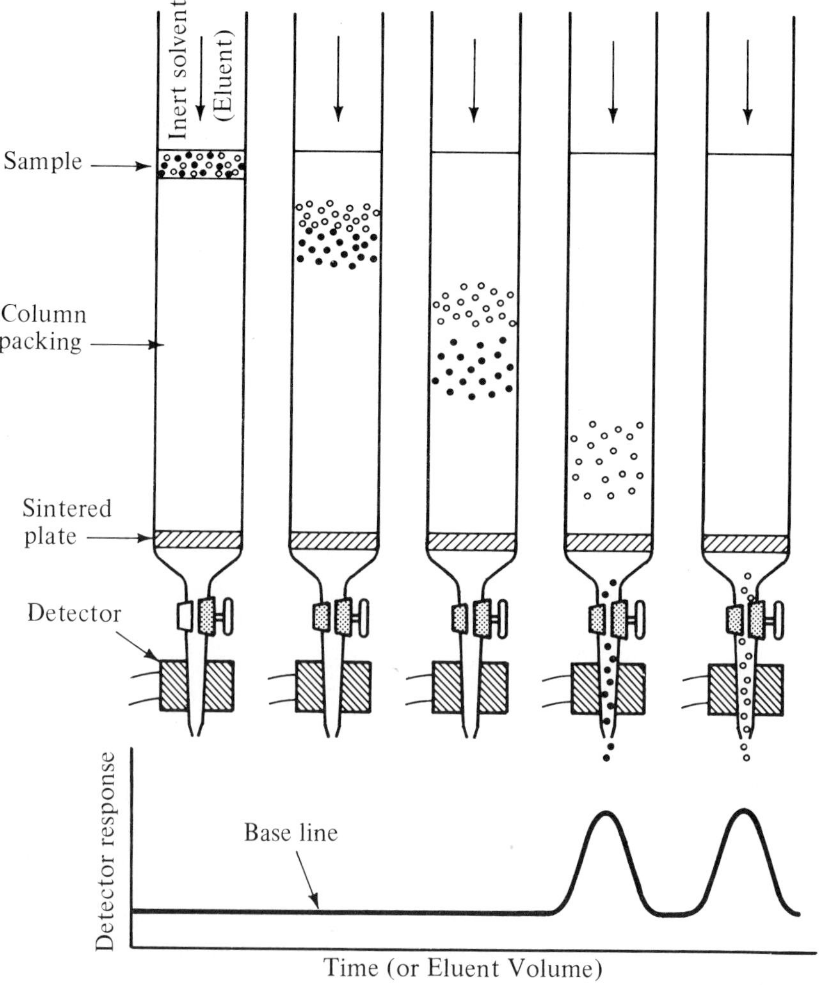

Fig. 46-1. (a) Simple arrangement for liquid-solid elution chromatography. (b) Chart recording of detector response.

The pictured arrangement consists of a glass tube, constricted at the lower end and fitted with a closure (for example, a buret). The tube contains a glass wool plug or a sintered plate on which the column of the stationary phase rests. This may be a pulverized insoluble adsorbent, such as calcium carbonate or aluminum oxide. Now a small amount of the sample solution containing two colored compounds is applied to the top of the column and the stopcock is opened briefly to permit the sample to "sink" into the column. Then the tube is filled with solvent, with care to avoid disturbing the sample layer. The stopcock is opened and the liquid is allowed to pass continuously; more solvent is added to the column top as needed. This operation is termed *elution;* the solvent, the *eluting agent,* or *eluent;* and the emerging liquid, the *eluate.*

During the elution, the chromatogram *develops* on the column; that is, two distinct zones form, which become progressively separated as they move down the column. Eventually the first zone is eluted and can be collected in one vessel, and then the second in another. In analytical practice, the sample components are usually not colored. Consequently, a detector is placed for "viewing" the eluate. For example, for compounds absorbing radiation in the visible or ultraviolet regions, a photometer can act as detector. Many physical and chemical properties can be used for detection (see further in Chapters 47 and 48). The electrical output of the detector can be fed to a chart recorder, and a tracing such as that at the foot of Fig. 46-1 might result. For given conditions (phases, temperature, column length, flow rate, and the like), the distance on the chart—which is proportional to either time or volume of eluent passed—between the start of the elution and the position of a peak maximum is characteristic of a compound. The area under the peak (or the height of a narrow peak) can be used for the measuring by means of calibration with known amounts of the compound.

Elution analysis has three important variants. *Simple elution* has already been described. If the compounds differ markedly in affinity for the stationary phase, it may be necessary to pass a substantial volume of solvent over a long period to elute the slower-moving compound. So that the process can be speeded, a second, more effective solvent can be passed after the first compound is eluted. This approach is termed *stepwise elution.* Alternatively, the two solvents can be mixed progressively and at a controlled rate as they flow onto the column. This process is called *gradient elution.*

46.5 Remarks

Because of the importance of gas chromatography, and because it involves a gas instead of a liquid as the mobile phase, this technique is considered separately (Chapter 47). Ion-exchange chromatography is considered part of the presentation of ion-exchange methods (Chapter 45). Diverse chromatographic techniques are grouped for study (Chapter 48).

47 Gas Chromatography

47.1 Introduction

Gas chromatography involves applying the principles considered in Chapter 46 to the separation and determination of substances volatile at the temperature used. The substances are transported by a carrier gas, under pressure, through the stationary phase, which is in the form of a column. The separated species are detected as they emerge from the column. In gas-solid chromatography (GSC), also known as gas adsorption chromatography, the stationary phase is a solid, like activated carbon or silica gel. In gas-liquid chromatography (GLC), also known as gas partition chromatography, the stationary phase is a supported nonvolatile liquid. Gas-liquid chromatography was not fully described until 1951-1952, when Martin and James published their findings; it has developed faster and has so far received more diverse application than any other chromatographic technique. Because of the importance of gas-liquid chromatography, we examine it in detail; however, most of the basic theory can be applied, with minor modification, to other chromatographic techniques subsequently touched on.

47.2 The Gas Chromatograph

Before we consider the principles and theory of gas chromatography, it will be helpful to delineate the basic components of a gas chromatograph used for analytical purposes. Such as instrument is shown as a block diagram in Fig. 47-1.

Gas Supply and Controls. The carrier gas is supplied from a high-pressure cylinder through a pressure reducer and valves. The pressure of the gas is usually displayed on a gauge. Common carrier gases include nitrogen, helium, argon, hydrogen, and carbon dioxide. A flow meter may be placed at the inlet or outlet side of the column to monitor the flow.

Injection Port. The sample—a gas, or liquid, or a solid dissolved in a solvent—is injected from a syringe capable of delivering microliter volumes by a hypodermic

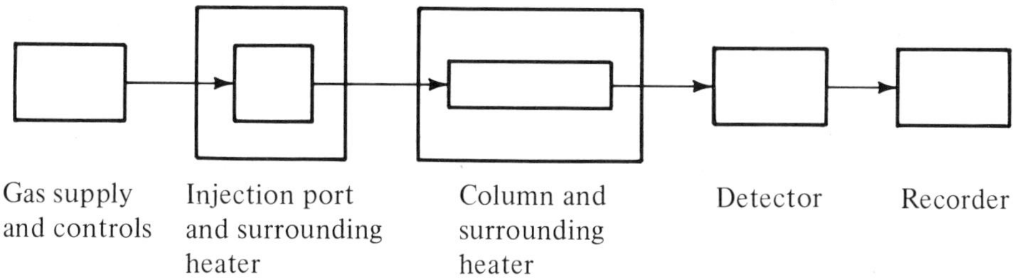

Fig. 47-1. Block diagram for gas chromatograph.

needle through a port having a self-sealing polymeric septum (disk). The port area is heated to a temperature that is higher than that of the column, to ensure prompt volatilization of the sample.

Detector. The detector signals the emergence of a species from the column and allows the amount to be estimated. Many physical and chemical properties of a substance in a gas phase can be made the basis for detection. (Some important detectors are considered in Section 47.6.)

Recorder. The electrical signal provided by the detector is amplified and fed to a chart recorder. The resulting graph, known as a *gas chromatogram,* is a tracing of signal intensity versus time. The area under a peak can be related to the amount or concentration of the associated species. For this purpose, integration is performed, either mechanically or by an electronic data processor. For some applications, the measurements and calculation of results can be based merely on the peak height.

Column. Gas chromatographic columns can differ markedly in the shape, dimensions, and construction material for the tubing and in the nature of the packing. The tubing is commonly of glass, copper, or stainless steel and may be coiled or in the form of a U-tube to save space.

In gas-liquid chromatography, the stationary phase is a nonvolatile, thermally stable liquid. For *capillary columns,* the liquid is applied as a thin coating to the inner wall of fine-bore tubing. Such columns may have lengths ranging from 3 to over 300 m. For *packed columns,* the liquid is applied as a thin, uniform coating to an inert, granular solid material, such as ground firebrick or diatomaceous earth or glass beads. Packing columns are often of about 4-mm bore and lengths of 1 to 4 m.

Many nonvolatile stationary phases are used in gas-liquid chromatography. Selection of a favorable one is somewhat empirical; guidelines, however, are often useful. The maxim "like dissolves like" is often appropriate. For example, for separating volatile aliphatic hydrocarbons, a higher hydrocarbon, such as squalane,

$C_{30}H_{62}$, is suitable as the stationary phase. For separating esters of fatty acids, a nonvolatile polyester is often used.

Column Heater. The column is maintained at a suitable temperature above room temperature by an electrically heated jacket, or it is placed within an electrically heated oven. In operation, the column either may be maintained at constant temperature (isothermal operation) or may have sections maintained at different temperatures, or the column temperature may be increased at a preselected rate to facilitate the elution of "slower-moving" species (temperature programming).

47.3 Some Terms and Concepts

Some underlying concepts and important terms are relevant to the theories of gas chromatographic separations.

Theoretical Plate. In plate theory, a chromatographic column is treated as an analog to distillation (Section 47.4). The separation is viewed as involving stages of equilibration that occur in a small part of the column known as a theoretical plate. In each theoretical plate, "average" equilibration is assumed between the species in the mobile and the stationary phases.

When a binary mixture of two substances, A and B, of which B has the higher boiling point, is distilled at constant pressure, ideally the boiling-point curve is of the type presented as Fig. 47-2. (You are probably familiar with vapor pressure and boiling-point curves for binary mixtures from earlier courses; if not, we suggest you review the relevant material in a textbook on elementary chemistry.) This figure is a plot of temperature versus composition in mole percentage. The lower-boiling component, A, is present in relatively greater proportion in the vapor phase; consequently, the upper curve corresponds to the composition of the vapor phase, and the lower curve to the composition of the liquid phase.

Consider a liquid mixture of composition m_1. It boils at temperature t_1; the corresponding vapor phase has the composition m_2, in which A is enriched. If this vapor phase is condensed, liquid of composition m_2 results. This liquid boils at temperature t_2 to give a vapor-phase composition of m_3, and so on. If the boiling takes place in a still pot fitted with a reflux column, then the compositions marked in Fig. 47-2 persist at certain heights in the reflux column once a steady state has been established. A reflux column that over its length brings about a change in composition, say of m_1 to m_4, would be described as having an efficiency of 3 theoretical plates (or a plate number n of 3), because three steps are involved, as indicated in the figure. Obviously, the shorter the column length over which the compositional change is effected, the greater the column efficiency. So that different columns can be compared, the term *height equivalent to a theoretical plate* (HETP, or simply H) is adopted:

$$H = \frac{\text{length of column}}{\text{total no. plates}} = \frac{L}{n} \qquad (47\text{-}1)$$

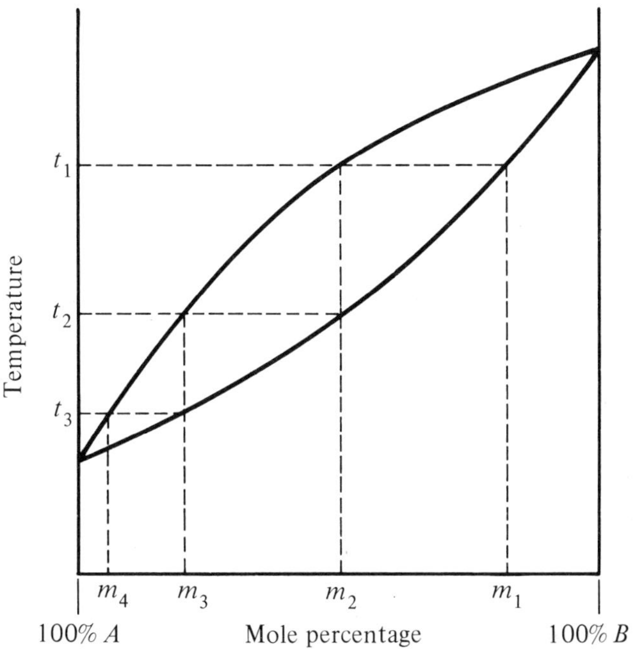

Fig. 47-2. Temperature-composition diagram for an ideal binary mixture of miscible liquids (see text).

You may understand this description better from what follows. Fractional distillation columns can be constructed having discrete plates. The principle is demonstrated in Fig. 47-3, with only two plates shown. The vapor from the mixture in the still pot enters the first plate through opening O_1; a portion is condensed to "flood" the plate and returns through opening R_1. The vapor above the liquid on the first plate enters the second plate through opening O_2, and so on. Most fractional distillation columns are not designed with discrete plates; then a plate becomes a theoretical concept, and the height of a theoretical plate can be established experimentally.

To advance the concept of a theoretical plate closer to chromatographic processes, we envision a multistep solvent extraction (see also Chapter 44). Operationally, a solute is allowed to equilibrate (partition) between two immiscible liquids in an initial vessel. Then the upper phase is transferred to a second vessel. Fresh upper phase is added to the first vessel and fresh lower phase to the second. After equilibration, the process is extended to a third vessel, and so forth. Each vessel then corresponds to a plate in a distillation column and the upper and lower phases to the mobile and stationary phases. A unit to perform such multistep extraction automatically has been constructed; it is often termed the *Craig countercurrent machine.* The Craig process has proved valuable for separating complex mixtures, such as those relating to the isolation of microamounts of antibiotics from experimental microbial cultures and of physiologically active substances from biological fluids.

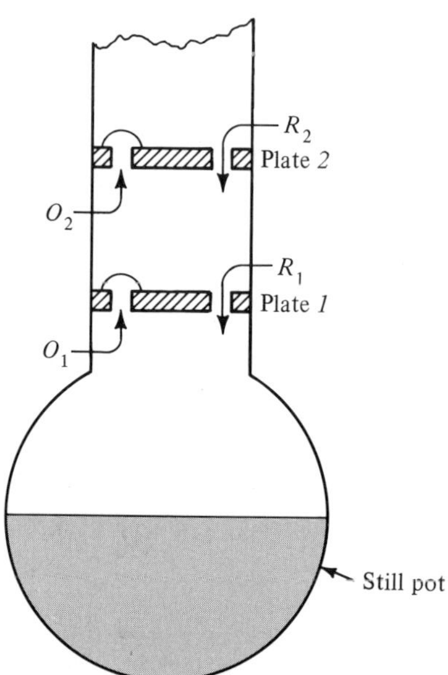

Fig. 47-3. Schematic representation of distillation "plate" column with two plates shown (see text).

Finally, the discussion must be extended to what pertains when the mobile phase is a gas and substances are distributed continuously along the column. Although it is not possible to establish actual plate heights, the concepts of theoretical plate and of height equivalent to a theoretical plate apply, for example, as parameters for expressing column efficiency. Although these concepts are abstract and largely mathematical tools, having examined them in terms of distillation and solvent extraction should help in your understanding them.

Partition Ratio. A substance (solute) in the sample introduced in a chromatographic column is distributed between the two phases to the extent described by the partition ratio K, given by

$$K = \frac{A_s}{A_m} \tag{47-2}$$

where A denotes the amount and the subscripts s and m denote the stationary and mobile phases. (Compare equation (44-2) in the consideration of solvent extraction.)

Retention Volume and Time. The retention volume V_R for a solute is the volume of the mobile phase passed through the column from the moment of sample injection to the appearance of the peak maximum for that solute band at the detector. The retention volume, and retention time t_R, are related to the flow rate F by

$$V_R = t_R F \tag{47-3}$$

Since the retention volume and time are proportional, they often can replace each other in mathematical relations. For gases, reduction of the volume and flow rate to standard conditions may be needed.

Pressure Gradient. In gas chromatography, the mobile phase (carrier gas) is forced through the column under pressure and usually emerges at atmospheric pressure. Consequently, a pressure gradient exists along the column. Since gases change in volume with pressure (Boyle's law), for some advanced considerations it is necessary to apply a pressure gradient (correction) factor, which can be calculated or taken from tables. (For chromatography with a liquid mobile phase, such correction is unnecessary because liquids have negligible compressibility.)

Peak Width. There are many possibilities for characterizing an elution band. If the conditions are close to ideal, the distribution around the peak maximum approaches a Gaussian one (see Section 2.8 and Fig. 2-1) and the standard deviation σ can be used to evaluate the peak width. As shown in Fig. 47-4, tangents are drawn at the inflection points to their intersection with the base line. The distance between these intersections, termed the peak-base width, W_B, corresponds to 4σ. The peak-base time is given the symbol, t_B.

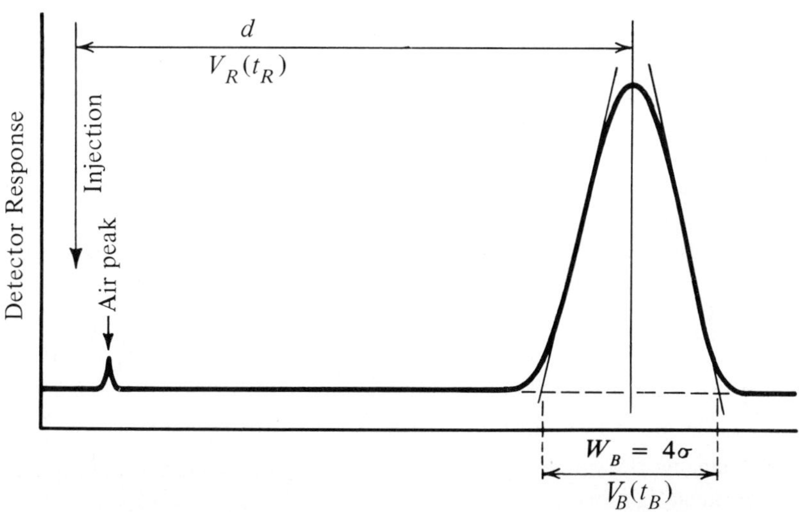

Fig. 47-4. Recorder tracing for elution of solute band with assessment of peak width (see text).

47.4 Plate Theory

In 1941-1942, Martin and Synge (who were later awarded the Nobel Prize for their contributions to chromatography) adopted a model for a chromatographic system involving the plate concept. They considered the hypothetical division of the column into plates and defined a single plate as a section or a zone of the column such that the concentration of a solute in the mobile phase contacting the plate would be in equilibrium with its average concentration in the stationary phase of the plate.

Ideally, a chromatogram then consists of a single narrow line for each solute; in reality, the lines are broadened to bands. The plate theory allows an accurate description of the peaks (bands), although some of the assumptions of the theory are unrealistic (see below). It is of interest that by inverse reasoning from the broadening of the bands, the number of plates and the plate height can be calculated, and the efficiency of the column thereby assessed. In relation to Fig. 47–4, the derivations by the plate theory allow expressing the plate height (actually, the height equivalent to a theoretical plate) H in terms of the variance, that is, σ^2, per unit length of the column:

$$H = \frac{\sigma^2}{L} \tag{47-4}$$

In relation to Fig. 47–4, all parameters can be expressed arbitrarily in the scale units of the recorder paper.

For a Gaussian peak, the following relations, involving ratios, can be established:

$$\frac{W_B}{d} = \frac{t_B}{t_R} = \frac{V_B}{V_R} = \frac{4\sigma}{L} \tag{47-5}$$

where d is the recorded distance between the injection point and the peak maximum and t_B and V_B are the time and volume corresponding to the peak-base width W_B.

Equations (47-1) and (47–4) apply even if the parameters are in scale units instead of, say, centimeters, because ratios are involved. Eliminating H between those equations gives

$$\frac{L}{n} = \frac{\sigma^2}{L} \tag{47-6}$$

Rearranging this equation and substituting equivalent ratios from Equation (47-5), after squaring provide

$$n = \frac{L^2}{\sigma^2} = 16\left(\frac{V_R}{V_B}\right)^2 \tag{47-7}$$

$$n = 16\left(\frac{t_R}{t_B}\right)^2 \tag{47-8}$$

There are some unrealistic assumptions in applying plate theory to chromatographic processes. The partition is assumed to take place in steps (as in the Craig process); actually the partition occurs continuously along the column. The mobile phase is added not in portions but continuously. For an ideal chromatogram (lines only), the base widths would be zero and the number of plates infinite (the "escape" from this difficulty is the definition of the plate as involving the *average* concentration of a solute in the stationary phase of the plate). Partition involves diffusion, which is a random process in all directions. Rapid equilibration means that diffusion is rapid, but with a restriction—it takes place only toward and away from the stationary phase in a perpendicular direction. Diffusion along the direction of the column length is not considered. Additionally, the plate theory does not take into account some dynamic factors, such as flow rate and the dependence of the partition ratio on solute concentration.

47.5 *Rate Theory*

Some shortcomings of the plate theory have been mentioned. Although the plate theory allows predicting—if the partition ratios and the plate number of the column are known—how feasible it is for two substances to be separated, it is not very helpful in suggesting how the performance of a column can be improved. In the treatment of band broadening, the plate theory suffers from the neglect of longitudinal diffusion and disregard of the flow rate of the mobile phase.

In the rate theory, to which various workers have contributed, a kinetic approach to examining gas-liquid chromatography is adopted, and band spreading is considered in terms of several rate factors; the concept of the height equivalent to a theoretical plate H is still retained, however.

The basic concepts of the rate theory can be understood from the "abbreviated" van Deemter equation:

$$H = A + \frac{B}{u} + Cu \tag{47-9}$$

Here u is the average linear velocity of the gaseous mobile phase in centimeters per second.* The parameters A, B, and C relate to the situation in the column as outlined below.

---

*The linear velocity should not be confused with the flow rate F, which is the volume of mobile phase passing through the column cross-section per unit time. An equivalent to equation (47–9) can be written involving flow rate, but the parameters A, B, and C would then have different values.

Parameter A, the Multipath Term. Parameter A is called the *eddy diffusion term,* or better, *the multipath term.* This parameter takes into account the different lengths that different solute molecules traverse as they pass between the particles of the column packing. In other words, it explains why all the solute particles do not arrive at the detector at one time. Since path lengths are independent of gas velocity, the first term in equation (47-8) contains only A and no u.

Parameter B, the Molecular Diffusion Term. The term B/u takes into account the effect of longitudinal diffusion of the solute molecules in the mobile phase. This effect is greater the longer the band of solute remains in the column, that is, the lower the gas velocity is.

Parameter C, Resistance to Mass Transfer. The term Cu takes into account that equilibration in or between the two phases is not instantaneous. The greater the gas velocity, the greater the value of this mass transfer term.

It is of interest to plot H versus u, as shown in Fig. 47-5. The lower thin lines show the individual contributions of the three parameters in the van Deemter equation and the bold curve is the plot for that equation. The contribution of the multi-

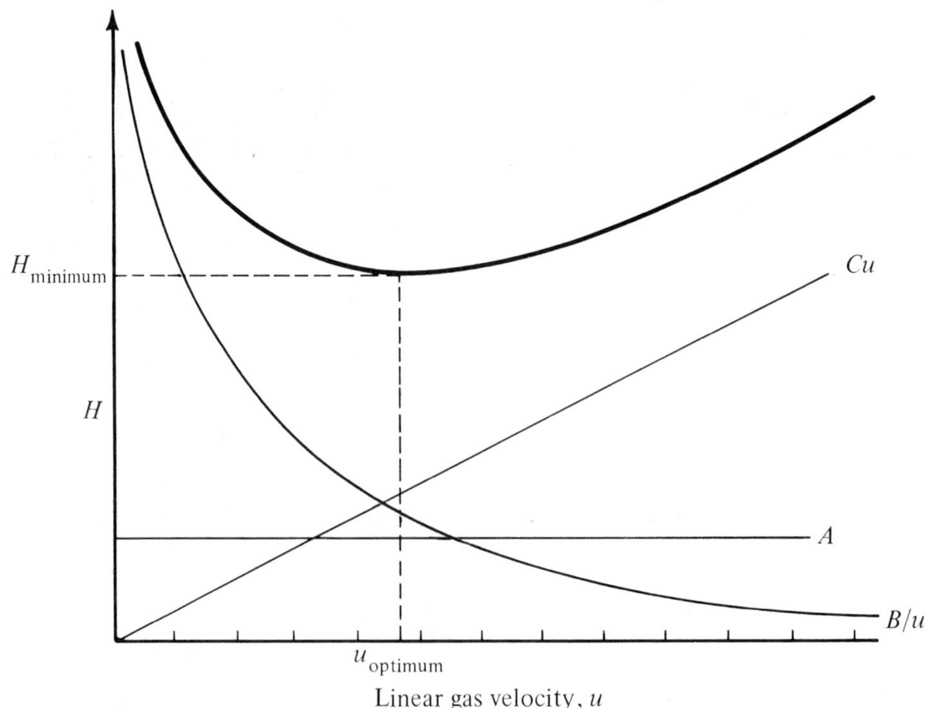

Fig. 47-5. Relation between plate height and linear gas velocity according to the van Deemter equation, with the contribution of the three parameters indicated (see text).

path term is independent of linear gas velocity and is therefore a horizontal line. The contribution of the B/u term is a hyperbola, and of the Cu term, an inclined line. It can be seen that a minimum value for H exists at some finite gas velocity, at which the efficiency of the column is the highest, as judged from the plate height. The multipath term does not depend on the flow rate, and the band broadening related to it can in practice be reduced by packing the column as uniformly as possible.

47.6 Detectors for Gas Chromatography

The detector in gas chromatography signals the emergence of a substance from the column and allows measuring the amount (or concentration) of the components in the column effluent. Many chemical and physical properties have been applied for such detection. The detector chosen depends largely on the samples and substances of interest, the sensitivity required, and the complexity and cost of the instrumentation. The ideal detector would respond to each compound specifically. A mass spectrometer (Section 42.5) coupled to a gas chromatograph approaches this ideal, but the arrangement is costly and receives only special, not general, application.

Detectors can be classified by whether they respond to the concentration of the detected compound in the gas stream or to the mass flow rate in that stream. Commonly, the first group leaves the emerging substances intact and the second group destroys them. To show the principles involved and the reasoning underlying their use, three of the most common detectors are here described: thermal conductivity, flame ionization, and electron capture detectors.

Thermal Conductivity Detector (TCD). The thermal conductivity detector, also called the katharometer, is based on the changes occurring in the thermal conductivity of gases with composition. Two electrical sensors, each consisting of a thin platinum or tungsten wire (or alternatively, a thermistor), are placed within chambers in a thermostated block of metal.

The "pure" carrier gas stream is passed through one sensor and the stream from the column through the other. The sensors are heated by an electric current and transfer heat to the gas flowing past. With a change in the gas composition, the rate of heat transfer changes; consequently, the temperature and thereby the electrical resistance of the sensor are altered. The two sensors are parts of a conductance bridge of the Wheatstone type. The bridge is initially balanced with pure carrier gas emerging from the column. When a substance comes out, the bridge becomes unbalanced and the electrical signal thereby created is amplified and fed to a chart recorder, which displays the chromatogram. Helium or hydrogen is the usual carrier gas because of the high thermal conductivity of each. Consequently, when a substance emerges, the heat loss by the sensor always decreases, and the chromatogram consists of peaks in only one direction from the base line. The conductivity detector has relatively poor sensitivity, and the response varies markedly toward various classes of compounds, with linearity maintained over only a limited range. The main advantages of this detector include simplicity, low cost, and its nondestructive nature toward the substances detected.

Flame Ionization Detector (FID). The flame ionization detector has as its basis that almost all organic compounds on decomposition in a hydrogen flame yield ions, if only as transient species. The ions migrate under the influence of an imposed electrical field and thereby cause a current to flow. Such a detector is shown schematically in Fig. 47-6. The negative side of a direct-current source, providing up to 500 V or even more, is connected to the metal tip of the burner and the positive side to a ring or a grid positioned above the flame. With a flame of hydrogen (freed of organic compounds), virtually no ions are produced, and any minute "background" current can be compensated for readily. When a substance emerges from the column and enters the flame, pyrolysis occurs, ions are produced, and current flows. The amplified electrical signal is fed to a recorder. The current produced depends largely on the number of carbon atoms in the compound with carbon-carbon and carbon-hydrogen bonds. The peak areas for hydrocarbons are roughly proportional to their weight concentration in the sample.

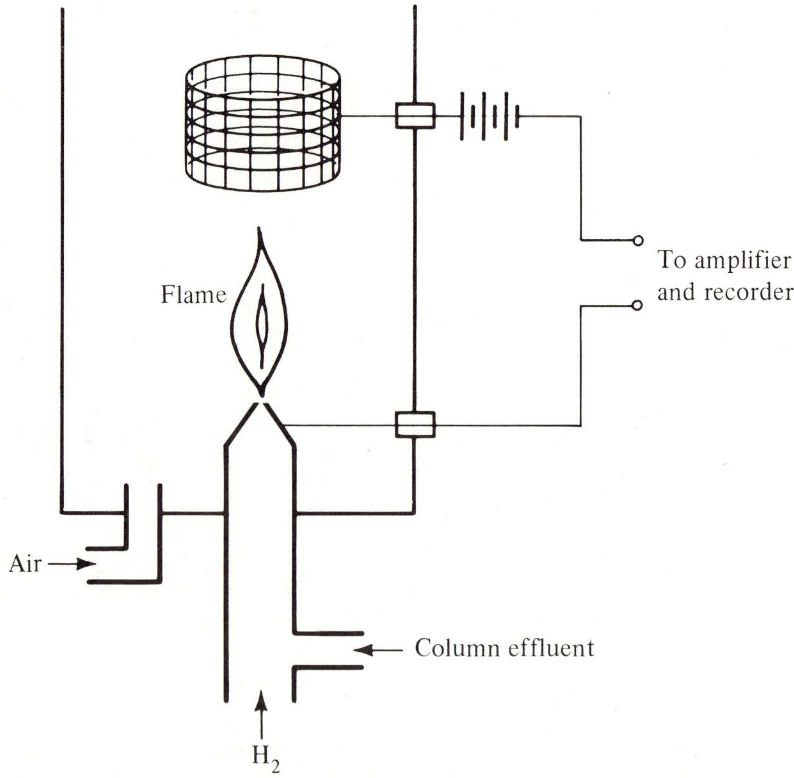

Fig. 47-6. Schematic diagram of a flame ionization detector.

The sensitivity of a flame ionization detector is favorable and the linearity range adequate. An important advantage is that many common inorganic gases (H_2, O_2, N_2, CO, CO_2, H_2O, and rare gases) do not yield ions and consequently do not interfere. This feature is especially important in analyzing environmental and biological samples. Since the electrical resistance of the flame is high, the currents produced are minute; consequently, the electronics involved are complex, making this detector expensive. This detector destroys the sample components.

Electron Capture Detector (ECD). The electron capture detector involves a source of beta rays—usually a nickel foil containing ^{63}Ni—and a collector for electrons. The beta particles ionize the carrier gas (N_2, Ar, or He with CH_4). A "standing" current then persists between the foil and the positive collector maintained at an applied potential of up to 100 V. When a compound enters the detector from the column that can capture electrons, the standing current is reduced. By appropriate electronic circuitry, this reduction in current is displayed by a chart recorder as a peak. The advantage of this detector is its extremely great sensitivity toward specific classes of organic compounds (such as halogen and nitro compounds) and low sensitivity to other classes (such as amines, alkanols, and hydrocarbons). The electron capture detector is nondestructive, and in some applications can be placed ahead of a flame ionization detector so that dual chromatograms are the outcome. The electron capture detector has a limited linearity in its response. An important area of application for gas chromatography with electron capture detection is in the detection of traces of halogenated pesticides in and on foods and in environmental samples. If large samples are used and there are preliminary separation and enrichment stops, pesticides can be detected at or below the nanogram-per-gram level (parts per billion).

47.7 Gas-Solid Chromatography

A distinction has been drawn (Section 47.1) between gas-solid chromatography (gas adsorption chromatography) and gas-liquid (gas partition chromatography). Emphasis has been on gas-liquid chromatography because of its far greater importance.

The instrumental and procedural details for gas-solid chromatography, are almost identical to the details for gas-liquid chromatography. The chief difference is that the stationary phase is a finely divided adsorbent packed into a column. Adsorbents used include activated carbon and alumina, silica gel, and molecular sieves. For a theoretical treatment, the sorption-desorption process for a substance between the adsorbent and the carrier gas replaces the partition between a liquid stationary phase and the gas. Gas-solid chromatography finds important special applications, notably in analyzing permanent ("fixed") gases and hydrocarbon mixtures.

47.8 Questions*

47–1 In gas chromatography, what parameters influence the shape of the elution peaks?

47–2 Why is it important in gas chromatography that on injection of a sample, the solvent and solutes volatilize almost instantaneously? What would be the effect on a gas chromatogram if a liquid sample evaporated only slowly?

47–3 Which would be the better way to improve a gas chromatographic separation: increasing the length of the column or increasing the number of plates for the given column length? Elaborate.

47–4 What is the advantage of using helium as the carrier gas when a thermal conductivity detector is used?

47–5 Compare the thermal conductivity, flame ionization, and electron capture detectors in underlying principles, sensitivity, relative advantages, and destruction of sample components.

47–6 In Fig. 47–4, a so-called air peak is shown. Since the components of the injected air have no affinity toward the stationary phase, the time between the injection point and the air peak allows the carrier gas flow rate to be estimated. Elaborate.

47–7 How is a liquid made the stationary phase in gas chromatography?

47–9 Problems

47–1 For a certain gas chromatographic system, the parameters A, B, and C have the numerical values 0.012, 0.25, and 0.0022 for the van Deemter equation, where u has the units of $cm \cdot s^{-1}$ and H is measured in centimeters.

(a) What are the units of the three parameters?

(b) What is the optimum velocity of the carrier gas for getting the smallest plate height? (*Hint:* Differentiate the equation with respect to u and apply conditions to find extreme values.)

(c) What is the height equivalent to a theoretical plate at the optimum gas velocity?

(d) If the actual length of the column were 20 cm, how many theoretical plates would be present?

Ans.: (b) $10._7$ $cm \cdot s^{-1}$; (c) 0.058 cm; (d) 352 plates

47–2 For the situation presented in Fig. 47–4, use a centimeter scale to estimate the relevant parameters and calculate the number of plates, applying Equation (47–7) or (47–8). Assume that the length of the column is 15 cm and calculate the HETP.

---

*Additional questions that relate gas chromatography to other chromatographic techniques appear at the end of Chapter 48.

47-3 A gas chromatographic column is 15 cm long, and for a certain compound the HETP value is 0.05 mm. How many theoretical plates are secured with that column?

47-4 A sample was subjected to gas chromatography. In addition to the peak for the main component, a peak for a single impurity *A* was recorded. For a quantitative determination, the following data were secured:

Wt of pure *A* injected into column	2.0 mg
Area of peak for *A*	3.29 cm^2
Wt of sample injected into column	25.0 mg
Area of peak for *A* in sample	0.49 cm^2

(a) What is the content of *A* in the sample in % *w/w*? (b) By difference, what is the "GC assay" of the sample, that is, the % *w/w* for the principal component?

48

Miscellaneous Chromatographic Methods

48.1 Introduction

In addition to gas-liquid and gas-solid chromatography, and ion-exchange chromatography, there are other chromatographic techniques. These include liquid column chromatography, liquid chromatography on paper and on thin layers, and gel permeation chromatography, which are considered in this chapter.

48.2 Liquid Column Chromatography

In our examination of the classical forms of liquid-solid (adsorption) chromatography and liquid-liquid (partition) chromatography, the movement of the mobile phase was assumed to be due to the pressure head provided by the liquid on top of the column or fed from a reservoir by gravity. Under those conditions, the technique is slow, with hours or even days needed for separating the components of practical samples. Dramatic improvements have now been effected, so that liquid column chromatography has become an important analytical tool. The improvements, in brief, have involved using pumps to reach higher pressures, improved and reusable column packings, and low dead-volume detectors of high sensitivity that allow direct measurement of the solutes in the column effluent.

For this improved technique, the terms *high-speed* and *high-pressure liquid chromatography* were introduced and given the acronyms HSLC and HPLC. The second of these has often come to stand for high-performance liquid chromatography. One proposal is to use the term *high-resolution liquid chromatography* (HRLC), because resolution is a key parameter, regardless of the pressure or speed used. Since the classical technique is now seldom used for quantitative analysis, some workers prefer the simple term *analytical liquid chromatography.*

In Fig. 48-1, the basic components of an analytical liquid chromatograph are indicated in a block diagram. The mobile liquid phase is fed from a reservoir through

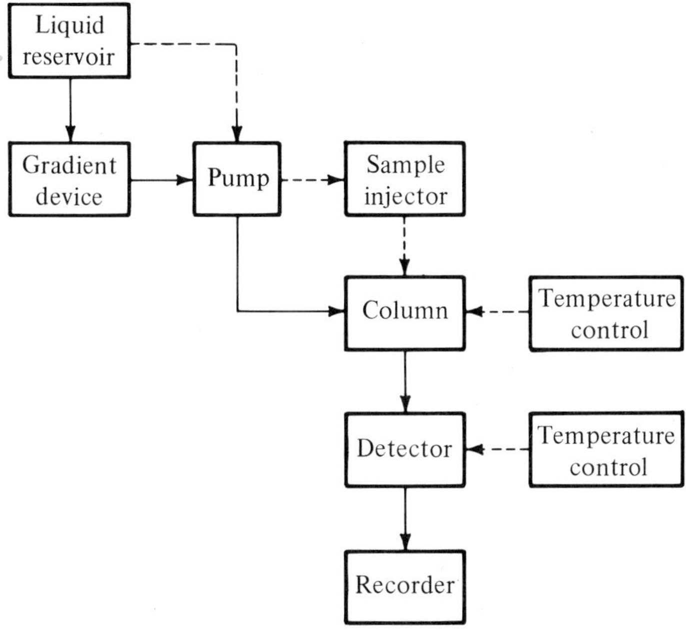

Fig. 48-1. Analytical liquid chromatography block diagram.

a pump to the column at typical pressures of 3.5×10^6 to 3.5×10^7 Pa (corresponding to 35 to 350 atm, or roughly 500 to 5000 lb$\cdot$in.$^{-2}$). When the composition of the mobile phase is to be changed progressively during the chromatographic analysis, a gradient device is also placed in the system (see Section 46.4 for the definition of gradient elution).

Direct injection of the sample into the column is possible, but at the high pressure involved a better arrangement is to inject the sample at low pressure into a by-pass tube, and then by a change in a valve position, sweep it into the column by a stream of the mobile phase under the high pressure.

As in gas chromatography, a detector is used to sense the presence of a solute in the column effluent and to measure its amount or concentration. Ideally, the detector should show high sensitivity and a wide linear response range, and it should respond to all kinds of compounds. Also, the detector should be "blind" to small fluctuations in flow rate or temperature. As you might expect, there is no ideal detector. Ultraviolet-visible photometers and differential refractometers (Section 42.1) with small flowthrough cells are widely used; fluorimeters serve as high-sensitivity detectors in special applications. The electrical signal provided by the detector is fed to a chart recorder.

The column is the heart of the analytical liquid chromatograph. It is usually packed in stainless steel tubing and fitted with a water jacket to maintain a selected temperature.

For liquid-liquid column chromatography, a stationary liquid is coated on an inert support. Liquids can be classified for polarity. Water, for example, is highly polar (large dipole moment); ethanol less so; chloroform, weakly polar; and carbon tetrachloride, nonpolar. If the stationary phase is a relatively polar liquid, then a relatively nonpolar liquid is used as the mobile phase; this is the "normal" mode of operation. If the stationary phase is nonpolar and the mobile phase polar, this mode is termed *reverse-phase* liquid-liquid column chromatography.

The inert supports used in liquid-liquid chromatography are relatively uniform, finely divided porous materials, such as diatomeceous earth or silica gel, or special materials of limited porosity, which have a hard core and a porous surface. The porous supports have a large surface area and provide large sample capacities for the column. The supports of limited porosity carry only a thin coating of stationary phase; this favors a more rapid separation but at the expense of the column sample capacity.

For liquid-solid column chromatography, two types of sorbents are useful. One consists of particles that are porous throughout and the other of particles that are porous only on their surface. Silicon dioxide and aluminum oxide are the most common materials.

A variant is the chemical bonding of the stationary phase to the support. The term *bonded-phase* liquid chromatography is then applicable. Since the bonded phase is not readily "stripped" from the support, the characteristics of the column are less apt to change with continued use.

The theoretical treatment of analytical liquid chromatography largely parallels the treatment of gas-liquid chromatography given in Chapter 47. Both plate and rate theories are applied.

The number of compounds that are soluble in one or more solvent systems is enormous; consequently, liquid chromatography is an important analytical tool, either actually or potentially, for a wide variety of materials. The technique is far less restricted than gas chromatography, because the compounds do not need to be thermally stable nor to have the capacity for being volatilized.

48.3 Paper Chromatography

Chromatography on paper first received extensive attention in the 1940s and rapidly became one of the most common chromatographic methods for analytical purposes. In this technique, the stationary phase is a liquid immobilized and held rigidly by the paper; the liquid is usually water. The mobile phase is another liquid that moves along the paper. The process is best viewed as a form of liquid-liquid partition chromatography, although the polar hydroxyl groups and carboxyl groups of the cellulose may be active in some cases and thereby introduce elements of adsorption and ion-exchange chromatography. Paper chromatography is of value for separating and subsequently identifying or determining micro amounts of both inorganic and organic substances.

Ascending Development. Ascending development can be performed by simple arrangements, such as the one shown in Fig. 48-2. The solvent system (for example, a mixture of 1-butanol and aqueous hydrochloric acid) is placed in the cylinder. The sample solution, often containing only microgram amounts of the substances to be separated, is placed on the paper strip as a spot. (The exact volume or amount spotted must be known if the separation is to be followed by a determination.) The strip, which is fixed to the holder at the bottom of the stopper and weighted (for example, by a glass angle or simply a paper clip) to prevent bending, is inserted into the cylinder so that the sample spot is well above the liquid level. By capillary action, the solvent mixture ascends the paper, showing a sharp solvent front. Water is preferentially retained by the hydrophilic cellulose of the paper; the mobile phase consequently has a higher concentration of the organic solvent. Various species, organic as well as inorganic, are hydrophilic to different degrees and are partitioned to different extents between the aqueous stationary phase and the organic-rich mobile phase. Consequently, the species move up the paper at different speeds. After some hours, the development will be complete and the separated species are located at different spots. Specially treated or impregnated paper can be used for improved separation of some substances.

The spots of colorless species, and even of colored ones when small amounts

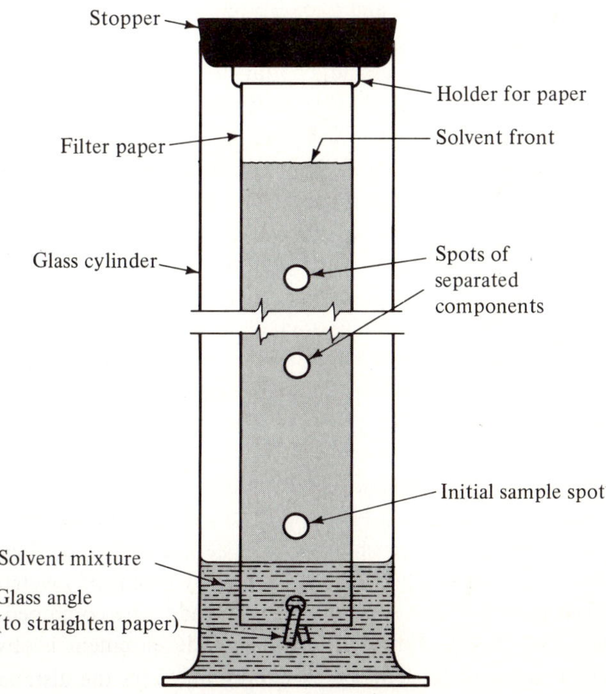

Fig. 48-2. Simple arrangement for ascending development in paper chromatography.

are involved, are not directly visible. The spots can be located ("visualized") by various means. Under ultraviolet radiation, substances that fluoresce strongly can be seen as bright, often colored, spots. Additionally, the dried paper can be exposed to ultraviolet radiation in the range 240 to 260 nm, a region in which many organic compounds absorb strongly. Such compounds then appear as dark areas on a faintly fluorescent background. A broader approach is treatment with one or more chromogenic or fluorogenic agents. A solution of the agent or agents can be sprayed on the paper, or the paper can be dipped briefly into the solution. Also the paper can be exposed to the vapors of a reagent, notably iodine. As a result of such treatments, one or more spots become blackened, colored, or fluorescent. Spots for metal ions that form colored sulfides are often made visible by spraying the dried paper with a hydrogen sulfide solution; the metal sulfides formed appear as brown or black spots.

Of special value for identifying substances separated by paper chromatography is the relative position of a spot on the paper, given by the R_F value:

$$R_F = \frac{\text{distance between center of spot and sample application line}}{\text{distance between solvent front and sample application line}} \tag{48-1}$$

For a given solvent mixture and paper, the R_F value is characteristic for a substance and is usually independent of its concentration in the sample. From measured R_F values, inferences are often possible about the substances present in the sample.

Spots can be evaluated in situ in various ways. An identified spot can be compared in color intensity (and area) with a dilution series of standards similarly spotted and developed. For photometric assessment, a densitometer is used, that is, a device in which the paper strip is drawn past a slit positioned in front of a photocell. The instrument output is fed to a chart recorder. Either the light reflected from the paper is measured or the light transmitted through the paper (often made transparent by saturation with mineral oil).

A substance can also be extracted from its spot and then determined by an appropriate method. So that larger amounts of separated substances will be available, the sample solution can be applied as a streak or a band across a wider strip or sheet of paper, rather than as a spot.

Descending Development. Descending development is also used with paper chromatography. For this procedure, a trough containing the solvent mixture is held at the upper portion of a chamber. The paper strip or sheet dips into the solvent, bends over the rim of the trough, and by means of a weight is kept stretched in a vertical position. The solvent descends under the combined influence of capillary action and gravity. For identical papers, development is obviously more rapid in the descending mode than the ascending one, and the distance that the solvent can travel is greater. The developed chromatogram can be visualized and evaluated as in the ascending development procedure.

48.4 *Thin-Layer Chromatography*

The advantages of paper chromatography include the relative simplicity and the potentiality it provides for separating, detecting, and even determining minute amounts of substances present in mixtures. It is often desirable, however, to use conditions that are either not provided or not tolerated by paper, and yet have the stated advantages. For such applications, thin-layer chromatography (TLC) has received prominence, beginning in the 1960s.

As the name suggests, a thin layer (up to 2 mm thick) of an adsorbent is coated on a supporting plate or sheet. Common adsorbents include silica gel, aluminum oxide, and powdered cellulose. For the coating of a glass plate, a water slurry of the adsorbent and a binding agent (such as calcium sulfate hemihydrate) is placed in an applicator, which is then drawn across the plate at constant speed. The coated plate is allowed to dry, and is placed in an oven for 30 to 60 min to allow full activation of the adsorbent. The dried, activated plate is placed in a desiccator until used.

Preparing a uniform layer on a large number of plates requires skill, but high quality TLC plates are purchased. Also available are flexible TLC sheets, which are prepared by coating a plastic film with the adsorbent, using a suitable binder.

For thin-layer chromatography, the application of samples, ascending development, and visualization are like their counterparts in paper chromatography. The development is often more rapid (10 cm in 20 to 30 min). In contrast with paper chromatography, corrosive reagents can be incorporated in the developing solvent or used in sprays for visualization. For example, spraying a developed thin-layer chromatogram on an inorganic adsorbent with sulfuric acid and warming leads to charring of many organic compounds so that they appear as black spots. The reproducibility of R_F values is generally better with thin-layer than with paper chromatography. Visual comparison or densitometry of developed and visualized spots allows quantitation. An inorganic fluorescent material can be incorporated into the slurry applied to the plate or sheet. Under ultraviolet illumination, spots of some organic substances (quenchers) then appear as darkened areas on a brightly fluorescent background.

48.5 *Gel Permeation Chromatography*

A gel has a three-dimensional network that markedly affects the diffusion of dissolved substances through it. This phenomenon allows separating substances of different molecular size. The larger the interstices, that is, the "pores," in the gel, the larger the molecular size of a substance that can enter. If a column is packed with small, swollen gel particles and a solution of various substances is passed, the solutes greater in molecular size than the interstices are excluded, and they move along the column at a rate close to that of the solvent. Solutes of smaller molecular size diffuse into the interstitial spaces and migrate down the column at different rates.

Separations based mainly on exclusion effects, such as differences in molecular size or shape, can be termed *exclusion chromatography*. When the stationary phase involves a swollen gel, the term *gel permeation chromatography* is widely used. It has the acronynm GPC. Other pertinent terms are *gel chromatography* and *gel filtration.*

Various materials can compose the gel. For work with aqueous solutions, the type of gel frequently used is a cross-linked dextran fraction. (Dextran is a polysaccharide produced by a fermentation of sugars by certain microorganisms.) This type of material is marketed under the trade name Sephadex®, in various pore sizes. Because of the hydroxyl groups, Sephadex gels are hydrophilic. When such gels are put to use, they are first treated with water to swell the granules. For use with organic solvents, cross-linked polystyrene is often applied. The gel is tailored by the degree of cross-linking to have an exclusion limit for molecular weight ranging from under one thousand to over a million.

A gel permeation chromatograph is somewhat like a liquid chromatograph. The solvent is pumped through the column and the sample solution is injected into the stream. Solutes can be detected (and measured) by various devices, including the differential refractometer, ultraviolet or infrared spectrophotometer, and flame ionization detector.

Sephadex gels are used in the separation and fractionation of biopolymers, including proteins and their degradation products. A Sephadex gel with a high degree of cross-linking allows a rapid desalting of aqueous solutions of large molecules. Gels of the cross-linked polystyrene type are applied to the study of synthetic polymers, including measurement of the average molecular weight and molecular-weight distribution.

48.6 Questions

48-1 What is meant by the expression "development" of a chromatogram?

48-2 What is meant by *elution* and *retention time*?

48-3 How is a liquid made the stationary phase in liquid chromatography? in paper chromatography and in thin-layer chromatography?

48-4 A column for liquid chromatography has a stationary phase that shows high affinity for polar compounds. If a mixture of carbon tetrachloride, ethanol, diethyl ether, and chloroform were chromatographed, in what order would you expect these compounds to emerge? Elaborate.

48-5 In paper chromatography does the paper function solely as the support? Elaborate.

48-6 For paper and thin-layer chromatography, define the term R_F *value*. Would a high affinity of a solute for the stationary phase correspond to a short retention time or a long one? a small R_F value or a large one?

48-7 What maximum and minimum does the R_F value have? What are the implications of these extreme values?

48-8 What is wrong with the statement, The R_F value of compound B is 1.6?

48-9 A student reported that the R_F of a compound is 0.85 cm. What is wrong with the statement?

48-10 Paper chromatograms are usually developed with a mixture of two solvents, for example, water and an alcohol. The development of some solutes can be effected with water alone, however. How do these two procedures differ in the principles underlying the separation of one or more substances from a mixture?

48-11 In paper chromatography, is it possible to obtain longer chromatograms by descending development rather than ascending? Explain.

48-12 Two-dimensional development in paper or thin-layer chromatography, involves spotting a square sheet or a plate near one corner and developing the chromatogram in solvent 1. The sheet or plate is dried and then developed with a different solvent 2 in a direction at right angles to the initial development. Draw your representation of the two-way chromatogram formed by a sample containing solutes A through E, each having the R_F values listed below in the two solvents, if each solvent front moves exactly 5 cm. Show each final spot as a small circle and mark the point of spotting the sample with an "X."

	Solute				
	A	B	C	D	E
R_F value, solvent 1:	0.15	0.35	0.40	0.75	0.85
R_F value, solvent 2:	0.70	0.15	0.40	0.25	0.50

What special merits of two-dimensional development can you see from this hypothetical example?

48-13 Radial development in paper chromatography involves spotting the sample at the center of a circular piece of filter paper, which is held horizontally; solvent is applied to the center, from which it spreads out radially. Speculate on the appearance of a developed and visualized radial paper chromatogram and on the special advantages the technique might have.

48-14 In thin-layer chromatography, what are the functions of the adsorbent that forms the layer?

48-15 In paper and thin-layer chromatography what limits the height to which the solvent can rise in ascending development?

48-16 For gel permeation chromatography, state what the stationary phase consists of and elaborate on the mobile-stationary phase relations involved in a separation.

48-17 List the various forms of *column* chromatography considered in Chapters 45 through 48. To the extent that the text information allows, note applications for each form.

48-18 In the various chromatographic techniques considered in Chapters 45 through 48, what is the stationary phase? the mobile phase?

48.7 Problems

48–1 In a thin-layer chromatographic separation, the solvent front was 8.2 cm from the starting line, and the spots for compounds A and B were at 5.4 and 3.8 cm, respectively. What are the R_F values for the two compounds? Where would the spots be located if the solvent front were at 9.6 cm? Would you view the separation achieved as favorable?

48–2 The R_F value of a certain compound is 0.55. At what distance from the starting line in a paper chromatogram would you find the compound when the solvent front has progressed to 8.6 cm? *Ans.:* 4.7 cm

49

Evolution Methods

The separation by volatilization of one or more components of a sample may be the method of choice for various reasons. Unwanted impurities or components may be removed, for example, ammonium salts and water by ignition of a precipitate. The weight of a sample before and after the volatilization of a single component may serve for the determination of the latter. Examples include the *indirect* gravimetric determination of water and carbon dioxide. Alternatively, the volatilized material may be recovered. Familiar examples include the preparation of distilled water and the distillation of ammonia. We shall also examine some other examples of evolution methods with particular emphasis on methods for water because of their practical importance.

49.1 Water in Solids

Before the possibilities determining water by evolution methods are considered, let us examine the various ways in which water may be associated with a sample. One broad classification is into *essential* and *nonessential* water.

Essential Water. Essential water is present in a stoichiometric ratio and is an essential part of the chemical composition of the substance. It may be present as *water of crystallization* or *water of constitution,* or both.

Water of Crystallization. Water of crystallization is held by different substances with different strengths. Some substances lose such water by mere exposure to dry air. For example, sodium carbonate decahydrate effloresces (that is, it withers) readily to the monohydrate. Other substances require warming or even heating at elevated temperatures. For example, copper(II) sulfate pentahydrate requires about 150°C or higher to be converted to the anhydrous salt. Some water is lost at lower temperatures and relations exist between vapor pressure and temperature in hydrate equilibria. It is frequently necessary to keep the ambient temperature as well as the humidity within a certain range if a specified hydrate is to be obtained and maintained.

Water of Constitution. Water of constitution does not show in the conventional formula of a substance as "H_2O" and is usually only volatilized at a high temperature. The loss of water by calcium hydroxide

$$Ca(OH)_2 \rightarrow CaO + H_2O \uparrow$$

offers a familiar example. The distinction between water of crystallization and water of constitution is made clear by the conversion of magnesium ammonium orthophosphate hexahydrate to the anhydrous salt and then to anhydrous magnesium pyrophosphate:

$$MgNH_4PO_4 \cdot 6H_2O \rightarrow MgNH_4PO_4 + 6H_2O \uparrow$$

$$2MgNH_4PO_4 \rightarrow Mg_2P_2O_7 + 2NH_3 \uparrow + H_2O \uparrow$$

Nonessential Water. Nonessential water is present in nonstoichiometric amounts and does not represent an essential part of the constitution of a substance. It may be classified as *occluded, sorbed,* or *adsorbed* water, or water present as a *solid solution.* Nonessential water loosely held by a material is often referred to as the "moisture" content of the sample.

Occluded Water. Occluded water is entrapped during the formation of crystals and cannot be removed by simple drying. An elevated temperature is needed; and when it is applied, the water is freed by rupture of the crystal. This rupture, known as decrepitation, is sometimes violent, and in quantitative operations care should be taken to avoid loss of material.

Sorbed Water. Sorbed water is held within the internal surfaces and crevices of materials that have a porous, capillary structure. Familiar examples include starch, active carbon, natural and synthetic zeolites (certain magnesium aluminum silicates), and colloidal substances (such as low-temperature dried hydrous aluminum oxide). Sorbed water in some cases is strongly held, and its complete removal may require heating a material to a temperature of several hundred degrees for a protracted period.

Adsorbed Water. Adsorbed water is usually held loosely as a monomolecular film on the surface of the material. Consequently, the amount thus retained increases with the degree of dispersion of the material. Such water can often be removed at room temperature merely by exposing the substance to a dry atmosphere.

Solid Solution. Solid solutions involving water are known. For example, glasses may hold water homogeneously dispersed.

Hygroscopicity. Some substances (including liquids—concentrated sulfuric acid, for example) take up water avidly from their surroundings. In some cases, a solid can take up so much water that it dissolves (magnesium chloride is a familiar example). The water taken up by a substance may be incorporated as essential or nonessential water, or both. For example, water taken up by dehydrated barium chloride goes to reestablish the dihydrate; in contrast, that taken up by an ignited zeolite (that is, a molecular sieve) merely establishes some degree of water sorption. Hygroscopic substances are often a nuisance in analytical work and may require special precautions in handling. However, some hygroscopic materials are useful as drying agents, that is, as desiccants.

49.2 Direct Gravimetric Determination of Water

If on drying (with or without the application of heat) of a sample, the weight loss is solely caused by the volatilization of water, an *indirect* gravimetric determination of water succeeds. However, when other volatilizable substances are present or if the sample reacts with the oxygen of the air (for example, iron(II) oxide in some minerals or ores is converted to iron(III) oxide), water must be determined *directly*. The water is volatilized in a suitable manner, collected selectively, and weighed. A simple assembly for this purpose is shown in Fig. 49-1. The material is weighed into an elongated porcelain or metal vessel known as a boat, which is then placed into the tube and heated. A stream of air or nitrogen, dried by being passed through a tower filled with a desiccant such as dried calcium sulfate or anhydrous magnesium perchlorate, carries the evolved water to a U-tube packed with the same desiccant. The difference in the weight of this tube before and after the "run" is the weight of water evolved from the sample.

49.3 Water Determination by Azeotropic Distillation

Some liquids form mixtures that boil at a temperature lower than any of the components separately. The mixture having the lowest boiling point is called the

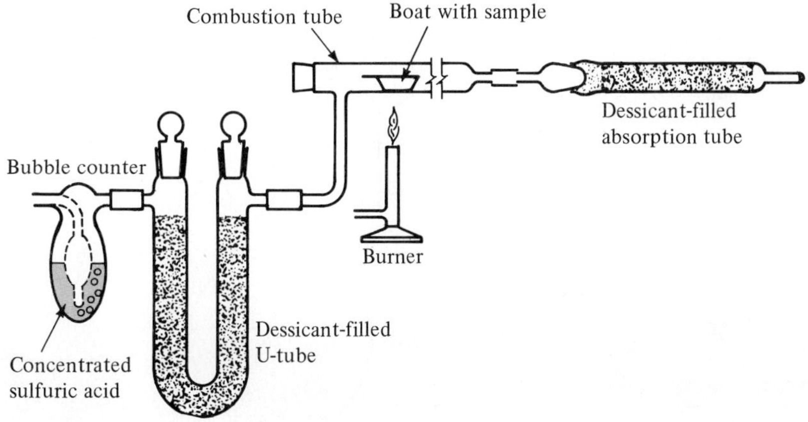

Fig. 49-1. Apparatus for the direct determination of water.

azeotropic mixture, in short, the azeotrope. The composition of the azeotrope is fixed at a given atmospheric pressure. The alcohol-water azeotrope (ca 96% v/v ethanol) and constant-boiling hydrochloric acid (Section 14.1) afford common examples. Azeotropic distillation is often applied to determine water firmly held by samples that would not tolerate extensive heating, such as coal and plant tissue. The assembly of Fig. 49-2 is suitable. The sample is placed in a distillation flask, xylene is added, a condenser is fitted, and the flask is heated. The xylene-water azeotrope distills; the condensate flows down the condenser walls and is collected in a vertical trap. When the condensate is cooled, the components of the azeotrope separate, and water, having the higher density, accumulates at the bottom of the trap. Xylene overflows into the sample flask and is recycled. The trap is graduated and the volume of water collected can be read. Alternatively, especially with small amounts of water, the water thus collected can be determined in other ways, notably by a Karl Fischer titration (Section 25.7).

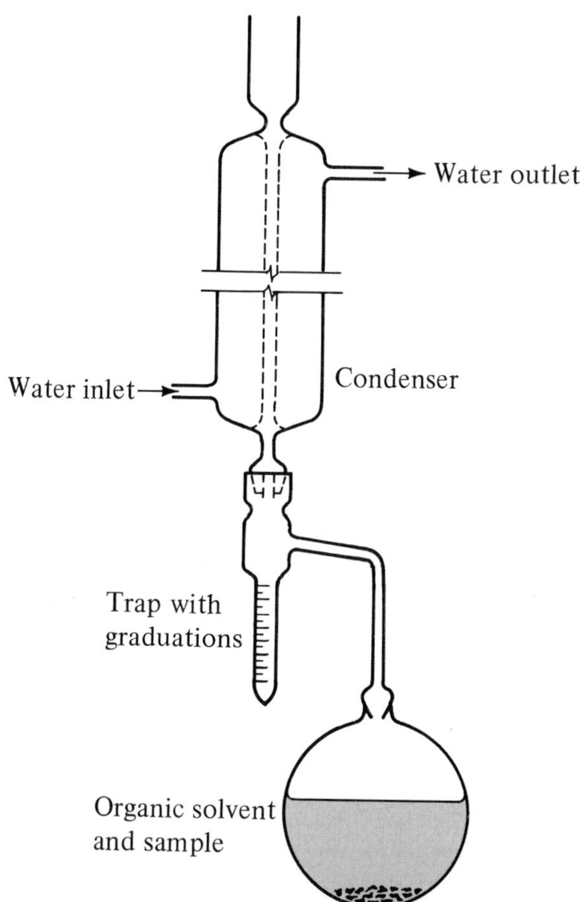

Fig. 49-2. Apparatus for the determination of water by azeotropic distillation.

49.4 Determination of Carbon Dioxide

The indirect determination of carbon dioxide suffers the same limitations as the analogous method for water, and direct evolution is necessary in many cases. For this purpose, the assembly shown in Fig. 49-1 may be used, with the U-tube filled with a soda-lime mixture, which absorbs carbon dioxide. In this case, the carrier gas is not only predried but also freed of carbon dioxide by an additional appropriate charge in the tower. If water is also evolved under the conditions of the run, it must be eliminated before the gas stream enters the soda-lime tube by allowing the stream to pass first through a tube containing a desiccant. Consequently, both water and carbon dioxide can be determined simultaneously.

49.5 C-H and N Analysis of Organic Samples

The possibility mentioned for the simultaneous determination of carbon dioxide and water allows determining the carbon and hydrogen contents of an organic material by its complete combustion. In essence, the assembly of Fig. 49-1 is used with two absorption tubes, one for water and the other for carbon dioxide. The water tube is packed with anhydrous magnesium perchlorate. The tube for carbon dioxide contains in the first portion sodium hydroxide coated on asbestos or glass fiber and in the second portion anhydrous magnesium perchlorate in order to retain water formed by the reaction

$$CO_2 + 2NaOH \rightarrow Na_2CO_3 + H_2O$$

The combustion tube contains a packing of oxidized copper wire, which is heated externally and assures the complete oxidation of evolved gases. Dry oxygen is often used as the carrier gas to facilitate the combustion. Interference by other gaseous substances that are formed is circumvented by filling the exit end of the combustion tube with various substances. For example, silver wool may be inserted to trap halogens and sulfur.

Nitrogen in organic samples, after Dumas, is determined as follows. The sample is mixed with copper(II) oxide and introduced into a tube that is already filled to about half its length with this oxide. The tube is heated, and combustion to water, carbon dioxide, and nitrogen occurs. The combustion products are swept by carbon dioxide as the carrier gas into a special gas buret, known as a nitrometer. This device is a graduated tube filled with concentrated potassium hydroxide solution. This solution absorbs all the carbon dioxide and water but not the nitrogen gas. The volume of nitrogen trapped above the solution is read and related to the amount of nitrogen in the sample.

49.6 Oxygen Flask Combustion

The oxygen-filled flask combustion technique (also known as the Schöniger method after its proponent) may not be strictly an evolution technique but is close enough for us to so classify it. This simple approach is widely applied to the microdetermination of sulfur, the halogens, phosphorus, and diverse metals in organic samples. The principle can be made clear by the description of the determination of sulfur. The organic sample, weighing a few milligrams, is placed on the square portion of filter paper cut in the form of a flag (*A* in Fig. 49-3), which is then folded, carefully to avoid losses, with the elongated strip left sticking out. This packet is then clamped into a V-shaped meshwork of platinum gauze, attached rigidly to a ground glass stopper by a wire (*B* in Fig. 49-3). Into a 250-mL conical flask, accommodating the ground-glass stopper, are placed about 10 mL of water and 3 to 5 drops of 30% hydrogen peroxide. A tube leading oxygen from a cylinder is inserted and the air in the flask is replaced by oxygen. Then the tip of the paper "fuse" is ignited and the stopper, with the attachment, is inserted into the flask (*C* in Fig. 49-3). In the oxygen atmosphere, the paper and the sample are combusted within 15 to 30 s; sulfur oxides are evolved, which are completely absorbed in the liquid when the flask is swirled and shaken. The sulfurous acid, formed from sulfur dioxide, is oxidized by the hydrogen peroxide to sulfuric acid. Any sulfur trioxide formed is converted directly to sulfuric acid. The flask is opened and the platinum gauze, the stopper, and the flask inner walls are rinsed with water. If no interfering (acid-forming) elements are present, the solution can be titrated with sodium hydroxide to the methyl red–methylene blue end point. Alternatively, the rinsing is conducted with ethanol so that the absorbing liquid becomes an ethanol-water mixture. Sulfate is then titrated with barium perchlorate. The indicator used is often Thorin, that is, *o*-[(2-hydroxy-3,6-disulfo-1-naphthyl)azo] benzenearsonic acid, disodium salt. In this medium, Thorin shows a sharp color change from pale yellow to pale pink (formation of the barium-Thorin complex). The barium sulfate persists

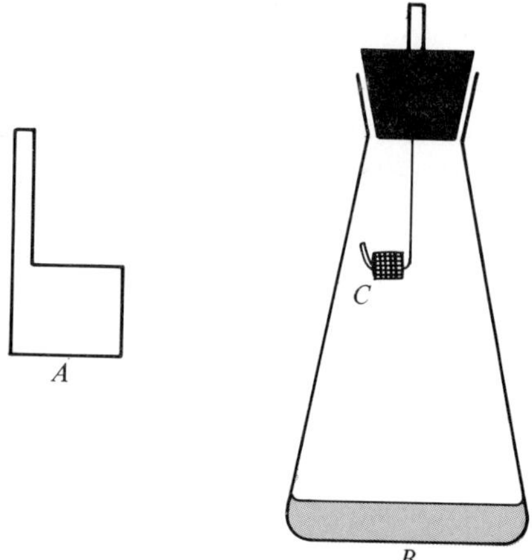

Fig. 49-3. Arrangement for oxygen-filled flask combustion. *A*, shape of sample holder cut from filter paper (not to same scale); *B*, sample packet clamped in platinum gauze; *C*, combustion flask.

as a colloidal suspension, and no substantial precipitate forms with the small amounts of sulfate involved.

Because of the small volume of the absorbing liquid required and the relative freedom from possible interfering species, many methods can be applied as a finish to oxygen flask combustion, including titrimetric, photometric, and electroanalytical approaches. Halogens are determined as is sulfur but with omission of the hydrogen peroxide. The absorbed chloride, bromide, or iodide can be titrated with silver nitrate. Chloride can also be titrated with mercury(II) nitrate. Metals that form stable complexes with EDTA can be titrated with that reagent.

In more elaborate arrangements, the sample may be placed in a capsule made of gelatin (or of high-purity methylcellulose, the best choice), which is placed in a small platinum spoon attached to the stopper. Ignition is accomplished within the closed, oxygen-filled flask by focusing a high-intensity light beam onto the capsule. Such a capsule can also be used in the simple arrangement of Fig. 49-3 when liquid samples are to be analyzed.

49.7 Distillation Methods

Separation by distillation is largely in the domain of organic chemistry but finds some application in inorganic analysis. The volatility of most inorganic substances is low. Consequently, only a few can be distilled at moderate temperatures, even at reduced pressure. The separation of water by azeotropic distillation has been considered (Section 49.3). The distillation of ammonia with water following a sample digestion by the Kjeldahl technique has also been described (Section 14.9). Fluoride can be distilled as hexafluorosilicic acid from an acidic medium containing silicic acid (Section 18.9). A further example is the distillation of the chlorides of arsenic, antimony, tin, and germanium. These chlorides distill from a concentrated hydrochloric acid solution at different temperatures and can thus be separated. Zinc metal is sometimes distilled at high temperature and thereby separated from less volatile components of alloys.

For fractional distillation, the concept of a theoretical plate and liquid-vapor equilibrium are important and are briefly considered in Section 47.3 in relation to the theory of gas-liquid chromatography.

49.8 Determination of Arsenic by Volatilization

When a material containing arsenic is boiled in the presence of zinc metal in an acidic solution, complete reduction to gaseous arsine, AsH_3, occurs (for example, ($As^{3+} + 3Zn + 3H^+ \rightarrow AsH_3 \uparrow + 3Zn^{2+}$). The arsine can be carried by a gas stream into a glass tube having a small-bore section 1 to 2 cm long. This portion is heated externally. Arsine is thermally decomposed, and elemental arsenic deposits as a mirror on the glass walls. The amount of arsenic in the sample is estimated from the length of the mirrored section, by comparison of the results with standards. Alternatively, the mirror is dissolved and the arsenic determined iodometrically (Section 25.7) or photometrically. This method, introduced by Marsh, is a standard procedure in forensic analysis when arsenic poisoning is encountered or suspected.

In the Gutzeit method, the reduction is performed in an assembly such as shown in Fig. 49-4. The hydrogen gas evolved in the reaction between zinc and acid carries the arsine into the upper tube. Droplets and any hydrogen sulfide formed are held back by a cotton plug impregnated with lead acetate. The upper tube contains a moistened strip of paper impregnated with mercury(II) bromide. This substance reacts with the arsine, and a coloration appears. The length of the colored portion is compared with that of strips obtained in the procedure with known amounts of arsenic under identical conditions. In this way trace amounts of arsenic can be determined rapidly and accurately.

A photometric finish to the determination of arsenic often replaces the older Marsh and Gutzeit methods. Usually, the generated arsine can be absorbed in a pyridine solution of silver diethyldithiocarbamate. The absorbance of the arsenic(III)-diethyldithiocarbamate species formed in this solution is measured.

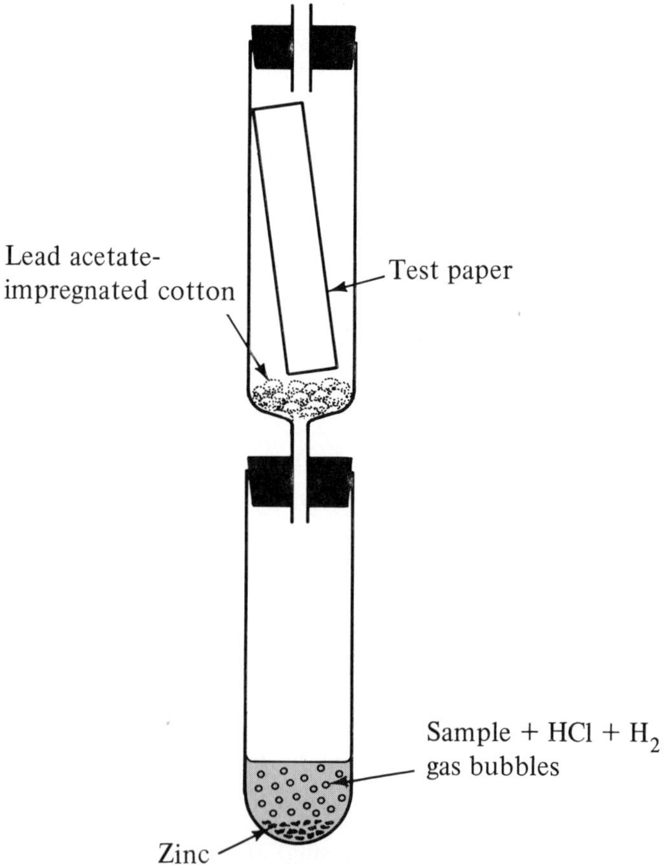

Fig. 49-4. Apparatus for determination of arsenic by the Gutzeit technique.

An atomic absorption spectrometric approach is based on the generation of arsine, usually by the action of sodium borohydride, $NaBH_4$, and passage into a hydrogen flame.

49.9 *Questions*

49-1 What is the difference between essential water and nonessential water? Describe in your own words and give other examples than those found in the text.

49-2 Is it possible for a substance to contain both essential and nonessential water? Give an example.

49-3 Define *hygroscopicity* and give examples of hygroscopic substances.

49-4 An air stream containing some water vapor is passed through an absorption tube filled with CaO. Write the equation for the reaction involved in the water absorption. Will the increase in weight give the correct amount for water?

49-5 A stream of dry air containing some carbon dioxide is passed through an absorption tube filled with CaO. Write the equation for the reaction involved in the absorption of the carbon dioxide. Will the increase in weight give the correct amount for carbon dioxide?

49-6 A stream of dry air containing some carbon dioxide is passed through an absorption tube filled with $Ca(OH)_2$. Write the equation for the reaction involved in the absorption of the carbon dioxide. Will the increase in weight necessarily give the correct amount for carbon dioxide? What else must be present in the absorption tube in order that the correct weight is obtained?

49-7 Name a solid substance, except calcium oxide or hydroxide, that is suitable for the absorption of carbon dioxide.

49-8 What is the distinction to be made between dry and water-free copper sulfate?

49-9 In the direct gravimetric determination of water (Section 49.4) and the C–H analysis of organic samples, the best practice is to fill the "guard" and water-absorption tubes with the *same* desiccant. Comment on the merit of this practice. (*Hint:* Is the vapor pressure of water the same over all desiccants?)

49.10 *Problems*

49-1 What amount of H_2O (*18.0*), in milligrams, is volatilized when 0.543 g of $BaCl_2 \cdot 2H_2O$ (*244.3*) is heated to constant weight? *Ans.:* 80.0 mg

49-2 A material contains 50.0% $CaCO_3$ (*100.1*) and 50.0% $MgCO_3$ (*84.3*). On ignition, CO_2 (*44.0*) is completely volatilized. Calculate the percentage loss on ignition for a 0.2500-g sample of that material. *Ans.:* 48.1%

49–3 A material containing 3.0% $CaCO_3$ (*100.1*) and 97.0% FeO (*71.8*) is ignited. Carbon dioxide (*44.0*) is evolved and oxygen (*16.0*) is absorbed, forming Fe_2O_3 (*159.7*). Calculate the percent loss on ignition for this mixture.
Ans.: –9.49%

49–4 When a 1.000-g sample of a material is dried, 16 mg of H_2O is lost. When 0.870 g of the same material (not dried) is *ignited*, 63 mg is lost. (a) Calculate the percent H_2O content and (b) the percent loss on ignition on a dry basis. *Ans.:* (a) 1.6%; (b) 5.7%

49–5 A material is known to contain 15.3% water. What will be the weight of an absorption tube when the water has been absorbed from a 1.520-g sample if the tube weighed 25.533 g before the start of the run? *Ans.:* 25.766 g

49–6 A substance known to contain *only* carbon, hydrogen, and nitrogen was analyzed by combustion. A microbalance precise to ±1 or 2 μg was used. The following data were secured: Combustion boat + sample = 0.053256 g; boat empty = 0.050231 g; absorption tube + water = 7.634614 g and tube empty = 7.632564 g; absorption tube + carbon dioxide = 7.193111 g and tube empty = 7.184535 g. Calculate the composition of the substance in percent of C, H, and N. Give the simplest empirical formula possible for the substance. *Ans.:* 77.3_9% C, 7.5_1% H, 15.1_0% N; C_6H_7N (aniline)

49–7 Arsenic can be determined in silicate rocks by fusing with sodium hydroxide and sodium peroxide, dissolving the melt in hot water, and treating the filtered solution with hydrochloric acid, potassium iodide, tin(II), chloride, and zinc dust. The evolved arsine is absorbed in exactly 3 mL of a silver diethyldithiocarbamate solution in pyridine and the absorbance of that solution is measured at 540 nm.

For a 0.525-g sample of a rock carried through this procedure, the absorbance was 0.542. Under similar conditions, 10.0 μg of arsenic, added as an arsenate solution to the evolution unit, gave an absorbance of 0.640. A 2.000-g sample of the rock was heated in an oven at 125°C to a constant weight of 1.790 g.

(a) Calculate the moisture content of the rock. (b) Calculate its arsenic content in micrograms per gram (ppm) on a "dry basis."
Ans.: (a) 10.5%; (b) 29.2 μg/g

49–8 An organic compound is analyzed for sulfur by the Schöniger oxygen flask technique. A 5.382-mg sample is combusted and the titration performed with 0.01045*F* barium perchlorate, using Thorin as the indicator. A volume of 3.27 mL is required. A blank using the paper "flag" alone with the absorbing liquid, performed repeatedly, had an average requirement of 0.02 mL for the titrant. (a) Calculate the percent of sulfur (32.06) in the compound. (b) If the sole source of sulfur and oxygen in the compound is a sulfonic acid group ($-SO_3H$), what is the percent of oxygen? (c) If the compound has a single sulfonic acid group in each molecule, what is the smallest molecular (formula) weight that can be assigned?
Ans.: (a) 20.25% S; (b) 30.3% O; (c) 158

50

Analysis of Gases

50.1 Introduction

Scientists in many areas of science and technology must often analyze mixtures of gases. Some common analytical problems include the evaluation of gaseous fuels, the analysis of combustion products, the study of respiration and metabolism, and the analysis of air pollutants and process gases from smelting and chemical operations.

Many approaches exist that allow resolving gas mixtures analytically. Some of the methods for individual gases considered in other sections have special application.

Water vapor may be determined gravimetrically by absorption in a drying tube packed with a desiccant (Section 49.2).

Sulfur dioxide may be absorbed in an aqueous medium in a gas scrubber bottle. It can then be determined iodometrically (Section 25.7), or after oxidation with hydrogen peroxide to sulfuric acid, gravimetrically as barium sulfate, by a precipitation titration with barium ion (Section 18.8), or by an acid-base titration.

Sulfur trioxide may be absorbed in an alkaline solution and be determined as the sulfate by any suitable method.

Ammonia may be absorbed in water or dilute acid and be determined by an acid-base titration (Sections 14.9 and 14.10) or photometrically by Nessler's method (Section 36.2).

Carbon dioxide may be absorbed in a tube suitably packed (for example, with sodium-hydroxide-coated asbestos or glass fiber) and determined gravimetrically. Alternatively, it may be absorbed in a barium hydroxide solution and determined by an acid-base titration or by measurements of the change in conductivity of the solution (Section 32.6).

Hydrogen chloride may be absorbed in water or in a base and determined by an acid-base titration. Alternatively, it may be absorbed in a silver nitrate solution, and the silver chloride precipitated then separated and weighed, or an excess of silver nitrate may be titrated with a standard chloride solution.

Physical measurements are valuable in analyzing gas mixtures, by infrared photometry (Section 35.19), for instance, or thermal conductivity or mass spectrometry. Gas chromatography has displaced some of the classical methods of gas analysis described below from key applications (Sections 50.2 through 50.3). In the biosciences, however, such methods continue to receive attention, for example, in the study of gases evolved and absorbed in cell respiration (Warburg apparatus) and the assessment of blood gases (Van Slyke technique).

50.2 Gas-Volumetric Methods

The most explored methods for the analytical resolution of many common gas mixtures are based on selectively removing one or more components of the mixture and measuring the resulting change in volume or pressure. When the term *gas analysis* is not further qualified, it usually implies these methods. In gas-volumetric methods, a known volume of the gas mixture is taken and a constituent is selectively removed, usually by an absorption process, and the new volume is measured at the same temperature and pressure. The process is repeated for other constituents using other absorbents. Since at constant pressure and temperature the amount of a gas is proportional to its volume, the composition of the gas in % v/v can be readily deduced. In manometric methods, the volume and temperature are kept constant and the change in pressure resulting from each absorption step is measured. The manometric method is capable of greater accuracy and is frequently used when a small sample must be analyzed.

The principles underlying gas-volumetric methods can be appreciated by considering the resolution of a simple gas mixture: nitrogen, oxygen, and carbon dioxide. Simple forms of a gas buret and an absorption bulb are illustrated in Fig. 50-1. Initially the gas buret, separated from the absorption bulb, is filled completely with a suitable confining liquid by opening both stopcocks and elevating the reservoir bulb; the lower stopcock is then closed. (For the assumed gas mixture, a sodium chloride solution acidified with sulfuric acid may serve as the confining liquid, since carbon dioxide and the other gases are only slightly soluble in it; mercury serves as the liquid of choice for the analysis of many gas mixtures.) The vessel containing the gaseous sample is next connected to the buret. The lower stopcock is then opened and a portion of the sample is drawn into the buret by lowering the reservoir bulb. The upper stopcock is then closed, and the reservoir is placed next to the gas buret and moved up and down until the level of the liquid in both is identical. The volume of the gas in the buret is read and recorded. An absorption bulb containing a concentrated potassium hydroxide solution (30 to 50%) is placed under suction until the solution is drawn through the capillary tubing to the inscribed mark and the stopcock is closed. This absorption bulb is then attached to the gas buret, as shown as Fig. 50-1. The three stopcocks are opened and the gas mixture is forced into the absorption bulb by elevating the reservoir until the confining liquid reaches the area of the upper buret stopcock, which is then closed. The absorption bulb is then shaken gently for 1 min or so. (The first of the two bulbs is often filled with glass beads so that when the solution is forced into the second

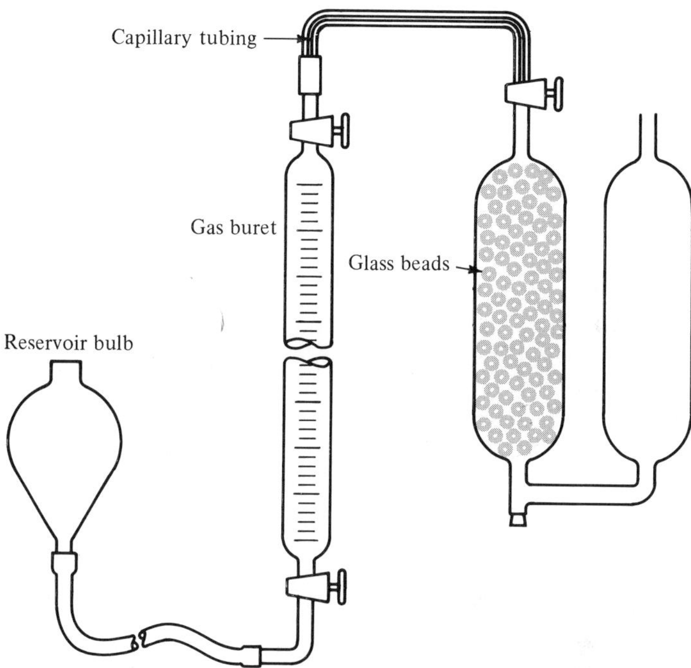

Fig. 50–1. Schematic diagram of simple assembly for gas analysis.

bulb by the gas, there is a greater area for the absorption of carbon dioxide, provided by the caustic-coated beads.) The stopcocks are then opened and the sample is sucked back by lowering the reservoir bulb until the caustic solution is again aligned with the mark. The buret stopcock is closed. The reservoir bulb is again placed next to the buret and moved up and down until the liquid level in both is identical. The gas volume is read. The difference between the initial volume and the new volume (at constant temperature) corresponds to the carbon dioxide in the sample.

The potassium-hydroxide-filled absorption bulb is replaced by a bulb containing an absorbent for oxygen such as a potassium hydroxide solution of pyrogallol, that is, 1,2,3-$C_6H_3(OH)_3$, or an ammoniacal copper(I) chloride solution containing a coil of copper gauze. The absorption operation is then conducted as described for carbon dioxide, and the change in volume related to oxygen. The nitrogen of the sample remains in the buret after the two absorption steps.

Example 50–1 In the gas-volumetric analysis of a mixture of CO_2, O_2, and N_2 as described, the volumes (at constant temperature) in the gas buret initially and after the absorption of CO_2 and O_2 are 90.0, 44.0, and 18.0 mL, respectively. Calculate the composition of the gas mixture in % v/v.

$$\% \, CO_2 = \frac{90.0 - 44.0}{90.0} \times 100 = \frac{46.0}{90.0} \times 100 = 51.1\% \text{ v/v}$$

$$\% \, O_2 = \frac{44.0 - 18.0}{90.0} \times 100 = \frac{26.0}{90.0} \times 100 = 28.9\% \text{ v/v}$$

$$\% \, N_2 = \frac{18.0}{90.0} \times 100 = 20.0\% \text{ v/v}$$

In more elaborate equipment (say the Orsat apparatus), the various absorption units are attached to one "manifold" line and any unit can be connected to the buret by opening the appropriate stopcock.

Hydrogen can be determined, usually after the absorption of the oxygen of the sample, by deliberately adding a known amount of oxygen. The mixture is then exposed to an electrically heated platinum spiral. A safe combustion occurs. The water formed condenses and the reduction in the volume of sample can be related to the hydrogen content of the sample.

50.3 Gasometric Methods

If a substance can be made to undergo a reaction that leads quantitatively to the liberation of a gas, collecting the gas and measuring it by volumetric or manometric means allows the determination of the substance. For example, it is sometimes necessary to determine the amount of a metal as such present together with its oxide. If the metal dissolves in an acid or a base with the liberation of hydrogen, a weighed amount of the material is treated with acid (or sodium hydroxide in the case of aluminum or zinc), and the hydrogen that is evolved, usually with boiling of the solution, is collected over water. When a gas-volumetric finish is applied, the volume of hydrogen is measured at a known temperature and pressure, and after correction is made for the vapor pressure of water and reduction to standard conditions, the amount of hydrogen is calculated.

Example 50–2 A 1.500-g sample of an aluminum oxide sample known to contain some metallic aluminum (*26.98*) is dissolved in 10% NaOH solution. The solution is boiled and a volume of 148 mL of hydrogen is collected over water at 23.0°C at an atmospheric pressure of 756.1 mm of Hg. What is the percentage of Al in the sample? The vapor pressure of H_2O at 23.0°C is 21.1 mm of Hg. One (gram) mole of a gas occupies 2.24×10^4 mL at 0.00°C and 760.0 mm of Hg. The "effective" pressure of H_2 at 23.0°C = 756.1 − 21.1 = 735.0 mm of Hg. The measured volume V_i is reduced to standard conditions V_{st} by applying the perfect gas law (where the subscripts i and st refer to the measured and standard conditions):

$$V_{st} = V_i \times \frac{T_{st}}{T_i} \times \frac{P_t}{P_{st}} = 148 \times \frac{273}{273 + 23} \times \frac{735.0}{760.0} = 132 \text{ mL}$$

From the reaction equation

$$2Al + 2OH^- + 2H_2O \rightarrow 2AlO_2^- + 3H_2 \uparrow$$

it can be seen that three moles of hydrogen gas corresponds to two moles of aluminum metal. Hence

$$\% \text{ Al} = \frac{132}{2.24 \times 10^4} \times \frac{2}{3} \times \frac{26.98}{1.500} \times 100 = 7.1\% \text{ Al in sample}$$

Other common gasometric methods include the determination of ammonia by its reaction with sodium hypobromite ($2NH_3 + 3OBr^- \rightarrow N_2 \uparrow + 3Br^- + 3H_2O$) and the determination of nitrite and nitrate by reaction with mercury and an excess of sulfuric acid (for example, $2H^+ + 2NO_2^- + 2Hg + H_2SO_4 \rightarrow 2NO \uparrow + Hg_2SO_4 + 2H_2O$). In the Dumas determination of nitrogen in organic compounds, the sample is burned in a carbon dioxide atmosphere with a metal oxide as the source of oxygen. Carbon dioxide and other acidic gases are absorbed in potassium hydroxide solution; the nitrogen gas formed is collected above that solution and the gas volume is measured (Section 49.5).

50.4 Questions

50-1 The gas laws discussed in previous courses are sometimes used in calculations relevant to gas analysis. What is stated by the laws of Boyle, Charles, Gay-Lussac, and Dalton?

50-2 Why should the water over which carbon dioxide is collected be previously saturated with that gas?

50-3 A material contains sulfur as the only active ingredient. A sample of this material is combusted and the gases passed through a measured volume of standard base, the excess of which is titrated with acid to the phenolphthalein end point. Would it matter whether the sulfur was combusted to only SO_2 or to only SO_3 or to a mixture? Explain.

50-4 A gas stream contains N_2, O_2, SO_2, and SO_3. Suggest a gas analytical method to determine the two sulfur oxides separately.

50-5 What are the "standard conditions" to which reference is made in measuring gases?

50-6 Why are pellets or rings of glass or another inert material often present in absorption bulbs used in gas analysis?

50-7 Do you expect the correction for water vapor pressure to be larger when measurements of gas volumes are made over pure water than when they are made over concentrated KOH? Explain.

50-8 The "mole volume" of 22.4 L is often used in gas analytical calculations. The values obtained are approximate. Why?

50-9 Gas analysis can be based on the following combustion technique. The known volume of a gas containing combustible components is mixed with a known volume of oxygen and the combustion initiated. The reaction mixture is then cooled and any CO_2 formed is determined. The data thus secured are used partly to calculate the composition of the sample. Write the reaction equations; indicate the volume of O_2 consumed, the contraction expected, and the volume of CO_2 produced for the combustion of one volume of each of the following substances: hydrogen; carbon monoxide; methane, CH_4; acetylene, C_2H_2; and ethane, C_2H_6.

Ans. for acetylene: $2C_2H_2 + 5O_2 \rightarrow 4CO_2 + 2H_2O$

O_2 consumption: 2.5 volumes

Contraction: 1.5 volumes [2 acetylene + 5O_2 volumes disappear and 4 volumes CO_2 appear per 2 volumes acetylene; thus $(2 + 5 - 4)/2 = 1.5$]

CO_2 produced: 2 volumes

50.5 Problems

50-1 A 0.320-g sample of limestone, containing $CaCO_3$ as the active ingredient and some inert material, is treated with acid and 78.5 mL of CO_2 developed is collected above water saturated with that gas. The temperature and pressure when the gas volume is measured are 23°C and 0.1128 kPa. Calculate the percent $CaCO_3$ (*100.1*) in the limestone. Vapor pressure of H_2O at 23°C is 2.786 kPa. *Ans.:* 97.9%

50-2 What minimum sample weight of a limestone containing 98% $CaCO_3$ (*100.1*) must be taken so that the CO_2 produced on ignition and collected dry over mercury will amount to at least 50 mL when measured at 20°C and 0.1006 kPa? *Ans.:* 0.21 g

50-3 What volume, in milliliters, of dry hydrogen gas will be evolved (measured at 23.5°C and 0.1002 kPa) when 86.2 mg of zinc metal (*65.4*) is treated with acid?

50-4 An organic material is combusted in air and the gas mixture containing CO_2, N_2, and O_2 after passing through a drying tube, is collected over mercury at 21°C and 752 mm of Hg. The gas volume is 60.2 mL. The gas is passed over concentrated K OH and the volume measured again and found to be 51.4 mL. What weight of carbon (*12.0*), in milligrams, was present in the combusted material?

51

Sampling

51.1 Sampling

In practice, a chemical analysis is often made of a small sample taken from a large consignment of a material. If the sample is to be representative of the entire consignment, it must be selected so that it possesses the characteristics of the bulk.* The manner in which this is accomplished is known as a sampling procedure. In the merchandising of an ore, it is often necessary to secure a representative sample from carloads, trainloads, or even shiploads. Obviously it is completely inadequate to take a single lump randomly. Since the random lump may be composed of accompanying rock (gangue) or of an extremely rich portion of the ore, the value obtained for an element of principal interest might range from zero to a percentage far above the average. In practice, establishing a representative sample often presents far more difficulties than the analysis itself. Law suits have resulted from inappropriate sampling. Indeed, today sampling schemes are sometimes specified in purchase contracts, especially when the material involved is extremely inhomogeneous or variable in composition.

In general, sampling involves three steps: (1) the establishment of a gross sample, (2) its reduction to a subsample that can be brought to the laboratory, and (3) the preparation of the sample actually taken by the analyst for chemical analysis. There is no uniform, general method of sampling. The best technique depends on the material of interest. The principal criterion is the degree of heterogeneity of the material.

---

*The discussion is here limited to establishing a representative "average" sample. However, this is not always the object of sampling. If the aim is to confirm, for example, that the active chlorine content of municipal water is great enough to prevent microbial growth, an average value is not sought nor is the sample taken at the chlorinator; rather, the sample is withdrawn from a far-distant, secondary distribution line. But even when a special sample is needed, an appropriate sampling procedure is necessary.

51.2 Establishing a Sample

Liquids are usually homogeneous, and if they are so, sampling them is simple. Be warned, however, that attaining homogeneity in thousands of gallons of a liquid product is far more difficult than preparing a simple solution in a beaker. Homogeneity cannot be blindly assumed in sampling a liquid. In practice, in the sampling of drums or tankcars, liquid is withdrawn at various levels and either a composite sample is established and analyzed or the samples from different levels are separately analyzed.

Establishing a representative sample in a solid product is far more difficult, especially if there is much variation in composition and the particle size ranges from large lumps several pounds in weight to a fine dust. Statistical treatment often helps in establishing the optimum sampling conditions and in determining the minimum size of the sample and subsample so that they represent the consignment adequately.

For bulk material, sample increments are often taken when the material is transferred. For example, with coal and ores, a shovelful or a wheelbarrow load may be taken from each truck or railway car or at regular intervals from the complete cross section of a conveyor belt. How often a sample increment should be taken and what size will depend on the inhomogeneity of the material. The increments must then be placed together, reduced to a relatively uniform size, for example by hammer or roll milling, and a subsample established.

Mechanical sampling devices are often used, but simple manual techniques are still widespread. In coning, the material is piled in a cone by placing each shovelful of material in the center of a ring. Finally, the cone is flattened by pulling down material equally in all directions, quarters are established by using a board, and two diagonally opposite quarters are rejected. The remaining quarters are piled into a second cone, and so on. In quartering, the material is laid out evenly in a flat square, diagonals are drawn, and opposite quarters are rejected. The simple device known as a riffle consists of a trough divided into an even number of transverse slots. The alternate slots discharge into a left and right collecting bin. Thus, when the sample is distributed evenly over the trough of a simple riffle and allowed to fall into the bins, the sample is divided in half. The contents of either the right or the left bin may then be riffled a second time, and so on. If one or more of these (or other) techniques are combined with one or more reductions in particle size, a portion of the gross sample is secured as a subsample, which is then sent to the laboratory.

Metals and alloys can be sampled by sawing or drilling a large consignment. The "sawdust" or turnings are "quartered," and so on, or when appropriate, may be melted together to form a homogenous subsample.

In the laboratory, the subsample may be further quartered or coned to obtain a sample of a few grams that is taken for analysis. Since many materials cannot be brought into solution readily, the sample may be further reduced in particle size. For this purpose, a ball mill or a mortar and pestle may be used.

A requirement of the sampling procedure is that the composition of the sample should remain unchanged or should change only in a known way. It is difficult to fulfill this requirement whenever the sample either contains loosely retained water (for example, damp coal), or as the particle size is reduced, absorbs water or is oxidized by the air. In such cases, a separate sample of a few pounds is usually taken and brought to the laboratory under protective conditions, and the component undergoing change is separately determined. For example, the moisture content is often established with a separate sample. The other determinations are effected with the representative sample on a dry basis. The results may then be recalculated to an "as received" or "as is" basis (see Section 6.4).

Even with liquid samples, possible changes in composition must be considered, and this is especially important in environmental analysis. For example, 1 to 50 L of water may be taken from a stream into one or more glass or plastic bottles (and often over several hours). The closed containers may then be transported some distance and be stored for hours or days before laboratory analysis. For sample integrity, points that must be considered include contamination from the bottle or closure or the entry of airborne contaminants, adsorption of trace elements on the container walls (for example, arsenic or mercury), escape of carbon dioxide, and changes in the concentration of some species (including dissolved oxygen) due to chemical reactions or microbial action.

51.3 Questions

51-1 Explain the term *quartering.*

51-2 An ore consists of lumps and fine dust. What will happen to a consignment of that ore during its transport by railroad car? Suggest how a representative sample could be secured from the car on its arrival at the smelter.

51-3 Explain why sampling liquids is simpler than sampling solids.

51-4 How might nonhomogeneity be revealed in the analysis of a laboratory sample?

51-5 Treated and filtered water from an industrial plant is discharged continuously for an 8-hr period each day into a river through a single pipe that is laid on the river bottom and opens at the river's center. The "average" concentration in the river of a stable chemical species of environmental concern is to be measured and the contribution of the plant discharge to that concentration is to be established. Elaborate on where and when samples should be taken.

51-6 Elaborate on the merits and purposes of the following sampling schemes.

(a) For a semicontinuous process, take a sample of the product for analysis at stated intervals of time.

(b) Take a sample whenever a process is started up (for example, after a lunch shutdown) or when a variable is adjusted.

(c) For a lot of a powdered product, take a sample from the first, last, and approximately middle drum as the material is withdrawn from the blender and analyze the three samples either separately or as a composite.

(d) For a hospital patient, combine the urine passed over a 24-hr period and submit a specimen for analysis.

52 Attack of the Sample

Most analytical methods require that the sample be dissolved in an aqueous medium. In practice, however, relatively few materials are completely or even substantially soluble in water. In the applications of "wet" methods mentioned in this textbook, it has been largely assumed that a sample solution has already been secured. Also, the student in parallel laboratory work mainly uses liquid samples, and possibly a few solid samples that are brought into solution by a simple treatment. In the everyday practice of inorganic quantitative analysis, the situation is different, and many samples in use are resistant to simple chemical attack.

Various measures for treating solid samples may have been brought to your attention when you undertook qualitative analysis. The measures commonly adopted in quantitative analysis are not basically different, but extreme care is needed to avoid losses of material from spattering or volatilization. Further, it is necessary during the entire preparation of the sample solution to avoid introducing any ions that will interfere in or otherwise complicate a subsequent separation or determination. In some cases, different dissolution techniques must be applied to separate samples so that different constituents of a complex material can be determined. When trace constituents of a sample are to be determined, it is imperative that significant amounts of these substances are not introduced from the reagents and vessels or as airborne contaminants.

52.1 *Acid Attack of the Sample*

Mineral acids, including hydrochloric, nitric, sulfuric, phosphoric, perchloric, and hydrofluoric acids, or mixtures of such acids, are frequently used to attack complex inorganic materials. Their use is advantageous, because any unwanted excess can usually be removed by volatilization. Any portion of the sample that is unattacked by suitable acid treatments is often separated by filtration and the washed residue is fused with a suitable flux (Section 52.2). The cooled fusion melt after solution may either be combined with the solution of the acid-soluble material or be analyzed separately. In the attack of a sample by heating with a readily volatilizable acid, such as hydrochloric or nitric acid, it is necessary to avoid excessive loss of the acid or to make up for losses by occasionally adding acid.

Hydrochloric acid is a good solvent for many metals, metal oxides, and metal carbonates. With carbonates, care must be exercised to avoid losing material from foaming or spattering during the evolution of carbon dioxide. The tendency for a metal to dissolve in a so-called nonoxidizing acid, such as hydrochloric acid, phosphoric acid, dilute perchloric acid, or dilute sulfuric acid, is relatable to the standard (or formal) electrode potential of the metal ion–metal redox couple (see Table G in the Appendix).

Nitric acid is known as an oxidizing acid, since its anion manifests oxidizing properties; and the acid attacks more metals than hydrochloric acid does. Some metals, however, notably aluminum and chromium, are not attacked by concentrated nitric acid. This so-called passivity is associated with the formation of an insoluble, protective layer of the oxide (or of a more complex entity) on the metal surface. Nitric acid is favored for dissolving many alloys and offers the special advantage that certain metals (tin, antimony, and tungsten) remain insoluble as oxides or acids and can be separated immediately by filtration.

Many metals and alloys, notably the noble metals, require an acidic medium with stronger oxidizing power than nitric acid alone provides. Various mixed acids receive attention, especially aqua regia, which is a mixture of one volume of concentrated nitric acid with three volumes of concentrated hydrochloric acid. The formation of chloro complexes may help in dissolving the noble metals by this mixture.

Sulfuric acid has certain advantages in attacking samples. It shows strong dehydrating properties and has a high boiling point (~340°C); the high boiling point allows removal of volatile acids by adding sulfuric acid and then heating the mixture.

Hot, concentrated perchloric acid is a powerful oxidant (Section 25.3) and attacks some alloys that are resistant to hydrochloric or nitric acid. Perchloric acid has the advantages of having a high boiling point (the azeotrope with water boiling at about 200°C) and of forming metal salts highly soluble in aqueous medium. Only potassium, rubidium, and cesium form sparingly soluble perchlorate salts (Section 7.2). Precautions must be taken to avoid contact of concentrated perchloric acid with organic matter, except under special, controlled conditions, because violent, even explosive reactions may result (see also Section 25.3).

Hydrofluoric acid receives special attention in the attack of silicate materials, including rocks, minerals, and glasses. When the acid is heated, fluosilicic acid and silicon tetrafluoride are volatilized (Section 7.3). The use of platinum or polytetrafluoroethylene (that is, Teflon®) ware is necessary. It is usually important to remove all the hydrofluoric acid before the analysis, because fluoride ion forms insoluble compounds or soluble complexes with many cations and also attacks glass. This removal is usually achieved by evaporation (in a hood!) of the sample solution with excess sulfuric acid or perchloric acid until white fumes appear.

Certain elements may be volatilized during the acid attack of a sample. This may be used to advantage as a means of separating the elements for their determination or to obviate their interference in subsequent determinations of interest. The loss in other cases may be unwanted and present a source of difficulty. What elements are volatilized depends on the acid used and the conditions existing, including the nature of other components of the sample. Carbon dioxide and hydrogen sulfide are evolved from carbonates and sulfides. Silicon as fluosilicic acid, H_2SiF_6, and silicon tetrafluoride, SiF_4 and boron as fluoboric acid, HBF_4 are volatilized from a hydrofluoric acid medium (see above). When a sample solution is boiled in hydrochloric acid, $AsCl_3$ and $GeCl_4$ are readily evolved and to some extent $SbCl_3$, $SnCl_4$, and $HgCl_2$, if adequate precautions are not taken (compare Sections 49.7 and 49.8). When a sample is heated with sulfuric acid, halides and also other anions forming volatile acids are evolved as the acid.

52.2 *Attack of the Sample by Fusion with a Flux*

Only a limited number of inorganic samples are brought completely into solution by acid attack. In such cases, the acid-insoluble residue or a further sample is subjected to fusion with a suitable flux. Fluxes usually melt in the region 180 to 850°C; however, with a few substances usually classified as fluxes, the attack involves sintering rather than melting. Depending on the flux and the sample, either the finely powdered sample is intimately mixed with the flux and placed in a crucible, or a layer of flux is placed in the crucible followed by the powdered sample and covered by additional flux. The crucible is covered with a lid and heated until the flux melts. The elevated temperature is maintained until the sample material is largely or completely dissolved in the melt. The color and appearance of the melt often allows the progress and success of the fusion operation to be assessed. The cooled melt is dissolved in water or treated with acids. The success of the fusion depends on the temperature, the time allowed for the attack by the flux, and frequently on the degree of dispersion of the sample material. Fluxes are usually classified as alkaline or acidic and oxidizing or reducing.

Acid-base behavior between solids or in the molten state at high temperature is explainable in terms of the Lewis theory (Section 16.4). Consider, for example, the fusion of silica with an alkali metal oxide; the oxide is classified as an alkaline flux. The silicon dioxide is an electron-pair acceptor and acts as a Lewis acid. The metal oxide has an ionic crystal lattice (for example, $Na^+O^{2-}Na^+$) and provides the oxide ion, which as an electron-pair donor acts as a Lewis base. The acid-base reaction in the fusion leads first to the metasilicate ion, SiO_3^{2-}:

$$
\begin{array}{l}
:\ddot{O}:Si:\ddot{O}: + :\ddot{O}:^{2-} \rightarrow :\ddot{O}:Si:O:^{2-} \\
\qquad\qquad\qquad\qquad\qquad\quad :\ddot{O}:
\end{array}
$$

Further analogous reaction with the oxide ion yields the orthosilicate ion, SiO_4^{4-}. Some metal oxides on fusion act as Lewis acids and react with alkaline fluxes. Others are amphoteric, notably aluminum, titanium, iron, and zinc, and react with both alkaline and acidic fluxes.

Alkaline fluxes are used more often than acidic fluxes. For the decomposition of silicates, fusion with anhydrous sodium carbonate is useful. The carbonate ion at high temperatures is a good source of the oxide ion; carbon dioxide is liberated in the process. The total reaction of an acid-insoluble aluminum metasilicate on fusion with sodium carbonate may be written

$$Al_2(SiO_3)_3 + 4Na_2CO_3 \rightarrow 3Na_2SiO_3 + 2NaAlO_2 + 4CO_2 \uparrow$$

The sodium metasilicate and sodium metaaluminate formed are soluble in water.

A mixture of potassium and sodium carbonate offers a flux melting lower than either carbonate alone. Other common alkaline fluxes include borax (sodium tetraborate), sodium or potassium hydroxide, and lithium metaborate.

Alkaline oxidizing fluxes convert metals to their higher oxidation states and thereby oxides that are more acidic and more readily undergo acid-base reactions with the alkaline component of the flux. Of such fluxes, sodium peroxide, Na_2O_2, is the most important. Milder oxidizing fluxes result from mixing sodium carbonate with sodium or potassium nitrate or chlorate.

Of acidic fluxes, anhydrous potassium pyrosulfate is the most common. Either the available reagent-grade form of this salt is used or it is prepared by heating potassium hydrogen sulfate until volatilization of water ceases:

$$2KHSO_4 \rightarrow K_2S_2O_7 + H_2O \uparrow$$

The pyrosulfate ion is a source of sulfur trioxide:

$$S_2O_7^{2-} \rightarrow SO_4^{2-} + SO_3$$

So that too much oxide is not lost by volatilization, the temperature should be just high enough to assure fusion of the salt, and the crucible should be covered.

The sulfur trioxide acts as an electron-pair acceptor (Lewis acid) in its reaction with metal oxides, including those of titanium, iron, chromium, and aluminum. In this way, soluble metal sulfates are formed.

Reducing fluxes receive special use. A flux of sodium carbonate and sulfur converts arsenic, antimony, and tin to water-soluble thiosalts, such as Na_3AsS_4, Na_3SbS_4, and Na_2SnS_3.

The material of the crucible used in the fusion should ideally not be subject to attack by the flux. With platinum, loss of an expensive metal is one reason. From the point of view of chemical analysis, introducing contaminants is more serious. Since fluxes generally act more vigorously at higher temperatures, it is beneficial to conduct the fusion at the lowest possible temperature at which the sample is attacked but the crucible left unaffected. Frequently, however, attack of the crucible is tolerable if the contaminants thus introduced are of no consequence in the determinations of interest. For example, an iron crucible is often used for the sodium peroxide fusion of chrome ores. The crucible is attacked severely, but the iron thereby introduced does not interfere in the subsequent determination of chromate by a redox titration. When the same ores are analyzed for their iron content, nickel crucibles can be used, since nickel does not interfere in the determination of iron.

For fusions with sodium carbonate, crucibles made of platinum are common, although zirconium is receiving increasing attention as a suitable material. For fusions with sodium hydroxide, potassium hydroxide, or sodium peroxide, using platinum is not feasible since it is severely attacked, unless the fusion is performed with strict control of the temperature in an electric oven. Nickel, silver, gold, and zirconium crucibles are appropriate for fusions with sodium or potassium hydroxide. Sodium peroxide fusions are best performed in zirconium crucibles, but nickel or iron ones may be serviceable. For fusions with potassium pyrosulfate, although platinum crucibles are appropriate, they are subject to attack if heated to too high a temperature. For this fusion, vitreous silica is a suitable material. Porcelain crucibles are severely attacked in alkaline fusions but may be used if the contaminants thus introduced do not interfere in the intended analysis.

52.3 *High-Temperature Attack by Halogens*

Some materials, notably sulfides, are attacked by a stream of chlorine or bromine gas at a high temperature. Sulfide is thereby converted to sulfur halides. Metals form either fused salts, or in the case of tin, antimony, arsenic, and certain other metals, are volatilized as their halides and may be absorbed in an appropriate solution and then determined.

52.4 *Attack of the Sample by Pyrohydrolysis*

When superheated steam is passed over a finely divided sample held in a tube within a furnace at a suitable temperature in the range 500 to 1300°C, a few elements can be volatilized, notably fluorine, chlorine, and boron. The hydrofluoric, hydrochloric, or boric acid formed is condensed with the steam and determined by conventional procedures. Pyrohydrolysis affords a rapid separation of fluorine from many samples that present difficulties to attack by acids or fusion or that contain elements that complicate the determination of fluorine.

52.5 *Questions*

52-1 Explain or elaborate on the terms *flux, oxidizing flux, nonoxidizing* versus *oxidizing acids, pyrohydrolysis,* and *aqua regia.*

52-2 Methods for attacking a sample are sometimes classified as either wet or dry. List the approaches in this chapter in terms of these two categories and elaborate.

52-3 State some factors that might deserve consideration in picking the method of attack for an inorganic material.

52–4 What difficulties might result in the fusion of an inorganic sample with sodium carbonate in a porcelain crucible? Of a silicate sample?

52–5 Iron(III) oxide on fusion with potassium pyrosulfate yields iron(III) sulfate and potassium sulfate. Write a balanced total reaction for this fusion. Elaborate on the possible acid-base reaction involved.

52–6 Magnesium mica (phlogopite), used in sheets for electrical insulation, has the approximate composition $KH_2Mg_3Al(SiO_4)_3$. When the ground mica sample is fused with sodium carbonate, the products remaining in the crucible are magnesium carbonate, potassium metaaluminate, and sodium metasilicate. The cooled melt is treated with dilute hydrochloric acid. Carbon dioxide is thereby evolved and silica precipitated. Write balanced equations for the reactions taking place during this attack of the mica sample and the subsequent acid treatment.

53

Classification of Analysis and Selection of Methods

At the beginning of this study, we did not try to classify analytical chemistry except as qualitative and quantitative analysis; and we commented only briefly on the distinction between a determination and an analysis. You are now in a position to view the subject with some perspective and to appreciate some general remarks on the various standpoints from which quantitative analysis may be examined.

53.1 *Organic versus Inorganic Analysis*

Inorganic analysis has been our central topic. Organic analysis, to which the same principles apply, is concerned with the detection, identification, and determination of organic compounds. Elemental analysis of organic samples for carbon, hydrogen, and nitrogen by combustion methods (Section 49.6) or for nitrogen via Kjeldahl digestion (Section 14.9) is very important in organic chemical research.

53.2 *Scale of Operations*

Analytical procedures can be differentiated according to the absolute amount of sample taken and the relative amount of the constituent or constituents to be determined. For sample size, that is, the scale of the analysis, the following classification is often used: macro analysis, 0.1 g or greater; semimicro analysis, 0.01 to 0.1 g; micro analysis, 0.001 to 0.01 g; and ultramicro (or submicro) analysis, less than 0.001 g. This classification scheme is a loose one, and since the limits of each category are really orders of magnitude rather than absolute numbers, the limits assigned by different workers will vary. This scheme has merit, however, because the techniques and methods that are best adopted may depend on the scale of analysis. For example, filtration by gravity is hardly possible in micro and ultramicro analysis, and the separation of a precipitate is accomplished either by filtration under suction or by centrifugation; in addition, the smaller the analytical sample, the greater the sensitivity needed for the balance used in weighing.

The scale of analysis selected depends on various factors, including the amount of material available, the constituents present and their relative amounts, the expense of reagents and the efforts required in their preparation, the instruments available in the case of physicochemical methods, the frequency with which the determinations are made, and the time available. Some operations may be effected more rapidly and more expeditiously with small samples, for example, the washing and drying of a precipitate.

The above statements are directed to the scale of analysis, that is, to the sample size. The relative amount of a constituent of a sample is also important, and the following classification is appropriate: major constituents, 1 to 100% of the sample, and minor constituents, 0.01 to 1%, which together may be termed macro constituents, and constituents less than 0.01% of the sample, which are spoken of as trace (or micro) constituents. Again it is stressed that the classification is loose, and based on orders of magnitude. Often a trace constituent is determined using a sample of the order of 10 g. Then one speaks of macro *analysis* with the particular trace *determination* performed by a micro or an ultramicro *technique.*

53.3 Determination versus Analysis

The distinction between a determination and an analysis has been briefly drawn in Chapter 1. Applications of the principles of quantitative analysis introduced in this textbook and by experiments performed by the student in parallel laboratory work are largely simple determinations. Usually a single constituent is determined under uncomplicated circumstances; no extensive separation of interfering substances is required and the sample is readily brought into solution. In a practical analysis, the situation is often far more complicated; establishment of a representative sample may present difficulties, many constituents are present, interference must be considered and resolved, the material may be refractory to dissolution, and more than one constituent must be determined.

53.4 Partial Analysis versus Complete Analysis

Sometimes only a few constituents of a complex material are of interest and need to be determined. This is known as partial analysis. In contrast, complete analysis ends only when all the constituents, as detected by suitable sensitive qualitative tests and methods, have been determined.

53.5 Proximate, Elemental, and Ultimate Analysis

Frequently two or more constituents show similar behavior in an analytical procedure and are determined together. This is known as proximate analysis. In the analysis of glass, for example, potassium and sodium may be weighed together as their sulfate salts and reported as total alkalis expressed as % Na_2O. Similarly, in the analysis of rocks and ores, the sum of the oxides of iron and aluminum may be determined together and expressed as % R_2O_3 (Section 7.2). The percent loss on ignition is usually proximate, because it includes moisture, carbonate, organic matter, and any other volatile substances present in the material. Other common proximate methods include the determination of the residue after evaporation (for volatile liquids), the determination of insoluble matter (for materials soluble in water or a designated solvent), and the "heavy metals" test (see Section 35.5 and Problem 35-21).

Determining the percentages of various elements present in a material is known as elemental analysis. The elemental analysis of a single substance may be termed *ultimate analysis.*

53.6 Nondestructive versus Destructive Analysis

Nondestructive analysis involves establishing the nature and content of the constituents of a material and at the finish having the sample present in its initial composition and amount. Such an ideal is closely approximated by some physical methods, notably methods based on the use of X rays or radioactivity. When emission spectrography involves only a minute portion of a specimen, it is also appropriately termed *nondestructive.* However, nondestructive methods usually fail to provide all the answers sought by analysis; consequently, additional methods that are more or less destructive must be relied on.

53.7 Manual Analysis versus Automated Analysis

Throughout this volume the implication has been largely that determinations are effected with close "manual" involvement by the analyst. The manual approach, however, in many situations is unsatisfactory and interest in automating the analysis of real samples is increasing progressively. The degree of automation may run from one or two steps of a procedure to total handling (even to establishing the sample and recording the final results). Reasons for introducing automated methods include saving of time, reduction of cost, possibilities for reducing human error, and the safe handling of radioactive and toxic materials. Finally, automation permits the continuous monitoring of a chemical system.

Automated analysis can be used to determine a single species or several concurrently. The clinical laboratory offers an instructive example. The manual determination of 12 different constituents in a single blood sample takes a long time even when many analysts are cooperating. In contrast, an automated analyzer using 12 "channels" can routinely determine these 12 constituents in 60 different blood samples within one hour. Such units save time and reduce costs. More importantly, the rapid turnaround of blood samples by the laboratory can aid in early diagnosis and the prompt start or modification of treatment.

Conceptually, the simplest type of automated analysis is the "transducer," mounted directly in a flowing stream of material or exposed to portions bled from that stream. Such transducers include a pH or other ion-selective electrode or a "flowthrough" cell mounted in a photometer.

Two extremes of automated analysis can be differentiated. The unit, much like a robot, merely simulates one or more steps that the human operator performs. For example, a unit can weigh a food sample placed in a tared vessel, pass it first to an oven and then to a cooling zone, and finally make a second weighing to establish the loss on drying. This approach fails to take full advantage of the potential for automation. The other extreme is illustrated by the blood analyzer just mentioned. Microliter portions of blood samples are sucked into plastic tubing in such a way that each sample drop is separated from the preceding one by an air bubble. The drops and bubbles are advanced through the tubing by a peristaltic pump. Chromogenic agents, buffer solutions, and the like can be added to the drops from joined side tubes. Heating, dialysis, and various other operations can be performed. Finally the sample drop, with color fully developed, passes to a photometer cell. The absorbance is measured as the peak height on a recorder tracing, or the detector signal is sent to a digital converter and the value is printed in the desired concentration unit.

An important aspect of any automated analytical system is the use of appropriate standards at appropriate intervals to ensure that the system is working properly and is correctly calibrated.

53.8 *Basic Steps in a Determination*

Generally, the determination of a constituent of a material involves four basic steps: (1) establishment of a sample suitable in character and amount; (2) its transfer by physical or chemical operations, or both, to a condition in which some property of the constituent can be measured; (3) the measurement; and (4) expression of the value in terms of the content of the constituent of interest by stoichiometric relations, physical laws, standards, and so on. Obviously the relative attention given to a particular step will vary from method to method, and two or more steps may overlap. You will benefit by examining the performance of typical gravimetric, titrimetric, and photometric determinations from the standpoint of these four basic steps. In these three types of analytical methods, the measurements establish the weight (that is, mass), volume, and absorption of light, respectively. These measurements are actually physical in principle. Chemistry enters when we transfer the sample to the form used in the measurement, when we interpret the result, and sometimes when we seek an understanding of the underlying phenomena.

53.9 Classification by Methods and Techniques

The study and practice of quantitative analysis is often conveniently divided in two broad categories. Full appreciation of the labels variously assigned to these categories requires appreciating many diverse methods. The first category, into which gravimetric and titrimetric methods are conventionally placed, is variously described as noninstrumental methods, classical analysis, chemical methods, or methods involving static properties. The second category, in which electrical and optical methods of analysis are the principal subjects, is variously labeled instrumental methods, physical or physicochemical methods, or methods involving dynamic properties. The implications of these labels will become clear from a brief discussion.

Gravimetric and titrimetric methods can be thought of as noninstrumental only by arbitrarily excluding the balance and the buret as instruments; hence this label is best retired from use. It is a carryover from earlier decades when these two types of methods were the principal ones at the command of the analytical chemist and electrical and optical instruments were recent in their introduction and still rare in practical analysis. This also explains why these two types of methods, along with gas-analysis procedures involving the measurement of pressures or volumes (Chapter 50), are frequently termed classical analysis. Both gravimetric and titrimetric determinations involve either a chemical (or an electrochemical) separation, metric determinations involve either a chemical or an electrochemical separation, and titration involves a chemical reaction proceeding virtually to completion. Static properties are possessed by a system itself and basically do not involve the passage of energy into, through, or out of an analytical system for external measurement. Common static properties include mass, volume, density, pressure, and chemical reactivity.

Electrical and optical methods can be thought of as instrumental ones, but the description is not recommended. If a crystal is placed in a light beam and the absorbance measured, this is clearly a physical procedure. In contrast, if the crystal is caused to react with an acid and the absorbance of the resulting solution is measured, the complete procedure is appropriately described as physicochemical. Dynamic properties are exhibited when energy (or particles) pass into, through, or out of an analytical system. Methods dependent on optical, electrical, and radioactive phenomena all involve dynamic properties.

Although potentiometric, amperometric, and photometric titrations are usually treated in the context of physicochemical methods, as we have done, they are still chemical methods, even though the end point is detected by a physicochemical technique. Electrogravimetric analysis involves an electrochemical separation, but a static property, mass, is still measured as in conventional gravimetric determinations.

53.10 Classification According to Application

It is also possible to examine analytical principles, methods, and techniques from the standpoint of their application to a restricted field, for example, metallurgical analysis or clinical analysis. Another possibility is restriction to a group of related substances or to a single important product of commerce, for example, steel analysis, cement analysis, and the analysis of copper metal. In all such treatments, the principles are the same, but the emphasis given particular topics differs.

53.11 *Selecting an Analytical Method*

The most critical question raised by the analytical chemist facing a new analytical problem is what method or methods should be selected. An adequate answer requires much knowledge and good judgment, usually gained only by long experience, and may involve some trials, often with prepared samples somewhat similar to the unknown. Such experience is not expected of the student and the nonspecialist in analytical chemistry, and, although they will not be faced immediately with choosing techniques, some remarks may be of general interest.

The scale of analysis selected depends on various factors, including the amount of material available and the value assigned to it. In practical analysis, time is often very important. The best analysis is wasted if the data become available only when they are no longer useful. This consideration applies, for example, when analyses are used to control chemical processing and metallurgical operations or to aid in diagnosing clinical conditions. In such circumstances, it is frequently necessary to curtail accuracy for speed.

Interferences must be considered in selecting a method; indeed, other factors being equal, a *selective* method may be preferred to a more *sensitive* one. Often we must assess how accurately and how precisely a result must be known to be useful. The attempt to achieve accuracy and precision over what is needed for the immediate determination or analysis may be costly in time and labor.

Sometimes selecting a method is based on other considerations besides establishing the optimum one. In many areas of analysis, so-called standard methods (actually standard procedures) are recognized either by their development or acceptance by professional societies and groups, by the force of purchase contracts, or by the requirements of industry or governmental codes. For example, in various countries, so-called pharmacopoeias provide the methods required in the identification and analysis of officially recognized drugs. Often standard methods are so devised as to avoid the use of equipment and instruments that are not readily available in modest laboratories. By their nature, such methods may lag in the adoption of the latest developments. However, they are usually presented in great detail and with attention given to even minor operations; consequently, regardless of the accuracy, the results obtained by different analysts usually agree favorably.

53.12 *Questions*

53-1 For liquid samples, analyses are sometimes placed on a volume basis, and so that the scale of analysis is fixed, 1 mL is taken as equal to 1 g. In volume units, what would be the approximate limits for macro, semimicro, micro, and ultramicro analysis? If a volume of 200 μL of blood is taken for a clinical analysis, what is the scale of analysis?

53-2 Is a component of 1 part per 1000 in a material a micro or a macro constituent?

53-3 Analytical methods are sometimes classified as "wet" or "dry," depending on whether or not a solution of the sample is a necessary prelude to or condition for the measurement. Namc some methods and determinations that might fall in these two categories.

53-4 Physicochemical methods are sometimes stated to depend on the measurement of a mass-dependent property rather than mass itself. Comment on the implications of this statement.

53-5 A material containing loosely held water is heated at 110°C to constant weight for establishing the percentage of water present. Is this a determination or an analysis? a proximate procedure? a chemical method? a nondestructive method? Elaborate your answers.

53-6 A nondestructive method of analysis provides the desired result and leaves the sample identical in weight and composition and available for further analysis or use. Elaborate on the extent to which various methods approach this ideal.

53-7 Some methods of analysis yield as the primary result the *amount* of the sought-for species and others, the *concentration* of that species in the final analytical preparation. Place in these categories various methods of analysis you are familiar with. For each method, elaborate on the information or calculations usually needed to express the result as the *content* of the sought-for species in the sample material.

Appendix

The data appearing in the following tables have been selected from various compilations and research publications. Principal sources include the following. The tables to which they pertain are given in brackets.

Sillén, L. G., and A. E. Martell. *Stability Constants of Metal-Ion Complexes.* Special Publication No. 17. London: Chemical Society, 1964. [Tables A, B, C, and D]

Kortüm, G., W. Vogel, and K. Andrussow. Dissociation constants of organic acids in aqueous solution. *Pure and Applied Chemistry* 1: 187–536 (1960). [Tables A and B]

Feitknecht, W., and P. Schindler. Solubility constants of metal oxides, metal hydroxides . . . in aqueous solution. *Pure and Applied Chemistry* 6: 130–199 (1963). [Table C]

Yatsimirskiĭ, K. Y., and V. P. Vasil'ev. *Instability Constants of Complex Compounds.* Translated by D. A. Paterson. Oxford: Pergamon Press, 1960. [Tables C and D]

ANSI/ASTM Standard E70–77 (*1977 Annual Book of ASTM Standards*); cf. Bates, R. G. *Determination of pH, Theory and Practice.* New York: Wiley, 2nd ed., 1973. Chapter 4. [Table E]

Charlot, G., D. Bezier, and J. Courtot. *Selected Constants, Oxydo-Reduction Potentials.* Oxford: Pergamon Press, 1958. [Table G]

Bishop, E. In *Comprehensive Analytical Chemistry,* eds. C. L. Wilson and D. W. Wilson. Amsterdam: Elsevier Publishing Company, 1960. Volume IB, pp. 151–184. [Tables A, B, and H]

Hickling, A., and F. W. Salt, *Transactions of the Faraday Society* 36: 1226 (1940). [Table I]

Page, C. H., and P. Vigoureux, eds. *The International System of Units (SI).* National Bureau of Standards Special Publication 330, 1972 ed. Washington: U.S. Government Printing Office, April 1972. 42 pp. [Table K]

Table A Dissociation Constants of Some Acids in Water at 25°C

Acid		K_a	*Acid*		K_a
Acetic	K_1	1.8×10^{-5}	Hypochlorous	K_1	2.8×10^{-8}
Arsenic	K_1	5.6×10^{-3}	Iodic	K_1	1.8×10^{-1}
	K_2	1.2×10^{-7}	Nitrous	K_1	5×10^{-4}
	K_3	3×10^{-12}	Oxalic	K_1	5.4×10^{-2}
Arsenious	K_1	1.4×10^{-9}		K_2	5.1×10^{-5}
Benzoic	K_1	6.3×10^{-5}	Phenol	K_1	1.1×10^{-10}
Boric	K_1	5.9×10^{-10}	Phosphoric	K_1	7.1×10^{-3}
Carbonic	K_1[a]	4.5×10^{-7}	(ortho)	K_2	6.3×10^{-8}
	K_2	5.6×10^{-11}		K_3	4.4×10^{-13}
Chloroacetic	K_1	1.4×10^{-3}	*o*-Phthalic	K_1	1.1×10^{-3}
Chromic	K_2	3×10^{-7}		K_2	3.9×10^{-6}
Citric	K_1	7.4×10^{-4}	Salicylic	K_1	1.0×10^{-3}
	K_2	1.7×10^{-5}		K_2	4×10^{-14}
	K_3	3.9×10^{-7}	Sulfamic	K_1	1.0×10^{-1}
Ethylene dinitrilo-	K_1	1×10^{-2}	Sulfuric	K_1	1.1×10^{-2}
tetraacetic	K_2	2.1×10^{-3}	Sulfurous	K_1	1.7×10^{-2}
	K_3	6.9×10^{-7}		K_2	6.3×10^{-8}
	K_4	7.4×10^{-11}	Tartaric	K_1	9.2×10^{-4}
Formic	K_1	1.8×10^{-4}		K_2	4.3×10^{-5}
Hydrocyanic	K_1	5×10^{-10}	Thiocyanic	K_1	1.4×10^{-1}
Hydrofluoric	K_1	6×10^{-4}			
Hydrogen sulfide	K_1	1.0×10^{-8}			
	K_2	1.2×10^{-14}			

[a] Apparent constant based on $C_{H_2CO_3} = [CO_2] + [H_2CO_3]$.

Table B Dissociation Constants of Some Bases in Water at 25°C

Base		K_b	*Base*		K_b
2-Amino-2-(hydroxymethyl)-1,3-propanediol	K_1	1.2×10^{-6}	Hydrazine	K_1	9.8×10^{-7}
			Hydroxylamine	K_1	9.6×10^{-9}
			Lead hydroxide	K_1	1.2×10^{-4}
Ammonia	K_1	1.8×10^{-5}	Piperidine	K_1	1.3×10^{-3}
Aniline	K_1	4.2×10^{-10}	Pyridine	K_1	1.5×10^{-9}
Diethylamine	K_1	1.3×10^{-3}	Silver hydroxide	K_1	6.0×10^{-5}
Hexamethylene-tetramine	K_1	1×10^{-9}			

Table C Solubility Products of Some Common Electrolytes in Water at 25°C

Substance	*Formula*	K_{sp}
Aluminum hydroxide (amorphous)	$Al(OH)_3$	6×10^{-32}
Barium carbonate	$BaCO_3$	5.5×10^{-10}
Barium chromate	$BaCrO_4$	1.2×10^{-10}
Barium fluoride	BaF_2	1.0×10^{-6}
Barium iodate	$Ba(IO_3)_2$	1.5×10^{-9}
Barium manganate(VI)	$BaMnO_4$	2.5×10^{-10}
Barium oxalate	BaC_2O_4	1.7×10^{-7}
Barium sulfate	$BaSO_4$	1.3×10^{-10}
Bismuth sulfide	Bi_2S_3	$\sim 10^{-97}$
Cadmium sulfide	CdS	$\sim 10^{-27}$
Calcium carbonate	$CaCO_3$	4.8×10^{-9}
Calcium fluoride	CaF_2	4×10^{-11}
Calcium hydroxide	$Ca(OH)_2$	3.7×10^{-6}
Calcium oxalate	CaC_2O_4	2.3×10^{-9}
Calcium phosphate	$Ca_3(PO_4)_2$	$\sim 10^{-26}$
Calcium sulfate	$CaSO_4$	1.2×10^{-6}
Chromium(III) hydroxide	$Cr(OH)_3$	$\sim 10^{-30}$
Cobalt(II) hydroxide (pink, inactive)	$Co(OH)_2$	4×10^{-16}
Cobalt(III) hydroxide	$Co(OH)_3$	$\sim 10^{-43}$
Cobalt(II) sulfide	CoS	$\sim 10^{-23}$
Copper(I) chloride	$CuCl$	1.2×10^{-6}
Copper(II) hydroxide (inactive)	$Cu(OH)_2$	2×10^{-19}
Copper(I) iodide	CuI	5.0×10^{-12}
Copper(I) sulfide	Cu_2S	$\sim 10^{-48}$
Copper(II) sulfide	CuS	$\sim 10^{-36}$
Copper(I) thiocyanate	$CuSCN$	1.9×10^{-13}[a]
Iron(II) hydroxide (inactive)	$Fe(OH)_2$	8×10^{-6}
Iron(III) hydroxide (amorphous, inactive)	$Fe(OH)_3$	8×10^{-40}
Iron(II) sulfide	FeS	$\sim 10^{-19}$
Lead carbonate	$PbCO_3$	1×10^{-13}[b]
Lead chloride	$PbCl_2$	1.6×10^{-5}
Lead chromate	$PbCrO_4$	1.8×10^{-14}[b]
Lead fluoride	PbF_2	2.7×10^{-8}
Lead hydroxide	$Pb(OH)_2$	$\sim 10^{-20}$
Lead iodide	PbI_2	7.1×10^{-9}
Lead sulfate	$PbSO_4$	1.7×10^{-8}
Lead sulfide	PbS	2.5×10^{-27}
Magnesium ammonium sulfate	$MgNH_4PO_4$	3×10^{-13}
Magnesium carbonate	$MgCO_3$	1×10^{-5}
Magnesium fluoride	MgF_2	6.6×10^{-9}
Magnesium hydroxide	$Mg(OH)_2$	1×10^{-11}
Manganese(II) carbonate	$MnCO_3$	1.8×10^{-11}
Manganese(II) hydroxide	$Mn(OH)_2$	1×10^{-13}
Manganese(II) sulfide	MnS	$\sim 10^{-13}$
Mercury(I) chloride	Hg_2Cl_2	1.3×10^{-18}
Mercury(I) chromate	Hg_2CrO_4	2.0×10^{-9}

Table C (continued)

Substance	*Formula*	K_{sp}
Mercury(II) hydroxide	$Hg(OH)_2$	$\sim 10^{-26}$
Mercury(I) iodide	Hg_2I_2	4.9×10^{-29}
Mercury(II) sulfate	Hg_2SO_4	6.8×10^{-7}
Mercury(I) sulfide	Hg_2S	$\sim 10^{-47}$
Mercury(II) sulfide (black)	HgS	$\sim 10^{-52}$
Nickel hydroxide (inactive)	$Ni(OH)_2$	6×10^{-18}
Nickel sulfide	NiS	$\sim 10^{-21}$
Potassium tetraphenylboron	$KB(C_6H_5)_4$	3.2×10^{-8}
Silver arsenate	Ag_3AsO_4	1×10^{-22}[a]
Silver bromide	$AgBr$	5.2×10^{-13}
Silver carbonate	Ag_2CO_3	8.1×10^{-12}
Silver chloride	$AgCl$	1.8×10^{-10}
Silver chromate	Ag_2CrO_4	1.5×10^{-12}
Silver cyanide	$AgCN$	2.3×10^{-16}
Silver dicyanoargentate	$Ag[Ag(CN)_2]$	1.3×10^{-12}
Silver hydroxide	$AgOH$	2.0×10^{-8}
Silver iodide	AgI	8.3×10^{-17}
Silver sulfate	Ag_2SO_4	2.1×10^{-5}
Silver sulfide	AgS	6×10^{-50}
Silver thiocyanate	$AgSCN$	1×10^{-12}
Strontium carbonate	$SrCO_3$	1.1×10^{-10}
Strontium chromate	$SrCrO_4$	3.6×10^{-5}
Strontium fluoride	SrF_2	2.5×10^{-9}
Strontium sulfate	$SrSO_4$	3.2×10^{-7}
Zinc carbonate	$ZnCO_3$	1.4×10^{-11}
Zinc hexacyanoferrate(II)	$Zn_2Fe(CN)_6$	4.1×10^{-16}
Zinc hydroxide (amorphous)	$Zn(OH)_2$	2.5×10^{-16}
Zinc sulfide	ZnS	$\sim 10^{-24}$

[a] At 20°C.
[b] At 18°C.

Table D Stability Constants of Some Metal Ion Complexes in Water

Unless it is otherwise indicated, the values are for 25°C and ionic strength μ of 0.1. The value given to the right of the formula of a complex is the *overall* stability constant of that complex. Where known, the *stepwise* stability constants are given as log K values beneath the formula.

Ammonia	
$Ag(NH_2)_2^+$	1.7×10^7
3.31, 3.91	
$Cd(NH_3)_4^{2+}$	5.8×10^6
2.56, 2.01, 1.35, 0.84	

Table D (continued)

Ammonia (contd.)	
$Cu(NH_3)_4^{2+}$	2.0×10^{12}
4.06, 3.41, 2.80, 2.04	
$Hg(NH_3)_4^{2+}$	2×10^{19}
8.8, 8.7, 1.0, 0.8	
$Ni(NH_3)_4^{2+}$	4.2×10^{7}
2.71, 2.16, 1.64, 1.11	
$Zn(NH_3)_4^{2+}$	8.0×10^{8}
2.23, 2.30, 2.36, 2.01	
Chloride	
$AgCl_4^{3-}$ ($\mu = 0.2$)	7.9×10^{5}
2.85, 1.87, 0.32, 0.86	
$HgCl_4^{2-}$ ($\mu = 0.5$)	1.2×10^{15}
6.74, 6.48, 0.85, 1.05	
Iodide	
CdI_4^{2-}	2.3×10^{5}
2.40, 1.26, 1.0, 0.7	
HgI_4^{2-} ($\mu = 0.5$)	7.2×10^{29}
12.87, 10.95, 3.67, 2.37	
Thiocyanate	
$Fe(SCN)_2^{+}$ ($\mu = 0.5$)	2.8×10^{3}
2.14, 1.31	
Cyanide	
$Ag(CN)_2^{-}$	1.1×10^{21}
$Cu(CN)_2^{-}$	1×10^{16}
$Fe(CN)_6^{4-}$	1×10^{24}
$Fe(CN)_6^{3-}$	1×10^{31}
$Hg(CN)_4^{2-}$	3.2×10^{41}
18.0, 16.70, 3.83, 2.98	
$Ni(CN)_4^{2-}$	1×10^{22}
Thiosulfate	
$Ag(S_2O_3)_2^{3-}$ ($\mu = 4$)	5.2×10^{12}
7.36, 5.36	
Ethylenedinitrilotetraacetate (= Y^{4-}) (values at 20°C)	
CaY^{2-}	5.0×10^{10}
CdY^{2-}	4×10^{16}
CuY^{2-}	6.3×10^{18}
FeY^{2-}	2×10^{14}
FeY^{-}	1×10^{25}
MgY^{2-}	4.9×10^{8}
ZnY^{2-}	3.2×10^{16}
Eriochrome Black T (= D^{3-}) (values at 20°C)	
CaD^{-}	1.9×10^{5}
MgD^{-}	7.4×10^{6}
ZnD^{-}	2×10^{12}

Table E Recommended Standard Values of pH_s for Six Standard Buffer Solutions

	Solution					
	A	B	C	D	E	F
°*C*	*Citrate*	*Phthalate*	*Phosphate*	*Phosphate*	*Borate*	*Bicarbonate*
10	3.863	4.003	6.984	7.534	9.464	10.317
10	3.820	3.998	6.923	7.472	9.332	10.179
20	3.788	4.002	6.881	7.429	9.225	10.062
25	3.776	4.008	6.865	7.413	9.180	10.012
30	3.766	4.015	6.853	7.400	9.139	9.966
35	3.759	4.024	6.844	7.389	9.102	9.925
40	3.753	4.035	6.838	7.380	9.068	9.889
50	3.749	4.060	6.833	7.367	9.011	9.828
60	...	4.091	6.836	...	8.962	...
70	...	4.126	6.845	...	8.921	...
80	...	4.164	6.859	...	8.885	...
90	...	4.205	6.877	...	8.850	...

Composition of the standard buffer solutions (25 °C):

A. 0.05 mol•kg^{-1} solution of KH_2 citrate. (Dissolve and dilute with distilled water to 1000 mL 11.41 g of potassium dihydrogen citrate, dried for 1 hr at 80°C.)

B. 0.05 mol•kg^{-1} solution of KH phthalate. (Dissolve and dilute with distilled water to 1000 mL 10.12 g of potassium hydrogen phthalate, dried for 1 hr at 110°C.)

C. Solution 0.025 mol•kg^{-1} in KH_2PO_4 and 0.025 mol•kg^{-1} in Na_2HPO_4. (Dissolve and dilute with distilled water to 1000 mL 3.388 g of potassium dihydrogen phosphate and 3.533 g of disodium hydrogen phosphate, both salts dried for 1 hr at 110°C.)

D. Solution 0.008695 mol•kg^{-1} in KH_2PO_4 -nd 0.03043 mol•kg^{-1} in Na_2HPO_4. (Dissolve and dilute with distilled water to 1000 mL 1.179 g of potassium dihydrogen phosphate and 4.302 g of disodium hydrogen phosphate, both salts dried for 1 hr at 110°C.)

E. 0.01 mol•kg^{-1} solution of $Na_2B_4O_7$. (Dissolve and dilute with carbon-dioxide-free distilled water to 1000 mL 3.80 g of sodium tetraborate decahydrate (borax), which should not be heated above room temperature.)

F. Solution 0.025 mol•kg^{-1} in $NaHCO_3$ and 0.025 mol•kg^{-1} in Na_2CO_3. (Dissolve and dilute with carbon-dioxide-free distilled water to 1000 mL 2.092 g of sodium bicarbonate, which should not be heated above room temperature, and 2.640 g of sodium carbonate, which should be heated for 1 hr at 270°C just before use.)

The solutions should be stored in polyethylene or chemical-resistant glass bottles and should be discarded if a visible change is noted or at the end of six weeks.

Table F Common Acid-Base Indicators

Common name	pK_{In}	pH visual transition interval	Color[a]	
			Acidic	Basic
Cresol red		0.2–1.8	Red	Yellow
Thymol blue	1.6	1.2–2.8	Red	Yellow
Methyl yellow	3.1	2.4–4.0	Red	Yellow
Bromophenol blue	4.2	3.0–4.6	Yellow	Blue
Methyl orange	3.5	3.2–4.4	Red	Yellow orange
Methyl orange + xylene cyanole FF, 40:56		(3.8–4.1)[b]	Violet	Green
Bromocresol green	4.9	3.9–5.4	Yellow	Blue
Methyl red	5.0	4.2–6.2	Pink	Yellow
Methyl red + methylene blue, 1:1		(~5.3)[b]	Red violet	Green
Chlorophenol red	6.2	4.8–6.4	Yellow	Red
Bromocresol purple	6.4	5.2–6.8	Yellow	Purple
Bromothymol blue	7.3	6.0–7.6	Yellow	Blue
Cresol red	8.4	7.2–8.8	Yellow	Red
Phenol red	8.0	6.8–8.2	Yellow	Red
Thymol blue	9.0	8.0–9.2	Yellow	Blue
Phenolphthalein	(8.7)	(8.0–9.8)[c]	Colorless	Red violet
Phenolphthalein + methylene green, 1:2		(8.8)[d]	Green	Violet
Thymolphthalein	(9.2)	(9.0–10.5)[c]	Colorless	Blue

[a]Colors in aqueous solution at lower and upper pH limits of the visual transition interval, respectively.

[b]Screened indicator, neutral gray at stated pH.

[c]Based on addition of 1 or 2 drops of a 0.1% indicator solution to 10 mL of aqueous solution.

[d]Screened indicator, pale blue at stated pH.

Table G Standard and Formal Electrode Potentials at 25°C (standard potentials are shown in boldface.)

Half-reaction equation	E^0 *or* E^f, *V*	*Solution conditions for formal potentials*
$S_2O_8^{2-} + 2e \rightleftharpoons 2SO_4^{2-}$	**+2.0**	
$H_2O_2 + 2H^+ + 2e \rightleftharpoons 2H_2O$	**+1.77**	
$MnO_4^- + 4H^+ + 3e \rightleftharpoons \underline{MnO_2} + 2H_2O$	**+1.69**	
	+1.70	$1F$ $HClO_4$
$Ce^{4+} + e \rightleftharpoons Ce^{3+}$	+1.60	$1F$ HNO_3
	+1.44	$1F$ H_2SO_4
	+1.28	$1F$ HCl
$NaBiO_3 + 6H^+ + 2e \rightleftharpoons Na^+ + Bi^{3+} + 3H_2O$	**~+1.6**	
$MnO_4^- + 8H^+ + 5e \rightleftharpoons Mn^{2+} + 4H_2O$	**+1.51**	
$2BrO_3^- + 12H^+ + 10e \rightleftharpoons Br_2 + 6H_2O$	**~+1.5**	
$\underline{PbO_2} + 4H^+ + 2e \rightleftharpoons Pb^{2+} + 2H_2O$	**+1.46**	
$Cl_2 + 2e \rightleftharpoons 2Cl^-$	**+1.359**	
$Cr_2O_7^{2-} + 14H^+ + 6e \rightleftharpoons 2Cr^{3+} + 7H_2O$	**+1.33**	
	+1.03	$1F$ $HClO_4$
	+1.00	$1F$ HCl
	+0.92	$0.1F$ H_2SO_4
$Tl^{3+} + 2e \rightleftharpoons Tl^+$	**+1.28**	
	+0.78	$1F$ HCl
$\underline{MnO_2} + 4H^+ + 2e \rightleftharpoons Mn^{2+} + 2H_2O$	**+1.23**	
$O_2(g) + 4H^+ + 4e \rightleftharpoons 2H_2O$	**+1.229**	
$ClO_4^- + 2H^+ + 2e \rightleftharpoons ClO_3^- + H_2O$	**+1.19**	
$2IO_3^- + 12H^+ + 10e \rightleftharpoons \underline{I_2} + 6H_2O$	**+1.19**	
$Br_2 + 2e \rightleftharpoons 2Br^-$	**+1.087**	
$VO_2^+ + 2H^+ + e \rightleftharpoons VO^{2+} + H_2O$	**+0.9994**	
$HNO_2 + H^+ + e \rightleftharpoons NO(g) + H_2O$	**+0.99**	
$NO_3^- + 3H^+ + 2e \rightleftharpoons HNO_2 + H_2O$	**+0.94**	
	+0.92	$1F$ HNO_3
$2Hg^{2+} + 2e \rightleftharpoons Hg_2^{2+}$	**+0.907**	
$Cu^{2+} + I^- + e \rightleftharpoons \underline{CuI}$	**+0.86**	
$Ag^+ + e \rightleftharpoons Ag$	**+0.7994**	
$Hg_2^{2+} + 2e \rightleftharpoons 2Hg$	**+0.792**	
$Fe^{3+} + e \rightleftharpoons Fe^{2+}$	**+0.771**	
	+0.75	$1F$ $HClO_4$
	+0.73	$1F$ HNO_3
	+0.71	$0.5F$ HCl
	+0.70	$1F$ HCl
	+0.68	$1F$ H_2SO_4
	+0.64	$5F$ HCl
	+0.53	$10F$ HCl
	+0.46	$2F$ H_3PO_4
Benzoquinone + $2H^+ + e \rightleftharpoons$ hydroquinone	**+0.6994**	
	+0.696	$1F$ HCl

Table G (continued)

Half-reaction equation	E^0 *or* E^f, *V*	*Solution conditions for formal potentials*
$O_2(g) + 2H^+ + 2e \rightleftharpoons H_2O_2$	**+0.69**	
$MnO_4^- + e \rightleftharpoons MnO_4^{2-}$	**+0.6**	
$MnO_4^- + 2H_2O + 3e \rightleftharpoons \underline{MnO_2} + 4OH^-$	**+0.57**	
$H_3AsO_4 + 2H^+ + 2e \rightleftharpoons H_3AsO_3 + H_2O$	**+0.559**	
	+0.577	$1F$ HCl
$I_3^- + 2e \rightleftharpoons 3I^-$	**+0.545**	
$\underline{I_2} + 2e \rightleftharpoons 2I^-$	**+0.536**	
$\underline{Ag_2CrO_4} + 2e \rightleftharpoons 2Ag + CrO_4^{2-}$	**+0.447**	
$UO_2^{2+} + 4H^+ + 2e \rightleftharpoons U^{4+} + 2H_2O$	+0.41	$0.5F\ H_2SO_4$
	+0.31	$1F$ HCl
$Fe(CN)_6^{3-} + e \rightleftharpoons Fe(CN)_6^{4-}$	**+0.356**	
	+0.71	$1F$ HCl
$VO^{2+} + 2H^+ + e \rightleftharpoons V^{3+} + H_2O$	**+0.337**	
	+0.360	$1F\ H_2SO_4$
$Cu^{2+} + 2e \rightleftharpoons Cu$	**+0.337**	
$Hg_2Cl_2 + 2e \rightleftharpoons 2Hg + 2Cl^-$	**+0.2680**	
	+0.3337	$0.1F$ KCl
	+0.2801	$1F$ KCl
	+0.2412	Sat'd KCl
$AgCl + e \rightleftharpoons Ag + Cl^-$	**+0.2224**	
$SbO^+ + 2H^+ + 3e \rightleftharpoons Sb + H_2O$	**+0.21**	
$SO_4^{2-} + 4H^+ + 2e \rightleftharpoons H_2\ O_3 + H_2O$	**+0.17**	
$Cu^{2+} + e \rightleftharpoons Cu^+$	**+0.153**	
$Sn^{4+} + 2e \rightleftharpoons Sn^{2+}$	**+0.14**	$1F$ HCl
$S + 2H^+ + 2e \rightleftharpoons H_2S$	**+0.14**	
$\underline{Hg_2Br_2} + 2e \rightleftharpoons 2Hg + 2Br^-$	**+0.1392**	
$TiO^{2+} + 2H^+ + e \rightleftharpoons$	+0.12	$2F\ H_2SO_4$
$Ti^{3+} + H_2O$	-0.01	$0.2F\ H_2SO_4$
$S_4O_6^{2-} + 2e \rightleftharpoons 2S_2O_3^{2-}$	**+0.09**	
$\underline{AgBr} + e \rightleftharpoons Ag + Br^-$	**+0.071**	
$2H^+ + 2e \rightleftharpoons H_2(g)$	**±0.0000**	
$Pb^{2+} + 2e \rightleftharpoons Pb$	**-0.126**	
$Sn^{2+} + 2e \rightleftharpoons Sn$	**-0.140**	
$\underline{AgI} + e \rightleftharpoons Ag + I^-$	**-0.152**	
$Ni^{2+} + 2e \rightleftharpoons Ni$	**-0.23**	
$V^{3+} + e \rightleftharpoons V^{2+}$	**-0.255**	
$Co^{2+} + 2e \rightleftharpoons Co$	**-0.28**	
$Tl^+ + e \rightleftharpoons Tl$	**-0.336**	
$Ti^{3+} + e \rightleftharpoons Ti^{2+}$	**-0.37**	
$Cd^{2+} + 2e \rightleftharpoons Cd$	**-0.402**	
$Cr^{3+} + e \rightleftharpoons Cr^{2+}$	**-0.41**	
	-0.38	$1F$ HCl
$Fe^{2+} + 2e \rightleftharpoons Fe$	**-0.440**	

Table G (continued)

Half-reaction equation	E^0 or E^f, V	Solution conditions for formal potentials
$Cr^{3+} + 3e \rightleftharpoons Cr$	**-0.74**	
$Zn^{2+} + 2e \rightleftharpoons Zn$	**-0.7628**	
$Mn^{2+} + 2e \rightleftharpoons Mn$	**-1.190**	
$Al^{3+} + 3e \rightleftharpoons Al$	**-1.66**	
$Mg^{2+} + 2e \rightleftharpoons Mg$	**-2.37**	
$Na^{+} + e \rightleftharpoons Na$	**-2.713**	
$Ca^{2+} + 2e \rightleftharpoons Ca$	**-2.87**	
$K^{+} + e \rightleftharpoons K$	**-2.925**	
$Li^{+} + e \rightleftharpoons Li$	**-3.03**	

NOTE: Standard potentials are shown in boldface.

Table H Redox Indicators

Common name	E^f, V	Color	
		Oxidized	Reduced
N-Phenylanthranilic acid	+1.08 at pH 0	Red violet	Colorless
1,10-Phenanthroline (iron(II) complex)	+1.06 at pH 0	Pale blue	Red
Xylene cyanole FF	+1.05 at pH 0	Orange	Green
Erioglaucine	+1.00 at pH 0	Orange pink	Green yellow
2,2′-Bipyridine (iron(II) complex)	+0.97 at pH 0	Pale blue	Red
Diphenylamine sulfonic acid (and its salts)	+0.84 at pH 0	Blue violet	Colorless
o-Dianisidine	+0.80 at pH 0	Red	Colorless
Diphenylamine and diphenylbenzidine	+0.76 at pH 0	Violet	Colorless
Variamine Blue B	+0.62 at pH 1.5	Blue	Colorless
Methylene blue	+0.52 at pH 3	Green blue	Colorless
Neutral red	-0.325 at pH 7	Red violet	Colorless

Table I Cathodic Overpotential, in Volts, of Hydrogen on Various Materials

Electrode material	*Current density, A/cm²*			
	0.001	*0.01*	*0.1*	*1*
Aluminum	0.58	0.71	0.74	0.78
Bismuth	0.69	0.83	0.91	1.01
Cadmium	0.99	1.20	1.25	1.23
Chromium			0.67	0.77
Copper	0.60	0.75	0.82	0.84
Iron	0.40	0.53	0.64	0.77
Lead	0.67	0.97	1.12	1.08
Mercury	1.04	1.15	1.21	1.24
Nickel	0.33	0.42	0.51	0.59
Platinized Pt	0.01	0.03	0.05	0.07
Platinum	0.09	0.39	0.50	0.44
Silver	0.46	0.66	0.76	
Tin	0.85	0.98	0.99	0.98
Tungsten	0.27	0.35	0.47	0.54

NOTE: In 1*F* HCl.

Table J Limiting Molar Conductivity

Cation	Λ_i^0 $S \cdot m^2 \cdot mol^{-1}$	*Anion*	Λ_i^0 $S \cdot m^2 \cdot mol^{-1}$
H^+	0.0350	F^-	0.0055
Li^+	0.0039	Cl^-	0.0076
Na^+	0.0050	Br^-	0.0078
K^+	0.0074	I^-	0.0077
NH_4^+	0.0073	ClO_4^-	0.0067
Ag^+	0.0062	OH^-	0.0198
Mg^{2+}	0.0106	NO_3^-	0.0071
Ca^{2+}	0.0120	HCO_3^-	0.0044
Sr^{2+}	0.0118	$C_2H_3O_2^-$	0.0041
Ba^{2+}	0.0128	$HC_2O_4^-$	0.0040
Cu^{2+}	0.0108	$B(C_6H_5)_4^-$	0.0018
Zn^{2+}	0.0106	SO_4^{2-}	0.0160
Pb^{2+}	0.0138	$C_2O_4^{2-}$	0.0148
La^{3+}	0.0210	CO_3^{2-}	0.0138

NOTE: Aqueous solution, 25 °C, rounded values.

Table K The International System of Units (SI)

Physical quantity	*Name of unit*	*Symbol of unit*	*Expression in SI base units*
Length	meter	m	[base]
Mass	kilogram	kg	[base]
Time	second	s	[basc]
Current, electric	ampere	A	[base]
Temperature, thermodynamic	kelvin	K	[base]
Amount of substance	mole	mol	[base]
Luminous intensity	candela	cd	[base]
Frequency	hertz	Hz	s^{-1}
Force	newton	N	$m \cdot kg \cdot s^{-2}$
Pressure	pascal	Pa	$m^{-1} \cdot kg \cdot s^{-2}$
Energy, work, quantity of heat	joule	J	$m^2 \cdot kg \cdot s^{-2}$
Power, radiant flux	watt	W	$m^2 \cdot kg \cdot s^{-3}$
Quantity of electricity, electric charge	coulomb	C	$s \cdot A$
Electrical potential, potential difference, electromotive force	volt	V	$m^2 \cdot kg \cdot s^{-3} \cdot A^{-1}$
Electrical resistance	ohm	Ω	$m^2 \cdot kg \cdot s^{-3} \cdot A^{-2}$
Electrical conductance	siemens	S	$m^{-2} \cdot kg^{-1} \cdot s^3 \cdot A^2$
Electrical capacitance	farad	F	$m^{-2} \cdot kg^{-1} \cdot s^4 \cdot A^2$
Magnetic flux	weber	Wb	$m^2 \cdot kg \cdot s^{-2} \cdot A^{-1}$
Magnetic flux density	tesla	T	$kg \cdot s^{-2} \cdot A^{-1}$
Inductance	henry	H	$m^2 \cdot kg \cdot s^{-2} \cdot A^{-2}$
Activity (radioactive)	becquerel	Bq	s^{-1}

NOTE: The General Conference of Weights and Measures has adopted this practical system of units of measurement. Listed are the SI base units and also some SI derived units having special names. For the derived units, the expression combines base units and centered dots.

SI Prefixes

Multiple or Submultiple	*Prefix*	*Symbol*
10^{12}	tera	T
10^{9}	giga	G
10^{6}	mega	M
10^{3}	kilo	k
10^{2}	hecto	h
10	deka	da
10^{-1}	deci	d
10^{-2}	centi	c
10^{-3}	milli	m
10^{-6}	micro	μ
10^{-9}	nano	n
10^{-12}	pico	p
10^{-15}	femto	f
10^{18}	atto	a

NOTE: Terms formed from prefixes listed, joined to the unit names, provide the multiples and submultiples in the International System. For example, the unit name *meter* with the prefix *kilo* added produces *kilometer,* meaning "1000 meters."

Index

Table of Base 10 Logarithms

No.	0	1	2	3	4	5	6	7	8	9	Proportional Parts 1	2	3	4	5	6	7	8	9
10	0000	0043	0086	0128	0170	0212	0253	0294	0334	0374	4	8	12	17	21	25	29	33	37
11	0414	0453	0492	0531	0569	0607	0645	0682	0719	0755	4	8	11	15	19	23	26	30	34
12	0792	0828	0864	0899	0934	0969	1004	1038	1072	1106	3	7	10	14	17	21	24	28	31
13	1139	1173	1206	1239	1271	1303	1335	1367	1399	1430	3	6	10	13	16	19	23	26	29
14	1461	1492	1523	1553	1584	1614	1644	1673	1703	1732	3	6	9	12	15	18	21	24	27
15	1761	1790	1818	1847	1875	1903	1931	1959	1987	2014	3	6	8	11	14	17	20	22	25
16	2041	2068	2095	2122	2148	2175	2201	2227	2253	2279	3	5	8	11	13	16	18	21	24
17	2304	2330	2355	2380	2405	2430	2455	2480	2504	2529	2	5	7	10	12	15	17	20	22
18	2553	2577	2601	2625	2648	2672	2695	2718	2742	2765	2	5	7	9	12	14	16	19	21
19	2788	2810	2833	2856	2878	2900	2923	2945	2967	2989	2	4	7	9	11	13	16	18	20
20	3010	3032	3054	3075	3096	3118	3139	3160	3181	3201	2	4	6	8	10	13	15	17	19
21	3222	3243	3263	3284	3304	3324	3345	3365	3385	3404	2	4	6	8	11	12	14	16	18
22	3424	3444	3464	3483	3502	3522	3541	3560	3579	3598	2	4	6	8	10	12	14	15	17
23	3617	3636	3655	3674	3692	3711	3729	3747	3766	3784	2	4	6	7	9	11	13	15	17
24	3802	3820	3838	3856	3874	3892	3909	3927	3945	3962	2	4	5	7	9	11	12	14	16
25	3979	3997	4014	4031	4048	4065	4082	4099	4116	4133	2	3	5	7	9	10	12	14	15
26	4150	4166	4183	4200	4216	4232	4249	4265	4281	4298	2	3	5	7	8	10	11	13	15
27	4314	4330	4346	4362	4378	4393	4409	4425	4440	4456	2	3	5	6	8	9	11	13	14
28	4472	4487	4502	4518	4533	4548	4564	4579	4594	4609	2	3	5	6	8	9	11	12	14
29	4624	4639	4654	4669	4683	4698	4713	4728	4742	4757	1	3	4	6	7	9	10	12	13
30	4771	4786	4800	4814	4829	4843	4857	4871	4886	4900	1	3	4	6	7	9	10	11	13
31	4914	4928	4942	4955	4969	4983	4997	5011	5024	5038	1	3	4	6	7	8	10	11	12
32	5051	5065	5079	5092	5105	5119	5132	5145	5159	5172	1	3	4	5	7	8	9	11	12
33	5185	5198	5211	5224	5237	5250	5263	5276	5289	5302	1	3	4	5	6	8	9	10	12
34	5315	5328	5340	5353	5366	5378	5391	5403	5416	5428	1	3	4	5	6	8	9	10	11
35	5441	5453	5465	5478	5490	5502	5514	5527	5539	5551	1	2	4	5	6	7	9	10	11
36	5563	5575	5587	5599	5611	5623	5635	5647	5658	5670	1	2	4	5	6	7	8	10	11
37	5682	5694	5705	5717	5729	5740	5752	5763	5775	5786	1	2	3	5	6	7	8	9	10
38	5798	5809	5821	5832	5843	5855	5866	5877	5888	5899	1	2	3	5	6	7	8	9	10
39	5911	5922	5933	5944	5955	5966	5977	5988	5999	6010	1	2	3	4	5	7	8	9	10
40	6021	6031	6042	6053	6064	6075	6085	6096	6107	6117	1	2	3	4	5	6	8	9	10
41	6128	6138	6149	6160	6170	6180	6191	6201	6212	6222	1	2	3	4	5	6	7	8	9
42	6232	6243	6253	6263	6274	6284	6294	6304	6314	6325	1	2	3	4	5	6	7	8	9
43	6335	6345	6355	6365	6375	6386	6395	6405	6415	6425	1	2	3	4	5	6	7	8	9
44	6435	6444	6454	6464	6474	6484	6493	6503	6513	6522	1	2	3	4	5	6	7	8	9
45	6532	6542	6551	6561	6571	6580	6590	6599	6609	6618	1	2	3	4	5	6	7	8	9
46	6628	6637	6646	6656	6665	6675	6684	6693	6702	6712	1	2	3	4	5	6	7	7	8
47	6721	6730	6739	6749	6758	6767	6776	6785	6794	6803	1	2	3	4	5	5	6	7	8
48	6812	6821	6830	6839	6848	6857	6866	6875	6884	6893	1	2	3	4	4	5	6	7	8
49	6902	6911	6920	6928	6937	6946	6955	6964	6972	6981	1	2	3	4	4	5	6	7	8
50	6990	6998	7007	7016	7024	7033	7042	7050	7059	7067	1	2	3	3	4	5	6	7	8
51	7076	7084	7093	7101	7110	7118	7126	7135	7143	7152	1	2	3	3	4	5	6	7	8
52	7160	7168	7177	7185	7193	7202	7210	7218	7226	7235	1	2	2	3	4	5	6	7	7
53	7243	7251	7259	7267	7275	7284	7292	7300	7308	7316	1	2	2	3	4	5	6	6	7
54	7324	7332	7340	7348	7356	7364	7372	7380	7388	7396	1	2	2	3	4	5	6	6	7
	0	1	2	3	4	5	6	7	8	9	1	2	3	4	5	6	7	8	9